changing the way the world learns

To get extra value from this book for no additional cost, go to:

http://www.thomson.com/wadsworth.html

thomson.com is the World Wide Web site for Wadsworth/ITP and is your direct source to dozens of on-line resources. *thomson.com* helps you find out about supplements, experiment with demonstration software, search for a job, and send e-mail to many of our authors. You can even preview new publications and exciting new technologies.

thomson.com: It's where you'll find us in the future.

Please look for our sociology.net home page with URL updates.

Other Titles of Related Interest

Edward P. Kardas and Tommy M. Milford, *Using the Internet for Social Science Research and Practice*

Daniel J. Kurland and Daphne John, *Internet Guide for Sociology*

Leonard Cargan and Jeanne H. Ballantine, *Sociological Footprints: Introductory Readings in Sociology*, 7th edition

Joel M. Charon, *Ten Questions: A Sociological Perspective*, 2nd edition

Judson R. Landis, *Sociology: Concepts and Characteristics*, 9th edition

sociology.net
Sociology on the Internet

Joan Ferrante
Northern Kentucky University

Wadsworth Publishing Company

I(T)P® An International Thomson Publishing Company

Belmont, CA • Albany, NY • Bonn • Boston • Cincinnati •
Detroit • Johannesburg • London • Madrid • Melbourne •
Mexico City • New York • Paris • Singapore • Tokyo •
Toronto • Washington

Sociology Editor: Eve Howard
Assistant Editor: Deirdre McGill
Editorial Assistant: Barbara Yien
Marketing Manager: Chaun Hightower
Project Editor: Jerilyn Emori
Print Buyer: Karen Hunt
Permissions Editor: Jeanne Bosschart
Interior and Cover Designer: Andrew Ogus/Book Design
Copy Editor: Thomas Briggs
Cover Photo: Copyright Barrie Rokeach 1996
Compositor: Thompson Type
Printer: Quebecor Printing/Fairfield

COPYRIGHT © 1997 by Wadsworth Publishing Company
A Division of International Thomson Publishing Inc.
I(T)P The ITP logo is a registered trademark under license.

 This book is printed on acid-free recycled paper.

Printed in the United States of America
1 2 3 4 5 6 7 8 9 10

For more information, contact Wadsworth Publishing Company, 10 Davis Drive, Belmont, CA 94002, or electronically at http://www.thomson.com/wadsworth.html

International Thomson Publishing Europe
Berkshire House 168-173
High Holborn
London, WC1V 7AA, England

International Thomson Editores
Campos Eliseos 385, Piso 7
Col. Polanco
11560 México D.F. México

Thomas Nelson Australia
102 Dodds Street
South Melbourne 3205
Victoria, Australia

International Thomson Publishing Asia
221 Henderson Road
#05-10 Henderson Building
Singapore 0315

Nelson Canada
1120 Birchmount Road
Scarborough, Ontario
Canada M1K 5G4

International Thomson Publishing Japan
Hirakawacho Kyowa Building, 3F
2-2-1 Hirakawacho
Chiyoda-ku, Tokyo 102, Japan

International Thomson Publishing GmbH
Königswinterer Strasse 418
53227 Bonn, Germany

International Thomson Publishing Southern Africa
Building 18, Constantia Park
240 Old Pretoria Road
Halfway House, 1685 South Africa

All rights reserved. No part of this work covered by the copyright hereon may be reproduced or used in any form or by any means—graphic, electronic, or mechanical, including photocopying, recording, taping, or information storage and retrieval systems—without the written permission of the publisher.

Library of Congress Cataloging-in-Publication Data

Ferrante, Joan M.
    Sociology.net : sociology on the Internet / Joan Ferrante.
      p.   cm.
    Includes bibliographical references and index.
    ISBN 0-534-52756-6
    1. Sociology.   2. Sociology—Computer network resources.
3. Internet (Computer network)   I. Title.
HM51.F465   1997
301—dc21   96-46313

to Eugene B. Gallagher

David C. Lundgren

P. Neal Ritchey

Brief Contents

1. Sociology on the Internet: An Introduction 1
2. The Sociological Imagination 8
3. Theoretical Perspectives 25
4. Research Methods and the Information Explosion 42
5. Culture 60
6. Socialization 94
7. Social Interaction and the Social Construction of Reality 114
8. Social Organizations 144
9. Deviance, Conformity, and Social Control 168
10. Social Stratification 187
11. Race and Ethnicity 211
12. Gender 233
13. Population and Family Life 259
14. Education 288
15. Religion 324
16. Social Change 348
17. Internet Home Library 369

Key Concepts 402

References 415

URL Index 435

Index 453

Detailed Contents

1 Sociology on the Internet: An Introduction 1
What Is the Internet? 2
What Is the World Wide Web? 2
What Is a Browser? 3
What Is a URL? 3
What Is a Search Engine? 4
What Is a Local Internet Access Provider? 5
Some Words of Advice 5

2 The Sociological Imagination 8
Troubles, Issues, and the Sociological Imagination 9
Social Relativity 11
Transformative Powers of History 12
The Industrial Revolution 13
The Nature of Work • The Nature of Interaction • Sociological Perspectives on Industrialization
The Focus of Sociology 17
Karl Marx (1818–1883) • Emile Durkheim (1858–1918) • Max Weber (1864–1920)
The Evolution of the Discipline of Sociology 20
Additional Reading 24

3 Theoretical Perspectives 25
The Functionalist Perspective 26
Overview • Critique of Functionalism • Merton's Concepts • The Functionalist Perspective on the Internet

The Conflict Perspective 32
Overview • Critique of Conflict Theory • The Conflict Perspective on the Internet

The Symbolic Interactionist Perspective 36
Overview • Critique of Symbolic Interactionism • The Symbolic Interactionist Perspective on the Internet

Discussion Question 41

Additional Reading 41

4 Research Methods and the Information Explosion 42

The Information Explosion 43

The Scientific Method 46

Step 1: Defining the Topic for Investigation 47

Step 2: Reviewing the Literature 48

Step 3: Identifying Core Concepts 48

Step 4: Choosing a Design, Collecting the Data, and Forming Hypotheses 49
The Population to Be Studied • Methods of Data Collection • Identifying Variables • Operational Definitions

Steps 5 and 6: Analyzing the Data and Drawing Conclusions 57

Discussion Question 59

Additional Reading 59

5 Culture 60

Material and Nonmaterial Components 62
Material Culture • Nonmaterial Culture

The Role of Geographical and Historical Forces 66

The Transmission of Culture 67
The Role of Language • The Importance of Individual Experiences

Culture as a Tool for Problems of Living 71
Cultural Formulas for Hunger • Cultural Formulas for Social Emotions

The Relationship Between Material and Nonmaterial Culture 75

Cultural Diffusion 76
Everyday Mingling • The State • The Marketplace • Social Movements

One Culture as the Standard 81

Culture Shock 83
Cultural Relativism 86
Subcultures 89
Discussion Question 92
Additional Reading 92

6 Socialization 94

Nature and Nurture 95

The Importance of Social Contact 96
Cases of Extreme Isolation • Less Extreme Cases

Individual and Collective Memory 99

The Role of Groups 101
Primary Groups • Ingroups and Outgroups

Symbolic Interactionism and Self-Development 104
The Emergence of Self-Awareness • Role-Taking • The Looking-Glass Self

Cognitive Development 107

Resocialization 109

The Unpredictable Elements of Socialization 112

Discussion Question 113

Additional Reading 113

7 Social Interaction and the Social Construction of Reality 114

The Context of Social Interaction 117
Mechanical Solidarity • Organic Solidarity

Zaire in Transition 120
Belgian Imperialism (1883–1960) • Independence (1960–Present) • HIV/AIDS and Zaire

The Content of Social Interaction 125
Social Statuses • Social Roles

The Dramaturgical Model of Social Interaction 130
Impression Management • Staging Behavior • The Back Stage of Blood Banks

Attribution Theory 134

Determining Who Is HIV-Infected 138

Television: A Special Case of Reality Construction 139

Discussion Question 142

Additional Reading 142

8 Social Organizations 144

The Multinational Corporation: Agent of Colonialism or Progress? 146

Features of Modern Organizations 147
Social Action • Means-to-End (Value-Rational) Thought and Action • Rationalization • The Concept of Bureaucracy

Factors That Influence Behavior in Organizations 153
Formal Versus Informal Dimensions • Trained Incapacity • Statistical Records of Performance

Obstacles to Good Decision Making 159
Expert Knowledge and Responsibility • The Problems with Oligarchy

Alienation of Rank-and-File Workers 162
Consequences of Worker Alienation • Criteria for Evaluating Multinationals

Discussion Question 166

Additional Reading 166

9 Deviance, Conformity, and Social Control 168

Deviance: The Violation of Norms 169
Folkways and Mores • Mechanisms of Social Control • The Functions of Deviance

Labeling Theory 173
Conformists and Deviants • The Falsely Accused

Rule Makers and Rule Enforcers 177

The Constructionism Approach 178
Claims Makers • Claims About the Dangers of Legal and Illegal Substances

Structural Strain Theory 181

Differential Association Theory 184

Discussion Question 185

Additional Reading 185

10 Social Stratification 187

Achieved and Ascribed Characteristics 187

Classifying People into Racial Categories 190

"Open" and "Closed" Stratification Systems 192
Caste Systems • Apartheid: A Caste System of Stratification • Class Systems • Does the United States Have a Class System?

Mixed Systems: Class and Caste 197

A Functionalist View of Stratification 200
The Functional Importance of Occupations • Critique of the Functionalist Perspective

Analyses of Social Class 203
Karl Marx and Social Class • Max Weber and Social Class • Occupational Structure and Poverty in the United States

Discussion Question 209

Additional Reading 210

11 Race and Ethnicity 211

Racial Classification and the Idea of Race 213

Minority Groups 215

Types of Assimilation 216
Absorption Assimilation • Melting Pot Assimilation

Stratification Theory and Assimilation 219
Racist Ideologies • Prejudice and Stereotyping • Discrimination • Institutionalized Versus Individual Discrimination

Social Identity and Stigma 227
Mixed Contact Between the Stigmatized and the Majority Population • Responses to Stigmatization

Discussion Question 230

Additional Reading 231

12 Gender 233

Feminism in the United States 234

Distinguishing Sex and Gender 235
Sex as a Biological Concept • Gender as a Social Construct

Gender Polarization 238

Sexual Stratification 241
Economic Arrangements • Household and Child-Rearing Responsibilities • Access to Agents of Violence Control • Hazards in the Workplace

Mechanisms of Perpetuating Gender Expectations 245
Socialization • Structural or Situational Constraints • Sexist Ideologies

Gender, Race and Ethnicity, and the State 252
Women as Biological Reproducers of Babies of a Particular Ethnicity or Race • Women as Biological Reproducers of the Boundaries of Ethnic or National Groups • Women as Transmitters of Social Values and Culture • Women as Signifiers of Ethnic and Racial Differences • Women as Participants in National, Economic, and Military Struggles

Gender and Date Rape 255

Discussion Question 257

Additional Reading 257

13 Population and Family Life 259

The Industrial Revolution and the Family 262

Mechanized Rich Versus Labor-Intensive Poor Countries 263

The Theory of Demographic Transition 265
Stage 1: High Birthrates and Death Rates • Stage 2: The Transition Stage • Stage 3: Low Death Rates and Declining Birthrates

The Demographic Transition in Labor-Intensive Poor Countries 269
Death Rates • Birthrates • Population Growth • Migration

Industrialization and Family Life 276
The Effects of Geographic Mobility on Family Life • The Consequences of Long Life • The Status of Children • Urbanization and Family Life • Women and Work

Discussion Question 285

Additional Reading 285

Appendix: How Demographers Measure Change 287

14 Education 288

What Is Education? 290

Social Functions of Education 291

Illiteracy in the United States 292
Illiteracy and Schools • Insights from Foreign Education Systems

The Development of Mass Education in the United States 297
Textbooks • Single-Language Instruction

Fundamental Characteristics of Contemporary American Education 300

The Availability of College • Differences in Curriculum • Differences in Funding • Education-Based Programs to Solve Social Problems • Ambiguity of Purpose and Value

A Close-Up View: The Classroom Environment 305
The Curriculum • Tracking • Tests • The Problems That Teachers Face

The Social Context of Education 316
Family Background • Adolescent Subcultures

Discussion Question 322

Additional Reading 322

15 Religion 324

What Is Religion? Weber's and Durkheim's Views 325
Beliefs About the Sacred and the Profane • Rituals • Communities of Worshipers • Critique of Durkheim's Definition of Religion

Functionalist and Conflict Perspectives on Religion 334
The Functionalist Perspective • The Conflict Perspective

Max Weber: The Interplay Between Economics and Religion 339

Two Opposing Trends: Secularization and Fundamentalism 341
Secularization • Fundamentalism

Discussion Question 346

Additional Reading 347

16 Social Change 348

Social Change: Causes and Consequences 349

Innovations 350
Innovations and Rate of Change • Cultural Lag • Revolutionary Ideas

The Actions of Leaders 354
Charismatic Leaders as Agents of Change • The Power Elite: Legal-Rational Authority and Change

Conflict 358
Consequences of Conflict • Structural Origins of Conflict

Capitalism 362
World System Theory • The Role of Capitalism in the Global Economy • The Role of Core, Peripheral, and Semi-Peripheral Economies

Discussion Question 366

Additional Reading 366

17 Internet Home Library 369

World Information 369
Population Profile of the United States 370
Information on Specific Population Groups 370
Country-Level Information 373
State-Level Information 376
City-Level Information (U.S. and World Cities) 377
County-Level Information 377
Electronic Books 378
Electronic Journals and Newsletters 379
Resources of Interest to Sociologists 382
Press Releases and Briefings 383
Social Issues 386
General Sources of Statistics 387
Daily News 388
Foreign Newspapers 389
U.S. Government Agencies 390
Historical Documents 391
Global Organizations 391
Dictionaries, Thesauruses, and Quotations 392
Encyclopedias and General References 393
Writing Resources 395
Handy Reference Guides 396
Study Abroad 397
Coping with College 398
Career Guides 399
The Best of the World Wide Web 400

Key Concepts 402

References 415

URL Index 435

Index 453

Preface

Overview

Sociology.net: Sociology on the Internet is first and foremost a brief introduction to the core principles of sociology. It introduces readers to key sociological concepts and theories that can be used to analyze a broad range of topics. But *Sociology.net* also incorporates internet-related applications throughout the book. Why internet applications? I decided to create an introductory sociology textbook with internet applications for several reasons. Anyone who has explored the internet realizes that within a few years this technology has the potential to put "all the representations of human ideas ever created"—books, journals, newspapers, magazines, statistics, interviews, photographs, art, police reports, news and radio broadcasts, and so on—at an individual's fingertips (Katz and Chedester 1992, p. 2). Because sociology offers a unique perspective for approaching any topic involving humans, the internet is an extraordinary resource that gives easy access to a rich variety of information on past and present human activities. For example, consider some of the internet sites in the culture chapter. After students are introduced to the concepts of norms, values, and beliefs, they have the opportunity to go to a Web site that describes the first euthanasia court case in China. After reading the information posted there, students are asked: (1) What are Chinese beliefs about a patient's rights in relationship to his or her family? (2) Which is the more important value in China—individual/patient autonomy or status of the family? (3) What is the norm in China for obtaining consent for a medical procedure such as an operation? (4) What norms do Chinese physicians follow in advising family members when medical treatment offers no hope for improving a patient's condition?

Similarly, after students read about the role of geographical and historical forces in shaping the character of culture, they are directed to a Web site that describes the emergency water conservation efforts implemented in the drought-stricken southwestern United States in 1996. They are then asked to give examples of conservation-oriented behaviors that resulted from drought-like conditions. Likewise, when students read about ethnocentrism they have an opportunity to go to a Web site and read some examples of "trite journalistic clichés and embarrassing or biased representations of Africa." They are then asked to list two examples of ethnocentric attitudes toward or portrayals of Kenya.

These examples show the pedagogical significance of the internet. The internet has another side, however: it can give people the feeling that they are drowning in a sea of information accumulating at a pace that overwhelms the brain's capacity to organize and evaluate it. To complicate the issue of quantity, users must determine whether information is accurate and useful. At the same time, these highly frustrating features of the internet give it an "untamed quality," one that allows skillful users to browse and discover information that they would not have considered looking for when they began an information-finding task.

Yvonne Katz and Gay Chedester (1992) maintain that the existence of this information-gathering and -distribution tool is expanding our conceptions of literacy to include *creative literacy* (the ability to create new meanings, make connections between ideas, and invent new concepts), *choice literacy* (the ability to evaluate and choose from a wide variety of resources), and *connectedness literacy* (the ability to communicate, collaborate, and cooperate with people who offer a variety of perspectives). As readers learn the core principles of sociology covered in this textbook, they are also exposed to these literacies. First, with regard to creative literacy, the sociological perspective gives new meanings to the human activities presented on the internet, and the internet applications bring sociological concepts and theories to life. Second, readers readily see that the sociological perspective enhances choice literacy as sociological concepts, theories, and methods become conceptual tools with which they can evaluate and organize the information on the internet. Third, because the information on the internet is a collaborative product of educators, government officials, public and private organizations, citizen groups, commercial interests, and so on, readers understand the importance of connectedness literacy as they come to see that their educational experience is enhanced through connections with people from across the United States and around the world.

I have structured the book so that it incorporates 10–15 Web sites (URLs) within each chapter and up to 20 or so additional Web sites at the end of each chapter. All URLs direct readers to sociologically relevant information—information that relates to a sociological topic, applies a sociological concept or theory, and/or gives insights into or presents a vivid example of that concept or theory. Instructors may choose to require each student to read all or only a portion of the internet sites in each chapter. Or instructors may divide up the Web sites among students and ask individual students to post their assessment of that site for the class as a whole to read or to make an in-class presentation.

Student Contributors

Eight students in my applied social research class browsed the World Wide Web for seven months to help me find and evaluate sites for this project. All eight are applied cultural studies (ACS) majors or minors at Northern Kentucky University. The ACS program integrates the disciplines of anthropology and sociology and offers a strong background in research methods and a framework for conceptualizing diversity in all its forms (age, race, gender, religion, country, culture).

Jenny DeBerry graduated in May 1996. She has done research on animals housed at a no-kill animal shelter. Jenny has recently moved with her dog, Snoopy, to

Montana to begin graduate studies at the University of Montana. She did the internet research for the culture and religion chapters.

Melissa Cox graduates in May 1997. Her research interests are in the area of gender and sexuality, with special emphasis on sexuality education in the public schools. Melissa coaches the youth volleyball team at her church and does volunteer work with the elderly. She did the internet research for the gender and sociological imagination chapters.

Lindsay Hixson graduates in May 1997. She has done research on youths who frequent roller rinks. If you are interested in *Moby Dick* and artwork inspired by that novel, check out Lindsay's home page on the internet (http://www.nku.edu/~hixson). Lindsay plans to go to graduate school to study sociology. She did the internet research for the social stratification and theoretical perspectives chapters.

Patricia Gaines is, in her own words, "a youthful-looking old person" who graduated in May 1996. Her research interests relate to the grieving process, especially the study of organizations and support groups dedicated to helping people through all varieties of loss. Patricia enjoys reading real-life crime stories, biographies, and Russian novels (in English), as well as painting, writing, and needlepoint. She is "surfing" the Help Wanted sections of the internet. Patricia did the internet research for the research methods chapter.

Ryan Huber graduates in May 1997. He is interested in all phases of the research process. Ryan enjoys playing competitive sports, reading novels (especially by Stephen King), and surfing the net. Ryan did the internet research for the socialization and social change chapters, and he created the URL index.

Laureen Norris graduated in May 1996. Laureen has a variety of interests including a desire to learn about foreign countries. She created the homepage for the International Student Center at NKU. Her hobbies include an interest in auto racing and adopting stray cats. Laureen did the internet research for the population and family life chapters.

Julie Rack graduated in May 1996 with a major in applied cultural studies and a minor in justice studies. Her current interests include writing about and studying foreign cultures. She plans to visit Kenya in the near future to learn about its society and culture. Julie did internet research for the deviance and social organizations chapters.

Jacob Stewart graduates in December 1996. His academic and research interests converge in that he does archaeological testing and surveying for the cultural resources management division of an environmental service company. Jacob did the internet research for the race and social interaction chapters.

Angela Vaughn joined the team in May 1996. She is an English major who loves sociology. Her research interests are broad, but she is drawn to topics related to current (yet timeless) social issues. Angela has played the piano for 14 years. She is a staff writer for *The Northerner* (NKU's student newspaper) and secretary of the literature and language club. Angela enjoys traveling and she is planning to visit Europe in May 1997. She coauthored the Internet Home Library (Chapter 17).

Ancillary Materials

The *Instructor's Manual* contains answers to questions asked about the internet sites in each chapter. We also have a Web page (found at **http://wadsworth.com**) that gives the latest updates on URL changes. One of the hazards of using the internet as a teaching tool is that sometimes Web sites move, change, or even disappear. Although we have made every effort to choose "stable" Web sites, instructors must prepare students for the possibility that some sites may not be there when they need them. In such an event they should check out **http://www.wadsworth.com** for advice and new information.

Acknowledgments

I was very fortunate to have worked on this project with the nine students named above. The untamed, unorganized, "you-never-know-what-you-will-get" quality of the internet gives it a special, although a frustrating, appeal. Each person played an important role in helping me with the time-consuming task of finding quality Web sites.

I give special thanks to my editor, Eve Howard, for her many suggestions and insights about how to approach this book. We have had a number of important conversations in which Eve's straightforward, resourceful style helped me clarify the direction and goals of this project. In addition, Jerilyn Emori, the project editor for this book, did an outstanding job of coordinating the various departments and attending to the many details involved in the book's production. Over the course of the production process, we talked on the phone many times a week (sometimes several times a day). While Jerilyn always tended to the business at hand, she also managed to convey an ease and sense of humor that left me smiling at the end of every conversation. I also had the good fortune to work with Tom Briggs, the copyeditor for this project. In a perfect world every writer would have a copyeditor as experienced and capable as Tom. Tom functioned as an advisor of sorts, suggesting important changes in wording that would make the writing and ideas clearer to readers.

I thank my father-in-law, Walter D. Wallace, for sending me a steady stream of information about the internet, including references to a number of useful internet sites. Special thanks goes to my mother, Annalee T. Ferrante, for helping me in every way. She maintains my files, acquires permissions, acts as my clipping service, and even cooks gourmet meals for my husband and me when time is tight (which is all the time).

I would also like to thank Dean Rogers Redding (College of Arts and Science, NKU) and my chair, Jim Hopgood, for supporting this project with two state-of-the-art laptop computers that students could check out for home use.

I dedicate this book to three people who were important in all phases of my academic career, but especially the early phases: Eugene B. Gallagher (University of Kentucky), David C. Lundgren (University of Cincinnati), and P. Neal Ritchey (University of Cincinnati). I also wish to acknowledge the influence of Horatio C Wood IV on my philosophy of education. Finally, I wish to express my love for the most important person behind the project, my colleague and husband, Robert K. Wallace.

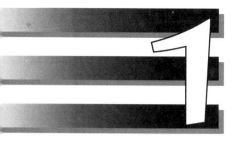

Sociology on the Internet: An Introduction

On the first day of class, when I announce that the course requires internet-related assignments, many students convey through their body language the impression that they might drop the class. Although they all have heard of the internet (more popularly known as the information superhighway), only a few have ever accessed it. I might also note that in a typical sociology class about 20 percent of the students have never used a computer, even as a word processor. Because many students have no experience with the internet and/or computers, they view the internet as something mysterious and incomprehensible. After less than an hour of instruction, however, everyone knows enough about the internet to do the assignments.

Over the course of the next few weeks, we learn that the most important internet-related skills are those that enable us to sift through, organize, comprehend, and evaluate the staggering and ever-increasing amount of information people from around the world have posted on the internet. To complicate matters, most of this information has not been evaluated in any formal way for accuracy or worth.

Sociologist Orrin Klapp offers a vivid metaphor to describe the dilemma of sorting through and keeping up with the massive amounts of information and the endless stream of new documents that can be accessed via the internet. Klapp envisioned a person "seated at a table fitting [together] pieces of a gigantic jigsaw puzzle. From a funnel overhead, pieces are pouring onto the table faster than one can fit them together. Most of the pieces do not match up. Indeed they do not belong to the same puzzle" (Klapp 1986, p. 110). The falling pieces represent research accumulating at such a pace that it interferes with people's ability to organize it into a comprehensible pattern.

Tim Berners-Lee, the inventor of the World Wide Web, predicts that in the next several years the internet will contain virtually everything that has ever been written, produced, filmed, photographed, recorded, painted, or otherwise created. Therefore the internet represents an extraordinary resource for information on a wide range of human activities, past and present. But a paradox exists. On the one hand, the internet has the power to put the most meaningful, thought-provoking, and/or influential of human creations at one's fingertips; on the other, the internet also gives easy access to the most trivial, inane, superficial, and/or inconsequential of human creations.

Sociology.net: Sociology on the Internet evolved as a reaction to this paradox. This book directs readers to some of the most sociologically useful information posted on the internet. Specifically this information covers sociological concepts and theories, applies a sociological perspective to a topic, and gives insight into or presents examples of sociological concepts or theories. As part of a class project, eight students enrolled in my Applied Social Research Methods class—Jennifer DeBerry, Melissa Cox, Lindsay Hixson, Patricia Gaines, Ryan Huber, Laureen Norris, Julie Rack, and Jacob Stewart—spent seven months searching for some of the most sociologically relevant material posted on the internet. Subsequently Angela Vaughn joined the team, creating Chapter 17, "Internet Home Library." This chapter lists URLs that take you to Web sites with information that can help you with most college assignments without leaving your desk. (This will help save wear and tear on you, your car, and the environment.)

Sociology.net: Sociology on the Internet is not a book about the internet. There are already many such books available for those who wish to learn the internet's extensive vocabulary and to acquire skills that go beyond those needed here. Consequently I cover only what you need to know to access the internet sites included in this book. Specifically you need to know the answers to these questions: (1) What is the internet? (2) What is the World Wide Web? (3) What is a browser? (4) What are Uniform Resource Locators (URLs)? (5) What is a search engine? (6) What is a local internet access provider? (Or, How can I gain access to the World Wide Web?)

What Is the Internet?

The **internet** is a vast network of computer networks. A computer network is a system of computers connected to one another via special software and phone, fiber-optic, or other types of lines. The internet links together tens of thousands of in-house computer networks maintained by businesses, libraries, government agencies, universities, and private organizations.

What Is the World Wide Web?

The **World Wide Web** is one of several internet-based services (others include e-mail and Usenet, a discussion/newsgroup service). The Web, which was invented in 1991, is a constantly changing and ever-expanding information-sharing tool that facilitates the exchange of text-, video-, and audio-based information. Because no one owns the Web or manages its content, it is impossible to calculate the number of computer networks involved or the amount of information available for exchange. According to one estimate some 500,000-and-counting documents are available through the Web (Steinberg 1996). Fortunately most of the documents available for access are formatted according to standard specifications known as *HTML* (hypertext markup language). Hypertext connects documents to one another and allows users to move quickly via links within and across documents located anywhere on the Web. *Links* are highlighted words or images that set them apart from the rest of a document's text. After users choose a link or click a link with a mouse, they move elsewhere. Any browser can read any basic HTML document, unless the HTML format has special features that can be read only by particular browsers.

What Is a Browser?

Users access information on the World Wide Web with a **browser.** There are two kinds of browsers: character-based client browsers and graphical client browsers. Character-based browsers read only letters, numbers, spaces, and nongraphical marks or signs; with the arrow keys and the space bar, users navigate through documents. The most popular character browser is Lynx. In contrast graphical browsers can process characters, images, and sounds. Two popular graphical browsers are Netscape Navigator and Microsoft Internet Explorer. To gain access to a browser, you need a local internet provider. (See "What Is a Local Internet Provider?")

What Is a URL?

Every document on the internet has an address called a **Uniform Resource Locator (URL).** Some examples of URLs include the following:

1. **http://www.usatoday.com/news/nweird.htm**
2. **http://library.whitehouse.gov/PressReleases-plain.cgi**
3. **gopher://gopher.etext.org/11/Politics/Disability.Rag**
4. **ftp://stats.bls.gov/pub/news.release/osh.text**
5. **http://www.wri.org**
6. **http://www.soc.surrey.ac.uk/socresonline**

The rules for URLs are very loose, but there is a basic structure for URLs. Most of the URLs listed in *Sociology.net: Sociology on the Internet* begin with http://, some begin with gopher://, and a very few begin with ftp://. "http" stands for hypertext transfer protocol. "ftp" stands for file transfer protocol. Protocol is a system of procedures to access information. "gopher" is the University of Minnesota's mascot; the University of Minnesota is the home of this internet navigational tool, which facilitates the search for information.

The *domain name system* or the host computer name appears between the double slashes (//) and the first single slash mark (/). It is called a domain name system because it makes reference to several domains or areas within an in-house computer network. Each domain is separated by a period. Notice that most of the domain name systems listed in our six sample URLs end with a three-letter code—.org, .com, .gov, or .edu. This tells us that the host computer is an organization (.org), company (.com), government agency (.gov), or educational institution (.edu). Sometimes the domain name system ends with a two-letter code known as a country code. For instance, in example 6 ".uk" stands for United Kingdom; other such codes include ".us" for the United States and ".tr" for Turkey.

To the left of these two-digit or three-digit codes are usually some clues about the identity of the organization, company, government agency, or educational institution. It's easy to see that the first URL refers to the company USAToday and the second URL to the White House, a department within the U.S. government. The domain name system in example 3 gives no clues to the identity of the organization. However, "etext" suggests that the URL leads to an electronic text. The "bls" in example 4 refers to the Bureau of Labor Statistics, the "wri" in example 5 to the World Resources Institute, and the "surrey" in example 6 the University of Surrey.

The information between the single slashes represents codes for the directories or hypertext paths one must follow to get to a particular document. Clues about the name or kind of document to which the URL takes you are to the far right of the URL address. If there are no paths, as in example 5, the URL takes you to the host institution's opening page or homepage. The documents for each of our six examples are as follows:

Path	Document
1. nweird.htm	Weird News Stories
2. PressReleases-plain.cgi	Press Releases
3. Disability.Rag	Disability Rag (a magazine that focuses on disability)
4. osh.txt	Workplace Injuries and Illnesses Report
5. No paths specified in URL	———
6. socresonline	Sociological Research Online (an online journal)

To gain access to these documents, you simply type in the URL at the appropriate place designated by the browser. In the case of the Lynx browser, type the letter "g" and begin typing the URL in the designated space. In the case of Netscape Navigator, delete the URL appearing in the "Location" box and type the URL you wish to locate in its place.

How do you find URLs? URL addresses can be found in many different places. There are URL "Yellow Pages" that are organized by subject with relevant URL addresses. Sometimes corporations and agencies list their URL address in an advertisement. For example, at the end of the Lehrer-Newshour, PBS lists its URL, which is http://www.pbs.org. Another way to find URLs is via search engines.

What Is a Search Engine?

A **search engine** allows users to submit a keyword or keywords to identify the topic on which they wish to find information. The search engine identifies Web sites and corresponding URLs (see the next section) that lead to information on that topic. Some names of search engines are InfoSeek, Yahoo, CUSI, Starting Point, Lycos, Web Crawler, and Excite. Because there is so much information on the Web and no central authority managing its contents, "organizing the Web is probably the hardest information science problem out there" (Young, in Steinberg 1996, p. 109). Search engines try "to bring order out of chaos in a frantic quest for the ultimate index of human knowledge" (Steinberg 1996, p. 109). Yahoo indexers, for example, claim to have catalogued more than 200,000 Web sites (and counting) into 20,000 different categories (Steinberg 1996).

What Is a Local Internet Access Provider?

Most people cannot afford their own direct connection to the internet. Consequently they rely on a **local internet access provider,** a nearby host institution that offers internet access via a local telephone call as a benefit of membership or for a fee. Many students can obtain access to the internet through their school or university. Some businesses and other organizations offer their employees or members access, and public libraries give access to patrons. In addition dozens of commercial internet service providers offer the connection for a fee (ranging from $10 a month and up depending on use). Examples include America Online, Prodigy, CompuServe, and perhaps your local telephone company.

To access the portion of the internet known as the World Wide Web from home, you must have a personal computer, a modem, and modem software (don't panic), which allows you to connect to a provider's host computer via dial-up telephone lines. You will have to ask your local internet service provider about software and other details for connecting to their service from your home.

If you are a student and do not have a personal computer and/or the necessary equipment to connect to the internet from your home, check out the availability of public computer labs located in dorms, academic departments, libraries, and elsewhere on campus.

Some Words of Advice

We close with some advice about the internet from some experts. The students who worked on this project have learned from experience how to handle the internet, and their suggestions can save you from a lot of headaches. You might want to try some internet sites before you read the advice.

Jenny DeBerry: "The internet presents a whole new way of thinking about information. My advice is to expect change and learn to be comfortable with it. People who post information on the internet can choose to take it off or revise it whenever they see fit. With books authors might want to change their ideas but have to wait until the next edition (if there is a next edition) to get it out there for the world to see. With the internet authors can revise their thinking whenever they see fit and make changes on the spot. So don't be surprised if you find that a document is gone or has changed in subtle or dramatic ways from one visit to the next."

Patricia Gaines: "My advice is not to wait until the last minute to do internet assignments. The internet might not be available an hour before your assignment is due. I read an article in *The New York Times* that says it all: 'More than a million computer users suffered interrupted or erratic Internet and on-line service connections last week because of a variety of planned and unplanned service shutdowns' (Lewis 1996, p. C1). Sometimes the 'system crashes' lasted as long as eight hours."

Ryan Huber: "My tip is to always double-check the URL you typed in against the one listed in the text. It must be typed *exactly* as listed in the text. For example, lowercase letters cannot be substituted for capital letters. This will prevent or solve many of the problems you might have in locating a site that you know is there but that the computer is telling you isn't. Also, when using a search engine such as the Web Crawler, Excite, or Lycos, try to think of more than one key phrase or search word. I usually try to think of eight different keywords/phrases for each topic I am researching. Be patient and explore. You may not always find what you are looking for right away. Don't be afraid to follow hypertexted links; sometimes the best information is found in the most unusual places."

Laureen Norris: "Using the internet requires you to have a great deal of patience, a low frustration level, and a detectivelike mentality. To get the best results for a search on the internet, be as specific as possible with your choice of words and phrases. The internet has so many sites that if your search word is too vague or general, the search engine may generate tens of thousands of sites. I found that the hypertexted words or suggested links within documents often are more valuable to my research than the document the search engine generated as directly related. Sometimes the highlighted words or links appear to have no relationship with my chosen topic."

Julie Rack: "Sometimes the internet can be very slow. You may have to wait 20 or 30 seconds (imagine that!) for the document to appear after you type in the URL. My advice is to try nonpeak-hour times before 9:00 A.M. and after 5:00 P.M. The fewer the people on the internet, the faster the response time."

Jacob Stewart: "The amount of information on the Web is massive. You can find something on almost any topic. However, you must be mentally and physically tough when searching for information on the Web. Sometimes you can feel totally overwhelmed to the point of panic. But don't worry; most people feel this way at one time or another. My advice is to relax and to not feel that you have to look or read over everything on a topic."

Angela Vaughn: "Be patient with the internet, but don't spend too much time at any one site. Remember, the internet has information on every subject; it's just a matter of finding it. Look through sites quickly, and if the information doesn't seem useful to the task at hand, keep going and don't look back. Above all, take an 'I'm in charge' attitude. Don't let the internet overwhelm you. Maneuvering around the internet is an acquired skill. The more time you spend at it, the more efficient you'll become."

Lindsay Hixson: "I could list a hundred tips about the internet, but I will offer just one. Many Web sites are rearranged or restructured on a regular basis so that the path one takes to get to a document might change. When that happens, the URL address listed in the textbook may not get you to the document. Don't be alarmed if you type in a URL and the message 'unable to connect to remote host' or 'a path does not exist' appears. The chances are very good that the document is still there but that the 'path' has changed. So what do you do? You could take the easy way out and tell your instructor that the URL didn't work so you couldn't do the assignment. Or with a little effort you could try to find the document yourself.

Here's what you do: Type in the entire URL and then erase the address back to the first slash. For example, if your URL is **http://www.cdinet.com/Rockefeller/Briefs/brief35.html,** erase /brief35.html and submit the revised address. If that revised address doesn't work, erase the address back to the next slash. In this case you will erase /Briefs. Keep doing this until you have only the domain name system to submit. If you gain access, search for hypertext links that match up with path code names in your original URL. Hopefully this will not happen too often, but if it does, you know what to do. Good luck."

Melissa Cox: "When I started this project, a friend suggested that I purchase a book covering the 'how to's' of the internet. I replied, 'Unless it is a book of URLs that lead to high-quality information, forget it.' The best way to learn about the internet is to talk to others who are also using the net. It is surprising how much you can pick up on your own and by sharing tips with others. Also don't be afraid to ask questions if you get stuck."

The Sociological Imagination

In 1884 the Pennsylvania Railroad company built a depot in Wilmerding, Pennsylvania, a town 13 miles east of Pittsburgh. Five years later George Westinghouse located an air brake manufacturing plant in Wilmerding. Under Westinghouse's influence Wilmerding "reached great importance. America's first company-planned community, Wilmerding stood shoulder-to-shoulder with any town in the valley and beyond." In "An Historical Small Town Treasure: A View from Our Front Porch," Richard Shumaker reminisces about life in Wilmerding.

http://www.lm.com/~rs7717/wilmerdi.html
Document: An Historical Small Town Treasure: A View from Our Front Porch

Q: How did the Westinghouse Air Brake Company and the train depot figure into the lives of the Shumaker family and the community as a whole? What was Wilmerding's population in 1890? In 1920? In 1960? What solution does Shumaker suggest to restore "Wilmerding"?

In his now classic book *The Sociological Imagination*, sociologist C. Wright Mills (1959) introduced the terms *troubles* and *issues*. Mills defined **troubles** as personal problems and difficulties that can be explained in terms of individual characteristics such as motivation level, mood, personality, or ability. The resolution of a trouble, if it can be resolved, lies in changing an individual's character or immediate relationships. Mills stated that when, in a city of 100,000, only one man or woman is unemployed, that is his or her personal trouble, and for its relief we properly look to that person's character, skills, and immediate opportunities (that is, "She is lazy," "He has a bad attitude," "She has always been a moody person," or "He didn't try very hard in school"). **Issues,** on the other hand, are matters that can only be explained by factors outside any one individual's control and immediate environment. Mills wrote, "But when in a nation of 50 million employees, 15 million men [and women] are unemployed, that is an issue, and we may not hope to find its solution within the range of opportunities open to any one individual. The very structure of opportunities has collapsed" (p. 9). In this situation unemployment cannot be explained simply in terms of personal shortcomings or character flaws.

The connection between troubles and issues is fundamental to sociology. As we will see in this chapter, perceiving this connection is a central component of the **sociological imagination**—the ability to link seemingly impersonal and remote historical forces to an individual's life or the ability to recognize when seemingly personal troubles are connected to issues.

Troubles, Issues, and the Sociological Imagination

Mills (1959) argued that people often feel "trapped" by issues because, even though they respond to their personal situations, they remain spectators to the larger forces that affect their lives. In other words they feel helpless because they seem to have no control over these forces. When a war breaks out, for example, an ROTC student becomes a rocket launcher, a teenager buys and wears a sweatshirt that announces his or her support (or lack of support) for the war, an automobile owner thinks about the role of oil in the war while pumping gas, and a child whose parents are in the military has to live without them for a while. "The personal problem of war, when it occurs, may be how to survive it or how to die in it with honor; how to make money out of it; how to climb into the higher safety of the military apparatus; or how to contribute to the war's termination" (Mills 1959, p. 9). The issues of war, on the other hand, concern its causes—for example, a threat to the resources (such as oil) on which a country depends. In a similar vein, when the railroad company went bankrupt and the Westinghouse Air Brake Company left Wilmerding, the personal troubles connected with this kind of economic collapse may have included filing for unemployment, choosing early retirement, finding a way to live on a reduced budget, fighting drowsiness while driving home from third-shift work, trying to stay cheerful while hunting for a job, or enrolling in college years after high school graduation. Such problems are not unique to one-industry towns such as Wilmerding, however. Since approximately the mid-1970s, mechanization, automation, and exportation of jobs to lower-wage cities within the United States or to low-wage countries have caused the collapse of many local economies. The Web document "An Historical Small Town Treasure" tells us that Wilmerding's population fell from a high of 6,441 in 1920 to 4,349 by 1960; more recent figures are provided by the Census Bureau.

http://tiger.census.gov/cgi-bin/gazetteer
Document: U.S. Gazetteer

Q: After you access this site, to find out the population size, type in the name of the town (Wilmerding), the state (PA), and the zip code (15148). What was the exact size of Wilmerding's population in 1990?

One indicator of the negative consequences of economic restructuring is the amount of out-migration, which suggests that people left for economic reasons. Between 1970 and 1990 the population of Allegheny County, of which Wilmerding is part, dropped 16.7 percent from 1,605,016 to 1,336,449 people. During this time there were also high rates of out-migration in the following states: Michigan, Illinois, New York, Texas, North Dakota, South Dakota, and Montana. Additional population statistics are available from the Census Bureau.

http://www.census.gov/ftp/pub/population/cencounts/
Document: Index

Q: Choose one of the states with a high rate of out-migration named in this document. How many counties in that state lost population between 1970 and 1990?

Moving may correct the personal troubles associated with economic collapse, but it does not change the issue of economic restructuring, which is determined largely by an economic system that measures success in terms of profit. Profit is achieved by lowering production costs, hiring employees who will work for lower wages, introducing labor-saving technologies (computerization), and moving operations out of high-wage zones inside or outside the country. Such strategies leave workers vulnerable to unemployment.

Mills asked, "Is it any wonder that ordinary people feel they cannot cope with the larger worlds with which they are so suddenly confronted?" (pp. 4–5). Is it any wonder that they cannot understand the connection between these forces and their own lives, that they turn inward and try to protect their own sense of well-being, that they feel trapped? Mills believed that to gain some sense of control over their lives, people need more than information, because we live in a time when information dominates our attention and overwhelms our capacity to make sense of all we hear, see, and read in a day. Thus we may be exhausted by the struggle to learn from this information about the forces that shape our daily lives.

According to Mills people need "a quality of mind that will help them to use information" in a way that they can think about "what is going on in the world and of what may be happening within themselves"(p. 5). Mills called this quality of mind the sociological imagination. Those who possess the sociological imagination can view their inner life and human career in the context of larger historical forces. The payoff for those who possess this quality of mind is that they can understand their own experiences and "fate" by locating themselves in history; they can recognize the responses available to them by becoming aware of all the individuals who share the same situation as themselves. Mills maintained that knowing one's place in the larger scheme of things is "in many ways . . . a terrible lesson; in many ways, a magnificent one" (p. 5).

This knowledge can be a terrible lesson because we learn that human agony, hatred, selfishness, sadness, tragedy, and hopelessness have no limits. It can be a magnificent lesson because we learn that human dignity, pleasure, love, self-sacrifice, and effort also are unlimited. According to Mills we will "come to know that the limits of 'human nature' are frighteningly broad. [We will] come to know that every individual lives, from one generation to the next, in some society; that [people] live out a biography, and that [they live] it out within some historical sequence" (p. 6). Yet, if only by living in a society, people shape that society, however minutely, even as the society shapes them.

Those who possess the sociological imagination can grasp history and biography and the connections between the two. If we understand the forces shaping our lives, we are able to respond in ways that benefit our own lives as well as those of others. Mills equated the historical forces shaping people's lives with the structure of opportunities or the chances available to accomplish a valued goal. That valued goal can be almost anything—the opportunity to attend a high school prom, the opportunity to become an effective reader, the opportunity to work, the opportunity to marry, the opportunity to live a long life. We must be able to see how opportunities—whatever they may be—are shaped by the way in which behavior is patterned or organized in society (that is, structured). The opportunity to attend a high school prom, for example, is constrained by the fact that we define a date to attend a school dance as a social occasion planned in advance between two people. Ideally one partner (the male) is taller than the other (the female). The opportunities might be widened if a date for a dance did not require people to pair off and if choice wasn't constrained by the height of the partner.

Sociology offers a perspective that permits people to define and analyze the structure of opportunities. It does not promise, however, that if people understand this structure, then their problems will be solved easily. On the other hand, the sociological imagination does decrease our chances of responding in inappropriate or misguided ways. The ability to see the structure of opportunities becomes most evident to people when they can grasp two interrelated concepts—social relativity and the transformative powers of history.

Social Relativity

Social relativity is the view that ideas, beliefs, and behavior vary according to time and place. People who have lived in more than one culture can teach us much about social relativity. Consider the following excerpts from diaries kept by some of my students, who were asked to interview foreign students about their adjustment to American life:

> I asked François (from France) what was the biggest adjustment he had to make coming to America. He answered in this way. When he came to America he took a job. The first day on the job he expected to take a lunch break around 12 o'clock.

What seemed like several hours passed before his boss told him to take a half-hour lunch break. Lunch is not taken so lightly in France. François was accustomed to a larger meal and to taking as long as two hours to eat.

I asked Irma if she had any problems readjusting to her native culture when she visited Indonesia. She said that she did have problems. Irma mentioned that in Indonesia it is customary to drop in on a relative or friend without advance notice. In the United States, on the other hand, she had to learn to call people first. When she was home over Christmas break, she began dialing the phone number of a relative to let the person know that she and her mother were coming over. Her mother asked her why she was behaving so strangely. Irma replied that she wanted to check to see if this was a convenient time. Her mother thought Irma was going crazy. In Indonesia calling before a visit is equivalent to asking someone if they will go out of their way to make special preparations. Indonesians view such a request as rude.

I asked Hannah what were some of the hardest things for her to get used to since coming to the United States. She explained that in Jordan a girl would never introduce a boy as a boyfriend. If that were to happen, the girl would be disowned not only by the family but by the whole society. In general, the nature of family ties is very different in the two countries. In Jordan people do not always do what they would like. They have a deep respect for family, as do most people in Jordan. This respect is so strong that family comes first and individual achievement second. Hannah described how marriage partners are chosen. The first and most important consideration is not the marriage partner per se, but the person's family. If someone was from a bad family, you would not marry the person for two reasons. First, the prevailing assumption is that if the family is bad, so is an individual raised in that family. Second, you would not disgrace the family by marrying a person of questionable character.

These accounts show that ideas about eating, visiting relatives, and dating vary across regions and cultures and that these ideas affect opportunities—in this case opportunities to relax in the middle of the day, to visit friends and relatives, and to select potential mates. Furthermore these accounts help us to recognize and understand that many of our ideas and behaviors are not personal responses, but rather are products of the environment into which we were born.

Transformative Powers of History

According to the concept of the **transformative powers of history,** the most significant historical events have dramatic consequences for people's thinking and behavior. As with social relativity, events must be seen in the context of time and place. That is, to understand the transformative power of an event, a person must have some idea of the way people thought or behaved before and after that event. Consider the Gutenberg printing press and its enormous transformative influence on every aspect of European culture.

http://www.digitalcentury.com/encyclo/update/print.html
Document: Printing: History and Development

How did the Gutenberg printing press help to expand church influence while at the same time helping to challenge its influence? What effect did the printing press have on the English language? How did the printing press encourage pursuit of personal privacy? What role did the printing press play in the American Revolution? (*Note:* These questions can be answered without accessing the hypertexted links in the document.)

As another example consider the tremendous transformative power of the Industrial Revolution of the nineteenth century. Before the Industrial Revolution people's lives varied little from one generation to the next. Thus parents could assume that their children's daily lives would be much like their own and that their children would face the same challenges and problems that they faced. Consequently societies embarked on long-term projects that extended to subsequent generations. For example, in the Middle Ages people built cathedrals that took several lifetimes to complete, "presumably never doubting that such edifices would be used and appreciated by their great-grandchildren when construction was complete" (Ornstein and Ehrlich 1989, p. 55).

During the Industrial Revolution, however, the pace of change increased to such a degree that parents could no longer assume that their children's lives would be similar to their own. "Imagine the reactions of an American today if asked to contribute to a building that would take 150 years to finish: 'We don't want to tie up our capital in something with no return for one hundred and fifty years!' 'Won't a new design and construction process make this one obsolete long before it's finished?'" (Ornstein and Ehrlich 1989, p. 55).

The sociological imagination permits those who adopt it to see how biography and history intersect—to understand that even some of the most personal experiences are shaped by time and place and by the transformative powers of history. This type of imagination is characteristic of all the great sociologists, including three of the most influential: Emile Durkheim, Max Weber, and Karl Marx. All had a tremendous impact on the discipline of sociology; all spent much of their professional careers in an attempt to understand the nature and consequences of one event—the Industrial Revolution.

The Industrial Revolution

Between 1492 and 1800 an interdependent world began to emerge. Europeans learned of, conquered, and/or colonized much of North America, South America, Asia, and Africa and set the tone of international relations for centuries to come. During this time colonists forced local populations to cultivate and harvest crops and to extract minerals and other raw materials for export to the colonists' home countries. When the Europeans' labor needs could not be met by indigenous populations, they imported slaves from Africa or indentured workers from Asia.

In light of colonization one can argue that the world has been interdependent for at least 500 years. The scale of social and economic interdependence changed dramatically, however, with the Industrial Revolution, which gained momentum in England around 1850 and soon spread to other European countries and to the United States. The Industrial Revolution drew people from even the most remote parts of the world into a process that produced unprecedented quantities of material goods, primarily for the benefit of the colonizing countries.

One fundamental feature of the Industrial Revolution was **mechanization**—the addition of external sources of power such as oil or steam to hand tools and to modes of production and transportation. Mechanization turned the spinning wheel into a spinning machine, the hand loom into a power loom, and the blacksmith's hammer into a power hammer. It replaced wind-powered sailboats with steamships, and horse-drawn carriages with trains. And it replaced hand-made paper with paper made by steam-powered machines.

http://www.ipst.edu/amp/machine.html

Document: The Advent of the Paper Machine

 How much paper could steam-powered machines produce every minute? Mechanization did more than speed up production, however. On a social level it changed the nature of work and the ways in which people interacted with one another.

The Nature of Work

Prior to the mechanization brought on by the Industrial Revolution, goods were produced and distributed at a human pace, as illustrated by the effort required to bake bread, to make glass, or to produce paper:

- Bakeries produced bread almost entirely by manual labor, the hardest operation being that of preparing the dough, "usually carried on in one dark corner of a cellar, by a man, stripped naked down to the waist, and painfully engaged in extricating his fingers from a gluey mass into which he furiously plunges alternately his clenched fists" (Zuboff 1988, pp. 37–38).
- In glassmaking, everything was done by hand, "the gatherers taking the metal from the furnace at the end of an iron rod, the blower shaping the body of the bottle with his breath, while the maker who finished the bottle off . . . tooled the neck with a light spring-handled pair of tongs. Each bottle was individually made no matter what household, shop, or tavern it was destined for" (Zuboff 1988, pp. 37–38).

For a description of manual labor involved in papermaking, see the documents "The Invention of Paper" and "Forerunners of Paper." As you read these documents, note that papermakers are referred to as skilled artisans in spite of the labor-intensive work they do.

http://www.ipst.edu/amp/inventn.html
Document: The Invention of Paper

http://www.ipst.edu/amp/forerun.html
Document: Forerunners of Paper

 Q: How did the Chinese make paper? What are some of the materials that have been used in the past to make paper? Based on the description and pictures (if you have Netscape) in "Forerunners of Paper," can you tell if there were any fast ways to make paper before mechanization? Explain.

Mechanization reduced the physical requirements needed to produce goods. Workers paid a price for the reduction in physical requirements, however, because skills and knowledge were intertwined with the physical effort. Before mechanization knowledge was

> inscribed in the laboring body—in hands, fingertips, wrists, feet, nose, eyes, ears, skin, muscles, shoulders, arms, and legs—as surely as it was inscribed in the brain. It was knowledge filled with intimate detail of materials and ambience—the color and consistency of metal as it was thrust into a blazing fire, the smooth finish of the clay as it gave up its moisture, the supple feel of the leather as it was beaten and stretched, the strength and delicacy of glass as it was filled with human breath.
> (ZUBOFF 1988, P. 40)

Consequently mechanization not only speeded up the production but also reduced the skills needed to create products. Industrialization transformed individual workshops into factories, craftspeople into machine operators, and hand production into machine production. Products previously designed as unique entities and assembled by a few people were now standardized and assembled by many workers, each performing only one function in the overall production process. This division and standardization of labor meant that no one person could say, "I made this; this is a unique product of my labor." The artisans yielded power over the production process to the factory owner because their skills were rendered obsolete by the machines. Now people with little or no skill could do the artisan's work, and at a faster pace.

The Nature of Interaction

Between 1820 and 1860 a series of developments—the railroad, the steamship, gas lighting, running water, central heating, stoves, iceboxes, the telegraph, and mass-circulation newspapers—transformed the ways in which people lived their daily lives. Although all of these innovations were important, the railroad and the steamship were perhaps the most crucial. These new modes of travel connected people to one another in reliable, efficient, less time-consuming ways. They permitted people to travel day and night; in rain, snow, or sleet; and across smooth and rough terrain. In sum these inventions caused people to believe they had "annihilated" time and space (Gordon 1989).

For example, before the steam engine train was introduced, a trip by coach between Nashville, Tennessee, and Washington, DC, took one month. Before the advent of the steamboat, a 150-mile trip on a canal took 32 hours (Gordon 1989), while an overseas trip from Savannah, Georgia, to Liverpool, England, took months. Each year following the development of steam power, transportation became faster and more reliable. For instance, in 1819, the year of the first ocean crossing by a steam-driven vessel, the trip took one month. By 1881 the time needed to make that trip had been reduced to seven days (*The New Columbia Encyclopedia* 1975).

Railroads and steamships increased personal mobility as well as the freight traffic between previously remote areas. These modes of transportation facilitated an unprecedented degree of economic interdependence, competition, and upheaval. Now people in one area could be priced out of a livelihood if people in another area could provide goods and materials at a lower price (Gordon 1989).

Sociological Perspectives on Industrialization

Industrialization did more than change the nature of work and interactions; it affected virtually every aspect of daily life. "Within a few decades a social order which had existed for centuries vanished, and a new one, familiar in its outline to us in the late twentieth century, appeared" (Lengermann 1974, p. 28). The changes triggered by the Industrial Revolution, and especially by mechanization, are incalculable—and they are still taking place.

Marx, Durkheim, and Weber all lived during the Industrial Revolution's most transformative decades. Although they wrote in the nineteenth and early twentieth centuries, their observations are still relevant today. In fact many insights into the character of contemporary society can be gained by reading their writings, because those who witness and adjust to a significant event are intensely familiar with its consequences in daily life. Because most of us living today know only an industrialized life, we lack a contrast to help us understand how industrialization shapes contemporary existence. Consequently the words of witnesses can be revealing, as the following anecdote suggests.

Recently a scientist was interviewed on the radio. He maintained that scientists were close to understanding the mechanisms that govern the aging process and that people might soon be able to live to be 150 years old. If the aging mechanisms in fact are controlled, the first people to witness the change will have to make the greatest adjustment. In contrast people born after this discovery is made will know only a life in which they can expect to live 150 years. If these postdiscovery humans are curious about how living to age 150 shapes their lives, they will have to look to those who recorded insights into life before the change and who described their adjustments to the so-called advancement.

So it is with industrialization: to understand how it shapes human life, we can look to some of the early sociologists, particularly Durkheim, Marx, and Weber. Because the Industrial Revolution (and all that it encompassed) was so important in the professional and personal lives of these early sociologists, the sociological perspective may be regarded as a "necessary tool for analyzing social life and the recurring issues that confront human beings caught up in the intensity and uncertainty of the modern era" (Boden, Giddens, and Molotch 1990, p. B1).

The Focus of Sociology

Because sociology emerged as an effort to understand the dramatic and almost incalculable effects of the Industrial Revolution on all phases of human life, the discipline is wide-ranging in its concerns. The value and the vastness of sociology are evident in the way we define the discipline (Bardis 1980). We might define **sociology** as the scientific study of the causes and consequences of human interaction. Human interaction can include an encounter between strangers on an elevator, a conversation between a physician and a patient, the interrelations among teammates, and the relationships among people of a large organization. Although Marx, Durkheim, and Weber would agree with this definition, each thinker had distinct ideas about what sociologists should emphasize (Lengermann 1974).

Karl Marx (1818–1883)

Karl Marx was born in Germany but spent much of his professional life in London, working and writing in collaboration with Friedrich Engels. For some interesting background information on Karl Marx, see the first five or so pages of the 1879 *Chicago Tribune* "Interview with Karl Marx."

http://csf.Colorado.EDU/psn/marx/Archive/Interviews/1879int1.htm

Document: Interview with Karl Marx

 Why was Marx living in London at the time of this interview? What intellectual qualities did Marx possess? How familiar was Marx with events taking place in the United States? Marx founded the International Society. What were the goals of the International Society?

Two of Marx and Engels's most influential treatises are *The Communist Manifesto* and *Das Kapital*. *The Communist Manifesto*, a pamphlet issued in 1848, outlines the principles of communism, among other things. It begins, "A spectre is haunting Europe," and closes with these lines: "The workers have nothing to lose but their chains; they have a whole world to gain. Workers of all countries unite." *Das Kapital*, a massive multivolume work, is critical of the capitalist system and predicts its defeat by a more humane and cooperative economic system, socialism. Marx's vision of the way to build a society as outlined in these treatises has profoundly influenced economic, social, and political life around the world, especially in the People's Republic of China, the former Soviet Union, and Eastern Europe—although Marx would not necessarily have approved of the effect.

According to Marx the sociologist's task is to analyze and explain *conflict*, the major force that drives social change. The character of conflict is shaped directly and profoundly by the **means of production,** the resources (land, tools, equipment, factories, transportation, and labor) essential to the production and distribution of goods and services. Marx viewed every historical period as characterized by a system of production that gave rise to specific types of confrontation between an exploiting class and an exploited class. For Marx class conflict was the vehicle that propelled people from one historical epoch to another. Over time free men and slaves, nobles and commoners, barons and serfs, guildmasters and journeymen—all have confronted one another. The emergence of machines as a means of production was accompanied by the rise of two distinct classes: the **bourgeoisie,** the owners of the means of production, and the **proletariat,** those who must sell their labor to the bourgeoisie to earn a living. Marx expressed profound moral outrage over the plight of the proletariat, who at the time of his writings were unable to afford the products of their labor and whose living conditions were deplorable. In fact every aspect of people's lives—their occupations, their religious beliefs, their level of education, their leisure activities—personifies the social class to which they belong. Marx devoted his life to documenting and understanding the causes and consequences of this inequality, which he linked to a fatal flaw in the organization of production (Lengermann 1974).

As one measure of the extent of Marx's influence, see the list of Marxist sites and biographies of famous Marxists posted on the Web. Keep in mind that people influenced by Marx may not have carried out his ideas in ways that Marx would have approved.

http://www.idbsu.edu/surveyrc/Staff/jaynes/marxism/websites.htm
Document: Other Marxist-Related Sites on the Web

 Select two of the worker/labor sites listed. How are Marx's ideas reflected in their mission/purpose statements?

Emile Durkheim (1858–1918)

The Frenchman Emile Durkheim believed that the sociologist's task is to analyze and explain **solidarity**—the ties that bind people to one another—and the mechanisms through which solidarity is achieved. Just as Marx viewed the means of production as a central sociological issue, Durkheim regarded solidarity as the essential concern. Durkheim observed that as society became industrialized, the nature of the ties that had bound people to one another changed in profound ways. Before the Industrial Revolution people were drawn together because they were similar to one another. Specifically most people made their living by agriculture and "spent their whole lives within a few miles of where they had been born . . . [and] nearly all lived much as their parents and grandparents had lived before them" (Gordon 1989, p. 106). Order and stability existed in society because of this continuity across generations.

Durkheim maintained that a new form of solidarity based on differences and interdependence is characteristic of industrialized society. In such societies there is considerable specialization, and the division of labor is complex. People are united by their inability to earn their livelihoods independent of others and by their need to cooperate with others to survive, even though they do not have emotional ties with most of the people they encounter in the course of a day.

Max Weber (1864–1920)

The German-born scholar Max Weber has had a monumental influence not only on sociology but also on political science, history, philosophy, economics, and anthropology. Although Weber recognized the significance of economic and material conditions in shaping history, he did not believe, as did Marx, that these conditions were the all-important forces in history. Weber regarded Marx's arguments as one-sided in that they neglected the interplay of economic forces with religious, psychological, social, political, and military pressures (Miller 1963).

According to Weber the sociologist's task is to analyze and explain the course and the consequences of **social action**—actions that people take in response to others—with emphasis on the subjective meaning that the involved parties attach to their behavior. Certainly Weber's preoccupation with the forces that move people to action was influenced by the fact that he saw "the thrust behind sociological curiosity as residing in the endless variety of societies. Everywhere one looks one sees variety. Everywhere one finds [people] behaving differently. . . . Scrutiny of the facts can show endless ways of dealing with the problems of survival, and an infinite wealth of ideas. The sociological problem is to make sense of this variety" (Lengermann 1974, p. 96).

In view of this variety Weber suggested that sociologists focus on the broad reasons people pursue goals, whatever those goals may be (for example, to make a profit, to earn a college degree, or to be recognized by others). He believed that social actions could be classified as belonging to one of four important types: (1) traditional (a goal is pursued because it was pursued in the past), (2) affectional (a goal is pursued in response to an emotion such as revenge, love, or loyalty), (3) value-rational (a goal is pursued because it is valued, with no thought of foreseeable consequences and often without consideration of the appropriateness of the means chosen to achieve it), and (4) instrumental (a goal is pursued after it has been evaluated in relation to other goals and after thorough consideration of the various means to achieve it. (Abercrombie, Hill, and Turner 1988; Coser 1977; Freund 1968).

An example will help clarify the distinctions between the four types of action. Consider, for instance, the goal of earning a college degree. If an individual pursues a college degree because everyone in her family for the past three generations is college-educated, the action can be classified as traditional. If she pursues a degree for the love or pure pleasure of learning, the action is affectional. If she decides to attend college because potential employers value and demand a diploma, the action is value-rational. Sometimes the people who approach college in this way view it only as a ticket to apply for jobs; they lose sight of all but this narrow reason for earning a diploma. As a result they may simply go through the motions in college and do only what they absolutely must do. They may take the easiest course, have others write their papers, and skip classes while relying on others to take notes for them.

Instrumental action is more complex. Suppose the individual considers other goals before settling on the goal of earning a college degree. She might contemplate traveling and seeing the world, enlisting in the military, or working for a few years. Once she has chosen the goal, she might consider the various means to obtain it, including taking out a loan, living at home, enlisting in the army to obtain money for college, getting a job and attending school at night, and so on. The individual also might consider the best approach to learning in the context of an increasingly interdependent world. Such a strategy might include taking classes with the most challenging professors; enrolling in mathematics, science, and foreign language classes; and participating in campus activities. All of these behaviors illustrate instrumental action, a well-thought-out, careful approach to defining and achieving goals. With instrumental action all angles and possibilities are considered.

Weber maintained that with industrialization behavior was less likely to be guided by tradition or emotion and was more likely to be value-rational. (He believed that instrumental action was rare.) Weber was particularly concerned about the value-rational approach; he believed it could lead to disenchantment, a great spiritual void accompanied by a crisis of meaning. Disenchantment occurs when people focus so uncritically on the means of achieving a valued goal that they lose sight of that goal.

The current system of education offers an example of disenchantment as a consequence of value-rational action. Multiple-choice and true-false tests, especially in large classes, are used often by teachers to measure how much knowledge students have acquired (the valued goal of education) by studying and attending classes. The danger is that students will come to associate learning with doing well on tests or that teachers will teach students only the material they need to know to do well on tests. In the end many students come to feel a spiritual void because they have lost sight of the real goals of education, such as personal empowerment and civic engagement.

The Evolution of the Discipline of Sociology

Marx, Durkheim, and Weber were wide-ranging and comparative in their outlook; they did not limit their observations to a single academic discipline, a single period in history, or a single society. They were particularly interested in the transformative powers of history, and they located the issues they studied according to time and place. All three lived at a time when Europe was colonizing much of Asia and Africa and when Europeans were migrating to the United States, Canada, South Africa, Australia, New Zealand, and South America. Sociologist Patricia M. Lengermann (1974) believes that overseas expansion had significant consequences for the discipline of sociology:

> Explorers, traders, missionaries, administrators, and anthropologists recorded and reported more or less accurately the details of life in the multitudes of new social groupings which they encountered. Westerners were deluged with the flood of ethnographic information. Never had man more evidence of the variety of answers which his species could produce in response to the problems of living. This knowledge was built into the foundations of sociology—indeed, one impulse behind the emergence of the field must surely have been Western man's need to interpret this evidence of cultural variation. (p. 37)

Even today the ideas of Marx, Durkheim, and Weber continue to influence the discipline of sociology because their ideas have survived the test of time. Over many decades people from a variety of backgrounds have found the ideas of these three scholars useful for thinking about a wide array of situations. (Ideas lose their usefulness if they cannot explain the situations and events that people consider important.)

Yet, despite the importance of Marx, Durkheim, and Weber, we must keep in mind that the discipline of sociology is always evolving and that tens of thousands of people have made it what it is today. The discipline evolves every time someone enters the profession. James M. Henslin, who edited a collection of essays entitled *Down to Earth*

Sociology (1993), asked the contributors to explain why they became sociologists. A sample of responses, which appear in the collection, shows how newcomers bring fresh insights and energy to the discipline:

Elijah Anderson: "I have always been interested in how individuals relate to society and how society relates to the individual. My interest in the social conditions that people experience—especially the marginality that so many blacks feel and how they relate to the wider social system—motivated me to go into sociology to look for some of the answers. I also had good teachers who inspired me. Later I found myself wanting to contribute in a meaningful way to correcting what I saw to be misrepresentations of reality in the academic literature about people who live in ghettos." (P. xxi)

Barrie Thorne: "A major reason I became a sociologist is that Everett C. Hughes, one of my teachers, used to observe that sociology continually shows us that 'It could be otherwise.' That is, social life—the way school, work, families, daily experience are organized—may feel permanent and given, but the arrangements are socially constructed, have changed over time, and can be changed." (P. xxxi)

William J. Chambliss: "I became a sociologist out of an interest in doing something about crime. I remained a sociologist because it became clear to me that until we have greater understanding of the political and economic conditions that lead some societies to have excessive amounts of crime we will never be able to do anything about the problem. Sociology is a beautiful discipline that affords an opportunity to investigate just about anything connected with human behavior and still claim an identity with a discipline. This is its strength, its promise, and why I find it thoroughly engaging, enjoyable, and fulfilling." (P. xxiii)

Unfortunately some voices and experiences have been left out of the record of sociology's history; still others were left out initially but were "discovered" later. The economist and novelist Harriet Martineau (1802–1836) is one example. This Englishwoman began writing in 1825, but only in recent years has her name been included in some sociology textbooks.

http://www.comet.chv.va.us/romance/htm/martbio.htm
Document: Harriet Martineau: A Short Biography

Q: What factors influenced Harriet Martineau to become a writer? What book by Harriet Martineau gained her fame and secured her future as a writer? Why was Martineau not well received in the United States?

From a sociological viewpoint Martineau's most important book is *Society in America*. Sociologists can learn much from the way she conducted her research on the United States. Martineau made it a point to see the country in all its diversity, and she believed it was important to hear "the casual conversation of all kinds of people" ([1837] 1968, p. 54). Martineau wrote:

I visited almost every kind of institution. The prisons of Auburn, Philadelphia, and Nashville; the insane and other hospitals of almost every considerable place; the literary and scientific institutions, the factories of the north, the plantations of the south, the farms of the west. I lived in houses that might be called palaces, in log-houses, and in a farm house. I traveled much in wagons, as well as stages; also on horseback, and in some of the best and worst of steamboats. I saw weddings, and christenings; the gatherings of the richer at watering places, and of the humbler at country festivals. I was present at orations, at land sales, and in the slave market. I was in frequent attendance on the Supreme Court and the Senate; and witnessed some of the proceedings of state legislatures.

I traveled among several tribes of Indians; and spent months in the southern States, with negroes. (PP. 52–53)

Perhaps most useful are the methods that Martineau chose to make sense of all this information. First, she wanted to communicate her observations without expressing her judgments of the United States. Second, she gave a focus to her observations by asking readers to compare the actual workings of the society with the principles on which the country was founded, thus testing the state of affairs against an ideal standard. Third, with this focus in mind, Martineau asked her readers "to judge for themselves, better than I can for them . . . how far the people of the United States lived up to, or fell below, their own theory" (pp. 48, 50).

Another voice that initially was ignored and later was "discovered" as important to sociology is W. E. B. Du Bois (1868–1963), who wrote in the "Forethought" to his book *The Souls of Black Folk:* "Herein lie buried many things which if read with patience may show the strange meaning of being black here in the dawning of the Twentieth Century. The meaning is not without interest to you Gentle Reader; for the problem of the Twentieth Century is the problem of the color-line." The following Web sites contain background on Du Bois's life and accomplishments.

http://www.gms.ocps.k12.fl.us/biopage/dubois.html
Document: W. E. B. Du Bois

http://ezinfo.ucs.indiana.edu/~jgreen/home16.html
Document: W. E. B. Du Bois

http://www.calpoly.edu/~clor/bois.html
Document: W. E. B. Du Bois

Q: What was Du Bois's scholarly contribution to the discipline of sociology? Which books are named as important to the discipline of sociology? What is Du Bois's most famous line?

Read the "Forethought" and "After-thought" to *The Souls of Black Folk.*

http://www.cc.columbia.edu/acis/bartleby/dubois/0.html
Document: The Forethought

http://www.cc.columbia.edu/acis/bartleby/dubois/15.html
Document: The After-thought

 Which sentences from the "Forethought" and "After-thought" suggest that Du Bois believed that the ideas in *The Souls of Black Folk* would not be heard?

Ignoring some people's ideas not only weakens the discipline but does a disservice to those who have been ignored. In *The Mismeasure of Man* Stephen Jay Gould (1981) writes about the agony of being denied the opportunity to participate: "We pass through this world but once. Few tragedies can be more extensive than the stunting of life, few injustices deeper than the denial of an opportunity to strive or even to hope, by a limit imposed from without, but falsely identified as lying within" (pp. 28–29). Gould's statements can be clarified by the case of Josh Gibson, an African-American baseball player whose talents have been equated to those of Babe Ruth. Legend has it that Gibson hit a thousand home runs while in the Negro Leagues, as they were known then—leagues for players of color who were not permitted to play in "organized baseball." (Baseball was integrated with the arrival of Jackie Robinson in 1947.) Spectators described Gibson's home runs as "quick smashing blows that flew off the bat and rushed out of the stadium" (Charyn 1978, p. 41). In 1943 Gibson suffered a series of nervous breakdowns and was institutionalized at St. Elizabeth's Hospital in Washington, DC, leaving to play baseball on the weekends. He began to hear voices; his teammates found him "sitting alone engaged in a conversation with [Yankee star] Joe DiMaggio," a man he was never permitted to play with or against: "'C'mon, Joe, talk to me, why don't you talk to me? . . . Hey, Joe DiMaggio, it's me, you know me. Why don't you answer me? Huh, Joe? Huh? Why not?'" (p. 41). The case of Josh Gibson reminds us that people's physical characteristics (race, sex, age) cannot be the criteria for judging the worth of their ideas and contributions.

One of the most troubling aspects of writing this introduction to the discipline was deciding whose ideas to include. Over the course of my career as a sociologist, I have read materials written by sociologists and nonsociologists that have enriched my understanding of the discipline. Yet I cannot hear, see, or read all of the significant ideas that exist. Nor is there enough space to include all of the important ideas I have encountered. Please keep in mind that I am limiting your exposure to the discipline of sociology to what I have been able to see, experience, read, and write. Remember that thousands of meaningful contributions exist. I hope you will gain an appreciation of the sociological perspective from what I have included in this text and will come to believe that sociology offers a valuable way of looking at personal, local, national, international, and global events. As sociologist Peter Berger (1963) wrote in *Invitation to Sociology*, "The fascination of sociology lies in the fact that its perspective makes us see in a new light the very world in which we have lived all our lives. It can be said that the first wisdom of sociology is this—things are not what they seem" (pp. 21, 23).

Additional Reading

For more on

- W. E. B. Du Bois
  http://www.swarthmore.edu/SocSci/History/Progs/Dubois.html
  Document: The Souls of Black Folk

- Emile Durkheim
  http://www.lang.uiuc.edu/RelSt/Durkheim/DurkheimHome.html
  Document: Durkheim Home Page

- Harriet Martineau
  http://miso.wwa.com/~jej/martinh.html
  Document: Harriet Martineau

- Karl Marx and Friedrich Engels
  http://csf.Colorado.EDU/psn/marx/index1.htm (for nongraphic browsers)
  Document: The Marx-Engels Internet Archives

  http://csf.colorado.edu/psn/marx (for graphic browsers)
  Document: The Marx-Engels Internet Archives

- The steam locomotive
  http://www.arc.umn.edu/~wes/misc/other.html
  Document: Other Web Sites Providing Steam Locomotive Information

  http://www-cse.ucsd.edu/users/bowdidge/railroad/rail-groups.html
  Document: Museums and Historical Societies

- The effects of economic restructuring on one-industry towns
  http://www.fivash.com/features/smltowns.htm
  Document: Washington Small Towns

- The effects of industrialization
  http://www.cadbury.co.uk/facts.htm
  Document: The Effects of Industrialization: The Case of Chocolate Easter Eggs

Theoretical Perspectives

Consider the following newspaper and magazine article headlines:

- "Drawing a Family History Out of Cyberspace"
- "Internet Is Becoming an Essential Tool"
- "Internet Snares First Criminal as FBI Traps Bank Robber in the Web"
- "Teachers Explore Online Options"
- "More Michigan Colleges Bring the Classroom to Cyberspace"
- "Parishioners Flock Back to the Fold As Some Churches Evangelize On-Line"
- "Judges Turn Back Laws to Regulate Internet Decency"

Most of us have seen these kinds of headlines, and even if we have not accessed the internet, we have heard many things about its potential and peril. Taken together such headlines suggest that the internet, "whatever it is . . . is coming straight at us with the speed and momentum of a locomotive at full steam. We can either prepare to jump on board or be mowed down by it" (Crawford 1995). Sociology offers several perspectives or theories that can help us put what we hear, read, or experience about the internet into a broader context.

In the most general sense a **theory** is a framework that can be used to comprehend and explain events. In every science theories serve to organize and explain events that can be observed through the senses (sight, taste, touch, hearing, and smell). A **sociological theory** is a set of principles and definitions that tell how societies operate and how people relate to one another. Three theories dominate the discipline of sociology: the functionalist, the conflict, and the symbolic interactionist perspectives.

This chapter outlines the basic assumptions and definitions of each of these theories and shows how each one offers a distinct framework that can be used to interpret any event or social phenomenon. Upon outlining each perspective, we show how each perspective can be used to think about the internet, a social phenomenon that is being billed by some as "Renaissance Two" or the "Second Coming of the Printing Press."

Each perspective offers a central question to help guide our thinking about a particular event and a vocabulary for answering the central question. Keep in mind that one theoretical perspective is not necessarily better than the others. Each simply gives a different angle from which to analyze an event. We turn first to an overview of the functionalist perspective. The central questions functionalists ask are, Why does a particular arrangement exist? and What are the consequences of this arrangement for society?

The Functionalist Perspective

Functionalists focus on questions of order and stability in society. They define society as a system of interrelated, interdependent parts. The parts are connected so closely that each one affects all the others as well as the system as a whole. Functionalists consider a **function** to be the contribution of a part to the larger system and its effect on other parts in the system.

Overview

To illustrate this complex concept, early functionalists used biological analogies. For example, the human body is made up of parts that include bones, cartilage, ligaments, muscles, a brain, a spinal cord, nerves, hormones, blood, blood vessels, a heart, a spleen, kidneys, lungs, and chemicals, all of which work together in impressive harmony. Each part functions in a unique way to maintain the entire body, but it cannot be separated from other parts, which it affects and which in turn help it to function. Consider eyelids: when they blink, they work in conjunction with tear fluid, tear ducts, the nasal cavity, and the brain to keep the corneas (the transparent coat over the eyes) from drying out and clouding over. Furthermore some scientists speculate that blinking functions in some way to activate the brain, which controls the body and performs thought processes (Rose 1988).

Functionalists see society, like the human body, as made up of parts, such as food-growing techniques, sports, medicine, bodily adornments, funeral rites, greetings, religious rituals, laws, language, modes of transportation, appliances, tools, and beliefs. Like the various body parts each of the social parts functions to maintain a larger system. For example, one of the many functions of the U.S. educational system is to transmit knowledge and skills from one generation to the next. Thus, for instance, young people learn about previous generations' solutions for meeting various environmental challenges and are not forced to start from scratch. As a second example consider sports teams, whether they be Little League, grade school, high school, college, Olympic, or professional teams. Sports teams function to unify people who are often extremely different from one another economically, culturally, linguistically, politically, religiously, and so on. Loyalty to a sports team transcends individual differences and helps to bind people to a school, a company, a city, or a country.

In the most controversial form of this perspective, functionalists argue that all parts of society, even those that seem not to serve a purpose, contribute to the system's stability. Functionalists maintain that a part would cease to exist if it did not serve some function. Thus, for example, such phenomena as poverty, illegal immigration, and the transfer of labor from the United States into Mexico contribute to the stability of the social system.

Herbert Gans (1972) argued this point in his classic analysis of the functions of poverty. To his own question, "Why does poverty exist?" he answered that poverty performs at least 15 functions, several of which are described here:

- The poor have no choice but to do the unskilled, dangerous, temporary, dead-end, undignified, menial work of society at low pay. Hospitals, hotels, restaurants, factories, and farms draw their employees from a large pool of workers who are forced to work at minimum or below-minimum wages. This hiring policy keeps the costs of their services reasonable and increases profits.

- Affluent persons contract out and pay low wages for many time-consuming activities such as housecleaning, yard work, childcare, and grocery shopping. This practice gives them time for other, more "important" activities. This function of poverty was brought to national attention in 1993 when President Clinton's first two nominees for U.S. attorney general (Zoe Baird and Kimba Wood) disclosed that they had employed illegal immigrants to care for their children. Nor is this childcare arrangement confined to the most prominent and most affluent Americans. Many middle-class and even lower-middle-class Americans make similar arrangements.

- The poor often volunteer for over-the-counter and prescription drug tests. Most new drugs must be tried on healthy subjects to determine potential side effects (for example, rashes, headaches, vomiting, constipation, drowsiness) and appropriate dosages. Money motivates subjects to volunteer. Because payment is relatively low, however, the tests attract a disproportionate share of poor people as subjects.

- The occupations of some middle-class workers—police officers, psychologists, social workers, border patrol guards, and so on—exist to serve the needs or to monitor the behavior of poor people. For example, about 3,300 U.S. immigration agents patrol the U.S.-Mexican border (Kilborn 1992). Similarly physicians and grocery store owners are reimbursed for serving poor people. The poor receive food stamps and medical cards; the providers receive cash.

- Poor people purchase goods and services that otherwise would go unused. Day-old bread, used cars, and secondhand clothes are purchased by or donated to the poor. In the realm of services, the labor of many less competent professionals (teachers, doctors, lawyers), who would not be hired in more affluent areas, is absorbed by low-income communities.

Gans outlined the economic usefulness of poverty to show how and what the parts of the society that everyone agrees are problematic and should be eliminated remain intact: these parts of the society contribute to the stability of the overall system. Based on this reasoning the economic system would be strained seriously if poverty were completely eliminated; industries, consumers, and occupational groups that benefit from poverty would be forced to adjust.

Although functionalists emphasize how parts contribute to the stability of the system, they also recognize that the system does not remain static: as one part changes, other parts adjust and change in ways that lessen or eliminate strain.

Critique of Functionalism

As you may have realized by now, the functionalist perspective has a number of shortcomings. First, critics argue that the functional theory is by nature conservative—"merely the orientation of the conservative social scientist who would defend the present order of things" (Merton 1967, p. 91). In other words, when functionalists identify how a problematic part of society such as poverty contributes to the system's stability, by definition they are justifying its existence and legitimating the status quo. Critics argue that so-called system stability is more often than not achieved at a cost to some segment of the society, such as those who are poor and powerless. Functionalists reject this criticism, claiming that they are simply illustrating why controversial practices or phenomena continue to exist despite efforts to change or eliminate them.

Second, critics argue that a part may not be functional when it is first introduced. Often practices and inventions find their usefulness only after they have come into existence. When the automobile was invented, for example, there were no paved roads, so the automobile's use was limited. It was not until some time later that it realized its transportation function:

> The availability of cheap energy and an inexpensive mass[-produced] car soon transformed the American landscape. Suddenly there were roads everywhere, paid for, naturally enough, by a gas tax. Towns that had been too small for the railroads were reached now by roads, and farmers could get to once-unattainable markets. Country stores that sat on old rural crossroads and sold every conceivable kind of merchandise were soon replaced by specialized stores, for people could drive off and shop where they wanted to. Families that had necessarily been close and inwardly focused in part because there was nowhere else to go at night became somewhat weaker as family members got in their cars and took off to do whatever they wanted to do. The car stimulated the expansiveness of the American psyche and the rootlessness of the American people; a generation of Americans felt freer than ever to forsake the region and the habits of their parents and strike out on their own. (HALBERSTAM 1986, PP. 78–79)

The examples of poverty and the automobile complicate the functionalist argument that a part exists because it contributes to order and stability in society. The automobile example also suggests that necessity is not always the mother of invention, but rather that the invention often is the mother of necessity.

Third, critics assert that the part of the system that fills a function is not necessarily the only way or the most efficient way to meet the function. Although the automobile functions to connect people with one another, it is not an environmentally sound means of doing so. In fact the widespread use of and reliance on the automobile hinders the development of more environmentally efficient modes of public transportation.

Reliance on automobiles also reduces socializing with others, something that public transportation demands and reinforces.

Finally, because the functionalist perspective assumes that a part's function must support the smooth operation of society, it has difficulty accounting for the origins of social conflict or other forms of instability. This assumption leads functionalists to overlook the fact that inventions may have negative consequences for the system or for certain groups within the society.

To address some of this criticism, sociologist Robert K. Merton (1967) introduced a few concepts that help us think about a part's overall effect on the social system and not just its contribution to stability. These concepts are manifest and latent functions and dysfunctions.

Merton's Concepts

Merton distinguished between two types of functions that contribute to the smooth operation of society: manifest functions and latent functions. **Manifest functions** are the intended, recognized, expected, or predictable consequences of a given part of the social system for the whole. **Latent functions** are the unintended, unrecognized, unanticipated, or unpredicted consequences. To illustrate this distinction, consider the manifest and latent functions associated with annual communitywide celebrations such as fireworks displays on the Fourth of July or Labor Day or concerts in the park. Often corporate sponsors join with the city to mount such events. Three manifest functions readily come to mind: (1) the celebration functions as a marketing and public relations event for the corporate sponsor or sponsors, (2) the event provides an occasion to plan activities with friends, and (3) the celebration unifies the community through a shared experience.

At the same time, however, several unexpected or latent functions are associated with community celebrations. First, such celebrations often give a visible role to public transportation systems as people take buses or ride trains to avoid traffic jams. Second, such events function to break down barriers across neighborhoods. People who do drive find they have to park some distance from the event, often in neighborhoods they would not otherwise visit. Consequently, after they park, people have the opportunity to walk through neighborhoods and observe life up close instead of at a distance.

Merton also pointed out that parts of a social system can have **dysfunctions** as well; that is, they can sometimes have disruptive consequences to the system or to some segment in society. Like functions dysfunctions can be manifest or latent. **Manifest dysfunctions** are the expected or anticipated disruptions to order and stability that a part causes in some segment of the system. Some predictable disruptions that seem to go hand in hand with communitywide celebrations are traffic jams, closed streets, piles of garbage, and shortage of clean public toilets.

Latent dysfunctions are unintended, unanticipated negative consequences. Communitywide celebrations often have some unanticipated negative consequences. Sometimes police departments choose to negotiate contracts with the host city just prior to the celebration, thereby using the event as a bargaining tool to secure a good contract. (Actually one might argue that this is a latent function for the police and a latent dysfunction for the city.) Another latent dysfunction of the communitywide celebration is that many people celebrate to the point that they miss class or work the next day in order to recover.

You can see from just this brief analysis of communitywide celebrations that the concepts of manifest and latent functions and dysfunctions make for a more balanced perspective than does the concept of function alone. The addition of these concepts eliminates many of the criticisms leveled at the functionalist perspective. But this broader functionalist approach also introduces a new problem: it gives us "no techniques . . . for adding up the pluses and the minuses and coming out with some meaningful over-all calculation or quotient or net effect" (Tumin 1964, p. 385). In regard to the communitywide celebration we are left with the question of whether the *overall* impact of this event has had a positive or negative effect on society.

The Functionalist Perspective on the Internet

To see how the functionalist perspective can be applied to a specific event, let's consider how functionalists would view the internet. From a functionalist viewpoint the internet can be described as "the convergence of all forms of electronic media into a 'neural net' of shared human consciousness at an international level" (Crawford 1995). The statement names at least two functions of the internet: (1) it functions as a "neural network" that permits access to ideas posted by people around the world and (2) access to this network is at an individual's fingertips.

We can apply Merton's concepts to analyze the overall consequences of the internet on society. The concepts of manifest, latent, function, and dysfunction remind us that in assessing the internet we must look for intended and unintended consequences that lead to both order and disorder.

To understand the intended, planned function of the internet, we need to understand why it was created in the first place. The internet is a Cold War invention. The Cold War is the name given to the political tension and military rivalry that existed between the United States and the former Soviet Union from approximately 1945 until its symbolic end on November 9, 1989, the day the Berlin Wall fell. The Cold War included an arms race, an intense and ongoing buildup by the Soviet Union and the United States in which each side competed to match and surpass any advances made by the other in the quantity and technological quality of weapons. Both countries created a vast, entrenched armaments industry; both made the research and development of weapons (especially nuclear weapons) their highest priorities. The U.S. government had pulled together scientists from three sectors—military, industrial, and academic—to coordinate their research and expertise in the Cold War effort. Because these scientists worked in offices and labs located all around the United States, Defense Department officials worried about what could happen if a military lab, defense contractor, or university site were attacked. Officials realized they needed a computer network that would allow information stored at one site to be transferred to another site in the event of an attack. At the same time, the computer network had to be designed such that if one or more parts of the network failed, the other parts could continue to operate. Such a design meant that there could be no central control over the network. After all, if central control is knocked out, the entire network crashes. The internet began as ARPANET (Advanced Research Projects Agency), and it linked together four universities: UCLA, UC Santa Barbara, the University of Utah, and Stanford University.

This brief overview of the history of the internet suggests that it was designed to achieve the following manifest functions: (1) to transfer information from one site to another quickly and efficiently in the event of war and (2) to create an information-sharing system absent central control. For insights about latent functions and manifest and latent dysfunctions, see "Renaissance Two: Second Coming of the Printing Press" by Jack Crawford.

http://www.lincoln.ac.nz/reg/futures/renaiss2.htm

Document: Renaissance Two: Second Coming of the Printing Press?

Crawford identifies a number of consequences associated with the introduction of the internet. For example, the internet will (1) speed up old ways of doing things, (2) give everyone access to the equivalent of a printing press, (3) represent a profound pedagogical challenge to traditional schooling, (4) change leadership and management styles, (5) challenge those who sit in positions of absolute power, (6) allow people to bypass the formalized hierarchy, and (7) change how students learn. How might you expand on the meaning of each of these seven functions? Would you classify each as a manifest function, manifest dysfunction, latent function, or latent dysfunction? Explain your rationale for each classification.

When making decisions about classification, keep in mind that the distinction between manifest and latent is not always clear-cut, nor is the distinction between function and dysfunction. For example, the absence of central control is a planned element of the internet; thus it is a manifest consequence. Depending on how you look at it, this consequence could be classified as a manifest function or a manifest dysfunction. The concepts are useful not because they allow for clear distinction, but because they remind us to look beyond the most obvious consequences. The "correctness" of your classification lies with the quality of your rationale.

It is virtually impossible to list all of the consequences associated with the internet because the internet, apart from the Defense Department's application, is still in its infancy. The following list of Web sites gives some ideas about the internet's functions and dysfunctions.

http://info.isoc.org/papers/truth.html

Document: Truth and the Internet

http://www.columbia.edu/~rh120/ch106.xpr

Document: Netizens: An Anthology: Preface

gopher://borg.lib.vt.edu/00catalyst/v22n3/katz.v22n3

Document: Redefining Success: Public Education in the 21st Century

http://www.sjcoe.k12.ca.us/employees.html

Document: Employee Internet Ethics and Acceptable Use Agreement

http://www.islandnet.com/%7Ercarr/oddy.html#mentornumbers

Document: Oddysey, Volume 3.2, May, 1995

Q: Select two documents from this list. Do they give insight into the net's manifest and latent functions or dysfunctions? Explain.

The Conflict Perspective

In contrast to functionalists, who emphasize order and stability, conflict theorists focus on conflict as an inevitable fact of social life and as the most important agent for social change. Conflict can take many forms besides outright physical confrontation, including subtle manipulation, disagreement, dominance, tension, hostility, and direct competition. The conflict theorists emphasize the role of competition in producing conflict. Dominant and subordinate groups in society compete for scarce and valued resources (access to material wealth, education, health care, well-paying jobs, and so on). Those who gain control of these resources strive to protect their own interests against the resistance of others. The fact-of-life viewpoint has motivated Michael and Rhonda Hauben to draft a Declaration of the Rights of Netizens, which argues that access to the internet should be considered a human right and not something reserved for a privileged elite.

http://www.columbia.edu/~rh120/netizen-rights.txt

Document: Proposed Declaration of the Rights of Netizens

Q: What rights are covered in the Declaration of the Rights of Netizens?

Overview

Conflict theorists draw their inspirations from Karl Marx, who focused on class conflict. Marx maintained that there are two major social classes and that class membership is determined by relationship to the means of production. The more powerful class is the bourgeoisie, or those who own the means of production (land, machinery, buildings, tools) and purchase labor. The bourgeoisie, motivated by the desire for profit, need constantly to expand markets for their products. In addition they search for ways to make production more efficient and less dependent on human labor (using machines, robots, and automation, for example), and they strive to find the cheapest labor and raw materials. These needs spread "the bourgeoisie over the whole surface of the globe. It must nestle everywhere, settle everywhere, establish connections everywhere" (Marx [1888] 1961, p. 531). According to Marx, in less than 100 years of existence, the bourgeoisie "has created more massive and more colossal productive forces than have all preceding generations together" (p. 531).

The less powerful class, the proletariat, consists of the workers who own nothing of the production process except their labor. The proletariat's labor is a commodity no different from machines and raw materials. Mechanization combined with the specialization of labor has left the worker with no skills; the worker is an "appendage of the machine, and it is only the most simple, most monotonous, and most easily acquired knack that is required of him" (p. 532). As a result workers produce goods that have no individual character and no sentimental value to either the worker or the consumer.

Conflict exists between the two classes because those who own the means of production exploit workers by "stealing" the value of their labor. They do so by paying workers only a fraction of the profits they make from the workers' labor and by pushing them to increase output. Increased output without a commensurate pay raise reduces wages to an even smaller fraction of the profit. In an editorial for the *Labour Standard* Friedrich Engels addressed the value of labor in "A Fair Day's Wage for a Fair Day's Work."

http://www.idbsu.edu/surveyrc/Staff/jaynes/marxism/fairwage.htm
Document: A Fair Day's Wage for a Fair Day's Work

Q: How is a "fair day's wage for a fair day's work" defined in modern capitalist society? Why is the idea of a fair day's wage for a fair day's work a "peculiar" kind of fairness?

The exploitation of the proletariat by the bourgeoisie is disguised by a **facade of legitimacy**—an explanation that members in dominant groups give to justify their actions—or by a justifying ideology. Conflict theorists define **ideologies** as fundamental ideas that support the interests of dominant groups. (The notion that poor people are poor because they are lazy rather than because they are paid low wages is an example of an ideology.) Because ideologies are backed by those in power, they are taken as accurate explanations of why things are the way they are.

On closer analysis, however, ideologies are at best half-truths, based on "misleading arguments, incomplete analyses, unsupported assertions, and implausible premises. . . . [A]ll ideologies foster illusions and cast a veil over clear thinking . . . that enable class divisions in society to persist" (Carver 1987, pp. 89–90). The capitalists' exploitation of the proletariat is justified by the argument that members of the proletariat are free to take their labor elsewhere if they don't like the arrangement. However, this is not the case. On the most basic level employers have considerably more leverage over workers than vice versa: if workers make too many demands, are unreliable, or do not produce—or if business is slow—employers can fire their workers. Workers have no comparable leverage against unreliable and overdemanding employers. Furthermore many workers must take what employers offer, because hundreds of other workers may be waiting to fill the jobs if they refuse. For the most part workers are an "incoherent mass scattered over the whole country, and broken up by their mutual competition" (Marx [1888] 1961, p. 533).

The ideas of Marx inspired most conflict theorists, and the fundamental theory has many variations. Yet, despite these variations, most conflict theorists ask this basic question: Who benefits from a particular pattern or social arrangement and at whose expense? In answering this question, conflict theorists strive to identify practices that the dominant groups have established, consciously or unconsciously, to promote and protect their interests. Exposing these practices helps to explain why there is unequal access to valued and scarce resources.

Most conflict theorists also examine the facade of legitimacy that supports existing practices. They observe how exploitive practices are justified logically by those in power. The most common methods of justification are (1) blaming the victims by proposing that character flaws impede their chances of success and (2) emphasizing that the less successful benefit from the system established by the powerful. Consider the argument a Denver woman gives "MacNeil/Lehrer Newshour" correspondent Tom Bearden (1993) for hiring an illegal immigrant to care for her children:

> *Mr. Bearden:* Does the employer of the undocumented worker have too much power over that person? There are some that believe that people who hire undocumented aliens gain an unfair power over them, it gives them influence over them because they're, in a sense, collaborating in something that's against the law. Do you agree with that, or have any thoughts about that?
>
> *Denver woman:* I guess I would disagree with that. The one thing that you get in undocumented child care or the biggest thing that you probably get, my woman from Mexico was available to me 24 hours a day. I mean, her cost of living in Mexico and quality of life in Mexico compared to what she got in my household were two extremes. When we hired her, she said, "I'll be available all hours of the day, I'll clean the house, I'll cook." They do everything. And if you hire someone from here in the States, all they're going to do is take care of your children. So not only do you have a differentiation in price, you have a differentiation in services in your household. I have to admit that was, at that point, with a newborn infant, was wonderful to have someone who was so available . . .
>
> *Mr. Bearden:* And it's not like the indentured servitude?
>
> *Denver woman:* That crossed my mind, and after she had been here for six months or so, we went to a schedule where she finished at 6 or 7 o'clock at night. And I don't think I ever really took advantage of her. Once a week I'd have her get up with the baby, so I didn't. . . . [S]he was available to me, but I don't feel like I really took advantage of her, other than the fact that I paid her less and she was certainly more available. But she got paid more here than she would have gotten paid if she'd stayed where she was. (p. 8)

Conflict theorists would take issue with the logic that this Denver woman uses to justify hiring an undocumented worker at a low salary. When it comes right down to it, the Denver woman is protecting and promoting her interests (having someone available all hours of the day to cook, clean, and provide childcare) at the expense of the Mexican worker.

As one final example of how capitalists use a facade of legitimacy to justify exploitive practices, let's look at how they explain the fate of jobless workers. More than 12 million American factory workers have lost their jobs because of plant relocations and shutdowns. Many of the companies from which factory workers have been laid off, such as General Electric and General Motors, have established job reentry programs. An article by Peter Kilborn in *The New York Times* highlighted the plight of 1,200 union workers laid off after General Electric moved refrigerator production from Cicero, Illinois, to Decatur, Alabama. There, nonunion workers are paid $9.50 per hour,

$4 less than the company paid the Cicero employees. In Kilborn's article Robert Jones, then the U.S. assistant secretary of labor for employment and training, described the laid-off workers as being "as dysfunctional as you can get." Others who were interviewed described the workers as "not able to meet entry-level requirements of other jobs" and "relatively old and unskilled." Many were described as functioning at fifth- or sixth-grade levels, and some Hispanics were said to "have never had to communicate in English" (Kilborn 1990, p. A12).

At face value these comments suggest that companies like General Electric, in conjunction with the U.S. Department of Labor, are concerned enough to be spending $4 billion in efforts to retrain laid-off employees who, employers believe, lack the intelligence, motivation, skills, and so on to take advantage of the potential employment opportunities available to them. Conflict theorists, however, would point out that the fate of these jobless employees represents the tragic outcome of a production process requiring so little from its workers (other than repetitive manual labor) that the workers have acquired no skills and have nothing marketable to show for decades of employment. From a conflict perspective the laid-off workers are not to blame for their fate. Those who used them as if they were machines to do mindless, repetitive work are to blame.

Critique of Conflict Theory

Like the functionalist perspective conflict theory has its shortcomings. A major criticism is that it overemphasizes the tensions and divisions between dominant and subordinate groups and underemphasizes the stability and order that exists within societies. It tends to assume that those who own the means of production are all-powerful and impose their will on workers who have nothing to offer except their labor. The theory also assumes that the owners exploit the natural resources and cheap labor of poor countries at will and without resistance. This is a somewhat simplistic view of the employer–employee relationship and of relationships between corporations and host countries. It also tends to ignore the real contributions of industrialization in improving people's standard of living.

Finally conflict theorists tend to neglect situations in which consumers, citizen groups, or workers use economic incentives to modify or control the way capitalists pursue profit. For example, the Environmental Defense Fund negotiated successfully with McDonald's to ban polystyrene packaging at its 8,500 restaurants across the United States. The Earth Island Institute Dolphin Project used lawsuits, court injunctions, and letter-writing campaigns to persuade Starkist to change its fishing techniques so as to make oceans safer for dolphins (Koenenn 1992). In a similar vein conflict theorists offer almost no ideas to help us conceptualize the internet as a tool that puts the individual at the center of information or that can be used to circumvent authority and challenge those in positions of absolute power.

The Conflict Perspective on the Internet

From a conflict perspective the internet is the perfect illustration of conflict as an agent of change because it is a product of the Cold War. According to Internet Society (1996) statistics, as of 1996 there were still more registered Defense Department networks on the internet than any other kind of site including educational, government, or commercial-based networks.

Conflict theorists would focus on the internet as part of the means of production, specifically its business (product sales, marketing, customer service, data interchange). Figures on business applications are available from the Commercenet/Nielsen Internet Demographics Survey.

http://www.commerce.net/work/pilot/nielsen_96/exec.html

Document: The CommerceNet/Nielsen Internet Demographics Recontact Study

Q: What percentage of persons surveyed have used the internet in the past three months? What percentage of *internet users* have internet connections at work? What percentage of *all those surveyed* have access to the internet at work? What are some examples of how the internet is used for business purposes?

Although the internet as a business tool is still in the early stage of development, its emergence can be traced to a central need of the capitalist system—"the need for an ever-expanding market for its products, the search for a technology that will allow it to nestle everywhere, settle everywhere, establish connections everywhere. [Capitalism] cannot exist without constantly revolutionizing the instruments of production, and thereby the relations of production, and with them the whole relations of society" (Marx and Engels 1848).

The Symbolic Interactionist Perspective

In contrast to functionalists, who ask how parts contribute to order and stability, and to conflict theorists, who ask who benefits from a particular social arrangement, symbolic interactionists ask, How do people define reality? In particular they focus on how people make sense of the world, on how they experience and define what they and others are doing, and on how they influence and are influenced by one another. These theorists argue that something very important is overlooked if an analysis does not consider these issues.

Overview

Symbolic interactionists have drawn much of their inspiration from American sociologist George Herbert Mead (1863–1931). Mead was concerned with how the self develops, how people attach meanings to their own and other people's actions, how people learn these meanings, and how meanings evolve. Consequently he focused on people and their relationships with one another. He maintained that we learn meanings from others, that we organize our lives around those meanings, and that meanings are subject to change (Mead 1934).

According to symbolic interactionists symbols play a central role in human life. A **symbol** is any kind of physical phenomenon—a word, object, color, sound, feeling, odor, movement, taste—to which people assign a meaning or value (White 1949). However, the meaning or value is not evident from the physical phenomenon alone. This is a deceptively simple idea that suggests that people *decide* what something means. In order to grasp this idea, you might think about how young children question the meaning of everything. As a parent once told me about children, "They don't understand anything. Everything is learned." Meaning for the child evolves through interactions with others.

Let's look at another example. Consider the colors black and white. *The Synonym Finder* (Rodale 1986) lists 160 synonyms for *black* including *threatening, menacing, treacherous,* and *sinful*. For the most part the color has no positive associations. On the other hand, 75 of the synonyms listed for *white* have positive associations, including *pure, stainless, innocent,* and *immaculate*. Although the color white has some negative associations (for example, *colorless, dull, dingy, bloodless, bland, nondescript*), it is not generally associated with ideas of danger.

Where do these symbolic associations come from? Some of my students argued that those for black are related to a universal fear of the dark. In other words the black of night is naturally threatening to all people. However, this interpretation may not be so universal. A Melanesian story, "Finding Night," from *In the Beginning: Creation Stories from Around the World,* shows that "blackness" can symbolize positive ideas such as rest, self-renewal, and rejuvenation. The following excerpt from this story illustrates the point. In this passage the god Quat is asked to do something to put an end to ceaseless daylight.

> In the beginning, there was light. It never dimmed, this light was over everything. It was bright all-light everywhere, and there was no rest from it. . . .
>
> "It's too light. . . . Quat, do something. We don't like the world so bright all the time. Make something to stop it, please, Quat."
>
> Quat looked everywhere for something. Something that was not light. He could find nothing. Light was everywhere. He'd heard about such a place at the far edge of the sky, and it was called *Qong,* Night. Quat . . . sailed over the sea toward the far edge.
>
> He sailed and sailed. Finally, Quat reached the edge where the sky came down and he could touch it. There lived Qong.
>
> Night was dark. It had no light anywhere in it. It touched Quat before his eyes. . . . It taught him to sleep, as well. And the great darkness, Night, gave him another piece of itself.
>
> So Quat went home, taking the piece of Night in his hand. . . . (HAMILTON 1988, PP. 10–13)

Consider a final example: the various meanings assigned to a suntan. In the United States a tan has at various times represented quite different ideas about social class, youthfulness, and health. Around the turn of the century, wealthy persons purposely avoided tanning to distinguish themselves from members of the working class (farmers and laborers). Pale complexions showed that they did not have to make their living outdoors, laboring under the sun. Then, as the basis of the U.S. economy changed from agriculture to manufacturing, a large portion of the population moved indoors to work. The meaning attached to a pale complexion changed accordingly to represent unrelieved indoor labor; a tan came to mean abundant leisure time (Tuleja 1987).

The presence or absence of a suntan also has reflected ideas about health and youthfulness. Many Americans describe a tan as something that makes them look good and feel better. But this meaning is likely to change with increasing reports about the connection between exposure to the sun and premature aging and skin cancer. In the face of such evidence, a tanned complexion symbolizes the skin's desperate attempt to protect the body from radiation. These changes in the meaning of a suntan underscore the fact that a physical form becomes a symbol because people agree on its meaning. Likewise they demonstrate that meanings of symbols change as conditions change.

Symbolic interactionists maintain that people must share a symbol system if they are to communicate with one another. Without some degree of mutual understanding, encounters with others would be ambiguous and confusing. The importance of shared symbols frequently is overlooked unless a misunderstanding occurs. Several TV situation comedies—including "Mork and Mindy," "Perfect Strangers," and "Beverly Hillbillies"—have depended on such misunderstandings by featuring characters who do not share the same symbols as do other characters. These comedies show us that problems arise when involved parties place different interpretations on the same event. They also show that during interaction the parties involved do not respond directly to the surroundings and to each other's actions, words, and gestures. Instead they interpret first and then respond on the basis of those interpretations (Blumer 1962). The interpretation–response process is taken for granted. Usually we are not conscious that the meanings we assign to objects, people, and settings make encounters understandable and shape our reactions. To make us aware of this fact, something must happen that challenges our interpretations.

Anthropologist Edward T. Hall (1992) describes the interpretation that guides how American white males shake hands: "The emphasis is on a firm strong handshake with direct and unblinking eye contact. One must demonstrate mutual respect, equality of status (for the moment at least), strength, sincerity, and dependability" (p. 105). This familiar interpretation, however, is challenged by the Navajo idea of a handshake. For the Navajo

> the emphasis is on proper feelings rather than image. One does not look the other in the eye (to do signals anger or displeasure). All that is necessary is to hold the other human being in one's peripheral vision field while grasping his hand gently, so as not to disturb the natural flow of feeling between his state of being and yours. These handshakes could be protracted because the Navajo . . . like to ease into things and are jarred by abrupt transitions. (pp. 105–6)

In order to understand what distinguishes the symbolic interactionist perspective from the functionalist and conflict, consider how the various theorists might answer the question, Why did the United States send 200,000 troops to the Persian Gulf region after President Saddam Hussein sent Iraqi troops into Kuwait, and why have more than 24,000 troops remained? From a conflict perspective the American military buildup and continued presence in the Persian Gulf can be analyzed in terms of (1) a conflict over a scarce and valued resource (oil), (2) the minority or working-class background of U.S. troops, especially as compared to the upper-class background of most government leaders who decide whether military presence is necessary, and (3) Arab resistance to the U.S. presence in the Middle East.

The buildup can also be analyzed in terms of its unifying function. That is, at the time it took place (August 1990 to January 1991), the military buildup functioned to divert the American people's attention from divisive domestic issues (for example, the savings and loan crisis, the huge budget deficit, relatively high levels of unemployment, and an impending recession). In addition the buildup and continued military presence functioned to protect the U.S. national interests related to oil. A functionalist would acknowledge the class differences between military personnel and government leaders but would emphasize that the military functions to offer working-class people, the unemployed or the poor, and those who want to serve their country a chance to earn money to go to college, to learn a skill, to gain on-the-job experience, or to fulfill a moral obligation.

Symbolic interactionists, in contrast, would focus on how the involved parties interpret the situation and construe each other's words and actions. Thus the symbolic interactionists would focus on interactions between Americans and their Middle Eastern hosts—specifically on situations in which one party adjusted its interpretations and actions to facilitate smooth interaction and situations in which interpretations and actions clashed. For example, symbolic interactionists would be interested in how American women adjusted to Saudi restrictions on dress. Some American women "chose to wear traditional Saudi *abaya,* or black shawl, in public, to avoid problems" (Erlanger 1996, p. Y5).

Critique of Symbolic Interactionism

Symbolic interactionists inquire into factors that influence how we interpret what we say and do, especially those factors that promote the same interpretations from significant numbers of people. Related topics of interest include origins of symbolic meaning, the way in which meanings persist, and the circumstances under which people question, challenge, criticize, or reconstruct meanings. Although symbolic interactionists are interested in these topics, they have established no systematic frameworks for predicting what symbolic meanings will be generated, for determining how meanings persist, or for understanding how meanings change.

Because of these shortcomings, the symbolic interactionist perspective does not give precise guidelines about where to focus one's attention. For example, whose interpretation should we focus on when we analyze an event? The cast of characters involved in these events is virtually endless. Even if we were able to consider every interpretation, we would still be left with the questions, What really happened? and Whose interpretation best captures the reality of the situation?

The Symbolic Interactionist Perspective on the Internet

As we have seen, symbolic interactionists explore the meanings that people assign to words, objects, actions, and human characteristics. Such a broad focus allows us considerable flexibility when it comes to applying it to the internet. For example, we can use this perspective to examine the vocabulary and rules of the internet. Of particular interest would be the rules and vocabulary that relate to and govern human interaction and behavior on the internet. Examples of rules and language can be found in the documents "On-line Survival Tips" and "Emoticons," which cover what is known as "nettiquette." *Nettiquette* is a play on the word *etiquette* and refers to a set of rules for behavior that have evolved over time to facilitate network-related communications.

http://heg-school.aw.com/bc/is/bclink/bclink1/Rules.html
Document: On-line Survival Tips

http://www.organic.com/1800collect/Emoticons/index.html

Document: Emoticons

Q: Why is it important to learn the nonverbal language of the network? What are some examples of nonverbal language? What do "flame on" and "flame off" mean? What are emoticons?

Symbolic interactionists would also be interested in media-generated images of the internet and its users, as well as any other efforts to characterize the internet and its users. Images are important to symbolic interactionists because people organize their behavior and reactions around them. As one example of the power of images, consider the reason that Commercenet/Nielsen gives for why the data they gather on internet users and nonusers is important to businesses and other organizations: "To guide the evolution of Internet commerce and to help businesses understand the opportunities, sound information is of critical importance. Understanding the demographics, attitudes and interests of internet users—and how they differ from those of nonusers—is essential to move the industry forward" (Commercenet/Nielsen 1995).

SRI International offers one profile of Web and non-Web users. Pay particular attention to the VALS segmentation system, which divides the general population into eight groups.

http://future.sri.com/vals/vals-survey.results.html

Document: Exploring the World Wide Web Population's Other Half

Q: What are the eight groups that make up the VALS segmentation system? Which group of Web users is the source of stereotypes about Web users and is the one to which the media devote the most coverage? What group constitutes "the other half" of the Web audience? Which groups are least likely to be part of the internet audience?

Note that SRI further profiles each of the eight groups according to the magazines they read, the packaged foods they purchase, and the kind of music they listen to. Keep in mind that we are not interested in critiquing the accuracy of the segmentation scheme. This profile system is of interest because it is a source of internet-user images for businesses, other organizations, and casual readers.

The symbolic interactionist perspective adds yet another dimension to the way we approach social events. It asks us to consider how the involved parties interpret an event. In this sense it enhances the functionalist and conflict perspectives, which focus on the origins (societal needs versus profit) and consequences (stability versus disruption and exploitation). Each perspective offers a unique set of questions and concepts to answer those questions and alerts us to the need to avoid making simple statements or generalizations about events.

Discussion Question

Based on the information presented in this chapter, how do the three theoretical perspectives help you to frame an analysis of almost any issue?

Additional Reading

For more on

- The global information infrastructure
  gopher://198.80.36.82/00s/usa/media/global/global.txt
  Document: Toward a Global Information Infrastructure

- Information technology
  http://www.beacham.com/essay_contents_848.html
  Document: Information Technology: Essays and Technology Criticism

- Symbolic interaction
  http://sun.soci.niu.edu/~sssi/
  Document: Society for the Study of Symbolic Interaction

- The internet
  http://info.isoc.org/speeches/interop-tokyo.html
  Document: The Present and the Future of the Internet: Five Faces

- Netizens
  http://www.columbia.edu/~rh120
  Document: Netizens: An Anthology

Research Methods and the Information Explosion

Research is a fact-gathering and fact-explaining enterprise governed by strict rules (Hagan 1989). **Research methods** are the various techniques that sociologists and other investigators use to formulate meaningful research questions and to collect, analyze, and interpret facts in ways that allow other researchers to check the results. In this chapter we outline the rationale behind and steps in social research methods.

We need to possess a working knowledge of research methods even if we do not plan to become sociologists or to do research of our own. One important reason is connected with a relatively new global phenomenon—the **information explosion.** This dramatic term describes an unprecedented increase in the volume of data due to the development of the computer and of telecommunications. Notice that we distinguish between data and information. **Data** consists of printed, visual, and spoken materials. Data becomes **information** after someone reads it, listens to it, or views it.

In addition to coping with large quantities of data, we also have to consider the data's quality. Most of the data that we hear, read, and see has been created by others. Therefore we can never be sure that it is accurate. In "Too Much of a Good Thing?: Dilemmas of an Information Society," Donald Michael (1984) argues that we cannot assume that more data will lessen uncertainty and increase feelings of control and security. In fact the opposite may be true: more data can overwhelm us to the point that we conclude we cannot believe anything we hear, read, or see. We need not accept

Michael's gloomy assessment, however, if we possess a working knowledge of social research methods. Such knowledge gives us the skills to identify and create high-quality data.

The Information Explosion

At least two technological innovations are responsible for the information explosion: computers and telecommunications. Both technologies help people create, store, retrieve, and distribute large quantities of data at mind-boggling speeds. Comparing the size and capabilities of the first computers to those of the present suggests why the volume of data has increased so rapidly over the past 50 years. The computers of the 1940s weighed five tons, stood 8 feet tall and 51 feet long, and contained 17,468 vacuum tubes and 5,000 miles of wiring (Joseph 1982). They performed simple calculations in a few seconds, but they tended to overheat and break down. They also used so much power that the lights in nearby towns often failed when the machines were turned on. And, because of their size and cost, computers were used only by the U.S. Defense Department and the Census Bureau. Today, in contrast, a single silicon chip a quarter-inch thick can process millions of bits of information in a second. The chip has reduced computer size and cost and has made possible the widespread use of the personal computer.

Similarly telecommunications have increased our ability to send data quickly across space. Although the telephone, radio, and television have existed in some form for up to 100 years, methods of rapidly transmitting clear signals have changed considerably. Fiber-optic cables have replaced wire cables as the means of transmitting images, voices, and data. In 1923 the cable connecting Britain and the United States contained 80,000 miles of iron and steel wire (enough to circle the earth three times) and 4 million pounds of copper. It could transmit the equivalent of 1,200 letters of the alphabet per minute across the ocean. In contrast, when the capabilities of fiber optics are exploited fully, a single fiber the diameter of a human hair can carry the entire telephone voice traffic of the United States and can transmit the contents of the Library of Congress anywhere in the world in a few seconds (Lucky 1985). This latter is no small feat if we stop to consider that the Library of Congress houses 100 million items on 532 miles of shelves (Thomas 1992).

One software tool that has helped to increase the amount of data available at our fingertips is hypertext. Before hypertext, reading was a linear matter. That is, a book, pamphlet, chapter, column, advertisement, or the like had a clear beginning and end. Most of us know no other way to read. The linear format, however, does not match the way we think, because the mind operates by association—"with one item in its grasp, it snaps instantly to the next by association" (Deemer 1994). Sometimes the associations lead us along unexpected paths. Hypertext is the opposite of the "start-and-end-here" approach in that readers are free to pick and choose among keywords (that is, important ideas) and follow links that enable them to learn more about those keywords. Readers can choose to wander off along tangential links, some of which may not be even remotely related to the topic at hand. Charles Deemer offers an excellent overview of hypertext.

http://www.teleport.com/~cdeemer/essay.html
Document: What Is Hypertext?

Q: How old is the idea of "hypertext" (not hyperspace)? When did the idea of hypertext catch on? What is the World Wide Web? What persistent question does hypertext "ask" readers? Give an analogy for what hypertext is like.

Such innovations as the personal computer and fiber optics have dramatically changed the form of information gathering and dissemination by putting data creation technologies into the hands of the general public and increasing the speed by which data is entered, edited, duplicated, stored, retrieved, and distributed. The brief introduction to a special issue of *Telecine* (Volume 3, Summer 1995), entitled "Turning the Page on Journalism: A Link to the Future of Information Acquisition and Distribution on the World Wide Web," explains how these technologies are transforming the methods for gathering and distributing news. Even though the special issue is directed at journalists, the information can be applied to understanding changes in any information-creating, -gathering, or -disseminating enterprise.

http://omnibus-eye.rtvf.nwu.edu/telecine/Tele-vol3.html#feature
Document: Turning the Page on Journalism

Q: What skills will future journalists need to possess? How did the internet affect the way the Oklahoma City bombing was reported?

As sociologist Orrin Klapp (1986) suggests, the speed with which data is produced and distributed overwhelms the brain's capacity to organize and evaluate it. Furthermore, when we want information on any subject, we must not only select from a large quantity of data but also often sift through distorted, exaggerated presentations. Klapp gives some reasons for these added burdens. New technologies permit large numbers of magazines, newspapers, radio stations, and television channels to exist. As a result message senders must compete for our attention. Reporters, producers, and others in the media often devise ways to entice us to read and listen. Common strategies include eye-catching headlines, misleading titles, and shocking stories (murders, car accidents, plane crashes). Too often such exaggerated headlines lure people into reading and listening to material that turns out to be trivial, repetitive, contradictory, and ultimately uninformative in any broader sense. The titles and headlines might catch our attention, but they usually mask rather than reveal a complex reality.

Klapp also cites **dearth of feedback** as a factor in creating poor-quality data. What does he mean by "dearth of feedback"? Much of the data that is televised and published is not subjected to honest, constructive feedback because there are too many messages

Keyword	Number of Documents, 2-26-96	Number of Documents, _-_-_	Increase in Number of Documents
Sociology	1,195	_____	_____
Sociological theory	92	_____	_____
Research methods	6,450	_____	_____
Culture	12,118	_____	_____
Socialization	105	_____	_____
Social construction of reality	309	_____	_____
Organizations	34,535	_____	_____
Deviance	57	_____	_____
Social stratification	31	_____	_____
Race	6,278	_____	_____
Gender	1,675	_____	_____
Population	6,632	_____	_____
Family	28,947	_____	_____
Education	32,508	_____	_____
Religion	5,193	_____	_____
Social change	3,579	_____	_____

and not enough critical readers and listeners to evaluate the data before it is released or picked up by the popular media. Without feedback the creators cannot correct their mistakes; thus the data they produce is diminished in quality. Klapp believes that data overload, coupled with distortion, exaggeration, and triviality, is as problematic as a lack of data.

With regard to the internet, it is discouraging to plug in a keyword and to confront the tremendous number of documents generated that have not been evaluated for accuracy or worth. Consider the number of documents each of the major sociological keywords shown in the accompanying chart generated on the search engine Webcrawler (**http://webcrawler.com**), as of February 26, 1996. Now plug in each keyword and record the number of documents generated by the search engine Webcrawler. For each keyword subtract the number of February 26 documents from the number generated by your search and record that number. This will give you an idea of how much the number of documents associated with each keyword has grown during this brief time period. Finally select one of the keywords and examine the first 20 to 30 documents. How many of these documents contain what you would consider useful information?

While it is easy to become pessimistic about the human ability to organize, evaluate, comprehend, and trust or question the growing quantity of data, the situation has another, brighter side. For one thing the information explosion also increases the chances that good and useful ideas will receive some exposure. For another the variety means that there is something for everyone. And even though the data is not well organized, researchers still can draw from it and organize it in new and unexpected ways.

The information explosion does not negate the need to be informed; it simply increases the need to be able to create, identify, and synthesize data and turn it into useful and worthwhile information. Decisions still must be made, actions still must be taken, and policies still must be formed. No constructive decision, action, or policy can be based on haphazard, misleading, or inadequate data. That situation would be equivalent to a physician's decision to perform heart surgery based only on the intuition that such action will solve the patient's health problems, without ordering medical tests, reviewing the patient's history, or interviewing the patient beforehand.

The larger point is that we live in a society in which people need to be more than computer literate (able to operate a computer and use it to input, access, and print out data). People also need to be **research-methods literate**—that is, they must know how to collect data that is worth putting into the computer and how to interpret the data that comes out of it. Unless computer-literate people also have research-methods literacy, they merely possess the skills to enter data. We turn now to basic techniques and strategies that sociologists (and all other researchers) use to evaluate and gather reliable data. We start with the guiding principle—the scientific method.

The Scientific Method

Sociologists are guided by the scientific method when they investigate human behavior; in this sense they are scientists. The **scientific method** is an approach to data collection guided by two assumptions: (1) knowledge about the world is acquired through observation and (2) the truth of the knowledge is confirmed by verification—by others making the same observations. Researchers collect data that they and others can see, hear, taste, touch, and smell. They must report the process by which they make their observations and must present conclusions so that interested parties can duplicate that process. If observations cannot be duplicated, or if upon duplication the results differ substantially from those of the original study, the study is considered suspect. Findings endure as long as they can withstand continued reexamination and duplication by the scientific community. When researchers know that others are critiquing and checking their work, the result is reinforcement of careful, thoughtful, honest, and conscientious behavior. Moreover, this "checking" encourages researchers to maintain **objectivity**—that is, to not let personal and subjective views about the topic influence the outcome of the research.

Because of continued reexamination and revision, research is both a process and a dialogue. It is a process because findings and conclusions never are considered final. It is a dialogue because a critical conversation between researchers and readers leads to more questions and additional research.

This description of the scientific method is an ideal one, because it outlines how researchers and reviewers *should* behave. In practice, though, questionable acts on both sides do occur sometimes. Some research may be dismissed as unimportant (often even before it is read) and as unworthy of examination simply because the topic is controversial or departs from mainstream thinking or because the results are reported by someone from a group that is considered "inferior." Moreover the scientific method works on the assumption that researchers are honest—that they do not manipulate data to support personal, economic, and political agendas. However, the extent to which researchers actually are honest is unknown. In one survey sponsored by the American Association for the Advancement of Science (AAAS), 25 percent of the 1,500 people surveyed reported that in the past 10 years they had witnessed some faking, falsifying, or plagiarizing of data (Marsa 1992).

The standards encouraged by the scientific method make research more than a fact-generating enterprise. As we are about to see, research is a carefully planned, multistep, fact-gathering and fact-explaining enterprise (Rossi 1988) that involves a number of interdependent steps:

1. Defining the topic for investigation
2. Reviewing the literature
3. Identifying core concepts
4. Choosing a research design, collecting data, and forming hypotheses
5. Analyzing the data
6. Drawing conclusions

Researchers do not always follow the steps in sequence, however. Sometimes they do not define the topic (step 1) until they have familiarized themselves with the literature (step 2). Sometimes an opportunity arises to gather information about a group (step 4), and a project is defined to fit the opportunity (step 1). Although the six steps need not be followed in sequence, all need to be completed to ensure the quality of the project.

In the sections that follow we will examine each stage individually, making reference to a variety of documents and research projects on the internet that illustrate some part of the research process.

Step 1: Defining the Topic for Investigation

The first step of a research project is choosing a topic. It would be impossible to compile a comprehensive list of the topics that sociologists study, because almost any subject involving humans is open to investigation. Sociology is distinguished from other disciplines not by the topics it covers, but by the perspectives it uses to study topics.

Good researchers explain to their readers *why* their chosen topic is significant. Explanation is vital because it clarifies the purpose and significance of the project, as well as the motivation for doing the work. If you don't know why you are conducting a study, it is unlikely to generate much personal or public interest.

Researchers choose their topics for a number of reasons. Personal interest is a common and often underestimated motive. Historical context and funding opportunities (which often reflect the historical context) are two often overlooked factors that affect the research topic selected. In "The Colonization of Eastern Europe Social Science" we gain further insights into the factors that influence researchers to study the topics they do.

http://szocio.tgi.bme.hu/replika/hozzaszol2.html

Document: Acquired Immune Deficiency Syndrome in Social Science in Eastern Europe: The Colonization of Eastern Europe Social Science

Q: How did the end of the Cold War affect the research agenda of those who specialized in Eastern European studies? Give an example. Who is setting the research agenda today? Why? How have relationships between Eastern European and Western researchers been affected by the post–Cold War agenda?

Step 2: Reviewing the Literature

All good researchers take existing research into account. They read what knowledgeable authorities have written on the chosen topic, if only to avoid repeating what has already been done. Even if researchers believe that they have revolutionary ideas, they must consider the works of past thinkers and show how they advance their ideas or correct errors or oversights. More important, reading the relevant literature can generate insights that the researcher may not have considered. Consider Oscar Grusky, Philip Bonacich, and Cynthia Webster's research on coalition structures in four-person families.

http://www.uiowa.edu/~grpproc/crisp/crisp.1.3.html
Document: The Coalition Structure of the Four-Person Family

 Q: Select "Current Research in Social Psychology." Then select "Prior Issues" and "Vol. 1, No. 3." Notice in their introduction that Grusky, Bonacich, and Webster refer readers to the work of at least 11 different authors, but especially the work of William Gamson and Theodore Caplow. How have the authors of "The Coalition Structure of the Four-Person Family" built upon and extended the work of Gamson and Caplow?

Step 3: Identifying Core Concepts

After deciding on a topic and reading the relevant literature (not necessarily in that order), researchers typically state their core concepts. **Concepts** are powerful thinking and communication tools that enable us to give and receive complex information in an efficient manner. When we hear or read a concept and are familiar with its meaning, it brings to our consciousness a host of associations stored in our brains.

A clear statement of core concepts enables researchers to focus their investigations. In clarifying the concept of family coalitions, Grusky, Bonacich, and Webster direct their attention to situations in which

> family members jointly use their power to control a decision. Coalitions are not the same as affective cliques of mutual attraction. Coalitions are not indicated by the absence of disputes among family members. Family members who do not argue are not in a coalition unless they support one another in disputes with other family members. . . . This definition meets the strict criterion stated by Gamson

(1961b:84) that "participation on the same side of an argument is sufficient justification for asserting that a coalition has been formed."

Step 4: Choosing a Design, Collecting the Data, and Forming Hypotheses

Once researchers have clarified core concepts, they decide on a **research design,** a plan for gathering data on a population they wish to study. Check out the interview with David Gotlib, a medical doctor who has worked with people claiming they have been abducted by UFOs, to learn why it is important that researchers, especially those involved with controversial subjects, work together to establish systematic and uniform procedures for gathering data.

http://www.myna.com/~davidck/gotlib.htm
Document: Transcripts-David Gotlib

What are the shortcomings of UFO research? What solution does Gotlib offer to correct these shortcomings?

A research design specifies the population to be studied and the **method of data collection,** the procedures used to gather relevant data. One research design in itself is not better than another. The population and the data-gathering procedures that are chosen for one study may not be appropriate for another study. Thus researchers choose a design to fit the circumstances of each study (Smith 1991).

The Population to Be Studied

Because of time constraints alone, researchers cannot study entire **populations**—the total number of individuals, traces (objects that yield information about human activity), documents, territories, households, or groups that could be studied. Instead, they study a **sample,** or a portion of the cases from a larger population.

Ideally a sample should be a **random sample,** with every case in the population having an equal chance of being selected. The classic, if inefficient, way of selecting a random sample is to assign every case a number, place the cards or slips of paper on which the numbers are written into a container, thoroughly mix the cards, and pull out one card at a time until the desired sample size is achieved. However, rather than follow this tedious system, most researchers use computer programs to generate their samples. If every case has an equal chance of becoming part of the sample, then theoretically the sample should be a **representative sample**—that is, one that has the same distribution of characteristics (such as age, gender, and ethnic composition) as the population from which it is selected. Thus, for example, if the population from which a sample is drawn is 12 percent Latino, then 12 percent of a representative sample will be Latino. Sometimes just by chance researchers draw samples that do not represent the population with regard to specified characteristics. In that case researchers may draw additional cases from the categories that are underrepresented in the sample. In theory, if the sample is representative, then whatever is true for the sample is also true for the larger population.

Obtaining a random sample is not as easy as it might appear. For one thing, researchers must begin with a **sampling frame**—a complete list of every case in the population—and each member of the population must have an equal chance of being selected. Securing such a complete list can be difficult. Campus and city telephone directories are easy to acquire, but lists of, say, American citizens, of adopted children in the United States, of U.S.-owned companies in Japan, or of Japanese-owned companies in the United States are not so easily obtained. Almost all lists omit some people (perhaps persons with unlisted numbers, members too new to be listed, or between-semester transfer students) and include some people who no longer belong (such as persons who have moved, died, or dropped out). What's important is that the researcher consider the extent to which the list is incomplete and update it before drawing a sample. Even if the list is complete, the researcher also must think of the cost and time required to take random samples and consider the problems of inducing all sampled persons to participate.

Researchers sometimes select samples to study that they know are not representative because the subjects are accessible. For example, researchers often sample from high school and college students because they are a captive audience. In addition, researchers may choose their subjects because little is known about them or because the subjects have special characteristics or their experiences clarify important social issues.

In "Strategies for Identifying and Interviewing 'Deviant' Informants: The Case of Marijuana Growers," Ralph Weisheit explains why it can be difficult to generate a sampling frame, especially when the population is composed of people believed to be "deviant." Weisheit, who studied people growing marijuana for profit in rural communities, describes the labor-intensive process by which he came to identify 71 growers.

http://sun.soci.niu.edu/~sssi/papers/ralphw.txt
Document: Strategies for Identifying and Interviewing "Deviant" Informants

Q: How did Weisheit find his cases and make contact with marijuana growers and other informants? How many marijuana growers did Weisheit interview? What resistance did Weisheit encounter from some government officials, neighbors, and growers? Does Weisheit's research yield an accurate picture of rural marijuana growers in the United States?

As another example of how nonrandom samples are generated, return to Grusky, Bonacich, and Webster's study.

http://www.uiowa.edu/~grpproc/crisp/crisp.1.3.html
Document: The Coalition Structure of the Four-Person Family

Q: By what process did the researchers identify 48 four-person families to interview?

Methods of Data Collection

In addition to identifying who or what is to be studied, the design also must include a plan for collecting information. Researchers can choose from a variety of data-gathering methods including self-administered questionnaires, interviews, observations, and secondary sources.

Self-Administered Questionnaire A **self-administered questionnaire** is a set of questions given (or mailed) to respondents, who read the instructions and fill in the answers themselves. This method of data collection is very common. The questionnaires found in magazines or books, displayed on tables or racks in service-oriented establishments (hospitals, garages, restaurants, groceries, physicians' offices), and mailed to households are all self-administered questionnaires. This method of data collection has a number of advantages. No interviewers are needed to ask respondents questions; the questionnaires can be given to large numbers of people at one time; and respondents are not influenced by an interviewer's facial expressions or body language, so they feel more free to give unpopular or controversial responses.

At the same time, there are some disadvantages to such questionnaires. Respondents are often self-selected in the sense that they choose to ignore or fill out the questionnaire. This leaves researchers wondering whether the people who do not fill out a questionnaire have different opinions than those who do. The results of a questionnaire depend not only on respondents' decisions to fill out and return it but also on the quality of the survey questions asked and a host of other considerations.

Interviews In comparison to questionnaires, **interviews** are more personal. They are face-to-face sessions or telephone conversations between an interviewer and a respondent in which the interviewer asks questions and records respondents' answers. As respondents give answers, interviewers must avoid pauses, expressions of surprise, or body language that reflect value judgments. Refraining from such conduct helps respondents feel comfortable and encourages them to give honest answers.

Interviews can be structured or unstructured, or some combination of the two. In a **structured interview** the wording and sequence of questions are set in advance and cannot be altered during the course of the interview. In one kind of structured interview, respondents are free to answer the questions as they see fit, although the interviewer may ask them to clarify or explain answers in more detail. In another kind of structured interview, respondents choose answers from a response list that the interviewer reads to them.

In contrast, an **unstructured interview** is flexible and open-ended. The question–answer sequence is spontaneous and resembles a conversation in that the questions are not worded in advance and are not asked in a set order. The interviewer allows respondents to take the conversation in directions they define as crucial. The interviewer's role is to give focus to the interview, to ask for further explanation or clarification, and to probe and follow up interesting ideas expressed by respondents. The interviewer appraises the meaning of answers to questions and uses what was learned to ask follow-up questions. Talk show hosts often use an unstructured format to interview their guests. However, sociologists have much different goals than talk show hosts. For one thing, sociologists do not formulate questions with the goal of entertaining an audience. In addition, sociologists strive to ask questions in a neutral way, and there is no audience reaction to influence how respondents answer the questions.

Select one of David Cherniack's transcripts of interviews.

http://www.myna.com/~davidck/transcr.htm

Document: Transcripts Index

Q: Study the interview carefully and try to determine Cherniack's interview plan or style. Is the interview structured, unstructured, or a combination of the two? Which questions do you think he prepared ahead of time and which were asked spontaneously? Can you tell what strategies he uses to keep the interview on track?

Observation As the term implies, **observation** involves watching, listening to, and recording behavior and conversations as they happen. This technique sounds easy, but observation is more than seeing and listening. The challenge of observation lies in knowing what to look for while still remaining open to other considerations; success results from identifying what is worth observing. "It is a crucial choice, often determining the success or failure of months of work, often differentiating the brilliant observer from the . . . plodder" (Gregg 1989, p. 53). Good observation techniques must be developed through practice to learn to know what is worth observing, to be alert to unusual features, to take detailed notes, and to make associations between observed behavior.

If observers come from a culture different from the one they are observing, they must be careful not to misinterpret or misrepresent what is happening. Imagine for a moment how an uninformed, naive observer might describe a sumo wrestling match: "One big, fat guy tries to ground another big, fat guy or force him out of the ring in a match that can last as little as three seconds" (Schonberg 1981, p. B9). Actually, for those who understand it, sumo wrestling is "a sport rich with tradition, pageantry, and elegance and filled with action, excitement, and heroes dedicated to an almost impossible standard of excellence down to the last detail" (Thayer 1983, p. 271).

Observational techniques are especially useful for studying behavior as it occurs, for learning things that cannot be surveyed easily, and for acquiring the viewpoint of the persons under observation. Observation can take two forms: participant and nonparticipant. **Nonparticipant observation** is detached watching and listening: the researcher only observes and does not interact or become involved in the daily life of those being studied. In contrast, researchers engage in **participant observation** when they join a group and assume the role of a group member, when they interact directly with those whom they are studying, when they assume a position critical to the outcome of the study, or when they live in a community under study.

In both participant and nonparticipant observation, researchers must decide whether to hide or to announce their identity and purpose. One of the primary reasons for choosing concealment is to avoid the **Hawthorne effect,** a phenomenon whereby research subjects alter their behavior when they learn they are being observed. If researchers announce their identity and purpose, they must give participants time to adjust to their presence. Usually, if researchers are present for a long enough time, the subjects eventually will display natural, uninhibited behaviors.

In addition to the Hawthorne effect, researchers must consider another issue: how to handle knowledge about "bad guys/girls." *Bad guys/girls* are people who have committed or who know about serious social transgressions. When researchers learn of such transgressions, they have acquired *dirty information*, information that if revealed would have serious repercussions for the researcher or for the research subject. The repercussions might range from professional discreditation to criminal prosecution to the death of an informant. When researchers acquire dirty information, they face the difficult question of what to do with that information. Jim Thomas and James Marquart discuss this issue in their paper "Dirty Information and Clean Conscience." They offer the concepts *honorable* and *ethical* as a way of helping researchers think through their response to dirty information dilemmas. However, their distinction between honorable and ethical is not clear.

http://sun.soci.niu.edu/~sssi/papers/dirty.data

Document: Dirty Information and Clean Conscience: Communication Problems in Studying "Bad Guys"

Q: Thomas and Marquart present three dirty-information scenarios researchers might encounter if they were observing prison life. How would you handle each? Why? Do you agree with Thomas and Marquart's recommendations?

Secondary Sources Another strategy for gathering data is to use **secondary sources.** This is data that has been collected by other researchers for some other purpose. Government researchers, for example, collect and publish data on many areas of life including births, deaths, marriages, divorces, crime, education, travel, and trade. Go to **gopher://jse.stat.ncsu.edu/11/jse/data** and explore the data sets the *Journal of Statistical Education* has archived. Or check out **http://www.census.gov**. Select "Subjects A-Z." Browse the documents to learn about the kinds of secondary data available at your fingertips.

Another kind of secondary data source consists of materials that people have written, recorded, or created for reasons other than research (Singleton, Straits, and Straits 1993). Examples include television commercials and other advertisements, letters, diaries, home videos, poems, photographs, artwork, graffiti, movies, and song lyrics. One example of this kind of secondary data is the *Life History Manuscripts from the Folklore Project,* a product of the Works Project Administration (WPA). The federal government established the WPA in the 1930s to give employment to those who could not find work during the Depression. Specifically the WPA Federal Writers' Project (1936–1940) gave work to unemployed writers or to anyone who could qualify as a writer. Approximately 300 writers from 24 states interviewed people across the country from all walks of life and circumstances. The product of their efforts was a collection containing 2,900 documents ranging from 2,000 to 15,000 words in length. The collection can be searched by keywords (such as textile workers, immigrants, or ex-slaves) and according to region or state.

http://lcweb2.loc.gov/wpaintro/wpahome.html

Document: Life History Manuscripts from the Folklore Project

Q: If you had the opportunity to study a particular group of people from the WPA Writers' Project, which group might it be? Why?

Identifying Variables

Acquiring a conceptual focus, identifying a population, and determining a method of data collection help researchers identify the variables they want to study. A **variable** is any trait or characteristic that can change under different conditions or that consists of more than one category. The variable "sex," for example, is generally divided into two categories: male and female. The variable "marital status" is often separated into six categories: single, living together, married, separated, divorced, and widowed.

Researchers strive to find associations between variables in order to explain and/or predict behavior. The behavior to be explained or predicted is the **dependent variable.** The variable that explains or predicts the dependent variable is the **independent variable.** Thus a change in the independent variable brings about a change in the dependent variable. A **hypothesis,** or trial explanation put forward as the focus of research, predicts how independent and dependent variables are related. This trial idea specifies what outcomes will occur as the independent variable varies.

In "The 'Unusual Episode' Exercise and Its History" Robert J. McG. Dawson presents data on the sex (male versus female), age (child versus adult), economic status, and fate (death versus escape) of passengers on the *Titanic* after it collided with an iceberg on April 15, 1912.

http://www2.ncsu.edu/ncsu/pams/stat/info/jse/v3n3/datasets.dawson.html

Document: The "Unusual Episode" Data Revisited

Q: In this case the independent variables are "sex," "age," and "economic status." The dependent variable is "fate." In predicting a passenger's fate, we might hypothesize that (1) the higher the economic status of the passenger, the greater the chances of escape from death, or (2) regardless of economic status, women and children were more likely to escape death than men (because of the norm "ladies first"). Can you think of a third hypothesis? (*Note:* We will revisit the *Titanic* later in this chapter.)

One major reason researchers collect data is to test hypotheses. In order for the findings to matter, other researchers must be able to replicate the study. For this reason researchers need to give clear and precise definitions and instructions about how to observe and measure the variables being studied.

Operational Definitions

In the language of research, such definitions and accompanying instructions are called **operational definitions.** An analogy can be drawn between an operational definition and a recipe. Just as anyone with basic cooking skills can follow a recipe to achieve a desired end, anyone with basic research skills should be able to replicate a study if he or she knows the operational definitions (Katzer, Cook, and Crouch 1991). For example, one operational definition of education is the number of years of formal schooling a person has completed. In the case of the *Titanic* study, the operational definition for the variable "passenger socioeconomic status" was the class by which that person traveled (first class, second cabin, or steerage cabin low in the ship's hull). The operational definition for "reasons people move from one dwelling to another" can be found in the document "Why Move?"

http://www.census.gov/ftp/pub/hhes/housing/ahs/tab2-11.html
Document: Why Move?

Q: Respondents are given a checklist of reasons for moving and asked to check the category that best explains the reason for their move. The categories include (1) private displacement, (2) government displacement, (3) disaster, (4) new job or job transfer, and (5) to be closer to work/school/other. What are the operational definitions for the variables "choice of present neighborhood" and "social mobility" (as measured by quality of home or neighborhood)?

Many more examples of operational definitions are available.

http://www.census.gov/ftp/pub/hhes/housing/ahs/tab2-3.html
Document: Big Homes?

Q: What are at least three operational definitions that could be used to observe the variable "home size"?

http://www.census.gov/ftp/pub/hhes/housing/ahs/tab2-2.html
Document: Good Buildings

Q: What is an operational definition for the variable "building quality"?

---

http://aspe.os.dhhs.gov/poverty/poverty.htm
Document: HHS Poverty Guidelines

Q: What operational definition does the U.S. government use to determine if a family lives in poverty?

If the operational definitions are not clear or do not indicate accurately the behaviors they were designed to represent, they are of questionable value. Good operational definitions are reliable and valid. **Reliability** is the extent to which the operational definition gives consistent results. For example, the question "How many magazines do you read each month?" may not yield reliable answers because respondents may forget some magazines. Thus, if you asked the question at two different times, the respondent likely would give two different answers. One way to increase the reliability of this question is to ask respondents to list the magazines that they have read in the past week. The act of listing forces respondents to think harder about the question, and shortening the amount of time to think back on makes it easier to remember.

Return to the *Titanic* study and read Dawson's account of the conflicting data on the number of people who perished.

---

http://www2.ncsu.edu/ncsu/pams/stat/info/jse/v3n3/datasets.dawson.html
Document: The "Unusual Episode" Data Revisited

Q: How does the *Titanic* data that Dawson reviewed at the STATS workshop differ from the data presented in the "Board of Trade Inquiry Report"? What factors might account for these differences?

Validity is the degree to which an operational definition measures what it claims to measure. Professors give tests to measure students' knowledge of a particular subject as covered in class lectures, discussions, reading assignments, and other projects. Students may question the validity of this measure if the questions on a test reflect only the material covered in lectures. In such instances students may argue that the test does not measure what it claims to measure.

---

http://www2.ncsu.edu/ncsu/pams/stat/info/jse/v3n3/datasets. dawson.html
Document: The "Unusual Episode" Data Revisited

Q: With regard to the *Titanic* study, is the class by which passengers travel a valid measure of the person's actual social status? Explain.

Steps 5 and 6: Analyzing the Data and Drawing Conclusions

When researchers get to the stage of analyzing collected data, they search for common themes, meaningful patterns, and/or links as well. Researchers may make use of graphs, frequency tables, photos, statistical data, and so on. The choice of presentation depends on what results are significant and how they might be best shown.

In the final stage of the research process, sociologists comment on the **generalizability** of findings, the extent to which the findings can be applied to the larger population from which the sample is drawn. The sample used and the response rate are both important when it comes to generalizability. If a sample is randomly selected, if all subjects agree to participate, and if the response rate is high, we can say that it is representative of the population and that the findings theoretically are generalizable to that population. If a sample is chosen for some other reason—perhaps because it is especially accessible or interesting—then the findings cannot be generalized to the larger population. Keep in mind that even though one goal of drawing conclusions is to make generalizations about the larger population, generalizations are not statements of certainty that apply to everyone. Because the generalizations do not apply to everyone, it is virtually impossible to claim that one independent variable causes a dependent variable. Consequently, instead of claiming causes, researchers search for independent variables that make significant contributions toward explaining the dependent variable.

At least three conditions must be met before a researcher can claim that an independent variable contributes significantly toward explaining a dependent variable. The first condition is that the independent variable must precede the dependent variable in time. Time sequence can be established easily when the independent variable is a predetermined factor such as sex, ethnicity, or birth date. These kinds of factors are fixed before a person is capable of any kind of behavior. Usually, however, time order cannot be established so easily.

A second condition that must be met before a researcher can claim that an independent variable contributes toward explaining a dependent variable is that the two variables must be **correlated.** The strength of this contribution is represented by a **correlation coefficient,** a mathematical representation of the extent to which a change in one variable is associated with a change in another (Cameron 1963). Correlation coefficients range in value from −1.0 to +1.0, with .00 representing no association between variables, −1.0 representing a perfect negative association, and +1.0 representing a perfect positive association. If, when the value of one variable (for example, number of fire trucks at the scene) increases there is a corresponding increase in the other variable (for example, dollar amount of fire damage), the correlation coefficient is a *positive* number. If, on the other hand, when one variable increases (number of fire trucks at the scene) the other decreases (dollar amount of fire damage), the correlation coefficient is a *negative* number.

Establishing a correlation is a necessary step but is not in itself sufficient to prove causation. A correlation shows only that the variables are related; it does not mean that one variable causes the other. For one thing, a correlation can be spurious. A **spurious correlation** is one that is coincidental or accidental; in reality some third variable is related to both the independent and the dependent variables. The presence of the third variable makes it seem that those two variables are related. A good example is the seeming correlation between the number of fire trucks at the scene and the amount of damage done in dollars (the more fire trucks, the greater the dollar amount of fire damage). However, common sense tell us that this is a spurious correlation. A third variable—the size of the fire—is responsible for both the number of fire trucks sent and the amount of damage done. Although the number of fire trucks called to the scene does help us to predict the amount of damage in dollars, that variable is not the variable of cause.

If a researcher is to claim that an independent variable helps to explain a dependent variable, there must be no evidence that another variable is responsible for a spurious correlation between the independent and the dependent variables. To check this possibility, sociologists identify **control variables,** variables suspected of causing spurious correlations. Researchers determine whether a control variable is responsible by holding it constant and reexamining the relationship between the independent and the dependent variables.

Even if researchers are able to establish these three conditions—that (1) the independent variable precedes the dependent variable in time, (2) there is a correlation between the independent and dependent variables, and (3) there is no evidence of a spurious correlation—it is extremely difficult for them to claim that the independent variable causes the dependent variable. Researchers use a **probabilistic model** of cause in which the hypothesized effect does not always result from a hypothesized cause. In the case of the *Titanic* study, for example, knowing a person's economic status does not mean that we can predict his or her fate with 100 percent accuracy. Under a probabilistic model, if researchers meet the three conditions, they can say that the independent variable makes a significant contribution toward explaining the dependent variable.

For additional information on the distinction among correlation, causation, and prediction, go to Allan J. Rossman's "Television, Physicians, and Life Expectancy." In this document Rossman describes a classroom exercise he uses to teach an important principle: "a strong association between two variables does not necessarily imply a cause-and-effect relationship between them." Note that some of the statistical concepts Rossman discusses are beyond the scope of this chapter. Focus on the information needed to answer the questions below.

http://www2.ncsu.edu/ncsu/pams/stat/info/jse/v2n2/datasets.rossman.html

Document: Television, Physicians, and Life Expectancy

Select "Parent Directory" and then "Rossman, A. J." What data-gathering strategy did Rossman use to find the data for this study? What is the population for this study? What criteria did Rossman use to select a sample from this population? What variables did Rossman consider? What is the dependent variable? Which variable is the best predictor of the dependent variable: "number of people per television set" or "number of people per physician"? Explain. Which one of these variables is clearly a spurious variable? Can you think of some reasons that the second variable might also be spurious?

Discussion Question

Based on the material presented in this chapter, how does a knowledge of the research process help you to evaluate research findings?

Additional Reading

For more on

- The information explosion
  http://www.educom.edu/educom.review/review.96/jan.feb/varion.html
  Document: The Information Economy: How Much Will Two Bits Be Worth in the Digital Marketplace?

  http://www.educom.edu/educom.review/review.96/mar.apr/peters.htm
  Document: Raison d'Net: Are You Ready for the Thing Called "Change"?

  http://www.cs.cmu.edu/afs/cs.cmu.edu/user/bam/www/numbers.html#ComputerInHomeWork
  Document: Computer Use at Home and Work

  http://www.educom.edu/educom.review/review.96/mar.apr/shapiro.html
  Document: Information Literacy as a Liberal Art

- Issues related to reliability
  gopher://justice2.usdoj.gov/00/fbi/November94/4prints.txt
  Path: /Federal Bureau of Investigations/November94
  Document: Best Foot Forward: Infant Footprints for Personal Identification

- Operational definitions (examples of)
  http://stats.bls.gov/flsfaqs.htm
  Document: FLS Frequently Asked Questions

  http://aspe.os.dhhs.gov/poverty/poverty.htm
  Document: Poverty Site

  http://stats.bls.gov/csxgloss.htm
  Document: Consumer Expenditure Surveys Glossary

- Data sets and analysis of data
  http://www2.ncsu.edu/ncsu/pams/stat/info/jse
  Document: Journal of Statistical Education

- Research design
  http://www.dnai.com/~children/report_guide.html
  Document: Step by Step Guide

Culture

The entry for "culture" in the *Cambridge International Dictionary of English* (1995, p. 334), which presents core definitions of words as well as examples, shows the most common usage of the word among English speakers. The entry reads as follows:

> **culture** WAY OF LIFE/*n* the way of life, esp. general customs and beliefs of a particular group of people at a particular time • *youth/working-class/Russian/Roman/mass culture* • *She's studying modern Japanese language and culture.* • *The cultures of Britain and Nigeria are very different.* • *Thatcher's enterprise culture* (= way of thinking and behaving) *of the 1980s brought many changes.* • *There's a* **culture gap** (= difference in ways of thinking and behaving) *between many teenagers and their parents.* • *It was a real* **culture shock** *to find herself in London after having lived on a small island* (= she felt alone and confused by the different way of life there). . . .

This entry shows that we use the word *culture* in conjunction with specific places (Russia, Rome, Nigeria, Britain, small islands) and with categories of people (the masses, teenagers, parents, Russians, Japanese). We also use the word in ways that emphasize differences ("the cultures of X and Y are very different"; "there is a cultural gap between X and Y"; "it is a culture shock to come from X and live in Y"). Our uses of the word suggest that we think of culture as having clear boundaries, as an explanation for behavior, and as a blueprint for living that people follow in mechanical ways. Our uses also suggest that we think of interaction between people of different cultures as problematic.

In light of the seemingly clear way we use the word, we may be surprised to learn that the real challenge and the sources of endless debate among people who study culture include the following:

- *Defining the term.* That is, is it possible to find words to define something so vast as culture?
- *Determining who belongs to a group designated as a culture.* For example, does a person who "looks" Japanese and who has lived in the United States most of his or her life belong to Japanese or American culture?
- *Identifying the distinguishing characteristics that set one culture apart from others.* For example, is eating rice for breakfast a behavior that makes someone Japanese? Is an ability to speak French a behavior that makes someone French?

Mary Hafford's essay on culture offers insights about how to answer these three questions.

http://www.loc.gov/folklife/cwc.html
Document: American Folklife a Commonwealth of Cultures

Q: Based on this description of folklife, can you see why defining culture, determining who belongs to a culture, and identifying distinguishing characteristics that set one culture apart from others represent challenges to the study of culture? Explain your answer.

David Levinson (1991) faced these three nagging questions when he set out to assemble the *Encyclopedia of World Cultures*, a nine-volume reference source (one volume for each of nine geographical regions of the world) that promised to give "accurate, clear, and concise descriptions of the cultures of the world" (p. xvii). As editor-in-chief Levinson chose an editor for each volume and gave these editors the task of selecting the cultures to be included the volume. He did not provide his editors with a working definition of a culture, nor did he offer them a fixed list of criteria for identifying a culture. Levinson did not furnish any of this information because there is no easy formula for identifying a culture or for determining who belongs. In other words, there are no distinctive features or **cultural markers** that can be used to clearly classify people into distinct cultural units. This challenges the common practices of using territory, race/ethnicity, country, and national origin and of identifying distinctive behaviors as markers.

Cynthia K. Mahmood and Sharon Armstrong's (1989) research represents one example of the challenges Levinson and his colleagues must have faced in creating the nine-volume encyclopedia. Mahmood and Armstrong traveled to Eastermar, a village in the Netherlands province of Friesland, to study the Frisian people's reactions to a book published about their culture. They found that the Frisian people were unable to agree on a single "truth" about them as described in the book. At the same time,

Frisian villagers could not come up with a list of features that would apply to all Frisians and that would distinguish them from other people living in Eastermar. Yet the Frisians were "convinced of their singularity," and the villagers reacted emotionally to the suggestion that perhaps they did not constitute a culture. The Frisian situation captures the conceptual challenges associated with the idea of culture: the paradox of recognizing a culture but being unable to define its boundaries, the characteristics determining where a culture begins and leaves off or the qualities marking some people off from others as a unified and distinctive group.

This chapter offers a framework for thinking about culture that considers both its elusive nature and its importance in shaping human life. In this regard sociologists agree on at least eight essential principles regarding the nature of culture.

Material and Nonmaterial Components

- *Principle 1:* Culture consists of material and nonmaterial components. In sociological terms material culture consists of objects or physical substances. Nonmaterial culture consists of those elements that cannot be directly observed or easily described.

At this point we are not concerned about identifying which people share material and nonmaterial culture. Instead, we focus on the kinds of objects, ideas, and behavior people *can* share.

Material Culture

Material culture consists of all the physical objects people have borrowed, discovered, or invented and to which they have attached meaning. Material culture includes natural resources such as plants, trees, and minerals or ores, as well as items that people have converted from natural resources into other forms for a purpose. Examples of the latter include cars and trucks to transport people, animals, and goods; microwave ovens to cook and heat food; computers to make calculations; video cameras equipped with devices that selectively soften facial features to make people look younger than they are; indoor plumbing to bring together "in one room the toilet from the outhouse or closet, the washbowl from the bedchamber, and the tub from the kitchen" (Nasaw 1991, p. 10); radios to entertain, inform, and provide background sound; and so on.

The significant feature of material culture is that people attach meanings to each item, including the purpose for which it is designed, the value placed on it, and the fact that some people are unhappy without it or unhappy about it and direct their energies toward acquiring or eradicating it (Kluckhohn 1949). In thinking about material culture, it is important to learn not only the most obvious and practical uses for which an object is designed but also the meanings assigned to that object by the people who use it (Rohner 1984). As an example consider the radio, a device for receiving and then broadcasting sound messages that travel through the air in the form of electromagnetic waves. For many people the radio takes on meanings beyond the obvious purposes mentioned above. People use it to fill the void that can accompany boring tasks, daily routines, or loneliness; to sustain or create a mood (for example, upbeat, romantic, relaxed); and to provide a social lubricant in that people can talk with one another about what they have heard on the air. The importance of the radio in some people's lives is evident in statements like these: "To me, when the radio is off, the house is empty";

"As soon as I get up in the morning, the first thing I do is turn on the radio"; "Radio puts me in a better mood"; and "It makes driving easier" (Mendelsohn 1964, pp. 242–43). Learning the meaning people assign to objects in the material culture helps us to grasp the significance of those objects in people's lives.

Nonmaterial Culture

Nonmaterial culture consists of intangible creations or things that we cannot identify directly through the senses. Four of the most important of these creations are beliefs, values, norms, and symbols.

Beliefs The first component of nonmaterial culture is **beliefs,** conceptions that people accept as true concerning how the world operates and where the individual fits in relationship to others. Beliefs can be rooted in blind faith, experience, tradition, or the scientific method. Whatever their accuracy or origins, they can exert powerful influences on actions and can be used to justify almost any kind of behavior, ranging from the most generous to the most violent. Here are some examples of beliefs that people can share:

- Interruptions or imbalances in the flow of *qi* (pronounced chee)—the energy that flows through the body—cause illness.
- Disturbances in social relationships cause illness.
- Very small organisms called germs cause disease.
- Continuous conversation, rather than silence, validates a relationship.
- Talent derives primarily from natural or inborn propensities to be good at something.
- Talent is essentially a product of hard work, practice, and persistence.
- Race is a clear-cut biological category.
- After death the human spirit returns to earth in a different form.

For an example of a situation in which beliefs affect behavior, see Dr. Cong Yali's description of the first euthanasia court case in China.

http://www.biol.tsukuba.ac.jp/~macer/EJ63/EJ63D.html
Document: The First Euthanasia Court Case in China—Cong Yali, M.D.

Q: What are the Chinese beliefs about patients' rights in relationship to their family?

Values The second component of nonmaterial culture is **values,** general, shared conceptions of what is good, right, appropriate, worthwhile, and important with regard to conduct, appearance, and states of being. Whereas beliefs are conceptions about how the world and people in it operate, values are conceptions about how the world *should* operate and how people *should* behave. Perhaps the most significant study on values was made by social psychologist Milton Rokeach (1973). Rokeach identified 36 values that people everywhere share to differing degrees, including the values of freedom, happiness, true friendship, broadmindedness, cleanliness, obedience, and national security. He suggests that societies are distinguished from one another not on the basis of which values are present in one society and not in another, but rather according to which values are more pervasive and more dominant. Americans, for example, place considerable value on the individual as an individual; they stress personal achievement and unique style (free choice). Return to the description of the first euthanasia court case in China.

http://www.biol.tsukuba.ac.jp/~macer/EJ63/EJ63D.html

Document: The First Euthanasia Court Case in China—Cong Yali, M.D.

Q: Which is the more important value: "individual/patient autonomy" or "status of the family"? Explain.

Norms The third component of nonmaterial culture is **norms,** written and unwritten rules that specify behaviors appropriate and inappropriate to a particular social situation. Examples of written norms are rules that appear in college student handbooks, on the backs of lottery tickets, on signs in restaurants ("no-smoking section"), and on garage doors of automobile repair centers ("honk horn to open"). Unwritten norms exist for virtually every kind of situation: wash your hands before preparing food, do not hold hands with someone of the same sex in public, leave a 15 percent tip for waiters and waitresses, remove your galoshes before entering the house. One unwritten rule is followed by approximately 80 percent of the women in the United States: shave or otherwise remove facial and body hair.

Some norms are considered more important than others, and so the penalties for violation are more severe. Depending on the importance of a norm, punishment can range from a frown to execution. In this regard we can distinguish between folkways and mores.

Folkways are norms that apply to the mundane aspects or details of daily life: when and what to eat, how to greet someone, how long the workday should be, how many times a day caregivers should change babies' diapers. Waitresses, waiters, and members of occupational groups that serve the public are expected to adhere to certain norms: "Oblige customers even if they treat you rudely . . . refrain from eating garlic and spicy foods, keep a toothbrush and toothpaste on hand, avoid nail biting, wear only clear nail polish and keep fingers out of mouths, ears and noses" (Murphy 1994, pp. A1, A14). As sociologist William Graham Sumner (1907) noted, "Folkways give us discipline and support of routine and habit"; if we were forced constantly to make decisions about these details, "the burden would be unbearable" (p. 92). Generally we go about everyday life without asking "why?" until something reminds us or forces us to see that other ways are possible.

Mores are norms that people define as essential to the well-being of a group. People who violate mores are usually punished severely—they are ostracized, institutionalized in prisons or mental hospitals, sentenced to physical punishment, condemned to die. In contrast to folkways, people consider mores to be unchangeable, regarding them as "the only way" and "the truth." In a 1994 case that received international attention, Singapore officials found Michael Faye, an 18-year-old American, guilty of violating their official mores, which place social order and citizens' general well-being ahead of individual rights. Faye was found guilty of vandalism because, over a 10-day period, he had damaged cars with eggs and spraypaint. Singapore officials defined this behavior as "a calculated course of criminal conduct" and sentenced him to caning and 60 days in jail. President Bill Clinton appealed the sentence; a Clinton administration official called the penalty excessive "for a youthful, nonviolent offender who pleaded guilty to reparable crimes against private property" (Wallace 1994, p. A5).

Again return to the description of the first euthanasia court case in China.

http://www.biol.tsukuba.ac.jp/~macer/EJ63/EJ63D.html
Document: The First Euthanasia Court Case in China—Cong Yali, M.D.

Q: What is the norm in China for obtaining consent for a medical procedure such as an operation? What norm do Chinese physicians follow in advising family members when medical treatment offers no hope for improving a patient's condition? Explain.

Symbols **The fourth component of nonmaterial culture is symbols,** physical phenomena—a word, object, color, sound, feeling, odor, gesture, taste, facial expression, behavior—to which people assign a meaning or value (White 1949). The meaning or value is not evident from the physical phenomenon alone. This deceptively simple idea suggests that people decide what something means. Unless people encounter a situation in which the meanings they assign are called into question, they act as if there can be no meanings other than those they know and have come to take for granted.

Without a shared system of symbols, life would be absurd, and efforts to communicate with others would be doomed to failure (Barnlund 1994). Language, whether signed or spoken, is a particularly important shared system of symbols because "it is only through language that we enter fully into our human estate and culture, communicate freely with our fellows, acquire and share information. If we cannot do this, we will be bizarrely disabled and cut off" (Sacks 1989, p. 8). No person can assign arbitrary meaning to words or other phenomena and expect to be understood. To be understood, people must use symbols that elicit the same general responses in others as in themselves. Otherwise no genuine communication would be possible without tremendous effort on the part of everyone involved. The movie *Nell*, in which Jodie Foster plays a woman isolated from all social interaction, gives us a sense of the time and effort needed to decipher words that have meaning for only one person. After learning of Nell's existence, a psychologist and a psychiatrist spend three months attempting to decipher the meanings of words such as "eviduh," "chickapap," and "ta" simply to learn something of Nell's background (Kempley 1994).

The Role of Geographical and Historical Forces

- *Principle 2:* Geographical and historical forces shape the character of culture.

Sociologists operate under the assumption that culture is "a buffer between [people] and [their] habitat" (Herskovits 1948, p. 630). That is, material and nonmaterial aspects of culture represent the solutions that people of a society have worked out over time to meet their distinctive historical and geographical challenges and circumstances:

> All mankind shares a unique ability to adapt to circumstances and resolve the problems of survival. It was this talent which carried successive generations of people into the many niches of environmental opportunity that the world has to offer—from forest, to grassland, desert, seashore, and icecap. And in each case, people developed ways of life appropriate to the particular habitats and circumstances they encountered. (READER 1988, P. 7)

Consider how the amount of natural resources available for human consumption affects energy conservation. Part of the reason U.S. manufacturers are still in the planning stages with regard to producing highly energy-efficient clothes washers, dishwashers, and refrigerators while Japanese manufacturers already are selling such machines has to do with the amount of natural resources in the two countries. Whereas Japan has no oil and virtually no other resources within its territory, the United States possesses abundant natural resources. Although the Japanese can import these resources, they face pressure to conserve unknown to most people in the United States—even as Americans have come to realize that resources are dwindling. This relative shortage reinforces the need to use resources sparingly and not to take them for granted. In sum conservation- and consumption-oriented values are rooted in circumstances of shortage and abundance. That is, resource abundance breaks down conservation-oriented behaviors while permanent shortage or dependence on others for resources promotes conservation.

To understand this connection, recall a time when your electricity or water was turned off. Think about the inconvenience you experienced after a few minutes and how it increased after a few hours. The idea that one must conserve available resources takes root. For example, you probably took care to minimize the number of times you opened the refrigerator door so as not to allow cold air to escape. Or consider how a long-term resource shortage affected Californians when their state experienced a six-year drought (1987–1992). In some water districts Californians cut their use of water by 25–48 percent. In Contra Costa County in the San Francisco Bay Area, for example, customers cut water consumption below the 280 gallons of water per day recommended under the voluntary rationing plan to an average of 165 gallons (Ingram 1992). Water conservation efforts also went into effect in the drought-stricken southwestern region of the United States in 1996.

http://www.mother.com/uswaternews/archive/96/conserv/swdrou.html

Document: **Emergency Water Conservation Measures Implemented in Drought-Stricken Southwest**

http://www.mother.com/uswaternews/archive/96/conserv/albuq.html
Document: Albuquerque Saves a Billion Gallons in '95 Usage

 Q: What are some examples of conservation-oriented behavior that resulted from droughtlike conditions in California? If Albuquerque saved a billion gallons of water in 1995, by what percentage did water use drop as a result of that savings?

For the most part people do not question the origin of the values they follow and the norms to which they conform "any more than a baby analyzes the atmosphere before it begins to breathe it" (Sumner 1907, p. 76). Nor are they aware of alternatives. This is because many values and norms that people believe in and adhere to were established before they were born. Thus people behave as they do simply because they know of no other way. And because these behaviors seem so natural, we lose sight of the fact that culture (in this case conservation and excessive consumption) is learned.

The Transmission of Culture

- *Principle 3:* Culture is learned.

Humans are born "with two endowments, or, more properly stated, with one and into one" (Lidz 1976, p. 5). Specifically we are born with a genetic endowment and into a culture. Our parents transmit via their genes a biological heritage at once common to all humans but uniquely individual. The genetic heritage that we share with all humans gives us a capacity for language development, an upright stance, and four movable fingers and an opposable thumb on each hand, as well as other characteristics. If these traits seem too obvious to mention, consider that they allow humans to speak innumerable languages, to perform countless movements, and to devise and use many inventions and objects. In fact "most people are shaped to the form of their culture, because of the enormous malleability of their genetic endowment" (Benedict 1976, p. 14).

Regardless of their physical traits (for example, eye shape and color, hair texture and color, skin color), babies are destined to learn the ways of the culture into which they are born and raised. The point is that our genes endow us with our human and physical characteristics, not our cultural characteristics. We cannot assume that someone comes from a particular culture simply because he or she looks like a person whom we expect to come from that culture.

An excerpt from a letter written by a first-generation Taiwanese-American mother to her daughter shows that even parents have a hard time accepting this idea when their children are raised in a country and a culture different from their own:

> To you, Taiwan is just a fun but humid place that you visited one summer, and your grandparents are just fuzzy voices over a telephone. To me, there is a lifetime that sits thousands of miles away, tucked inside the navy blue suit you see on me now. And to me, Taiwan is still home. HOME. Yes, our white stucco house with the orange door is home, too, but part of my blood still flows toward Taiwan. Can

you understand that? Maybe that's why sometimes I expect you to understand Chinese culture without having experienced any of it firsthand. I think that you have the same blood, and that it pulses to the same beat. You ARE Chinese still, and I know that you have some interest and even some pride in it, but there's so much you don't know. It is your right to know. It is your right to know your family's experiences, even if you don't care about them. Maybe someday, you will care. (YEH 1991, P. 2)

The development of language illustrates the relationship between genetic and cultural heritages. Human genetic endowment gives us a brain that is flexible enough to allow us to learn the language(s) that we hear spoken by the people around us.

The Role of Language

Language is an important nonmaterial component of culture. As young children learn words and the meanings of words, they learn about their culture. They also acquire a tool that enables them to think about the world—to interpret their experiences, to establish and maintain relationships, and to convey information. Anyone who speaks only one language might not realize this property of language until he or she learns another language. To become fluent in another language is not merely a matter of reading and conversing in that language, but of actually being able to think in that language. Similarly, when young children learn the language of their culture, they acquire a thinking tool.

The following characteristics of language show the relationship between learning the meaning of words and learning the ways of a culture:

- *Language conveys important messages above and beyond the actual meaning of words.* Words have two levels of meaning—denotative and connotative. *Denotation* is literal definition; *connotation* is the set of associations that a word evokes. The connotation of a word is as important, and sometimes more important, to understanding meaning as the literal definition. Idioms help us to see the distinction between denotation and connotation. An *idiom* is a group of words that when taken together have a meaning different from the internal meaning of each word understood on its own. For examples of idioms, see "The Weekly Idiom," a free internet-based service provided by the Comenius group to assist students of English.

    http://www.comenius.com/idiom/index.html
    Document: The Weekly Idiom

    Q: Read the idiom for the week and look over the idioms posted in the "weekly idiom index." How would you distinguish between the literal meanings and the meanings assigned to the phrase?

- *Words refer to more than things; they also describe relationships.* The word *adoption*, for example, as in the adoption of a child, refers to more than the child or the taking in and raising of that child. It also implies the presence of biological and adoptive parents and evokes assumptions about the relationship among the child, the biological parents, and the adoptive parents. The norms that guide these relationships are important features of the word. In Korea, for example, most people think of adoption as "a system whereby a sonless couple may receive a son from one of the husband's brothers or male cousins" (Peterson 1977, p. 28). An adopted son cannot come from the wife's family or from a sister of the husband. In contrast, in the United States adoption usually involves a situation in which a child's biological parents release him or her to responsible adults who agree to raise the child as their own. The adoptive parents usually have no connection with the biological parents.

- *Words mirror cultural values.* Language embodies values considered important to the culture. For instance, in Korean society age is an exceedingly important measure of status: the older a person is, the more status, or recognition, he or she has in the society. Korean language acknowledges the importance of age by its use of special age-based hierarchical titles for everyone. In fact, it is nearly impossible to carry on a conversation, even among siblings, without taking age into consideration. Every word referring to one's brother or sister acknowledges his or her age in relation to the speaker. Even twins are not equal, because one twin was born first. Furthermore, norms that guide Korean forms of address do not allow the speaker to refer to older brothers or sisters by first name. A boy addresses his older brother as *hyung* and his older sister as *nuna;* a girl addresses her older brother as *oppa* and her older sister as *unni.* Regardless of gender, however, people always address their younger siblings by their first name (Bo-Kyung and Kirby 1996).

  Among many non-Western cultures the importance of the group over the individual is reflected in rules governing the writing and speaking of one's name. Peter K. W. Tan considers Western and non-Western name conventions.

  gopher://liberty.uc.wlu.edu/00/library/human/eashum/nameconv
  Document: Non-Western Name Conventions

  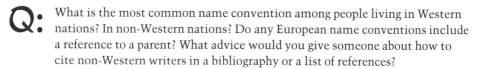 What is the most common name convention among people living in Western nations? In non-Western nations? Do any European name conventions include a reference to a parent? What advice would you give someone about how to cite non-Western writers in a bibliography or a list of references?

- *Common expressions embody the preoccupations of the culture.* Frequently used phrases and words serve as indicators of cultural preoccupations—stresses, strains, and values. For instance, Americans use the word *my* to express "ownership" of persons or things over which they do not have exclusive rights: my mother, my school, my country. The use of *my* reflects the American preoccupation with the needs of the individual over those of the group. In contrast, Koreans express possession as shared: our mother, our school, our country. The use of the plural possessive reflects the Korean preoccupation with the needs of the group over individual interests.

  Another example of cultural preoccupations is the Korean response to a full moon—"Isn't that sad?"—or to singing birds—"They are weeping!" These comments reflect patterns of response rooted in centuries of invasions and wars. In part because of its geographical location, Korea has had a continuous history of invasions—by the Japanese, the Chinese, and the Russians—that caused severe hardship, substantial loss of life, and widespread devastation. Generations of warfare and occupation by foreigners have created in the Korean people a sadness that is reflected in their responses to many natural phenomena.

These four characteristics of language demonstrate that children do more than learn words and meanings. They also acquire a perspective that reflects what is important to the culture. Nevertheless, learning about and acquiring a cultural perspective does not make people cultural replicas of one another.

The Importance of Individual Experiences

- *Principle 4:* People are products of cultural experiences but are not cultural replicas of one another.

Why is this the case? A baby enters the world and, by exposure to an already established set of human relationships, is introduced to many versions of the culture. Virtually every event the child experiences—being born, nursing, being cleaned, being talked to, weaning, toilet training, talking, playing, and so on—involves people. The people involved in the child's life at any one time include various combinations of father, mother, grandparents, brothers, sisters, playmates, other adult relatives, neighbors, babysitters, and others (Wallace 1952). All of these people expose the child to their own versions of culture, which they have acquired in the same way and which they pass on to the child in modified forms. The following excerpt from the essay "Faculty Brat: A Memoir," by Emily Fox Gordon (1995), captures this selective and interpretive dynamic:

> My father's background was Jewish, my mother's Presbyterian. Both of them were agnostic rationalists, and I grew up hearing almost nothing of belief or doctrine. My mother preserved the aesthetic parts of her Christian heritage. We spent two weeks before Christmas, my mother, sister, brother and I at the kitchen table mixing food coloring into vanilla icing in small glass dishes—pale green, pink, a shade I called chocolate blue. We used toothpicks to paint striped frosting trousers on the rudimentary legs of gingerbread men, buttoned up their blurred pastel waistcoats with silvery sugar balls. We also collected pine cones and sprayed them, over newspaper, with silver and gold (the wonderful toxic reek of those cans, which were also preternaturally cold to the touch!); we saved the tops and bottoms of tin cans and used metal shears to cut them into stars and spirals for the Christmas tree. We made Santas, gluing triangles of cotton on the chins of walnuts and red felt hats on their foreheads. . . .
>
> We children learned nothing of Judaism, except a vague understanding that the pickles and corned beef sandwiches my father loved, and the demonstrative relatives from New York and Philadelphia we occasionally visited, were things from the Jewish side of the universe. . . . We understood as Jewish the sometimes jarring jokes and epigrams my father loved to repeat. A dog goes into a bar, they typically began. Maybe I should have said DiMaggio, they ended. We children looked quickly toward our mother when my father told these stories, to catch the quiver of distaste which flickered over her lip. . . .
>
> I am the only one of my siblings to marry a Jew, and from her birth my daughter has always been Jewish to me. My brother and sister consider themselves and their children to be unaffiliated, but they celebrate Christmas and Easter. I'm not sure what I call myself, but now, having a child, I find I cannot celebrate the Christian holidays, even though the memory of some carols—"It Came upon the Midnight Clear," and "Lo, How a Rose Ere Blooming"—brings tears to my eyes when

I find myself humming them in December. I know more about Judaism now, and I have a great abstract respect for it, but my mother's holiday Christianity, its sweetness, the memories of food and music and the surfaces of familiar things embellished and glittering, is like a beloved country from which I have exiled myself. (PP. 6–8)

This excerpt shows that individuals are products and carriers of cultural experiences and that they pass on those experiences selectively with varying degrees of clarity and confusion. The people to whom they transmit these experiences then repeat the process. Because individuals perceive, interpret, select, manipulate, revive, and create culture, they cannot be viewed as passive agents who absorb one version of culture. And culture cannot be viewed as a singular force causing people to behave and think as they do. Consider the case of the Louisiana Creole language.

http://www.teleport.com/~napoleon/louisianafrenchcreole/introtolouisianafrenchcreo.html
Document: Louisiana Creole

Q: What is the function of language? How does Louisiana Creole illustrate that function?

Still we cannot disregard culture when considering any behavior or thought (Kroeber and Kluckhohn 1952). L. K. Frank (1948) offers a useful analogy considering the shared nature of cultural traits while acknowledging the idiosyncratic versions that each person possesses: "We can abstract the regularities and uniformities and likewise observe the personality distortions and skewings, as we have learned to observe the statistical regularities of gas but also recognize and acknowledge the irregular and nonconforming behavior of individual molecules of that gas" (p. 98).

Culture as a Tool for Problems of Living

- *Principle 5:* Culture is the tool that enables the individual to adjust to the problems of living.

Although our biological heritage is flexible, it presents all of us with a number of challenges. As noted previously, we are dependent on others for a relatively long time. We also feel emotions; we experience organic drives such as hunger, thirst, and sexual desire; and we age and eventually die. These biological inevitabilities have given rise to functional requisites, arrangements, or "formulas" necessary for the survival of all societies. There are formulas for caring for children; for satisfying the need for food, drink, and sex; for channeling and displaying emotions; for segmenting the stages and activities of the life cycle; and eventually for departing this world. In this section we will focus on the differing cultural formulas for dealing with two biological events—hunger and the social emotions.

Cultural Formulas for Hunger

All people become hungry, but the factors that stimulate and satisfy appetite vary considerably across cultures. One indicator of a culture's influence is that people define only a portion of the potential food available to them as edible. Culture determines not only what is defined as edible but also who prepares the food, how the food is served and eaten, what the relationship is among those eating together, how many meals are eaten in a day, and when during the day meals are eaten. For example, dogs and snakes are among the foods defined by many Asians as edible, but they are not defined as such by Americans. Americans cannot understand why some Koreans eat dog meat. This reaction should not be surprising when we consider that more than one-third of American households include at least one dog. The United States has more than 10,000 pet shops, 19,000 dog food vendors, 11,000 grooming shops, 7,000 kennels, and 300 pet cemeteries, as well as 200 products for dogs, ranging from feeding dishes to raincoats and sunglasses (Rosenfeld 1987). In addition consider the number of documents—13,206—that the keyword "Dog" generated on the search engine Webcrawler as of June 1, 1996. Many internet sites, such as "Pet Loss and Grieving Resources," focus solely on helping grieving pet owners whose pets have died.

http://www.primenet.com/~meggie/bridge.htm
Document: Informational Dog-Related Web Sites

 Choose the files "Dogs Remembered" and "More Dogs Remembered." Read some of the dog owners' tributes to their deceased pets, and briefly profile the relationship between owner and pet. Do you believe that these characteristics apply to most dog owners?

Among other factors these differences in attitudes toward dogs are rooted in historical and environmental factors. Whereas the United States uses an abundance of fertile, flat land for grazing cattle, many Asian countries such as Korea and China with limited space use available land to grow crops, not to graze cattle. The few existing cattle are important to the agricultural system as a source of labor—to pull plows. Even today cattle are more efficient than tractors in tilling steep inclines. The agricultural importance of cattle, combined with the lack of land to support a cattle industry, discourages the widespread practice of eating beef and encourages the consumption of dogs and snakes as alternative food sources.

Cultural Formulas for Social Emotions

Culture also influences the expression of emotion, just as it influences people's responses to food needs. Social emotions are internal bodily sensations that we experience in relationships with other people. Grief, love, guilt, jealousy, and embarrassment are a few examples of social emotions. Grief, for instance, is felt at the loss of a relationship; love reflects the strong attachment that one person feels for another person; jealousy can arise from fear of losing the affection of another (Gordon 1981). People do not simply express social emotions directly, however. They also interpret, evaluate, and modify their internal bodily sensations upon considering "feeling rules" (Hochschild 1976, 1979).

Feeling rules are norms that specify appropriate ways to express these internal sensations. They define sensations that one should feel toward another person, object, or behavior. As one example of how culture channels expressions of emotion, consider the Japanese practice of memorializing animals used for research and human bodies donated for medical research.

http://www.biol.tsukuba.ac.jp/~macer/EJ51A.html
Document: Editorial—Why a New Journal?

 Q: How do the Japanese acknowledge the experimental and research contributions of research animals and human bodies?

As a second example of how culture channels emotional expression, consider that in the dominant culture of the United States same-sex friends are supposed to like each other but not to feel romantic love toward each other. We come to learn these feeling rules through a complex process that evolves through interactions with others.

In her novel *Rubyfruit Jungle* Rita Mae Brown (1988) describes a situation in which feeling rules are articulated. The central character, Molly, who is about seven years old at the time, wonders whether girls can marry each other. She approaches Leota, a girlfriend whom she likes very much, about this possibility:

"Leota, you thought about getting married?"
"Yeah, I'll get married and have six children and wear an apron like my mother, only my husband will be handsome."
"Who you gonna marry?"
"I don't know yet."
"Why don't you marry me? I'm not handsome, but I'm pretty."
"Girls can't get married."
"Says who?"
"It's a rule."
"It's a dumb rule. Anyway, you like me better than anybody, don't you? I like you better than anybody."
"I like you best, but I still think girls can't get married." (p. 49)

In another scene in the same novel, Molly walks in on her father, Carl, while he is comforting his friend Ep, whose wife has just died. In this passage Molly reflects on the feeling rules that apply to men:

I was planning to hotfoot it out on the porch and watch the stars but I never made it because Ep and Carl were in the living room and Carl was holding Ep. He had both arms around him and every now and then he'd smooth down Ep's hair or put his cheek next to his head. Ep was crying just like Leroy. I couldn't make out what they were saying to each other. A couple of times I could hear Carl telling Ep he had to hang on, that's all anybody can do is hang on. I was afraid they were going to get up and see me so I hurried back to my room. I'd never seen men hold each other. I thought the only things they were allowed to do was shake hands or fight. But if Carl was holding Ep maybe it wasn't against the rules. Since I wasn't sure, I thought I'd keep it to myself and never tell. I was glad they could touch each other. Maybe all men did that after everyone went to bed so no one would know the toughness was for show. Or maybe they only did it when someone died. I wasn't sure at all and it bothered me. (p. 28)

As these passages suggest, people learn norms that specify how, when, where, and to whom to display emotions. Another example of the role culture plays in channeling expressions of emotion has to do with laughter. Laughter is *not* something that happens only when people are amused. Instead laughter can be an expression of emotions such as anxiety, sadness, nervousness, happiness, or despair. Check out some of the many reasons people need to laugh.

http://rampages.onramp.net/~ejunkins/
Document: Laughter Therapy Homepage

Q: What is the function of laughter?

No matter what emotion laughter releases or defuses, however, it occurs when something is awry in a given situation or when behavior in a specific circumstance differs from what is expected. Because culture provides the guidelines for what is expected in a specific set of circumstances, this observation about laughter implies that the situations that make someone laugh in one culture may not be funny in another. In fact communication specialists generally agree that jokes do not translate well. The following anecdote, in which a Japanese translator decides not to translate a foreign speaker's joke because the audience will not understand the cultural context that makes the joke funny, represents one way of handling this dilemma.

I began my speech with a joke that took me about two minutes to tell. Then my interpreter translated my story, and about 30 seconds later the Japanese audience laughed loudly. I continued with my talk, which seemed well received . . . but at the end, just to make sure, I asked the interpreter, "How did you translate my joke so quickly?" The interpreter replied: "Oh, I didn't translate your story at all. I didn't understand it. I simply said, 'Our foreign speaker has just told a joke, so would you all please laugh.'" (MORAN 1987, p. 74)

To this point we have discussed a number of principles about culture. We have emphasized that culture is a tool that people learn and draw from to meet the challenges of living. The discussion may have led you to believe that the nonmaterial components of culture have a greater influence on behavior than do the material components, but that is debatable. In fact the material and nonmaterial components of culture are interrelated: the nonmaterial shapes the material, and the material—

particularly the introduction of some new object or technological advance—shapes the nonmaterial.

The Relationship Between Material and Nonmaterial Culture

- *Principle 6:* It is difficult to separate the effects of nonmaterial and material cultures on behavior.

To illustrate how nonmaterial components of culture shape the material, consider how the design of the American flag reflects the nonmaterial culture of the United States. Each of the 13 alternating red and white stripes represents one of the 13 original colonies. Each of the 50 stars on the blue field represents a state. Although the United States began as a British colony, the flag does not acknowledge that influence (except in the red, white, and blue colors). Nor do the stars and stripes overlap. This separateness reflects the value placed on independence or freedom from a strong central power. The absence of any symbol that represents Britain or Europe also reflects the value placed on shedding the influence of the past to make a clean start—evidence of the norm that rejects the role of the past in defining a person.

We can see the influence of nonmaterial culture on material culture in Masahiro Morioka's analysis of barriers to organ transplantation in Japan.

http://www.biol.tsukuba.ac.jp/~macer/EJ54E.html

Document: Bioethics and Japanese Culture: Brain Death, Patients' Rights

Q: What are Japanese beliefs about the body, and how do they represent barriers to organ transplantation? What norms govern patient–physician relationship in Japan? How are they related to the practice of organ transplantation from brain-dead donors?

Just as nonmaterial components of culture shape material culture, material components also shape the nonmaterial. Consider the microwave oven. Approximately 70 percent of American households have microwave ovens, which have eliminated one of the incentives for families to eat together. Before the microwave was available, it was more efficient for one member of the household to cook for everyone because of the time—whether calculated in person time or in oven time—required to cook a meal. Now that meals take only minutes to prepare, it is no longer considered inefficient for family members to cook and eat separately. The microwave also enhances independence in that a person is not tied to a particular meal schedule or even family group:

> The old dining room table required each individual to give up some personal autonomy and bow to the dictates of the group and the social system. If we stop eating together, we shall save time for ourselves and achieve mealtime self-

sufficiency. . . . The communal meal is our primary ritual for encouraging the family to gather together every day. If it is lost to us, we shall have to invent new ways to be a family. It is worth considering whether the shared joy that food can provide is worth giving up. (VISSER 1989, P. 42)

The microwave oven not only changed the time and effort required to cook a meal but also the way family members relate to one another. Indeed inventions often have consequences that go beyond their intended purpose. As this example shows, material components affect values and norms and can have a profound influence on the ways in which people in a society relate to one another.

Most people tend to think that their material and nonmaterial culture is self-created. In reality, however, any culture is susceptible to outside influences to a varying extent.

Cultural Diffusion

- *Principle 7:* People borrow ideas, materials, products, and other inventions from other societies.

Most people tend to underestimate, ignore, or distort the extent to which familiar ideas, materials, products, and other inventions are connected in some way to outside sources or are borrowed outright from those sources (Liu 1994). The process by which an area, an invention, or some other cultural item is borrowed from a foreign source is called **diffusion.** The term *borrow* is used in the broadest sense; it can also mean to usurp, pirate, steal, imitate, learn, plagiarize, purchase, or copy. The opportunity to borrow occurs whenever two people who have different cultural traits make contact, whether face-to-face, by phone or fax, or via the internet.

Instances of cultural diffusion are endless and can be found by skimming the newspaper headlines. Here are some examples:

- "School for Japan's Executives in the United States"
- "In China, Beauty Is a Big Western Nose"
- "Thai Publisher Plans to Expand Empire in U.S."
- "Invasion of the Discounters: American-Style Bargain Shopping Comes to the United Kingdom"
- "Global Goliath: Coke Conquers the World"
- "Japan's Favorite Import from America: English"
- "Skilled Asians Leaving U.S. for High-Tech Jobs at Home"

Notice that each headline suggests some form of cultural diffusion. This ranges from some Chinese undergoing cosmetic surgery to "copy" some conception of a Western nose to an estimated 195,000 skilled Asians, after living for a while in the United States, deciding to return home, leaving behind a significant yet unknown effect on the United States and taking with them the effects of living in the United States (Dunn 1995).

People from one society do not usually borrow cultural components outright from foreign sources. Borrowing is usually selective: people decide which features of a foreign cultural trait they will adopt or modify that trait to fit their own values and beliefs. Even the simplest invention is really a complex of elements, including various associations and ideas about how to use it. Thus, not surprisingly, people borrow the most tangible elements and then develop new associations and reshape the items to serve new purposes or to fit their existing assumptions (Linton 1936).

Consider, for example, how the Suzuki method of teaching very young children to play the violin has been altered to fit American assumptions about talent and the learning process. The Japanese violinist Shinichi Suzuki invented this method in 1946. As a result of the diffusion process, today an estimated 300,000 students in 23 countries study violin according to this method. Approximately half of these students live in the United States and Canada.

In the process of being adopted by and transplanted to other societies, the method, as Suzuki originally conceived it, has been modified to fit the educational assumptions and practices of the borrowing cultures. In the United States, for example, modifications have been made in the kinds of homework that Suzuki instructors assign to students (Taniuchi 1986). Specifically U.S. instructors do not emphasize repetition in the same way as their Japanese counterparts:

> [In Japan] repetition is central to this activity of diligent, persistent practice. Although the word in English connotes doing the same thing again and again in the same way, when a [Japanese] Suzuki instructor tells a student to repeat a particular passage 100 or 1,000 times, they expect that each time the student will strive to play it better than the last. [In Japan] repetition is continued long past the point at which reasonable competence is reached, until correct performance becomes automatic . . . the American attitude is that repetition is usually boring, whereas learning new material is fun, [which] means that Suzuki students and parents [in the United States] become oriented toward rate of progress through the material rather than quality of the performance. [U.S.] students are less willing to continue to practice and study earlier material, and teachers meet resistance from both parents and students in spending lesson time in continued study of "easy" songs. American Suzuki teachers are also less likely to prescribe their Japanese counterparts' favorite prescription for difficult passages: "practice it 1,000 times this week and then see if it still seems hard." (TANIUCHI 1986, PP. 130–31)

A second example of how diffusion is selective involves the musician M. C. Solaar. One of the most popular music stars in France in 1994, Solaar modified American rap music to appeal to French listeners. When Solaar released his first album in 1991, rap music imported from the United States was not very popular in France. Therefore, instead of using rap to boast or brag, as Solaar believes (rightly or wrongly) most American rappers do, he used it to show the strength and versatility of the French language and in the process to make listeners think, "to give them the envy of knowing more" (National Public Radio 1994, p. 31). Solaar incorporates complicated words and names of philosophers and old singers and movies into his lyrics because, as he puts it, "I know that my little cousin, she will listen to that, she will open a dictionary, she will open a storybook, she will open an art book, and she will know more. For her, it's to get through school, for me it's something good not to be simply basic but to make people learn stuff by listening to music" (National Public Radio 1994, p. 31). Rap is also very popular in Asia, but like the French, Asian rappers have modified rap to appeal to their fans. For example, Japanese rappers have borrowed the beat and hip-hop persona of

American rappers (baggy pants, baseball caps, heavy jewelry), but their lyrics express social concerns in a much less direct way than do American lyrics (Kristof 1996).

Anthropologist Ulf Hannerz (1992) identifies at least four interrelated kinds of relationships through which cultural diffusion or flow occurs: (1) between people as they mingle with one another every day, (2) between state officials and the masses, (3) between sellers and buyers in the marketplace, and (4) between believers and nonbelievers in social movements. In naming these four sources of cultural flow, Hannerz did not intend to be exhaustive or to imply that each source is independent of the other. His purpose was to identify some of the major relationships through which diffusion occurs.

Everyday Mingling

Everyday mingling encompasses people's routine talking, looking, and/or listening as they live their lives. The settings in which people mingle are endless; they include the workplace, the home, the neighborhood, and the school. According to Hannerz one characteristic of everyday mingling is that people acquire cultural traits simply by observing and participating in their daily routines. This learning occurs as people do the same things over and over again and as they see and hear others doing the same things repeatedly. Consider that when Shinichi Suzuki visited Germany, he formulated his ideas about the Suzuki method.

http://www.intrnet.net/~jbailey/suzuki/whois-ss.html
Document: Who Is Shinichi Suzuki?

: How did Suzuki's visit to Germany influence his theories about how to play the violin?

Another example is the experience of Dr. Makoto Kondo, a Japanese radiologist who lived for a year in the United States. Kondo observed that the oncologists he worked with told their patients the truth about their condition—that is, that they had cancer. In Japan physicians tend to withhold or mask that diagnosis; only about 25 percent of doctors tell their patients that they have cancer. According to a 1989 ruling from the highest court in Japan, "the extent of disclosure is at the discretion of the physician." Thus Japanese physicians are under no legal obligation to tell cancer patients their true condition. Apparently Kondo was greatly affected by his daily mingling with U.S. physicians and came to realize that "if doctors practicing in the United States could tell the truth then he could do the same in Japan" (Kristof 1995, p. Y6). Consequently, on his return to Japan, Kondo became an outspoken critic of the Japanese medical system, especially as it relates to the physician–patient relationship.

The State

A state is a political entity recognized by foreign governments, with a civilian and military bureaucracy to carry out its policies, to enforce its rules, to regulate other activities within its borders, and to control activities outside its borders that have the potential to affect its citizens. Among other things states control the flow of cultural traits by regulating the movement of goods, services, and people across their borders. The Schengen Agreement, which went into effect in March 1995, is one example of state control over cultural flow. Seven European countries—Germany, Belgium, the Netherlands, Luxembourg, France, Spain, and Portugal—created a free-travel zone by eliminating all border controls, allowing people to journey from one country to another *inside* the zone without passports. This agreement, in conjunction with the elimination of customs checks in 1993, removed barriers to the free flow of goods, services, and people across the borders of these seven countries. These governments, however, established stricter controls over movement *into and out of* the free-trade zone, creating barriers to the free flow of material and nonmaterial cultural traits *into* the zone from countries outside the zone (Marshall 1995).

In addition states affect the free flow of cultural information when their officials institute policies to influence others outside their borders. The United States Information Agency (USIA) can be seen as an agent of cultural flow.

http://www.usia.gov/abtusia/factsht.htm

Document: About the U.S. Information Agency and the U.S. Information Service

Q: How many posts does the USIA have around the world? In how many countries? What is the general mission of the USIA? What is the USIA's best-known cultural and educational exchange program?

The Marketplace

A **market** involves transactions between sellers and buyers. **Sellers** conduct one or more of the extensive transactions (producing, advertising, shipping, storing, selling) needed to sell a commodity or service (oriental rugs, insurance, automobiles, MTV). **Buyers** are the persons in a targeted territory to which commodities are marketed and sold. The target market can encompass a territory ranging in size from a neighborhood to the entire world, and/or it can encompass a certain demographic group.

In an interesting trend associated with advances in communications technology, companies in one country market goods and services to their compatriots who have taken up permanent, long-term, or even short-term residence in other countries. In the 1994 winter issue of the *Philippine American Quarterly*, for example, FILSAT ran an advertisement offering readers a one-year free subscription to 24-hours-a-day coverage of top Filipino TV programs (news, dramas, variety shows, sitcoms, celebrity gossip, movies) with the purchase of a satellite dish. This example shows how the market functions both as an agent of cultural flow and as a link between expatriates and events in their home countries.

Social Movements

Social movements are organized, deliberate efforts by believers to transform, reform, or replace some element of culture, and in the process to convert nonbelievers to their position. Examples of social movements include efforts to change the laws (that is, norms) prohibiting the use of illegal drugs; to outlaw alcoholic beverages; to translate the Bible into the estimated 6,000–9,000 distinct languages spoken in the world; to establish norms governing political or human rights; to end racial segregation in employment, housing, and public facilities; to abolish the death penalty; to end social and legal discrimination against gays; to win voting rights; and to ban abortions.

Consider the work of Wycliff Bible Translators and its sister organization, the Summer Institute of Linguistics. Together they employ 6,000 translators, literary specialists, and support staff members, who work on six continents. The organization's aim is to translate the Bible into languages spoken by fewer than 10,000 people each. To meet this goal, translators must live (often for as long as 25–30 years) with the people who speak the languages (Vanderknyff 1995). Most of these languages have never been studied by outsiders and do not exist in written form (Weiss 1995, p. 36). See the document "A Historical Perspective" for more on the organization.

http://www.gospelcom.net/ibs/who.html#HISTORIC

Document: A Historical Perspective

Q: What groups, areas of the world, and physical places has Wycliff managed or attempted to reach?

Cultural flow is a two-way process of exchange, but the details of the exchange are usually sacrificed to a focus on the most obvious trait in the exchange. After decades of interaction between Wycliff Bible Translators and the people who speak the language the translators are trying to learn, both parties are at risk of change. Yet in this case the focus is on the most visible product of the exchange—the translated Bible—and on the party doing the translating. Most people do not consider how the new language transforms the Bible. The point is that any contact sets up an opportunity for the involved parties to alter each other, if only in subtle ways. But how are we to determine what each side gains or loses as a result of the contact? Anthropologist Michael Agar (1994) expresses this idea as follows: "The thing I wonder about is: What do you learn, what is its relationship to what you already know, and who are you after you learned it?" (p. 248).

One of the problems in trying to assess cultural flow is determining who owns the idea, behavior, or product that is exchanged. The issue of "ownership" was one of the themes of an enlightening 10-part PBS television series entitled "Connections," produced and narrated by James Burke in the late 1970s. Burke traced the history of eight inventions that many Americans would probably attribute to homegrown ingenuity—the atomic bomb, the telephone, the computer, the assembly line, aircraft, plastics, the guided missile, and television. His underlying thesis was that each of the eight is a synthesis of ideas, tools, materials, and other inventions extending from ancient times to the present.

The computer, for example, is the "offspring" of the printing press, with its system of movable type. According to Burke's convincing logic the computer is related to a

Korean invention, movable metal type, which dates from 1313. The Koreans borrowed the idea from the Chinese, who had developed a movable wood-and-clay type. Because wood and clay tend to wear down quickly and break easily, the Koreans developed metal type, which probably was carried to Europe by Arab traders. The Chinese-Korean-Arab-European connection is supported by the fact that the fifteenth-century typesetting system "invented" by Gutenberg is almost identical to the Korean system developed 200 years earlier (Burke 1978). The point is that if we take the time to learn the history of any cultural trait, we will find it very difficult to attribute ownership to any one culture. Unfortunately most people do not take the time. If they did, they would realize that cultures are not distinct entities; they are, in fact, interdependent. And such knowledge would protect against ethnocentrism.

One Culture as the Standard

- *Principle 8:* The home culture is usually the standard that most people use to make judgments about another culture.

People with an attitude of **ethnocentrism** use their home culture as the standard for judging the worth of foreign ways. From this viewpoint there are no acceptable alternatives to the home culture; "one's group is the center of everything, and all others are scaled and rated with reference to it" (Sumner 1907, p. 13). Other cultures are regarded as inferior, strange, exotic, inferior, even subhuman.

The attitude is exemplified in the attitudes and actions of a Tavares, Florida, woman who led a fight to institute a policy mandating that the county's 22,526 students be "taught that the United States is superior to any other society in all of human history." According to *New York Times* reporter Larry Rohter, the woman, who describes herself as a "patriot, a Christian, and a Republican," did not object to students' learning about other cultures so long as they never forgot that the United States is the best of the best. The woman acknowledged that she had never been outside the United States and that she spoke no foreign language, but she maintained that she didn't "need to visit other countries to know that America is the best country in the world." It was enough for her just to know that "thousands of people risk life and limb to come to America because they know this is the land of the free" (Rohter 1994, p. Y12).

Ethnocentrism has different levels and various consequences. Two relatively harmless types are defining foreign ways as peculiar and making jokes about the people from that culture. One example is a popular joke about people from the United States: What do you call someone who knows three languages? Answer: trilingual. What do you call someone who knows two languages? Answer: bilingual. What do you call someone who knows only one language? Answer: an American. Jokes are problematic because they misrepresent reality, take things out of context, or otherwise mask the true state of affairs. The language joke, for instance, misrepresents reality in that more than 32 million people over age 5 in the United States speak another language at home in addition to English. Of these 32 million, 56 percent speak English very well (U.S. Bureau of the Census 1993). Or consider the "trite journalistic clichés and embarrassing or biased representations of Africa."

http://www.lawrence.edu/dept/anthropology/classics_april.html
Document: East Africa: Colonized Minds

Q: What are two examples of ethnocentric attitudes toward or portrayals about Kenya?

One of the most extreme and destructive forms of ethnocentrism is **cultural genocide,** wherein the people of one society define the culture of another society not only as offensive but also as so intolerable that they attempt to destroy it. When Turkey invaded Cyprus in 1970, Turkey's authorities and its military embarked on a campaign of cultural genocide against everything Greek.

http://daedalus.ee.ic.ac.uk/~cyprus/Cyprus_Problem/destruction2.html
Document: Destruction of the Cultural Identity of the Occupied Area

Q: Which elements of Greek culture were destroyed when Turkey invaded Cyprus?

Sociologist Everett Hughes (1984) identifies yet another, less obvious type of ethnocentrism: "One can think so exclusively in terms of his own social world that he simply has no set of concepts for comparing one social world with another. He can believe so deeply in the ways and the ideas of his own world that he has no point of reference for discussing those of other peoples, times, and places. Or he can be so engrossed in his own world that he lacks curiosity about any other; others simply do not concern him" (p. 474).

This attitude is demonstrated in many ways. For example, people may meet someone from a foreign country and expect him or her to speak English without questioning their own inability to speak that person's language—or any language other than English. Or Americans may ask someone from another country, living temporarily in the United States, "'You'll be staying here, won't you?' . . . unshakably convinced that anything else would be abnormal" (Ugresic 1994, p. 80). This uninterested, self-absorbed attitude is evident in the answers that respondents in three U.S. cities gave to a Japanese sociologist who asked them to name the most famous Japanese people they know. The three most common answers were (1) Yoko Ono, (2) Bruce Lee, and (3) Godzilla (National Public Radio 1995). The problem is not so much that these Americans cannot name famous Japanese people as that so little thought seemingly was given to these answers, especially the third one.

Another relatively unknown type of ethnocentrism is **reverse ethnocentrism,** whereby the home culture is seen as inferior to a foreign culture. People who engage in this kind of thinking often regard other cultures as utopias. For example, the former

Soviet Union is labeled as the model of equality, Japanese culture as the model of human connectedness, the United States as the model of self-actualization, India as the model of otherworldliness, Israel as the model of pioneering spirit, Nigeria as the model of family values, and Native Americans as the original environmentalists (Hannerz 1992). People who engage in reverse ethnocentrism not only idealize other cultures but also reject any information suggesting otherwise. For example, some people in the United States view the various peoples encompassed by the term *Asian American* as model minorities—as people who work hard, do well in school, and have or cause few problems. Consequently, if someone such as Bong Hwan Kim, director of the Korean Youth Center in Los Angeles, tells them that he works "with juvenile delinquents in Koreatown, they get a blank look on their face and the conversation ends right there" (Schoenberger 1992, p. A24).

Because most people come to learn and accept the ways of their home culture as natural, they can experience a mental and physical strain known as culture shock when they encounter people from other cultures.

Culture Shock

The term **culture shock** describes the strain that people experience when they encounter foreign cultural traits and/or must orient themselves to new traits. In particular they must adjust to the idea that the behaviors and responses they learned in their home culture and came to take for granted do not apply in the foreign setting. People experience "shock" when they confront "new demands, new food, new language, new weather, new transportation, new philosophy, new harassments, and new pleasures, too; not all bad, not all good, but different" (Anderson 1971, p. 1123).

The intensity of culture shock depends on an array of factors including (1) the extent to which the home and the foreign cultural traits differ from one another; (2) one's knowledge about what to expect; (3) the circumstances—research, study abroad, cultural exchange, vacation, job transfer, forced migration, or war—surrounding the encounter; (4) the affinity one expects to have with the hosts; (5) one's ability to speak the language; (6) the extent to which living accommodations function as a shelter, separating one from the culture; and (7) one's personal characteristics—openness to new experiences, ability to deal with unfamiliar situations, patience, tolerance for ambiguity.

The intensity of culture shock also depends to some degree on the hosts' characteristics. Obviously it is easier to learn a new language and acquire other cultural traits if the hosts attempt to understand guests, make allowances for mistakes, and constructively point out "errors." Conversely it is more difficult to adapt if the hosts refuse to make allowances, point out errors of interpretation relentlessly and critically, or refuse to help guests learn the language (Gullick 1990).

Some cases of culture shock can become so intense and unsettling that people feel ill. Among the symptoms are "obsessive concern with cleanliness, depression, compulsive eating and drinking, excessive sleeping, irritability, lack of self-confidence, fits of weeping, nausea" (Lamb 1987, p. 270). Other symptoms include complaining continually, finding fault with the host country, playing up the perceived superiority of the home culture, craving food from home (for instance, longing for a hamburger and a milk shake), and assigning disproportionate significance to symbols and mementos from home encountered in the course of travel (for example, the home country's flag or an item in a department store marked as made in that country).

People may respond to culture shock in a number of ways, including (1) retreating from the host culture into self-imposed isolation or into an expatriate community, (2) "going native" or overidentifying with the host culture (perhaps by marrying someone from the host culture after living there for only a short time), (3) resisting the host culture's ways and/or seeking to convert hosts to the ways of the home culture, or (4) overcoming the shock by learning from differences and accepting them as just another way of facing the human condition (Gullick 1990).

Culture shock is not a reaction restricted to exotic behaviors. A series of small and (when considered individually) inconsequential differences can be enough to make someone uneasy and uncertain (Smalley 1963). Anthropologists Waud Kracke (1987) describes his difficulties in adjusting to one aspect of Kagwahiv culture, namely the norms associated with gift giving:

> Another cultural pattern I had difficulty dealing with (as other anthropologists working with South American indigenes have also noted) was the recurrent demand for gifts, and the lack of expression of gratitude when receiving one. My reaction to this pattern—and the matter-of-fact way in which gifts are received, without the obligatory expression of gratitude required in our culture—was to feel that they regarded what I gave them in a purely utilitarian light, without the emotional bond established or expressed in our own culture by a gift; and I particularly resented the pressure on me to give, which always mounted during the last few days before I left a place. In retrospect I realize that my reaction was based on a failure to appreciate fully what giving means in Kagwahiv culture. At one level, of course, sharing is quite matter-of-fact and expected (I have noted elsewhere that there is no term designating the virtue of "generosity" in their culture, so expected is it of any decent person, while the word for "stingy" . . . carries a heavy weight of opprobrium); but at another level, giving is for them a paramount expression of emotional bond, even more than in our culture. Gratitude would be inappropriate in response to a gift, since it is an expression of love, or relatedness; the demands for gifts on my departure, which I experienced as meaning that I was nothing but a source of material goods for them, in fact meant quite the opposite: it expressed their attachment to me, their sense of being bereft at my departure. (P. 75)

Kracke questions whether the term *culture shock*—with its connotations of sudden and violent disturbance, intense surprise and disgust, and strong blows to the senses—captures the experience of coming to terms with or orienting oneself to another culture. Instead Kracke prefers *encounter with another culture*, which he defines as "an alternation of excitement, discovery, frustration, embarrassment, liberation, depression, elation, puzzlement, and only occasionally the kind of abrupt and disconcerting surprises implied in the term 'shock'" (p. 60). The point is that pathological reactions are not the only outcomes of exposure to foreign ways. Most discussions of culture shock tend to simplify the array of emotions people feel and to overlook the self-knowledge, enlightenment, and other kinds of intellectual enrichment they acquire.

One does not adjust to foreign ways simply by abandoning the old ways of thinking and behaving and by adopting the new ways. Anthropologist Barbara Gallatin Anderson (1971) maintains that the "home culture never lets go of us" (p. 1121). During the initial stages of adjustment, for example, one may rely on an interim cultural framework, a hybrid support that combines elements of the host culture with whatever old cues from the native culture are useful in the new setting.

Sometimes the home culture's influence on people may not be obvious; it may be repressed or displaced. Yet it still "communicates" with people, even if only in their dreams. While doing fieldwork in India, Anderson and her colleagues kept records of their dreams and noticed definite themes that corresponded to their stage of adjustment to the new culture. During the first few weeks their dreams were filled with persons from their past, most of whom they had not thought of in years. Approximately one month later the dreams took a different form: family members and contemporaries appeared in the dreams, but at a distance and with themes of mixed identifications (for example, Indians in the dream displayed Americanlike qualities, and vice versa). Anderson describes what their dreams were like at that point:

> After more than a month in India, and much movement, there was a change reported in the dream patterning. Families were entering the context of dreams but almost shyly. One man reported dreaming for the first time since he left the United States of his wife, "but she simply talked to me from the doorway." Another dreamed he was having lunch with a colleague but the table was broad and they could not converse easily. A third watched as his children fly kites from a considerable distance. Some dreamed of new Indian acquaintances but often they were doing particularly un-Indian things: playing cards, riding in a sports car, or sitting in a conventional though unfamiliar Western house. One professor wrote remembering that in his dream he touched his hand to his head and felt the folds of an Indian turban. In one our Indian woman guide was smoking and spoke English without an accent at all, though actually she did not smoke and her speech was difficult to follow with the rolling Kerala accent. (p. 1122)

As the time to return home approached, the dreams changed themes again. Now Anderson and her colleagues reported no mixed identification or distant family members and friends. In fact, when significant others from the home culture appeared in dreams, they interacted and mingled easily with people from the host culture.

Do not assume that culture shock is limited to experiences with foreign cultures. People can experience **reentry shock**, or culture shock in reverse, upon returning home after living in another culture (Koehler 1986). In fact some researchers have discovered that many people find it surprisingly difficult to readjust to the return after spending a significant amount of time elsewhere. As in the experience of culture shock, they are in a situation in which differences jump in the forefront.

As with culture shock the intensity of reentry shock depends on an array of factors including the length of time lived in the host culture and the extent of the returnee's immersion in the everyday lives of people in that culture. Symptoms of reentry shock are essentially the mirror image of those associated with culture shock. They include panic attacks ("I thought I was going crazy"), glorification of the host culture, nostalgia for the foreign ways, panic, a sense of isolation or estrangement, and a feeling of being misunderstood by people in the home culture. These comments by Americans returning from abroad illustrate these reactions:

> "People pushed and shoved you in New York subways; they treated you as if you simply don't exist. I hated everyone and everything I saw here and had to tell my-

self over and over again: 'Whoa, this is your country; it is what you are part of.'" (WERKMAN 1986, P. 5)

"I find people fascinated with the fact that we have lived abroad for eleven years—fascinated for about 30 seconds." (HUNTER 1986, P. 186)

"America was a smorgasbord. But within two weeks, I had indigestion. Then things began to make me angry. Why did Americans have such big gas guzzling cars? Why were all the commercials telling me I had to buy this product in order to be liked? Material possessions and dressing for success were not top priorities in the highlands. And American TV? I missed the BBC." (SOBIE 1986, P. 96)

Although many people expect to have problems in adjusting to a stay in a foreign culture and even prepare for such problems, most do not expect trouble upon returning home. Because reentry shock is unexpected, many people become anxious and confused and feel guilty about having problems with readjustment ("How could I possibly think the American way was anything but the biggest and the best?"). In addition they are apprehensive about how their family, friends, and other acquaintances might react to their critical views of the home culture; for example, they are afraid others might view them as unpatriotic.

The experience of reentry shock points to the transforming effect of an encounter with another culture (Sobie 1986). The fact that the returnees go through reentry shock means that they have experienced up close another way of life and have been exposed to new norms, values, and beliefs. Consequently, when they come home, they see things in a new light. Strengths, but especially weaknesses, are more obvious. A position that runs counter to ethnocentrism and neutralizes cultural shock and reentry shock is cultural relativism.

Cultural Relativism

Cultural relativism means two things: (1) that a foreign culture should not be judged by the standards of a home culture and (2) that a behavior or way of thinking must be examined in its cultural context in terms of that society's values, norms, beliefs, environmental challenges, and history. Anthropologist Mary Catherine Bateson (1968) observed that when people first make contact with others who do not share their cultural traits, their initial reaction is usually ethnocentric. People notice only the cultural traits that differ from the ones they possess, and they generally describe the differences in negative and highly oversimplified terms. For example, "they" (whoever they are) have no conception of individuality, have no sense of privacy, do not really grieve, are money-driven, are not creative, are stuck in the past, have no sense of responsibility, do not know how to act genuine, and so on. Even when differences are perceived as positive, they are described in oversimplified terms, as in the claim that Native Americans treasure the earth and never did anything to hurt it (Johnson 1995) or when Japanese tell Americans, "You don't hold things in like we do. You express how you really feel. You don't have to hide your true feelings" (Berry 1995, p. 8). In any event, whether differences are described in positive or negative terms, one of the parties is described as lacking in some quality in relation to the others.

When Bateson (1968) did fieldwork in the Philippines, she noticed that Filipinos she encountered and Americans she knew each held very simplistic views about how the other grieves when someone close to them dies. Filipinos she knew regarded Americans as people who do not really grieve. The Americans she knew, in contrast, viewed Filipinos as insensitive and tactless because they seemed to go out their way to evoke expressions of grief from the survivors. Bateson recounted several experiences in which she observed the Filipino style, including a personal experience that occurred only a few hours after the premature stillbirth of her first child in a Manila hospital:

> On the afternoon of that day I was able to describe, so that my husband and I would be prepared, the way in which Filipinos would express sympathy. They show concern, in this as in many other contexts, by asking specific factual questions, and the primary assumption about those who have suffered a loss is that they should not be left alone. Rather than a euphemistic handling of the event and a denial of the ordinary course of life, one should expect the opposite. Whereas an American will shake hands and nod his head sadly, perhaps murmuring, "We were so sorry to hear [that]," and beat a swift retreat, a Filipino will say "We were so sorry to hear that your baby died. How much did it weigh? How long was labor? Etc. Etc." (p. 611)

On one level—an ethnocentric level—Bateson could simply have labeled the Filipino behavior she observed as intrusive or as an invasion of privacy. Yet even in this stressful situation she possessed the intellectual discipline to see differences as sources of both insight and unexpected comfort. Bateson argues that if people look hard enough, they will find not that "they" lack a certain quality, but that "they" usually have different notions of how to display that quality—in this case grief. According to Bateson's experiences and observations as an American, American norms governing interactions with the bereaved are intended to not remind them of the death so as not to arouse emotional outbreaks. The norms include not talking about the death except in euphemistic and indirect terms, not mentioning the name of the deceased in conversation, and glossing over the details of the death. From Bateson's perspective "Americans treat grief almost like an embarrassing disease" (p. 612).

From a Filipino's perspective such norms can be interpreted as evidence that Americans really don't grieve and/or that they are superficial. Such an interpretation embodies an ethnocentric viewpoint because, in arriving at this judgment, a Filipino applies his or her norms as the standard for judging American behavior. On closer examination it becomes clear that Americans are following norms that function to help survivors maintain their composure and control expressions of "real" grief.

Bateson offers at least two suggestions for comparing cultural traits, suggestions that correspond to a perspective of cultural relativism. First, one should focus on points of similarity and parallelism between the two cultures, indeed across all cultures, in the behavior or idea of interest. In the case of bereavement, it is important to point out that all cultures must confront the facts of death and the pain felt by survivors, and all cultures have developed strategies to make death bearable to the survivors. Some societies make an organized effort to help the bereaved control themselves and forget; others help survivors express and live out their grief. When the analysis is viewed in this light, it is difficult to support ethnocentric judgments such as that Americans are superficial or unemotional or that Filipinos are insensitive to survivors' needs.

A second suggestion for comparing cultural traits is to keep in mind "ways in which both cultures succeed and fail in solving problems common to all human beings" (Bateson 1968, p. 617). American norms governing how to interact with survivors may succeed in helping them avoid breakdowns in public, but they may fail when survivors believe that openly grieving for more than a few days is pathological. Filipino norms may succeed in helping people express grief, but they fail by denying survivors opportunities to be alone.

As this example shows, cultural relativism aims to understand foreign behavior and thinking, not to condone or to discredit it. People who take a position of cultural relativism would control quick, judgmental reactions to seemingly strange cultural elements, such as cow manure figurines made by Amish farmers, until they acquired information to reach an informed judgment.

http://www.poopets.com/cgi-bin/dbml.exe?Template=/mall/poopets/abouttewks.dbm&sid=1&cid=-1406890265

Document: All About Poopets

Q: What are some of the uses to which cow manure is put around the world? What environmental payoffs are there for finding uses for manure?

Cultural relativism, however, does not entail an "anything goes" position. Such a position allows every cultural trait—even some of the most harmful and violent ones (for example, infanticide, human sacrifice, foot binding, death threats against those holding controversial ideas, witch-hunts, scientific experimentation on naive human subjects)—to escape judgment or criticism. Unfortunately uninformed critics often interpret cultural relativism as a search to justify any behavior and as support for one or more of the following attitudes: "Whatever they do is fine"; "That's just the way that culture is"; "It's none of my business what others do"; or "Everything is relative. What's good in one culture is bad in another, and vice versa."

Anthropologist Clifford Geertz (1995a) maintains that if one is serious about addressing a problematic cultural trait, the first step is to ask the persons affected (supporters, critics, victims, beneficiaries) about the practice. Investigators should not assume that they know the truth of the situation and proceed as if they have the solution. As a rule of thumb one should avoid making judgments and recommendations that affect another culture when these are based on superficial knowledge and/or derive from ethnocentric thinking. With regard to education, for example, Iranian education critic, reformer, and scholar Samad Behrangi maintains that "unless we have seen the school's environment and surrounding community, unless we have lived among the people, unless we have been friends with the people, we have not heard their voice and have not known their desires, it is not even proper to have sympathy for the environment, or impose unrelated educational policies, or even to write stories or textbooks for them" (Behrangi, in Fereshteh 1994, p. 28).

Understanding a cultural trait or taking a position of cultural relativism does not mean that one accepts the trait unconditionally, nor does it mean that one has no values of one's own. As Clifford Geertz (1995b) argues, the challenge of taking this position is finding "a way to keep one's values and identity while living with other values—values you can neither destroy or approve" (Berreby 1995, p. 47). In Geertz's words, "I hold democratic values, but I have to recognize that a lot of people don't hold them. So it doesn't help much to say, 'This is the truth.' That doesn't mean I don't believe anything. You can't assert yourself in the world as if nobody else was there. Because this isn't a clash of ideas. There are people attached to those ideas. If you want to live without violence, you have to realize that other people are as real as you are" (p. 47).

Cultural relativism is especially useful when people must make a decision supporting one cultural trait over another or must contemplate taking action to change, modify, or even eliminate a cultural trait that "others" share. In all cases cultural relativism facilitates decision making and/or negotiation by making the cultural elements in question intelligible to everyone involved. Certainly a decision is more appropriately made by people who understand the traits than by those who do not.

Studying and learning about other cultures teaches us that morality is both relative and universal (Redfield 1962a). Morality is relative in that norms, values, and beliefs about rightness and wrongness vary across time and place. It is also universal, however, in that every culture has conceptions of morality. Although mores do exist that can make virtually any idea or behavior seem right or wrong, "mores have a harder time making some things right than others" (Redfield 1962b, p. 451).

Here I am reminded of a scene from the film *Schindler's List* featuring a high-ranking Nazi officer and his Jewish housekeeper and cook, Helen. One evening he walks down to the cellar where she sleeps and has the following conversation with himself, during which Helen looks straight ahead and says nothing. As a Nazi the officer believes that the human race can be improved by careful selection of those who mate and produce offspring. At the same time, he finds himself attracted to his Jewish housekeeper.

> "I came to tell you that you are a wonderful cook and a well-trained servant. I mean it. If you need a reference after the war, I'll be happy to give you one. It must get lonely down here . . . doesn't it? You can answer. . . . Yes, you're right, sometimes we are both lonely. I, I mean, I would like so much to reach out and touch you in your loneliness. What would that be like? What would be wrong with that? I mean, I realize you are not a person in the strictest sense of the word. But maybe you are right about that, too. Maybe what's wrong is not us, it is this. When they compare you to vermin, to rodents, to lice. You make a good point, a very good point. Is this the face of a rat? The eyes of a rat? Hath not a Jew's eyes? I feel for you, Helen. [He reaches for her and then withdraws.] No, I don't think so. You Jewish bitch. You nearly talked me into it, didn't you?"

This scene supports Redfield's idea that mores have an easier time making actions "right" when they prescribe behavior that supports equal opportunity, life, liberty, dignity, self-determination, and well-being than when they deny human dignity and rights. It is difficult to imagine that people anywhere could be troubled, as is this Nazi officer, by actions that would have positive and constructive effects on people's lives.

Subcultures

- *Principle 9:* In every society certain groups possess distinctive traits that set them apart from the main culture.

Groups that share in some parts of the dominant culture but have their own distinctive values, norms, language, or material culture are called **subcultures.** A subculture that conspicuously challenges, rejects, or clashes with the central norms and values of the dominant culture is referred to as a **counterculture.** All of the cultural principles discussed thus far apply to subcultures (which include countercultures).

Often we think we can identify subcultures on the basis of physical traits, ethnicity, religious background, geographic region, age, sex, socioeconomic or occupational status, dress, or behavior defined as deviant by society. However, determining which people constitute a subculture is a complex task that requires careful thought; it must go beyond simply including everyone who shares a particular trait. For example, using broad ethnic categories as a criterion for identifying the various subcultures within the United States makes little sense. Realistically a biologically and culturally intermixed population numbering hundreds of millions cannot be divided neatly into white, African-American, Hispanic, Native American, and Asian subcultures (Clifton 1989). The broad categorization of Native American, for example, ignores the fact that the early residents of North America "practiced a multiplicity of customs and lifestyles, held an enormous variety of values and beliefs, spoke numerous languages mutually unintelligible to the many speakers, and did not conceive of themselves as a single people—if they knew about each other at all" (Berkhofer 1978, p. 3). The point is that the presence or absence of a single trait cannot be the only criterion for classifying someone as part of a subculture. Consider persons of Korean ancestry who live in the United States.

http://www.koma.org/ktown_about.html

http://www.koma.org/ktown_15.html

Document: All About Koreatown

Q: What percentage of Southland's Korean population live in Koreatown? How many Korean Americans live in Southern California? Select the document "Koreatown History." Which Asian groups have the longest history in the United States? When did these groups arrive in relation to Koreans? Who are the 1.5 generation? What does the information tell us about using a single trait—in this case Korean—as a criterion for placing someone in a subculture?

Sociologists determine whether a group of people constitutes a subculture by learning whether they share a language, values, norms, or a territory and whether they interact with one another more than with people outside the group. One characteristic central to all subcultures is that their members are isolated in some way from other people in the society. This isolation may be voluntary, it may result from an accident of geography, or it may be imposed consciously or unconsciously by a dominant group. Or it may be a combination of all three. Whatever the reason, subcultures are cut off in some way. The cutoff may be total or may be limited to selected segments of life such as work, school, recreation and leisure, dating and marriage, friendships, religion, medical care, or housing.

The Old Order Amish are an example of a subculture within the United States whose members voluntarily cut themselves off from many elements of mainstream culture. A geographically isolated subculture within the United States is that of the Aleuts, natives of the Aleutian Islands, which stretch in a chain from Alaska across the northern Pacific Ocean. Their language, an Eskimo-Aleut dialect, is incomprehensible to other American Eskimo groups. The plight of other Native American groups is an example of isolation imposed by a dominant group. These groups represent numerous subcultures within the United States that were forced onto reservations by the federal government. About one third of Native Americans live on reservations, and many are geographically, economically, and socially isolated from mainstream American culture.

Some subcultures within the United States are isolated in different ways or to different degrees. Their members integrate into mainstream culture when possible but voluntarily or involuntarily accept a segregated role in other areas of life. For example, African Americans who work or attend school primarily with whites often are excluded from personal and social relationships with their white colleagues. This exclusion forces them to form their own fraternities and sororities, study groups, support groups, and other organizations.

The number and variety of affiliations between members of a subculture and people outside their group are a rough indicator of the subculture's relationship to mainstream culture. In general, the greater the number and the more diverse the affiliations with "outsiders," the more likely the subculture shares a significant portion of the mainstream culture. Some subcultures are **institutionally complete** (Breton 1967); their members do not interact with anyone outside their subculture to shop for food, attend school, receive medical care, or find companionship because these needs are provided for by the subculture. In the United States retirement communities are a typical example of a setting in which residents' needs to affiliate with the mainstream culture are minimized.

Often there is a clear association between institutional completeness and language differences. People who cannot speak the language of the dominant culture are likely to live in institutionally complete ethnic communities (for example, Little Italies, Chinatowns, Koreatowns, Mexican barrios). For a complete description of one potentially institutionally complete community, see the Koreatown homepage on the internet.

http://www.koma.org/ktown_dir.html

Document: Koreatown Directory

: What businesses and services made Koreatown institutionally complete?

Whether an immigrant chooses to belong to a community of like persons depends on a variety of factors: circumstances of migration (forced or voluntary), age, sex, status of mother country relative to host country, knowledge of the host country's language, level of skills, education, and social class.

Discussion Question

Based on the information presented in this chapter, what conclusions can you draw about the importance of culture in people's lives?

Additional Reading

For more on

- Cultural relativism (see especially section IA, Relational Relativism)
  http://www.law.indiana.edu/glsj/vol1/green.html
  Document: **Cultural Identities and Global Political Economy from an Anthropological Vantage Point**

- Ethnocentrism
  http://www.univie.ac.at/voelkerkunde/theoretical-anthropology/godina.html
  Document: **What Is Wrong with the Concept of Human (Social) Evolution?**

- Cultural flow/cultural diffusion
  http://www.pangea.ca/~trade
  Document: **The Global Connections Project**

  http://www.carleton.ca/npsia/cfpj/john.html
  Document: **Culture and Foreign Policy**

  http://www.perseus.tufts.edu/GreekScience/Students/Ellen/EarlyGkAstronomy.html
  Document: **Early Greek Astronomy**

- Asian values
  http://ifrm.glocom.ac.jp/DOC/k02.001.html#5
  Document: **Can "Asian Values" Be Unique?**

- Defining culture
  http://www.wsu.edu:8001/vcwsu/commons/

  http://www.wsu.edu:8000/vcwsu/commons/topics/culture/culture-definitions/whose-text.html
  Document: **The Culture Debate in the U.S.: Whose Culture Is This Anyway?**

- The Somali alphabet
  http://www.teleport.com/~napoleon/somali/alphabet.html
  Document: **Brief History of the Somali Alphabet**

- Water conservation efforts (establishing norms)
  http://www.mother.com/uswaternews/archive/96/conserv/consort.html
  Document: **Consortium Promotes Washing Machines That Use Less Water, Energy**

  http://www.innovativ.com/waterwise.html
  Document: **Be Water-Wise: 7 Ways in 7 Days**

  http://ianrwww.unl.edu/ianr/waterctr/wctriv.html
  Document: **Water Trivia**

http://www.mbnet.mb.ca/wpgwater/welcome.html
Document: Welcome to the Waterfront—Winnipeg's Water Conservation Information Source

- Languages/symbols
  http://www.travlang.com/languages/indextext.html
  Document: Foreign Languages for Travelers

  http://mendel.mbb.sfu.ca/berg/breden.lab/dotw.html
  Document: Days of the Week

- Subcultures
  gopher://alishaw.ucsb.edu/00/.thresholds/.tvc8/.histpunk/.steigner.txt
  Document: Do-It-Yourself Punk Rock and Hardcore

- Cultural issues
  http://www2.uchicago.edu/jnl-pub-cult
  Document: Public Culture Home Page

Socialization

At birth the human cerebral cortex (the seat of complex thought) is not sufficiently developed to permit a sophisticated awareness of self and others or reflection on the rules of social life, which we call norms. These capacities develop, however, as children mature biologically and as they interact with others.

Most 2-year-olds are biologically ready to show concern for what adults regard as the rules of life. They are bothered when rules are violated: paint peeling from a table, broken toys, small holes in clothing, and persons in distress raise troubling questions. From a young child's point of view, when something is "broken," then somebody somewhere has done something very wrong (Kagan 1988a, 1988b, 1989).

To show this kind of concern with standards, however, 2-year-olds must first be exposed to information that leads them to expect behavior, people, and objects to be a certain way (Kagan 1989). They develop these expectations in the course of their social relations with adults. Children go to adults with their questions and needs. Adults respond in different ways: they may offer explanations, express concern, try to help, show no concern, or pay no attention. Through many such simple exchanges, children learn how to think about objects and people. They learn about the social group to which they primarily belong and about other groups to which they do not belong. In addition through these exchanges children are acquiring basic skills such as the ability to talk, walk upright, and reason. This learning from others about the social world is part of a complex lifelong process called **socialization.**

Socialization begins immediately after birth and continues throughout life. It is a process by which newcomers develop their human capacities and acquire a unique personality and identity. This process also enables **internalization,** by which people take as their own and accept as binding the norms, values, beliefs, and language needed to participate in the larger community. And socialization is the process by which culture is passed on from generation to generation.

This chapter explores the socialization process and its significance for both societies and individuals. Keep in mind that socialization is just one factor of many that help us understand the relevant strategies people everywhere use to teach newcomers how to participate in the society into which they are born. The newcomers, however, do not become carbon copies of their teachers. They learn about the environment they inherit and then come to terms with it in unique ways.

Nature and Nurture

No discussion of socialization can ignore the importance of nature and nurture to physical, intellectual, social, and personality development. **Nature** is the term for human genetic makeup or biological inheritance. **Nurture** refers to the environment or the interaction experiences that make up every individual's life. Some scientists debate the relative importance of genes and environment, arguing that one is substantially more important than the other to all phases of human development. But most consider such a debate futile, because it is impossible to separate the influence of the two factors or to say that one is more important. Both are essential to socialization. Trying to distinguish the separate contributions of nature and nurture is analogous to examining a tape player and a cassette tape separately to determine what is recorded on the tape rather than studying how the two work together to produce the sound (Ornstein and Thompson 1984).

Consider the search for the cause of alcoholism. Many researchers frame the question in either/or terms: Is alcoholism a learned behavior or a genetic trait? Few researchers consider the possibility that both genes and environment may be important factors. The following is an example of research that focuses on alcoholism as learned behavior.

http://home.navisoft.com/aapa/aca.htm
Document: Adult Children of Alcoholics

Q: Notice that the researcher focuses on the environment in which children of alcoholics are raised and their reaction to that environment. What are two or three characteristics of the "alcoholic" environment?

Here is an example of research that focuses on alcoholism as a genetic or inherited condition.

http://www.niaaa.nih.gov/publications/aa18.htm

Document: The Genetics of Alcoholism

 Although "The Genetics of Alcoholism" begins by acknowledging the importance of the environment, it focuses on the genetic component. Why do researchers do genetic research? What research design did the scientists use to study the genetic component? Instead of identifying one factor as the cause, perhaps we should consider a third possibility: that both genetics and environment interact in a complex way to "create" the alcoholic. Can you think of an example of an argument that incorporates both factors?

The development of the human brain illustrates rather dramatically the inseparable qualities of genes and environment. As part of our human genetic makeup, we possess a cerebral cortex—the thinking part of the brain—which allows us to organize, remember, communicate, understand, and create. The cortex is made up of at least 100 billion neurons or nerve cells (Montgomery 1989); the fibers of each cell form thousands of synapses, which make connections with other cells. The number of interconnections is nearly infinite; it would take 32 million years, counting one synapse per second, to establish the number (Hellerstein 1988; Montgomery 1989).

Perhaps the most outstanding feature of the human brain is its flexibility. Scientists believe that the brain may be "set up" to learn any of the more than 6,000–9,000 known human languages. In the first months of life, babies are able to babble the sounds needed to speak all of these languages, but this enormous potential is reduced by the language (or languages) that the baby hears and eventually learns. Evidence suggests that the brain's language flexibility begins to diminish when the child reaches one year of age (Ornstein and Thompson 1984; Restak 1988). The larger implication is that genetic makeup provides essential raw materials but that these materials can be shaped by the environment in many different ways.

The human genetic makeup is flexible enough to enable a person to learn the values, beliefs, norms, behavior, and language of any culture, but a multitude of experiences must combine with genetic makeup to create that person. If there is no contact with others, a person cannot ever become a normally functioning human being, let alone learn to become part of society.

The Importance of Social Contact

Cases of children raise in extreme isolation or in restrictive and unstimulating environments show the importance of social contact (nurture) to normal development. Some of the earliest and most systematic work in this area was done by sociologist Kingsley Davis, psychiatrist Rene Spitz, and sociologist Peter Townsend. Their work shows how neglect and lack of socialization influence emotional, mental, and even physical development.

Cases of Extreme Isolation

In two classic articles, "Extreme Isolation of a Child" and "Final Note on a Case of Extreme Isolation," sociologist Kingsley Davis (1940, 1947) documented and compared the separate yet similar lives of two girls, Anna and Isabelle. Each girl had received a minimum of human care during the first six years of her life. Both were illegitimate children and for that reason were rejected and forced into seclusion. When authorities discovered the girls, they were living in dark, atticlike rooms, shut off from the rest of the family and from daily activities. Although both girls were 6 years old when authorities intervened, they exhibited behavior comparable to that of 6-month-old children. Anna "had no glimmering of speech, absolutely no ability to walk, no sense of gesture, not the least capacity to feed herself even when food was put in front of her, and no comprehension of cleanliness. She was so apathetic that it was hard to tell whether or not she could hear" (Davis 1947, p. 434).

Like Anna, Isabelle had not developed speech; she communicated with gestures and croaks. Because of a lack of sunshine and a poor diet, she had developed rickets: "Her legs in particular were affected; they 'were so bowed that as she stood erect the soles of her shoes came nearly flat together, and she got about with a skittering gait'" (Davis 1947, p. 436). Isabelle also exhibited extreme fear of and hostility toward strangers.

Anna was placed in a private home for retarded children until she died four years later. At the time of her death, she behaved and thought at the level of a 2-year-old child. Isabelle, on the other hand, was placed in an intensive and systematic program designed to help her master speech, reading, and other important skills. After two years in the program, Isabelle had achieved a level of thought and behavior normal for someone her age.

On the basis of Anna and Isabelle's case histories, Davis concluded that extreme isolation has a profound and negative effect on mental and physical development. At the same time, Davis concluded, Isabelle's case demonstrates that extreme "isolation up to the age of six, with the failure to acquire any form of speech and hence failure to grasp nearly the whole world of cultural meaning, does not preclude the subsequent acquisition of these" (Davis 1947, p. 437).

In addition Davis speculated on the question of why Isabelle did so much better than Anna. He offered two possible explanations: (1) Anna may have inherited a less hardy physical and mental constitution than Isabelle's and (2) Anna's condition may have resulted from not being exposed to the intensive and systematic therapy that Isabelle received. The case history comparisons are inconclusive, though. Because Anna died at age 10, researchers will never know whether she eventually could have achieved a state of normal development.

In drawing these conclusions, Davis overlooked one important factor. Although Isabelle had spent her early childhood years in a dark room, shut off from the rest of her mother's family, she spent most of this time with her deaf-mute mother. Davis and the medical staff who treated Isabelle seemed to equate deafness and muteness with feeble-mindedness (not an unusual association in the 1940s). In addition they seemed to assume that being in a room with a deaf-mute is equivalent to a state of isolation. A considerable body of evidence, however, now indicates that deaf-mutes have rich symbolic capacities (see Sacks 1989). The fact that Isabelle was able to communicate through gestures and croaks suggests that she had established an important and meaningful bond with another human being. Although the bond was less than ideal, it gave her an advantage over Anna. In view of this possibility we must questions Davis's conclusion that children may be able to overcome the effects of extreme isolation during the first six years of life, provided they have a "good enough" constitution and systematic training. We use the phrase "good enough" because researchers do not know the exact profile of a person or training regime that would allow someone to overcome such effects.

Less Extreme Cases

Other evidence of the importance of social contact comes from less extreme cases of neglect. Rene Spitz (1951) studied 91 infants who were raised by their parents during their first three to four months but who later were placed in orphanages because of unfortunate circumstances. At the time of their entry into the institution, the infants were physically and emotionally normal. At the orphanages they received adequate care with regard to bodily needs—good food, clothing, diaper changes, clean nurseries—but little personal attention. Because there was only one nurse for every 8–12 children, the children were starved emotionally. The emotional starvation caused by the lack of social contact resulted in such rapid physical and developmental deterioration that a significant number of the children died. Others became completely passive, lying on their backs in their cots. Many were unable to stand, walk, or talk (Spitz 1951). The experiences and information the Parent Network for Post-Institutionalized Children posts on the internet reinforce Spitz's findings. The network functions as a resource center for U.S. and Canadian families who have adopted institutionalized children from Eastern European countries and the former Soviet Union.

http://www.cyfc.umn.edu/Adoptinfo/institutionalization.html

Document: The Long-Term Effects of Institutionalization on the Behavior of Children from Eastern Europe and the Former Soviet Union

Q: What are the characteristics of institutional life that affect children's emotional, psychological, and behavioral development?

Such cases teach us that children need close contact with and stimulation from others in order to develop normally. Adequate stimulation means the existence of strong ties with a caring adult. The ties must be characterized by a bond of mutual expectation between caregiver and baby. In other words there must be at least one person who knows the baby well enough to understand his or her needs and feelings and who will act to satisfy them. Under such conditions the child learns that certain actions on his or her part elicit predictable responses: getting excited may cause Dad to be equally excited; crying may get Mom to soothe the child. When researchers set up experimental situations in which a parent fails to respond to his or her infant in expected ways (even for a few moments), the baby suffers considerable tension and distress ("Nova" 1986).

Meaningful social contact with and stimulation from others are important at any age. Strong social ties with caring people are linked to overall social, psychological, and physical well-being. British sociologist Peter Townsend (1962) studied the effects of minimal interaction that can characterize life for the elderly in nursing homes. The consequences for the institutionalized elderly are strikingly similar to those described by Spitz in his studies of institutionalized children:

> In the institution people live communally with a minimum of privacy, and yet their relationships with each other are slender. Many subsist in a kind of defensive shell of isolation. Their mobility is restricted, and they have little access to general society. Their social experiences are limited, and the staff leads a rather separate existence from them. They are subtly oriented toward a system in which they submit to orderly routine and lack creative occupation, and cannot exercise much self-determination. They are deprived of intimate family relationships. . . . The result for the individual seems to be a gradual process of depersonalization. He may become resigned and depressed and may display no interest in the future or things not immediately personal. He sometimes becomes apathetic, talks little, and lacks initiative. His personal and toilet habits may deteriorate. (TOWNSEND 1962, pp. 146–47)

The work of Kingsley Davis, Rene Spitz, and Peter Townsend supports the idea that a person's overall well-being depends on meaningful interaction experiences with others. On a more fundamental level social interaction is essential to a developing sense of self. Sociologists, psychologists, and biologists agree that "it is impossible to conceive of a self arising outside of social experience" (Mead 1934, p. 135). Yet, if the biological mechanisms involved in remembering or learning and recalling names, faces, words, and the meaning of significant symbols were not present, people could not interact with one another in meaningful ways: "You have to begin to lose your memory, if only in bits and pieces, to realize that memory is what makes our lives. Life without memory is no life at all. . . . Our memory is our coherence, our reason, our feeling, even our action. Without it we are nothing" (Bunuel 1985, p. 22).

Individual and Collective Memory

Memory, the capacity to retain and recall past experiences, is easily overlooked in exploring socialization. Yet without memory individuals and even whole societies would be cut off from the past. On the individual level memory is what allows people to retain their experiences; on the societal level memory preserves the cultural past.

How memory works is still largely a mystery. The latest neurological evidence suggests that some physical trace remains in the brain after new learning takes place, stored in an anatomical entity called an engram. **Engrams,** or memory traces as they are sometimes called, are formed by chemicals produced in the brain. They store in physical form the recollections of experiences—a mass of information, impressions, and images unique to each person:

> It may have been a time of listening to music, a time of looking in at the door of a dance hall, a time of imagining the action of robbers from a comic strip, a time of waking from a vivid dream, a time of laughing conversation with friends, a time of

listening to a little son to make sure he was safe, a time of watching illuminated signs, a time of lying in the delivery room at childbirth, a time of being frightened by a menacing man, a time of watching people enter the room with snow on their clothes. (PENFIELD AND PEROT 1963, P. 687)

Scientists do not believe that engrams store actual records of past events, like films stored on videocassettes. More likely engrams store edited or consolidated versions of experiences and events, which are edited further each time they are recalled.

As we noted previously, memory has more than an individual quality; it is strongly social. First, no one can participate in society without an ability to remember and recall such things as names, faces, places, words, symbols, and norms. Second, most "newcomers" easily learn the language, norms, values, and beliefs of the surrounding culture. We take it for granted that people have this information stored in memory. Third, people born at approximately the same time and place are likely to have lived through many of the same events. These experiences, each uniquely personal and yet similar to one another, remain in memory long after the event has passed. We will use the phrase **collective memory** to describe the experiences shared and recalled by significant numbers of people (Coser 1992; Halbwachs 1980). Such memories are revived, preserved, shared, passed on, and recast in many forms, such as stories, holidays, rituals, and monuments. Read the first part of the document by Carla Armstrong through her analysis of the National Vietnam Memorial in Washington, DC.

http://www.saed.kent.edu/Architronic/v2n2/v2n2.05.html
Document: Ritual and Monument

How do rituals and monuments help to facilitate the socialization process? With regard to the design of the National Vietnam Memorial, what response did the architect hope to evoke in those who view it?

The point is that socialization is not possible without memory. Memory is the mechanism by which group expectations become internalized in individuals and, by extension, in the whole society and by which the past remains an integral, living part of the present. The picture that any one individual holds cannot be a complete record of the past. Yet memories include the perceived reasons that things are the way they are.

Memory of past experiences allows individuals to participate in society and shapes their viewpoint. In his essay "The Problem of Generations," sociologist Karl Mannheim (1952) maintained that first impressions or early childhood experiences are fundamental to a person's view of the world. In fact Mannheim believed that an event has more biographical significance if it is experienced early in life than if it is experienced later in life. Many of our most significant early experiences take place in groups, some of which leave a powerful and lasting impression.

The Role of Groups

In the most general sense a **group** is two or more persons who (1) share a distinct identity (the biological children of a specific couple; members of a gymnastics team, military unit, club, or organization; or persons sharing a common cultural tradition), (2) feel a sense of belonging, and (3) interact with one another in direct and/or indirect, but broadly predictable, ways. Interaction is broadly predictable because norms govern the behavior expected of members depending on their position (mother, coach, teammate, sibling, employee) within the group. Groups vary according to a whole host of characteristics including size, degree of intimacy among members, member characteristics, purpose, duration, and the extent to which the members socialize newcomers and one another. However, sociologists identify primary groups and ingroups and outgroups as particularly powerful socialization agents.

Primary Groups

Primary groups, such as the family or a high school sports team, are characterized by face-to-face contact and strong ties among members. Primary groups are not always united by harmony and love; they can be united by hatred for another group. But in either case the ties are emotional. A primary group is comprised of members who strive to achieve "some desired place in the thoughts of [the] others" and who feel allegiance to the others (Cooley 1909, p. 24). A person may never achieve the desired place but may still be preoccupied with that goal. In this sense primary groups are "fundamental in forming the social nature and ideals of the individual" (Cooley 1909, p. 23). The family is an important primary group because it gives the individual his or her deepest and earliest experiences with relationships and because it gives newcomers their first exposure to the "rules of life." In addition the family can serve to buffer its members against the effects of stressful events or negative circumstances, or it can exacerbate these effects.

Moving to a new community is one example of a stressful event, especially for children. The American Academy of Child and Adolescent Psychiatry's "Children and Family Moves" identifies actions parents can take to buffer against that stressful event.

http://www.psych.med.umich.edu/web/aacap/factsFam/fmlymove.htm
Document: Children and Family Moves

 Q: What actions can parents take to reduce the stress of moving for children?

One very clear example of a primary group is a military unit. A unit's success in battle depends on strong ties among its members. Apparently soldiers in the primary group fight for one another, rather than for victory per se, in the heat of battle (Dyer 1985). Military units train their recruits always to think of the group before the self. In fact the paramount goal of military training is to make individuals feel inseparable from their unit. Some common strategies to achieve this goal include ordering recruits to wear uniforms, to shave their heads, to march in unison, to sleep and eat together, to live in isolation from the larger society, and to perform tasks that require the successful participation of all unit members: if one member fails, the entire unit fails. Another key strategy is to focus the unit's attention on fighting together against a common enemy. An external enemy gives a group a singular direction and thus increases its internal cohesiveness. All types of primary groups, however, have boundaries—a sense of who is in the group and who is outside of the group.

Ingroups and Outgroups

Sociologists use the term **ingroup** to describe those groups with which people identify and to which they feel closely attached, particularly when that attachment is founded on hatred from or opposition toward another group. Ingroups exert their influence on our social identity in conjunction with outgroups. An **outgroup** is a group of individuals toward which members of an ingroup feel a sense of separateness, opposition, or even hatred. Outgroups also can make us conscious of where we belong. Obviously one person's ingroup is another person's outgroup. The very existence of an outgroup heightens loyalty among ingroup members and magnifies characteristics that distinguish the ingroup from the outgroup. An outgroup can unify an ingroup even when the ingroup members are extremely different from one another.

Loyalty to an ingroup and opposition to an outgroup are accompanied by an us-versus-them consciousness. Boundaries between the two groups are sharp; they are reinforced by residential, occupational, educational, and/or religious segregation. Because there is little interaction between ingroup and outgroup members, they know little about one another. This lack of firsthand experience deepens and reinforces misrepresentations, mistrust, and misunderstandings between members of the two groups. Members of one group tend to view members of the other in the most stereotypical of terms. Often one of the groups has superior status, material conditions, and facilities.

Fortunately many programs are designed to weaken the boundaries that separate ingroups from outgroups. Consider a Rhode Island police department program designed and implemented with the goal of improving relations between police officers and the many elderly people who live in that state. The size of the elderly population and its special needs increase the probability of contact between police and elderly.

gopher://justice2.usdoj.gov/00/fbi/January95/jan6.txt
Document: Focus on Training: Teaching Officers to Serve Seniors

Q: Which myths and misconceptions about the elderly could turn the elderly into an "outgroup" for police officers? What features of the program are intended to offset these misconceptions?

A second example can be found in the Virginia Beach Police Department. The Police Equestrian Program seeks to bridge the gap between officers and children from low-income neighborhoods characterized by drug and crime problems.

gopher://justice2.usdoj.gov/00/fbi/January95/jan2.txt
Document: Police Practice: Horseplay Brings Officers Closer to Community

 Q: What strategy does the Virginia Beach Police Department use to develop understanding between children and officers and to promote constructive relations?

Often an ingroup and an outgroup clash over symbols—objects or gestures that are clearly associated with and valued by one group. These objects can be defined by members of the other group as so threatening that they seek to eliminate them: destroying the objects becomes a way of destroying the group. Danny Rubenstein, an Israeli reporter covering the West Bank, observed that most of the clashes between Palestinian youths and Israeli soldiers are over symbols:

> Of the hundreds of clashes I have witnessed pitting Palestinian youths against Israeli military and administrative authorities in the West Bank and Gaza, most have involved symbols. Thus, for example, an ongoing battle is being waged over the Palestinian flag. Arabs hoist the flag (which is very much like the Jordanian flag), while Israeli soldiers bring it down and attempt to catch and punish the perpetrators. At times the situation takes a ridiculous turn. Some time ago, in Bethlehem, I heard an Israeli officer issue an order to close down for a week all shops on a certain street where a Palestinian flag had been hoisted on the corner utility pole the night before. I saw schoolgirls in Hebron knitting satchels modeled on the Palestinian flag and clothing stores with window displays arranged to fit its color and pattern.
>
> The flag is but one example. Military censors crack down on anti-Israeli expression in Arabic language newspapers, textbooks and plays. Every day Israeli soldiers remove pro-Palestinian graffiti from walls in Arab towns and villages. The warfare over symbols is accompanied by demonstrations of young people, rock throwing, protest marches and the burning of automobile tires—routine . . .
> (RUBENSTEIN 1988, P. 24)

To this point we have examined how socialization is a product of nature and nurture. We have discussed how genetic makeup provides each individual with potentials that are developed to the extent made possible by the environment. We also have considered the importance of stimulation from caregivers in developing our genetic potential and the connection between group membership and self-awareness. Even groups to which we do not belong can have powerful influences on our sense of self. For example, an outgroup makes us clearly aware of who "they" are, and this in turn reminds us of who "we" are. Next we will examine the theories of two symbolic interactionists, George Herbert Mead and Charles Horton Cooley, regarding some specific ways in which the self develops and in which information is transmitted to newcomers.

Symbolic Interactionism and Self-Development

Humans are not born with a sense of self; it evolves through regular interaction with others. The emergence of a sense of self depends on our physiological capacity for **reflexive thinking**—stepping outside the self and observing and evaluating it from another's viewpoint. Reflexive thinking allows individuals to learn how they come across to others and to adjust and direct behavior in ways that meet others' expectations. In essence self-awareness emerges hand in hand with awareness of others and of their evaluations of one's behavior and appearance.

The Emergence of Self-Awareness

According to George Herbert Mead significant symbols and gestures are the mechanisms that allow an individual to interact with others and in the process to learn about the self. (For a discussion of symbolic interaction theory, see Chapter 3.) A **significant symbol** is a word, gesture, or other learned sign that is used "to convey a meaning from one person to another, and that has the same meaning for the person transmitting it as for the person receiving it" (Theodorson and Theodorson 1979, p. 430). Language is a particularly important significant symbol because "it is only through language that we enter fully into our human estate and culture, communicate freely with our fellows, acquire and share information. If we cannot do this, we will be bizarrely disabled and cut off" (Sacks 1989, p. 8).

Symbolic gestures or signs are nonverbal cues, such as tone of voice, inflection, facial expression, posture, and other body movements or positions that convey meaning from one person to another. Bernard Hibbitts identifies a range of symbolic gestures that are and were important to various legal transactions.

http://www.law.pitt.edu/hibbitts/re_mem.htm

Document: Re-membering Law: Legal Gesture in the Past, Present, and Future

Q: Hibbitts states that legal gestures serve at least eight broad functions. Which two functions do you consider to be the most interesting? What is an example of a gesture that illustrates each function?

As people learn significant symbols—language and symbolic gestures—they also acquire the ability to do reflexive thinking and to adjust their presentation of self to meet other people's expectations. Mead believed, however, that humans do not adhere mechanically to others' expectations. Instead a dialogue goes on continuously between two aspects of the self—the *I* and the *me*.

The *me* is Mead's term for the self as the internalized expectations of others. Before an individual acts, the *me* takes others into account by assessing the appropriateness of the act and anticipating the responses. The *I* is the spontaneous, autonomous, creative self, capable of rejecting expectations and acting in unconventional, inappropriate, or unexpected ways. For example, a student recently expressed disappointment at receiving a failing grade on an examination. Upon seeing her score she blurted out,

"A 50! I skipped two classes to study for this stupid test!" In making this comment, she failed to anticipate its effect on the professor. Presumably the unexpected grade came as such a shock that her spontaneous *I* overwhelmed her calculating *me*.

Although Mead does not specify how the *I* emerges, we know that a spontaneous, creative self must exist; otherwise human life would never change and would stagnate. Mead is more specific about how the *me* develops: through imitation, play, and games, all of which give the developing child practice with role-taking.

Role-Taking

Mead assumed that the self is a product of interaction experiences. He maintained that children acquire a sense of self when they become objects to themselves. That is, they are able to imagine the effect of their words and actions on other people. According to Mead a person can see him- or herself as an object after learning to role-take. **Role-taking** involves stepping outside the self and viewing its appearance and behavior imaginatively from an outsider's perspective.

Researchers have devised an ingenious method for determining when a child is developmentally capable of role-taking. A researcher puts a spot of rouge on the child's nose and then places the child in front of a mirror. If the child shows no concern with the rouge, he or she presumably has not yet acquired a set of standards about how he or she ought to look; that is, the child cannot role-take or see him- or herself from another person's viewpoint. But if the child shows concern over the rouge, he or she presumably has formed some notion of self-appearance and therefore can role-take (Kagan 1989).

Mead hypothesized that children learn to take the role of others (1) through imitation, (2) through play, and (3) through games. Each of these stages involves a progressively sophisticated level of role-taking.

The Preparatory Stage In this stage children have not yet developed the mental capabilities that allow them to role-take. Although they mimic or imitate people in their environment, they have almost no understanding of the behaviors they are imitating. Children may imitate spontaneously (by mimicking a parent writing, cooking, reading, and so on), or they may repeat things that adults encourage them to say and reward them for saying. In the process of imitating, children learn to function symbolically; that is, they learn that particular actions and words arouse predictable responses from others.

The Play Stage Mead saw children's play as the mechanism by which they practice role-taking. **Play** is a voluntary and often spontaneous activity, with few or no formal rules, that is not subject to constraints of time (for example, 20-minute halves, 15-minute quarters) or place (for example, a gymnasium, a regulation-size field). Children, in particular, play whenever and wherever the urge strikes. If there are rules, they are not imposed on participants by higher authorities (for example, rule books, officials). Participants undertake play for their amusement, entertainment, or relaxation. These characteristics make play less socially complicated than organized games (Corsaro 1985; Figler and Whitaker 1991).

In the play stage children pretend to be **significant others:** people or characters who are important in their lives—important in that they have considerable influence on a child's self-evaluation and encourage the child to behave in a particular manner. Children recognize behavior patterns characteristic of these significant persons and incorporate them into their play. For example, when a little girl plays with a doll and pretends she is the doll's mother, she talks and acts toward the doll the same way her mother does toward her. By pretending to be the mother, she gains a sense of the mother's expectations and perspective and learns to see herself as an object. Similarly two children playing doctor and patient are learning to see the world from viewpoints other than their own and to understand how a patient acts in relation to a doctor, and vice versa.

The Game Stage In Mead's theory the play stage is followed by the game stage. **Games** are structured, organized activities that almost always involve more than one person. They are characterized by a number of constraints, including one or more of the following: established roles and rules, an outcome toward which all activity is directed, and an agreed-upon starting time and place. Through games children learn (1) to follow established rules, (2) to take simultaneously the role of all participants, and (3) to see how their position fits in relation to all other positions.

When children first take part in games such as organized sports, their efforts seem chaotic. Instead of making an organized response to a ball hit to the infield, for example, everyone tries to retrieve the ball, leaving nobody at the base to catch the throw needed to put the runner out. This chaos exists because children have not developed to the point at which they can see how their role fits with the roles of everyone else in the game. Without such knowledge a game cannot have order. Through playing games children learn to organize their behavior around the **generalized other**—that is, around a system of expected behaviors, meanings, and viewpoints that transcend those of the people participating. "The attitude of the generalized other is the attitude of the whole community. Thus, for example, in the case of such a social group as a baseball team, the team is the generalized other insofar as it enters—as an organized process or activity—into the experience of [those participating]" (Mead 1934, p. 119). In other words, when children play organized sports, they practice fitting their behavior into an established behavior system.

The following documents present daycare activities designed to encourage the development of various social, emotional, and physical skills.

gopher://gopher-cyfernet.mes.umn.edu:4242/00/ChildCare/ChildDevel/cynet15

Document: Preschooler Development

gopher://gopher-cyfernet.mes.umn.edu:4242/00/ChildCare/ChildDevel/cynet17

Document: Toddler Development

gopher://gopher-cyfernet.mes.umn.edu:4242/00/ChildCare/Curriculum/cynet21

Document: Play Activities

http://people.delphi.com/punkyhaake/preschool.htm

Document: Preschool for Day Care

Q: What examples can you locate of activities that would be appropriate for children in the preparatory stage? The play stage? The game stage? At what age are children likely to pretend to be significant others? When do children begin acting around the concept of a generalized other?

As we have learned, George Herbert Mead assumed that the self develops through interaction with others. Mead identified the interaction that occurs in play and games as important to children's self-development. When children participate in play and games, they practice at seeing the world from the viewpoint of others and gain a sense of how others expect them to behave. Sociologist Charles Horton Cooley offered a more general theory about how the self develops.

The Looking-Glass Self

Like Mead, Cooley assumed that the self is a product of interaction experiences. Cooley coined the phrase **looking-glass self** to describe the way in which a sense of self develops: people act as mirrors for one another. We see ourselves reflected in others' reactions to our appearance and behaviors. We acquire a sense of self by being sensitive to the appraisals of ourselves that we perceive others to have: "Each to each a looking glass, / Reflects the other that [does] pass" (Cooley 1961, p. 824). As we interact, we visualize how we appear to others, we imagine a judgment of that appearance, and we develop a feeling somewhere between pride and shame: "The thing that moves us to pride or shame is not the mere mechanical reflection of ourselves but . . . the imagined effect of this reflection upon another's mind" (Cooley 1961, p. 824).

Cooley went so far as to argue that "the solid facts of social life are the facts of the imagination." According to this logic one person's effect on another is defined most accurately as what one person imagines the other will do and say on a particular occasion (Faris 1964). Because Cooley defined the imagining or interpreting of others' reactions as critical to self-awareness, he believed that people are affected deeply even when the image they see reflected is exaggerated or distorted. One responds to the perceived reaction rather than to the actual reaction.

Both Mead's and Cooley's theories suggest that self-awareness derives from an ability to think—to step outside oneself and view the self from another's perspective. Although Cooley and Mead described the mechanisms (imitation, play, games, and other people) by which people learn about themselves, neither theorist addressed how a person acquires this level of cognitive sophistication. To answer this question, we must turn to the work of Swiss psychologist Jean Piaget.

Cognitive Development

Piaget is the author of many influential and provocative books on how children think, reason, and learn. The titles of some of his many books—*The Language and Thought of the Child* (1923), *The Child's Conception of the World* (1929), *The Moral Judgement of the Child* (1932), *The Child's Conception of Time* (1946), and *On the Development of Memory and Identity* (1967)—give some clues about the many categories of childhood thinking that Piaget investigated.

Piaget's influence reaches across many disciplines: biology, education, sociology, psychiatry, psychology, and philosophy. His ideas about how children develop increasingly sophisticated levels of reasoning stem from his study of water snails (*Limnaea stagnalis*), which spend their early life in stagnant waters. When transferred to tidal water, these lazy snails engage in motor activity that develops the size and shape of the shell to help them remain on the rocks and avoid being swept away (Satterly 1987).

Building on this observation, Piaget arrived at the concept of **active adaptation,** a biologically based tendency to adjust to and resolve environmental challenges. The theme of active adaptation runs through almost all of Piaget's writings. He believed that learning and reasoning are rooted in active adaptation. He defined logical thought, another biologically based human attribute, as an important tool for meeting and resolving environmental challenges. Logical thought emerges according to a gradually unfolding genetic timetable. This unfolding must be accompanied by direct experiences with persons and objects; otherwise a child will not realize his or her potential ability. On the basis of cumulative experiences, a child constructs and reconstructs his or her conceptions of the world.

Piaget's model of cognitive development includes four broad stages—sensorimotor, preoperational, concrete operational, and formal operational—each characterized by a progressively more sophisticated reasoning level. A child cannot proceed from one stage to another until the reasoning challenges of earlier stages are mastered. Piaget maintained that reasoning abilities cannot be hurried; a more sophisticated level of understanding will not show itself until the brain is ready.

- *Sensorimotor stage (from birth to about age 2).* In this stage children explore the world with their senses (taste, touch, sight, hearing, smell). The cognitive accomplishments of this stage include an understanding of the self as separate from other persons and the realization that objects and persons exist even when they are out of sight. Before this notion takes hold, very young children act as if an object does not exist when they can no longer see it.

- *Preoperational stage (from about ages 2 to 7).* Piaget focused most of his attention on this stage. Children in this stage typically demonstrate three characteristic types of thinking. First, they think anthropomorphically; that is, they assign human feelings to inanimate objects. Thus they believe that objects such as the sun, the moon, nails, marbles, trees, and clouds have motives, feelings, and intentions (for example, dark clouds are angry; a nail that sinks to the bottom of a glass filled with water is tired). Second, they think nonconservatively, a term Piaget used to signify an inability to appreciate that matter can change form but still remain the same in quantity. Third, they think egocentrically in that they cannot conceive how the world looks from another point of view. Thus, if a child facing a roomful of people (all of whom are looking in his or her direction) is asked to draw a picture of how another person in the back of the room sees the people, the child will draw the picture as he or she sees the people. Related to egocentric thinking is centration, a tendency to center attention on one detail of an event. As a result of this tendency, the child fails to process other features of a situation.

- *Concrete operational stage (from about ages 7 to 12).* By the time children enter this stage, they have mastered these preoperational tasks but have difficulty in thinking hypothetically or abstractly without reference to a concrete event or image. For example, a child in this stage has difficulty envisioning a life without him or her in it. One 12-year-old struggling to grasp this idea said to me, "I am the beginning and the end; the world begins and ends with me."

- *Formal operational stage (from adolescence onward).* At this point people are able to think abstractly. For example, they can conceptualize their existence as a part of a much larger historical continuum and a larger context.

Diane Clark Johnson offers an alternative version of Piaget's theory that further subdivides each of the four major stages to create an eight-stage model of development.

http://www.fishnet.net/~pparents/johnson1.html
Document: What Is Normal Development?

Q: How can Piaget's theory help adults interact in effective and constructive ways with children?

As far as we know, this progression by stages toward increasingly sophisticated levels of reasoning is universal, but the content of people's thinking varies across cultures.

The theories of Mead, Cooley, and Piaget all suggest that the process of social development is multifaceted and continues over time. It is important to realize that socialization is a lifelong process in that people make any number of transitions over a lifetime: from single to married, from married to divorced or widowed, from childlessness to parenthood, from healthy to disabled, from one career to another, from civilian status to military status, from employed to retired, and so on. In making such transitions, people undergo resocialization.

Resocialization

Resocialization is the process of being socialized over again. In particular it is a process of discarding values and behaviors unsuited to new circumstances and replacing them with new, more appropriate values and norms (standards of appearance and behavior). A considerable amount of resocialization happens naturally over a lifetime and involves no formal training; people simply learn as they go. For example, people marry, change jobs, become parents, change religions, and retire without formal preparation or training. However, some resocialization requires that, to occupy new positions, people must undergo formal, systematic training and demonstrate that they have internalized appropriate knowledge, suitable values, and correct codes of conduct.

Such systematic resocialization can be voluntary or imposed (Rose, Glazer, and Glazer 1979). It is voluntary when people choose to participate in a process or program designed to "remake" them. Examples of voluntary resocialization are wide-ranging—the unemployed youth who enlists in the army to acquire a technical skill, the college graduate who pursues medical education, the drug addict who seeks treatment, the alcoholic who joins Alcoholics Anonymous. The 12-step program of AA offers concrete examples of behavior alcoholics must learn to perform if they are to free themselves from alcohol dependency.

http://www.shore.net/~tcfraser/blurblst.htm

Document: Blurbs from the BigBook: Chapter 5

Q: What are the key features of AA's resocialization program? What is the "spiritual" component of this resocialization program?

Resocialization is imposed when people are forced to undergo a program designed to train them, rehabilitate them, or correct some supposed deficiency in their earlier socialization. Military boot camp (when a draft exists), prisons, mental institutions, and schools (when the law forces citizens to attend school for a specified length of time) are examples of environments that are designed to resocialize but that people also enter involuntarily.

In *Asylums: Essays on the Social Situation of Mental Patients and Other Inmates*, sociologist Erving Goffman wrote about a setting—total institutions (with particular focus on mental institutions)—in which people undergo systematic socialization. In **total institutions** people surrender control of their lives, voluntarily or involuntarily, to an administrative staff and, as inmates, carry out daily activities (eating, sleeping, recreation) in the "immediate company of a large batch of others, all of whom are [theoretically] treated alike and required to do the same thing together" (Goffman 1961, p. 6). Total institutions include homes for the blind, the elderly, the orphaned, and the indigent; mental hospitals; jails and penitentiaries; prisoner-of-war and concentration camps; army barracks; boarding schools; and monasteries and convents. Their total character is symbolized by barriers to social interaction, "such as locked doors, high walls, barbed wire, cliffs, water, forests, or moors" (p. 4).

Goffman was able to identify the general and standard mechanisms, despite their wide range, that the staffs of all total institutions employ to resocialize "inmates." When the inmates arrive, the staff strips them of their possessions and their usual appearances (and the equipment and services by which their appearances are maintained). In addition the staff sharply limits interaction with people outside the institution in order to establish a "deep initial break with past roles" (p. 14). Goffman observed:

> We very generally find staff employing what are called admission procedures, such as taking a life history, photographing, weighing, finger-printing, assigning numbers, searching, listing personal possessions for storage, undressing, bathing, disinfecting, haircutting, issuing institutional clothing, instructing as to rules, and assigning to quarters. The new arrival allows himself to be shaped and coded into an object that can be fed into the administrative machinery of the establishment. (p. 16)

Goffman maintained that the admission procedures function to prepare inmates to shed past roles and assume new ones by participating in the various enforced activities that staff members have designed to fulfill the official aims of the total institution—whether to care for the incapable, to keep inmates out of the community, or to teach people new roles (for example, to be a soldier, priest, or nun).

The Human Rights Watch Children's Rights Project describes the life of child soldiers in Sudanese and Liberian military camps. Some of these children have been forcibly recruited while others have volunteered, but even the volunteers do so because they see no other way to survive.

gopher://server.gdn.org/00/Human_Rights_Reports/Human_Rights_Watch/Child_Soldiers_in_Sudan

Document: Human Rights Watch/Africa

gopher://server.gdn.org/00/Human_Rights_Reports/Human_Rights_Watch/Child_Soldiers_in_Liberia

Document: Human Rights Watch

Q: What qualities of the military camps make them total institutions?

In general it is easier to resocialize people when they want to be resocialized than when they are forced to abandon old values and behaviors. Furthermore resocialization is likely to be easier if acquiring new values and behaviors requires competence rather than subservience (Rose, Glazer, and Glazer 1979). A case in point is the resocialization that takes place in medical school. Theoretically medical students learn (among other things) to be emotionally detached in their attitudes toward patients, to not prefer one patient over another (that is, patients of a particular ethnicity, sex, or age, or even level cooperation), and to provide medical care whenever it is required (Merton 1976). These attitudes are necessary for proper diagnosis and treatment, as the following accounts make clear.

An emergency room physician, Elisabeth Rosenthal, reported on an unkempt man who had been brought into the emergency room. He was uncooperative, his speech was slurred and did not make sense, he did not know the date or the name of the president of the United States, and he could not decide whether he was in a hotel or a hospital. On the basis of his appearance and behavior, he seemed to be drunk; the best treatment seemed to be to let him sleep it off. However, physicians are trained (in theory at least) not to be influenced by stereotypes and to look beyond commonplace interpretations associated with certain physical traits.

In this case subsequent tests revealed that the man had a massive kidney infection: "His incoherence and his lack of cooperation were caused not by intoxication but by a metabolic disturbance resulting from his infection" (Rosenthal 1989, p. 82). For the physician professional competence is demonstrated by learning to abandon or hold in check widely shared misconceptions about the meaning of behavior, especially the behavior of persons in different ethnic, sex, and age groups. Consider the case of Dr. Steven Baumrucker and his relationship with a patient, "Mr. G." Mr. G asked to be kept alive at all costs, even though his death was inevitable. Baumrucker noted that Mr. G's desperation seemed to go far beyond a fear of death.

http://www.ilinkgn.net/commercl/author/news.htm
Document: News from the Front Lines of Hospice Care

Q: What did Baumrucker learn about Mr. G that helped him become a more effective physician to his patient?

The Unpredictable Elements of Socialization

Socialization is a multifaceted, lifelong process. Among other things it is a process through which newcomers acquire a sense of self and internalize the norms, values, beliefs, and language needed to participate in the larger community. Many of the important social theorists—George Herbert Mead, Charles Horton Cooley, and Erving Goffman—are considered to be symbolic interactionists. Consequently this chapter centers around the process by which newcomers acquire the skills needed to become social beings, including the ability to understand the symbol system of their culture, to internalize the pattern of expected behaviors, and to interpret how they fit into it.

At the same time, as the examples of the effects of extreme social isolation and limited association with other people demonstrate, socialization goes beyond the needs of individuals. It functions to link people to one another in orderly and predictable ways. Without the benefits of social interaction, newcomers suffer physically and fail to learn the skills they need to achieve meaningful connections with others. In addition, without meaningful contact between the generations, culture (solutions to the problems of living) cannot be passed on from one generation to the next. Without this link a culture ceases to exist.

In addition to being a process by which newcomers acquire a symbol system, solutions to problems of living, and the skills to interact in orderly and predictable ways, socialization is a process by which newcomers learn to think and behave in ways that reflect the interests of the teachers and to accept and fit into a system that benefits some groups more than others.

Whatever its consequences, socialization consists of a gradual unfolding of one's genetic endowment and an extraordinary number of social experiences. We are just beginning to discover how experiences (nurture) and genes (nature) combine to enlarge human potential, to bestow identity, and to shape thought and behavior. Clearly it is impossible to separate the influences of nature and nurture or to say that one is more important than the other for development: "Genetic determination is like the blueprint of a beautiful house. But the house itself is not there; you can't sleep in a blueprint. The kind of building you eventually have will depend on the choice of which bricks, which wood, which glass are used—just as the virgin brain will be shaped by what is given to it from the environment" (Delgado 1970, p. 170).

Without stimulation from other people, we cannot learn language, appropriate behaviors, and the skills and information we need to live with other people. Without the operation of biological mechanisms, we cannot learn or develop, no matter how great the stimulation. Yet biology is not destiny, and people are more than the sum of their experiences. Our biological materials can be shaped in innumerable ways. For example, as noted previously, at birth we have the flexibility to learn any of the world's 6,000–9,000 known languages and to learn the ways of any culture into which we are born. One might argue that, although we are flexible biologically, we have little control over how the raw materials are shaped—that we are the products of our upbringing and other circumstances beyond our control. Yet our biological makeup gives us a brain that is capable of thinking and creating and that allows us to think about our environment and act to change it. In this vein George Herbert Mead believed that humans do not adhere mechanically to others' expectations. He believed, rather, that there is a continuous dialogue between two aspects of the self: the *I* and the *me*.

What we know about the socialization and resocialization processes suggests that life does not have to unfold in a seemingly predictable fashion. Human genetic and social makeup contains considerable potential for change. First, people are not born with preconceived notions about standards of appearance and behavior. To develop standards, people must be exposed to information that leads them to expect people, behavior, and objects to be a certain way. Second, the cerebral cortex allows people to think reflexively—to step outside the self and observe and evaluate it from another viewpoint. Third, the mechanisms that teach prejudice and hatred for another group—mechanisms such as imitation, play, and games—also teach respect and understanding. The problem is that they often are used to teach children respect and understanding after the children already have learned prejudice and hatred through the same mechanisms. Finally, people can be resocialized to abandon one way of thinking and behaving for another. Preferably, however, people will choose to abandon old habits and, in doing so, will gain a feeling of competence.

Discussion Question

Based on the information presented in this chapter, why is socialization critical to human growth and development?

Additional Reading

For more on

- Adult views of what skills young people should acquire
  gopher://gopher.undp.org/00/ungophers/popin/unfpa/speeches/1995/youth95.gen
  Document: Geneva Youth Forum–Panel on Health Issues

- Children's reactions to the environment they have inherited
  http://www.unicef.org/voy/
  Document: Voices of Youth Home Page

- Lifelong socialization
  gopher://borg.lib.vt.edu/00/catalyst/v21n4/cross.v21n4
  Document: The Renaissance in Adult Learning

- The looking-glass self
  http://www.dnai.com/~children/media/content_study.html
  Document: The Reflection on the Screen: Television's Image of Children

Social Interaction and the Social Construction of Reality

Margrethe Rask, a Danish surgeon, was exposed to the virus now known as human immunodeficiency virus, or HIV, while working in a small village clinic in Zaire. The excerpt that follows highlights some of the final events in Dr. Rask's life, her last interactions with colleagues and close friends.

Grethe Rask gasped her short, sparse breaths from an oxygen bottle.... "I'd better go home to die," Grethe had told [her friend] Ib Bygbjerg matter-of-factly. The only thing her doctors could agree on was the woman's terminal prognosis. All else was mystery. Also newly returned from Africa, Bygbjerg pondered the compounding mysteries of Grethe's health. None of it made sense. In early 1977, it appeared that she might be getting better; at least the swelling in her lymph nodes had gone down, even as she became more fatigued. But she had continued working, finally taking a brief vacation in South Africa in early July.

Suddenly, she could not breathe. Terrified, Grethe flew to Copenhagen, sustained on the flight by bottled oxygen. For months now, the top medical specialists of Denmark had tested and studied the surgeon. None, however, could fathom why the woman should, for no apparent reason, be dying. There was also the curious array of health problems that suddenly appeared. Her mouth became covered with yeast infections. Staph infections spread in her blood. Serum tests showed that something had gone awry in her immune system; her body lacked T-cells, the [essential parts of] the body's defensive line against disease. But biopsies showed she was not suffering from a lymph cancer that might explain not only the T-cell deficiency but her body's apparent inability to stave off infection. The doctors could only gravely tell her that she was suffering from progressive lung disease of unknown cause. And, yes, in answer to her blunt questions, she would die.

Finally, tired of the poking and endless testing by the Copenhagen doctors, Grethe Rask retreated to her cottage near Thisted. A local doctor fitted out her bedroom with oxygen bottles. Grethe's longtime female companion, who was a nurse in a nearby hospital, tended her. Grethe lay in the lonely whitewashed farmhouse and remembered her years in Africa while the North Sea winds piled the first winter snows across Jutland.

In Copenhagen, Ib Bygbjerg, now at the State University Hospital, fretted continually about his friend. Certainly, there must be an answer to the mysteries of her medical charts. Maybe if they ran more tests. . . . It could be some common tropical culprit they had overlooked, he argued. She would be cured, and they would all chuckle over how easily the problem had been solved when they sipped wine and ate goose on the Feast of the Hearts. Bygbjerg pleaded with the doctors, and the doctors pleaded with Grethe Rask, and reluctantly the wan surgeon returned to the old Rigshospitalet in Copenhagen for one last chance. On December 12, 1977, just twelve days before the Feast of the Hearts, Margrethe P. Rask died. She was forty-seven years old. (SHILTS 1987, PP. 6–7)

We can visualize some of the interactions between Grethe Rask and her friends and colleagues. For example, we can visualize Dr. Rask telling her friend Ib Bygbjerg, "I'd better go home to die," or asking the Copenhagen doctors whether she would die after they tell her that she is "suffering from progressive lung disease of unknown cause." Finally, when Dr. Rask decides to leave the hospital and die at home, we can imagine Ib Bygbjerg pleading, "Please come back to the hospital for more tests; maybe there is still hope."

Sociologists looking at this situation would agree that Dr. Rask's illness is the obvious and immediate reason for these **social interactions**—everyday events in which at least two people communicate and respond through language and symbolic gestures to affect one another's behavior and thinking. In the process the parties involved define, interpret, and attach meaning to the encounter. Sociologists also assume that any social interaction reflects forces beyond the obvious and immediate. Hence they strive to locate the interaction according to time (history) and place (culture). From a sociological viewpoint history and culture limit the range of potential experience and point people "towards certain definite modes of behaviour, feeling, and thought" (Mannheim 1952, p. 291).

When sociologists study social interaction, they seek to understand and explain the forces of context and content. **Context** consists of the larger historical circumstances that bring people together. **Content** includes the cultural frameworks (norms, values, beliefs, material culture) that guide behavior, dialogue, and interpretations of events. In the case of Dr. Rask, sociologists would determine the context by asking what historical events brought Dr. Rask to Africa in the first place and what further events put her in direct contact with a deadly virus. To understand the content of her interactions, sociologists would ask how the parties involved are influenced by their cultural frameworks as they strive to define, interpret, and respond meaningfully to her condition.

In this chapter we explore the sociological theories and concepts that sociologists use to analyze any social interaction in terms of context and content. We give particular focus to social interaction as it relates to the transmission of HIV and the treatment of AIDS (see **http://www.safersex.org/hiv/howisaidstransmitted.html** for the Centers for Disease Control definition of AIDS). In doing so, we look closely at the central African country of Zaire (formerly the Belgian Congo) for two important reasons. First, focusing on Zaire and its relationship to other countries helps us connect the transmission of HIV to a complex set of intercontinental, international, and intrasocietal interactions. Specifically these interactions involve unprecedented levels of international and intercontinental air travel of the privileged for pleasure and business, as well as legal and illegal migrations of the underprivileged from villages to cities and from country to country (Sontag 1989).

Second, focusing on Zaire highlights evidence that HIV existed as early as 1959—evidence in the form of an unidentified blood sample frozen in that year and stored in a Zairean blood bank. Although this hardly proves that AIDS originated in Zaire, this hypothesis has received considerable support from government and health officials in Western countries. HIVNET posts the document "An Introduction to AIDS: What, Where, Who, How, etc."

http://gopher.hivnet.org:70/0/hivtext1/aids101

Document: An Introduction to AIDS

 Skim down to the question "What is AIDS?" What is the prevailing scientific opinion about the origin of HIV?

Whether Zaire is actually the country of origin of HIV is irrelevant to our purposes. Far more important is the idea that reality is a social construction. That is, people give meaning to phenomena (events, traits, objects), meanings that almost always emphasize some aspect of a phenomenon and ignore other aspects. For example, to say that HIV traveled from Zaire to the United States ignores the possibility that it traveled from the United States to Zaire. When we compare Western values and beliefs related to the origin of the virus and the treatment of AIDS with African values and beliefs, we realize that the Western framework is only one way of viewing the AIDS phenomenon. Moreover, by comparing the two frameworks, we can see that the meanings people give to an event have enormous consequences for the individuals involved (medical personnel, infected persons and those close to them, and noninfected persons). These meanings influence how people interact with one another and what decisions they make and actions they take to deal with HIV infection and AIDS.

Continue to think about Dr. Rask as you read this chapter. As we explore issues of context and content, we will see that an individual's seemingly unique and personal interactions are affected by history and culture, just as they are affected by genetics (Tuchman 1981). We begin by exploring the context of social interactions—the unprecedented mixing of the world's peoples and the large-scale social disruptions that have accompanied the emergence of worldwide economic and social interdependence. If we can understand these social forces, we can understand more about the transmission of viruses in general and the transmission of HIV in particular.

The Context of Social Interaction

Emile Durkheim was one of the first sociologists to provide insights into the social forces that contributed to the rise of a "global village." In *The Division of Labor in Society* ([1933] 1964) Durkheim gave us a general framework for understanding both global interdependence and conditions that can cause large-scale social upheaval, leaving people vulnerable to phenomena like AIDS. More specifically Durkheim's ideas provide a framework for understanding how Zaire was transformed, in less than 200 years, from a land of isolated and independent nations to a country characterized by immense social disruptions and participation in the world economy. (A *nation* is a geographical area occupied by people who share a culture and a history; a *country* is a political entity, recognized by foreign governments, with a civilian and military bureaucracy to enforce its rules.)

Durkheim observed that an increase in population size and density intensifies the demand for resources. This in turn stimulates the development of more efficient methods for producing goods and services. As population size and density increase, society "advances steadily towards powerful machines, towards great concentrations of forces and capital, and consequently to the extreme division of labor" (Durkheim [1933] 1964, p. 39). As Durkheim described it, **division of labor** refers to work that is broken down into specialized tasks, with each task performed by a different set of persons. Not only are the tasks themselves specialized, but the parts and materials needed to manufacture products come from many geographical regions.

The West vigorously colonized much of Asia, Africa, and the Pacific in the late nineteenth and early twentieth centuries because of the growing demand for resources. Western countries forced local populations to cultivate and harvest crops and to extract minerals and ores for export. The Belgian government claimed territory in central Africa, named it the Belgian Congo, and forced the people living there to extract rubber and mine copper. As industrialization proceeded in Europe, so did the demand for various raw materials. Over time the world grew to depend on Zaire as a source of cobalt (needed to manufacture jet engines), industrial diamonds, zinc, silver, gold, manganese (needed to make steel and aluminum dry-cell batteries), and uranium (needed to generate atomic energy and to fuel the atomic bomb). The worldwide division of labor now included the indigenous people of Zaire, who mined the raw materials needed for products in distant parts of the world.

Durkheim noted that as the division of labor becomes more specialized and as the sources of materials for products become more geographically diverse, a new kind of solidarity or moral force emerges. Durkheim used the term **solidarity** to describe the ties that bind people to one another in a society. He referred to the solidarity that characterizes preindustrial society as mechanical, and the solidarity that characterizes industrial societies as organic.

Mechanical Solidarity

Mechanical solidarity is social order and cohesion based on a common conscience or uniform thinking and behavior. In this situation everyone views the world in much the same way. A person's "first duty is to resemble everybody else, [and] not to have anything personal about one's beliefs and actions" (Durkheim [1933] 1964, p. 396). Such uniformity derives from a simple division of labor and the corresponding lack of specialization. (In other words everyone is a jack-of-all-trades.) When everyone in a society does the same thing, they have common experiences, possess similar skills, and hold similar beliefs, attitudes, and thoughts. Therefore a simple division of labor means that people are more alike than different. People are bound together because similarity gives rise to consensus and to a common conscience. In societies characterized by mechanical solidarity, the ties that bind people to one another are based primarily on kinship and religion.

As one of about 200 nations in Zaire, each of which has a distinct language and belief system, the Mbuti pygmies, a hunting-and-gathering people who live in the Ituri Forest (an equatorial rain forest) of northeastern Zaire, exhibit this type of solidarity. Their society represents the way of life that many people were forced to abandon after colonization began.

The Mbuti share a forest-oriented value system. Their common conscience derives from the fact that the forest gives them food, firewood, and materials for shelter and clothing. Anthropologist Colin Turnbull has written extensively about the Mbuti and their value system in three books, *The Forest People* (1961), *Wayward Servants* (1965), and *The Human Cycle* (1983). Excerpts from these books show the extent to which Mbuti forest-centered values permeate their life:

> For them the forest is sacred, it is the very source of their existence, of all goodness. . . .
>
> Young or old, male or female . . . the Mbuti talk, shout, whisper, and sing to the forest, addressing it as mother or father or both. (1983, P. 30)

> It is not surprising that the Mbuti recognize their dependence upon the forest and refer to it as "Father" or "Mother" because as they say it gives them food, warmth, shelter, and clothing just like their parents. What is perhaps surprising is that the Mbuti say that the forest also, like their parents, gives them affection. . . . The forest is more than mere environment to the Mbuti. It is a living, conscious thing, both natural and supernatural, something that has to be depended upon, respected, trusted, obeyed, and loved. The love demanded of the Mbuti is no romanticism, and perhaps it might be better included under "respect." It is their world, and in return for their affection and trust it supplies them with all their needs.
> (1965, P. 19)

Turnbull provides several examples of the intimacy between the Mbuti and the forest. In one instance he came upon a youth dancing and singing by himself in the forest under the moonlight: "He was adorned with a forest flower in his hair and with forest leaves in his belt of vines and his loin cloth of forest bark. Alone with his inner world he danced and sang in evident ecstasy" (1983, p. 32). When questioned as to why he was dancing alone, he answered, "'I am not dancing alone, I am dancing with the forest'" (1965, p. 253).

In a second instance Turnbull asked a Mbuti pygmy if he would like to see a part of the world outside the forest. The pygmy hesitated a long time before asking how far they would go beyond the forest. Not more than a day's drive from the last of the trees, replied Turnbull, to which the Mbuti pygmy responded with disbelief, "No trees? No trees at all?" He was highly disturbed about this and asked if it was a good country. From the Mbuti perspective people living without trees must be very bad to deserve that punishment. In the end he agreed to go if they took enough food to last them until they returned to the forest. "He was going to have nothing to do with 'savages' who lived in a land without trees" (1961, p. 248).

Finally Turnbull notes that the pygmies are aware of the ongoing destruction of the rain forest by companies that push them farther into the forest's interior. By consensus the pygmies do not wish to leave the forest and become part of the modern world: "The forest is our home; when we leave the forest, or when the forest dies, we shall die. We are the people of the forest" (1961, p. 260).

Organic Solidarity

A society with a complex division of labor is characterized by **organic solidarity**—social order based on interdependence and cooperation among people performing a wide range of diverse and specialized tasks. A complex division of labor increases differences among people, and common conscience in turn decreases. Yet Durkheim argued that the ties that bind people to one another can be very strong nonetheless. In societies characterized by a complex division of labor, these ties are no longer based on similarity and common conscience but on differences and interdependence. When the division of labor is complex and when the materials for products are geographically scattered, few individuals possess the knowledge, skills, and materials to permit self-sufficiency. Consequently people find themselves dependent on others. Social ties are strong because people need each other to survive.

Specialization and interdependence mean that every individual contributes a small part in creating a product or delivering a service. Because of specialization, relationships among people take on a transitory, limited, impersonal, and abstract character. We relate to one another in terms of our specialized roles. We buy tires from a dealer; we interact with a sales clerk by telephone, computer, and fax; we fly from city to city in a matter of hours and are served by flight attendants; we pay a supermarket cashier for coffee; and we deal with a lab technician for only a few minutes when we give blood. We do not need to know these people personally to interact with them. Nor do we need to know that the rubber in the tires, the cobalt from which the jet engine is built, and the coffee we purchase come from Zaire. Similarly we do not need to know whether the blood we give is kept in the United States or is exported elsewhere.

When we interact in this manner, we can ignore personal differences and can treat those who perform the same tasks as interchangeable. Yet members of society "are united by ties which extend deeper and far beyond the short moments during which the exchange is made. Each of the functions that they exercise is, in a fixed way, dependent upon others. . . . [W]e are involved in a complex of obligations from which we have no right to free ourselves" (Durkheim [1933] 1964, p. 227). In other words, because everyone is dependent on everyone else, each individual has a stake in preserving the system.

A curious feature of organic solidarity is that although people live in a state of interdependence with others, they maintain little awareness of it, possibly because of the fleeting and impersonal nature of the relationships. Because the ties with most of the people with whom we come in contact during a day are instrumental (we interact with them for a specific reason) rather than emotional, we seem to live independently of one another.

Durkheim hypothesized that societies become more vulnerable as the division of labor becomes more complex and more specialized. He was particularly concerned with the kinds of events that break down the ability of individuals to connect with one another in meaningful ways through their labor, a process we take for granted until something disrupts those connections. Such events include (1) industrial and commercial crises caused by such occurrences as plant closings, massive layoffs, crop failures, technological revolutions, and war; (2) workers' strikes; (3) job specialization, insofar as workers are so isolated that few people grasp the workings and consequences of the overall enterprise; (4) forced division of labor to such an extent that occupations are based on inherited traits (race, sex) rather than on ability, in which case the "lower" groups aspire to the positions that are closed to them and seek to dispossess those who occupy such positions; and (5) inefficient management and development of workers' talents and abilities so that work for them is nonexistent, irregular, intermittent, or subject to high turnover. For example, a country might not develop enough workers (teachers, scientists, nurses) or too many workers (athletes and entertainers) for available positions, or it might fail to retrain people whose positions are vulnerable to layoff or obsolescence. These events are particularly disruptive when they arise suddenly.

In his 1986 book *The Reckoning* David Halberstam profiles the life of one man—Joel Goddard—after he is laid off from Ford. Goddard is married and has two children. Halberstam chronicles the changes in Goddard's life that affect his ability to connect in meaningful ways with others. For example, after Goddard loses his job, his daily contacts shift from colleagues at work to contacts with those at the unemployment office; moreover Goddard loses the structure of his life that comes with the routine of his job. Instead he watches TV, fishes, or reads want-ads. His ties are further disrupted when some of his friends from work move to Texas to find employment. Goddard eventually takes a job selling insurance but finds himself selling to his acquaintances. Eventually he quits, and then his wife decides to go to work. However, her success at work strains their marriage as Joel is reminded of his failures.

As a second example of how disruptions to the division of labor affect people's ties with one another, consider how war-induced industrial and commercial crises break down people's ability to connect with one another in meaningful ways. Specifically consider the refugee crisis that resulted from civil war in Rwanda, with 1.2 million Tutsi and Huti refugees fleeing Rwanda to the North Kivu area of Zaire and swelling the local population from 300,000 to 1.5 million.

http://www.visions.net/zaire.html
Document: Medissage in Zaire: The Biodiversity Emergency Team

 Q: What are at least five consequences of the massive influx of people into North Kivu?

Zaire in Transition

From 1883 to the present, at least one of the five disruptive situations that Durkheim postulated has existed in Zaire. A brief summary of Zaire's history in the past 100 years shows the extent to which these disruptions have created a social order that severs the connections that people have to one another.

Belgian Imperialism (1883–1960)

Before 1883 inhabitants of what is now Zaire lived in villages characterized by common conscience and a simple division of labor. In 1883, however, King Leopold II of Belgium claimed the land as his private property, and millions of people were forced from their villages to work the land and mine raw materials (forced division of labor). Leopold's personal hold over the land was formally legitimized in 1885 by leaders of 14 European countries attending the Berlin West Africa Conference. The purpose of this conference was to carve Africa into colonies. The continent was divided without regard to preexisting national boundaries, so that friendly nations were split apart and hostile nations were thrown together.

For 23 years Leopold capitalized on the world's growing demand for rubber. His reign over Zaire was the "vilest scramble for loot that ever disfigured the history of human conscience and geographical location" (Conrad 1971, p. 118). The methods he used to extract rubber for his own personal gain involved atrocities so ghastly that in 1908 international outrage forced the Belgian government to assume administration of the Belgian Congo. The Belgian government operated more humanely than Leopold; but it too forced the indigenous peoples to build roads so that minerals and crops could be transported from mines and fields across the country for export. And Africans still were forced to leave their villages to work the mines, to cultivate and harvest the crops, and to live alongside the roads and maintain them. The Belgians introduced a cash economy, imported goods from Europe that eventually became essential to native life, established a government, and built schools, hospitals, and roads. Under this system the African people could acquire cash in one of two ways: growing cash crops or selling their labor. They were no longer allowed to be self-sufficient, as they were before European colonization. In addition under European colonization the African people were denied access to most professional-level and high-skilled occupations (forced division of labor; inefficient management and development of workers' talents).

The introduction of European goods and a cash economy pulled the people who inhabited the Belgian Congo into a worldwide division of labor and created a migrant labor system within the country. Since colonization people have moved continuously from the villages to the cities, mining camps, and plantations (Watson 1970). In addition family members have endured prolonged separations as a result of the migrant labor system:

> The migrant labor system affected Africans' lives in many fundamental ways. . . . [M]ale workers were typically recruited from designated labour supply areas great distances from the centres of economic activity. This entailed prolonged family separations which had serious physical and psychological repercussions for all concerned. The populations of African towns "recruited by migration" were characterized by a heavy preponderance of men living in intolerably insecure and depressing conditions and lacking the benefits of family life or other customary supports.
> (DOYAL 1981, P. 114)

The Belgians did not anticipate the Africans' anger about the exploitation of their land, minerals, and people and were not prepared for the violent confrontations that took place in the late 1950s. Those in power termed the revolutions "savage" and claimed that the Africans did not appreciate the "benefits" of colonialism. When the Belgians pulled out suddenly in 1960, the Belgian Congo became an independent country without trained military officers, businesspeople, teachers, doctors, or civil servants. In fact there were only 120 medical doctors in a country of 33 million people (Fox 1988).

Independence (1960–Present)

In the vacuum left by the Belgians, the various African ethnic groups that had been forced together to form the Belgian Congo now fought one another to obtain power (industrial and commercial crises). Civil wars raged until 1965, when Sese Mobutu took power with the help of the United States. Since that time several power struggles have occurred between various nations within Zaire, especially in the later 1970s. To stop the rebellions, Mobutu called on mercenary forces from Morocco, Belgium, and France. To make up for the lack of skilled personnel, Mobutu invited French, Danes, Haitians, Portuguese, Greeks, Arabs, Lebanese, Pakistanis, and Indians to work as civil servants, teachers, doctors, traders, businesspeople, and researchers. It was through this invitation that Dr. Rask arrived in Zaire.

In addition to problems posed by civil war, several other major problems have made life difficult for the people of Zaire since independence. First, many cash crops were priced out of competition in a growing world economy, and a technological revolution in synthetic products reduced the demand for African raw materials (industrial and commercial crises). Second, civil wars raging in neighboring countries have caused thousands of refugees from Sudan, Rwanda, Angola, Uganda, Congo, and Burundi to flee to Zaire. Meanwhile Zaireans suffering from their own civil wars and economic problems sought refuge in those same countries (Brooke 1988a; U.S. Bureau for Refugee Programs 1988). Finally Mobutu has diverted much of Zaire's wealth to European banks and has invested it in property outside of Africa (Brooke 1988b; Kramer 1993). Some people estimate Mobutu's personal fortune to be as much as $5 billion (Kramer 1993).

The local populations have suffered greatly and are still suffering from the ongoing massive upheaval in Zaire. These events disrupted the division of labor, and people lost an important social connection to one another. Such "change of existence, whether it be sudden or prepared, always brings forth a painful crisis" (Durkheim [1933] 1964, p. 241). Out of economic necessity, a desire for a higher standard of living, and a need to escape war, many people have left their villages for the cities and for industrial, plantation, and mining sites. Women, children, and the elderly left behind in the villages have had little choice but to change to higher-yield and less labor-intensive crops to survive. Unfortunately the new crops, such as cassava, are low in protein and high in carbohydrates. (Low-protein diets compromise the human immune system, making people more vulnerable to infection.) To further complicate matters, significant numbers of single women with no means of supporting themselves in the villages and rural areas have migrated to labor sites in search of employment. Because there are few employment opportunities for women at the labor sites, many are forced to survive through prostitution (inefficient management and development of workers' talent). In the meantime, when the men and women who had migrated out of the villages to find employment became sick and could no longer work, they returned home to their villages and infected an already vulnerable population with whatever diseases they carried (Hunt 1989).

The magnitude of these migrations is reflected in the population increase of Kinshasa, the capital city, which grew from 390,000 in 1950 to 3.5 million in 1988. A large portion of its population (almost 60 percent) lives in squatter slums, the largest of which is named the Cite. Here the "streets [are] stuffed with children and families living under cardboard roofs held down by rocks. Kinshasa [is] a wasteland of flooded streets and cracked sidewalks, smoldering garbage and bars catering to whores and lonely white men. There [is] an end-of-civilization atmosphere, with survivors finding shelter in the rubble" (Clarke 1988, pp. 175, 178).

As a result of this upheaval and mismanagement, Zaire fell from its position as one of the wealthiest colonies in Africa to become the poorest independent country and one of the 12 poorest countries in the world. Its annual per capita income is $220. As another indicator of how dire the situation has become, in 1960 Zaire had 90,000 miles of passable highway; by 1988 that number had dwindled to only 6,000 miles. Obviously this impedes the transportation of medical supplies, food, and fertilizer to villages that have become dependent on these commodities (Noble 1992).

These historical events are the contextual forces that brought Dr. Margrethe Rask to a small village clinic in Zaire. Amid such chaos and poverty Dr. Rask had to perform operations on less than a shoestring budget, with only minimal supplies. "Even a favored clinic would never have such basics as sterile rubber gloves or disposable needles. You just used needles again and again until they wore out; once gloves had worn through you risked dipping your hands in your patient's blood because that was what needed to be done" (Shilts 1987, p. 4).

HIV/AIDS and Zaire

The importance of considering Zaire's history and global connections to understand the origins of AIDS is supported by the fact that disease patterns historically are affected by changes in population density and transportation patterns, both of which bring together previously isolated groups (McNeill 1976). In "Travel and the Emergence of Infectious Diseases" physician Mary E. Wilson explains why travel is a potent force in the emergence of disease.

http://www.cdc.gov/ncidod/EID/vol1no2/wilson.htm

Document: Travel and the Emergence of Infectious Diseases

Q: What do the following ideas mean: (1) human activities are the most potent factors driving disease emergence, (2) to determine the consequences of travel both the traveler and the population visited must be considered, and (3) the current volume, speed, and reach of travel are unprecedented?

Leading AIDS researchers believe that the transmission of the HIV infection is indeed linked to changes in population density and transportation. Interestingly "the medical condition which was later to be called AIDS began to be noticed in the late 1970s and early 1980s in several widely separated locations, including Belgium, France, Haiti, the United States, Zaire, and Zambia" (The Panos Institute 1989, p. 72). Before this time the virus may have survived in a dormant state in an isolated population with a tolerance to the virus but may have been activated when this population came into contact with another population with no tolerance. Or two harmless retroviruses, each existing in a previously isolated population, may have interacted to produce a third, lethal virus. The point is that

> a major dislocation in the social structure—love, hate, peace, war, urbanization, overpopulation, economic depression, people having so much leisure [or having no alternative source of income] they sleep with five different people a night—whatever it is that puts a stress on the ecological system, can alter the equilibrium between [people] and microbes. Such great dislocations can lead to plagues and epidemics. (KRAUSE 1993, P. xii)

In view of this information about the global context of interaction, it is difficult to say who is responsible for triggering and transmitting the virus that can cause AIDS. Clearly "the foreigners introduced hitherto unknown diseases and probably aggravated some previously endemic diseases to epidemic proportions by the facilitation of transportation, forced migration of rural populations to work sites, and the creation of congested cities" (Lasker 1977, p. 280). For example, the destruction of the Ituri rain forest in Zaire by companies from around the world brings developers into contact with previously isolated populations (such as the Mbuti) and forces many people to migrate to the city because they have lost their homes. In addition to disrupting people's lives, the commercial activities in the rain forest cause climatic changes (such as the greenhouse effect) that can alter the structure of viruses. What's important is not to determine who started the transmission (because that is impossible to ascertain), but to recognize the extent to which the world's people are interacting with one another and to become aware that the actions of one group can affect other groups.

Placing Dr. Rask's interactions in this global context helps us see how historical events bring people into interaction with one another. The opportunity for Dr. Rask to go to Zaire and practice medicine, the circumstances that placed her in direct contact with patients' blood, and her subsequent illness arose from a unique sequence of historical events.

The point is that somewhere in this unprecedented mixing of people from all over the world are the conditions that facilitated both the activation and the transmission of HIV. The importing and exporting of blood (which brings people into contact with one another in indirect ways), the development of wide-body jet aircraft, and large-scale migrations are cited as events that increased opportunities for large numbers of people from different countries and from different regions of the same country to interact with one another and to transmit the HIV infection through unprotected sexual intercourse, needle sharing, and other activities that involve blood and blood products (De Cock and McCormick 1988).

For one indicator of the amount of "indirect" interaction between blood donors and receivers, see the American Red Cross homepage, which includes an overview of that organization's biomedical services (specifically blood and tissues services).

http://www.crossnet.org/biomed/bio-fact.html
Document: Biomedical Services 1995–96

 When did the Red Cross begin its blood collection program? How many volunteer blood donations are made to the Red Cross each year? How much of the nation's blood does the Red Cross supply? What international services does the Red Cross provide?

In addition to the context, the content of social interactions is important. When people interact, they identify the social status of the people they interact with. Once they determine another person's status in relation to their own, people proceed to interact on the basis of role expectations.

The Content of Social Interaction

As the division of labor has become more specialized and the sources of labor and raw materials have become more geographically diverse, the ties that connect people to one another have shifted in character from mechanical to organic, and the number of interactions people have with strangers has increased. How can people interact smoothly with people about whom they know nothing? They eliminate "strangeness" by identifying the social positions or social status of the strangers with whom they interact. Knowing a person's social status gives us some idea of the behaviors we can expect from someone in that status. It also affects how we will interact with that person. To grasp this principle, think about when you meet someone for the first time. How do you start the interaction? You ask the stranger a question to determine his or her social status. You might ask, "What do you do?" That question sets the interaction into motion.

Social Statuses

In everyday language people use the term *social status* to mean rank or prestige. To sociologists **social status** refers to a position in a social structure. A **social structure** consists of two or more people interacting and interrelating in specific expected ways, regardless of the unique personalities involved. For example, a social structure can consist of the two statuses of doctor and patient and their relationship. Other familiar examples of statuses in a two-person social structure are husband and wife, professor and student, sister and brother, and employer and employee.

Examples of multiple-status social structures are a family, an athletic team, a school, a large corporation, and a government. Again the common characteristic of all social structures is that it is possible to generalize about the behavior of people in each of the statuses, no matter who occupies them. Just as the behavior of a person occupying the status of a football quarterback is broadly predictable, so too is the behavior of a person occupying the status of nurse, secretary, mechanic, patient, or physician. Once we know a person's status, we think we know how to interact with the person.

Types of Status Statuses can be of two kinds—ascribed or achieved. An **ascribed status** is a position that people acquire through no fault or virtue of their own. Examples include sex, age, ethnic, and health statuses—for example, male, female, African American, Native American, adolescent, senior citizen, retired, son, daughter, and disabled. An **achieved status** is a position earned (or lost) by a person's own actions or abilities. Occupational, educational, parental, and marital statuses include athlete, senator, secretary, physician, high school dropout, divorcé, and single parent.

The distinction between ascribed and achieved statuses is not always clear-cut. Often achieved statuses such as financial, occupational, and educational positions are related to ascribed statuses such as sex, ethnicity, and age. For example, in the United States physicians are disproportionately male (80 percent) and white (92 percent), whereas registered nurses are overwhelmingly female (86 percent) and white (90 percent) (U.S. Bureau of the Census 1992b).

Stigmas Some ascribed and achieved statuses are such that they overshadow all other statuses that a person occupies. Sociologist Erving Goffman calls such statuses **stigmas** and classifies them into three broad varieties: (1) physical deformities, (2) character blemishes due to factors such as sexual orientation, mental hospitalization, or imprisonment, and (3) stigmas of ethnicity, nationality, or religion. When a person possesses a stigma, he or she is reduced in the eyes of others from a multifaceted person to a person with one tainted status. To illustrate this point, Goffman opens *Stigma: Notes on the Management of Spoiled Identity* (1963) with a letter written by a 16-year-old girl born without a nose. Although she is a good student, has a good figure, and is a good dancer, no one she meets can get past the fact that she has no nose. Or consider the stigma of "character blemish" attached to those scientists engaged in "sex research."

http://math.ucsd.edu/~weinrich/theScientist2.html
Document: SEX: Still a Bad Word for Some People

 Scroll down to the second article. How does being classified as a sex researcher affect access to funding opportunities? Why?

Every person occupies a number of statuses. For example, Dr. Rask was middle-aged, a female, a physician, a patient, and Danish. A status has meaning, however, only in relation to other statuses. For instance, the status of a physician takes on quite different meanings depending on whether the interaction is with someone who occupies the same status or a different status such as patient, spouse, or nurse. This is because a physician's role varies according to the status of the person with whom he or she interacts.

Social Roles
Sociologists use the term **role** to describe the behavior expected of a status in relationship to another status (for example, professor to student). The distinction between role and status is subtle but noteworthy: people *occupy* statuses and *enact* roles.

Associated with every status is a **role set,** or an array of roles. For example, the status of physician entails among other roles the role of physician in relationship to patient, to nurse, to other doctors, and to a patient's family members. The sociological significance of statuses and roles is that they make it possible for us to interact with other people without knowing them. Once we determine another person's status in relation to our own, we interact on the basis of role expectations attached to that status relationship.

Role expectations are socially prescribed and include both rights and obligations. The **rights** associated with a role define what a person assuming that role can demand or expect from others depending on their status. For instance, teachers have the right to demand and expect that students will come prepared for class. The **obligations** associated with a role define the appropriate relationship and behavior that the person enacting that role must assume toward others occupying a particular status. For example, teachers have an obligation to their students to come to class prepared.

One of the best-known descriptions of a role and its accompanying rights and obligations was given by sociologist Talcott Parsons (1975). According to Parsons we assume a **sick role** when we are sick. Ideally sick persons have an obligation to try to get well, to seek technically competent help, and to cooperate with a treatment plan. Sick persons also have certain rights: they are exempt from "normal" social obligations and are not held responsible for their illness.

Social Roles and Individual Behavior Roles set general limits on how we think and act but do not imply that behavior is totally predictable. Sometimes people do not meet their role obligations, as when professors come to class unprepared or when patients do not cooperate with their physicians' treatment plans or when physicians blame their patients for getting sick (for example, thinking they have AIDS because they engaged in promiscuous and perverse sexual activity or thinking they have lung cancer because they don't have the will to quit smoking). By definition, when people fail to meet their role obligations, other people are not accorded their role rights. When professors are unprepared, students' rights are violated; when patients do not follow treatment plans, physicians' rights are violated; when physicians blame their patients, patients' rights are violated. Moreover the idea of role does not imply that all people occupying the same status enact the roles of that status in exactly the same way. Roles are enacted differently because of individual personalities and interpretations of how the roles should be carried out. Finally roles are enacted differently because people resolve role strain and role conflict differently.

Role strain is a predicament in which contradictory or conflicting expectations are associated with the role that a person is occupying. For example, military doctors, as physicians, have an obligation to preserve life. At the same time, they are employed to care for patients who have been placed deliberately in situations that threaten their health and lives. **Role conflict** is a predicament in which the expectations associated with two or more roles in a role set are contradictory. For example, a sick person has an obligation to want to get better and to comply with treatment plans. This obligation, however, can interfere with other roles that the person holds, as in the case of a woman who finds that the side effects of a prescribed drug prevent her from being alert enough to work or care for her children.

Socially prescribed rights and obligations notwithstanding, there is always room for improvisation and personal style. Yet, despite variations in how people enact roles, role is still a useful concept because an appreciable degree of predictability generally exists, "sufficient to enable most of the people, most of the time to go about [the] business of social life without having to improvise judgments anew in each newly confronted situation" (Merton 1957, p. 370). In other words the variations usually fall within "a certain range of culturally acceptable behavior—if the performance of a role deviates very much from the expected range of behavior the individual will be negatively sanctioned" (p. 370).

Cultural Variations in Roles: The Patient–Physician Interaction   The behaviors expected of one status in relation to another status vary across cultures. Role expectations are intertwined with norms, values, beliefs, and nonmaterial culture. In the United States the major objective of the patient–physician interaction is to determine the exact physiological malfunction and use the available material culture and technology (tests, equipment, machines, drugs, surgery, transfusions) to treat it. This objective is shaped by a profound cultural belief in the ability of science to solve problems. Practitioners of Western medicine are expected to use all of the tools of science at hand to establish the cause, combat the disease for as long as possible, and return the body to a healthy state. When physicians violate this norm, the violation is almost always accompanied by intense public debate. In view of this cultural orientation, it is not surprising that in the United States physicians and their patients rely heavily on technological elements such as X-rays and CAT scans to diagnose the condition and on vaccines, drugs, and surgery to cure it. This reliance is reflected in the fact that the United States, with a population of 226 million—5 percent of the world's population—consumes 23 percent of the world's pharmaceutical supply (Peretz 1984). Given this emphasis it is not surprising that tremendous effort is devoted to finding a technological solution to the AIDS problem.

http://gpawww.who.ch/whademo/approach.htm
Document: New Approach to Fighting AIDS

What is the new approach to fighting AIDS? How will the proposed technology work? What "side effects" should *not* result from the new technology? Should a new technology be abandoned if it affects fertility?

We can contrast the U.S. physician–patient relationship with the traditional African healer–patient relationship. Although Zaire has modern health-care facilities, the majority of people go to traditional healers. The social interaction between the healer and patient is very different from the U.S. physician–patient social interaction.

Just as Western physicians do, traditional healers recognize the organic and physical aspects of disease. But they also attach considerable importance to other factors—supernatural causes, social relationships (hostilities, stress, family strain), and psychological distress. This holistic perspective allows for a more personal relationship between healer and patient. Another significant difference from Western medicine is that healers concentrate on providing symptom relief instead of searching for a total cure.

African healers, for example, focus on treating the debilitating symptoms of AIDS such as diarrhea, headaches, fevers, and weight loss, and they employ remedies with few side effects so as not to make people sicker (Hilts 1988). When traditional methods fail, Africans may make a trip to a hospital or clinic to consult with a doctor of Western medicine.

Obviously each system has its attractions and its weaknesses, as evidenced by the fact that Westerners suffering from incurable diseases sometimes turn to alternative treatments and medicines. See the account by Brent Boyd, who describes his decision to abandon Western medicine for alternative healing. Keep in mind that we are using this case to illustrate situations in which people decide to switch from one healing system to another, not to endorse one healing technique over another.

http://www.talamasca.org/avatar/alt-healing.html
Document: Alternative Healing

 Q: What factors motivated Brent Boyd to abandon Western medicine? What did the literature on Chinese medicine say about its healing powers? What treatment did Boyd eventually choose? Did it work for him?

Sociologist Ruth Kornfield observed Western-trained physicians working in Zaire's urban hospitals and found that success in treating patients was linked to the foreign physician's ability to tolerate and respect other models of illness and include them in a treatment plan. For example, among some Zairean ethnic groups, when a person becomes ill, the patient's kin form a therapy management group and make decisions about administering treatments. Because many people in Zaire believe that illnesses are caused by disturbances in social relationships, the cure must involve a "reorganization of the social relations of the sick person that [is] satisfactory for those involved" (Kornfield 1986, p. 369; also see Kaptchuk and Croucher 1986, pp. 106–8).

The intent here is not to evaluate the quality or outcome of either culture's patient–practitioner interaction. Rather, by comparing the two cultures, we can see more clearly that people think and behave in largely automatic ways because they are influenced by norms, values, beliefs, and nonmaterial culture. Those involved with Dr. Rask automatically assumed a scientific framework to define the origin and treatment of her condition. Such a conceptual framework defines illness as "a state of disease and dysfunction 'impersonally' caused by microorganisms, inborn metabolic disturbances, or physical or psychic stress" (Fox 1988, p. 505). On this basis we can begin to understand why Dr. Rask, other physicians, and her close friends all defined her illness as a condition contracted in Africa, requiring hospitalization, having biological origins, and related to direct contact with a patient's blood. The interactions and dialogue between Dr. Rask, other medical personnel, and friends would have been quite different if they had believed that her conditions were caused by spiritual beings, conflict with family, failure to observe religious practice, or the forces of nature, gods, ancestors, or other beings.

The Dramaturgical Model of Social Interaction

A number of sociologists have compared roles attached to statuses with dramatic roles played by actors. Erving Goffman is a sociologist associated with this **dramaturgical model** of social interaction. In this model social interaction is viewed as though it were theater, people as though they were actors, and roles as though they were performances presented before an audience in a particular setting. People in social situations resemble actors in that they must be convincing to others and must demonstrate who they are and what they want through verbal and nonverbal cues. In social situations, as on a stage, people manage the setting, their dress, their words, and their gestures to correspond to the impression they are trying to make or the image they are trying to project. This process is called **impression management.**

Impression Management

On the surface the process of impression management may strike us as manipulative and deceitful. Most of the time, however, people are not even aware that they are engaged in impression management because they are simply behaving in ways they regard as natural. Women engage in impression management when they remove hair from their faces, legs, armpits, and other areas of their bodies and present themselves as hairless in these areas. From Goffman's perspective, even if people are aware that they are engaged in impression management, it can be both a constructive and a normal feature of social interaction because smooth interactions depend on everyone's behaving in socially expected and appropriate ways. If people spoke and behaved entirely as they pleased, civilization would break down. Goffman (1959) also recognized the dark side of impression management that occurs when people manipulate their audience in deliberately deceitful and hurtful ways.

Impression management often presents us with a dilemma. If we do not conceal inappropriate and unexpected thoughts and behavior, we risk offending or losing our audience. Yet, if we conceal our true reactions, we may feel that we are being deceitful, insincere, or dishonest or that we are "selling out." According to Goffman in most social interactions the people involved weigh the costs of losing their audience against the costs of losing their integrity. If keeping our audience is important, concealment is necessary; if showing our true reactions is important, we may risk losing the audience.

In the United States people who test positive for HIV antibodies face the dilemma of impression management. If they disclose the test results, they risk discrimination and the loss of their jobs, insurance coverage, friends, and family. If they keep this information to themselves, they may feel they are being untrue to themselves and to others who care about them. Rarely are interaction situations "either/or." Usually people compromise between the two extremes.

The dramaturgical model and the specific idea of impression management are useful concepts for understanding the dilemma that one partner may face if he or she suggests using a condom as a precautionary condition of sexual intercourse. This dilemma is quite different in the United States than in Zaire. In both countries health officials recommend condom use as a way to reduce substantially the risk of HIV infection. Risk of infection extends to new sexual partners, members of high-risk groups, and even currently monogamous partners if they were sexually active in the past, used intravenous drugs, or received a blood transfusion in the past 10 years. (At the time of this writing, the interval between HIV infection and the onset of AIDS is known to be as long as 10 years. In essence an individual is going to bed not only with the other person but also with that person's sexual partners of the past 10 years.)

In both Zaire and the United States, the subject of condom use is a sensitive one because of the message that the condom conveys to potential sexual partners. In Zaire condom use is associated with birth control, not with disease prevention. The taboo against birth control is strong because many Africans measure their spiritual and material wealth by the number of offspring. If children survive, they become economic assets to the family as well as links to ancestors (Whitaker 1988). In view of the strong pressures to have children, many Zaireans believe that condom use virtually deprives them of the approval of their families and ancestors.

Researcher Kathleen Irwin (1991) and 15 colleagues interviewed healthy factory workers and their wives from Kinshasa, Zaire. They found that, although many respondents had heard of condoms, few actually used them. The researchers also found that among these respondents it is the men who decide whether to use and purchase condoms.

Because of the widespread availability of female contraceptives for many people in the United States, the contraceptive value of condoms is a secondary concern. In the United States, if a person suggests using a condom during sex, he or she is implying that the partner's sexual orientation or sexual history is suspect. Advertisers try to package and market condoms in ways that counteract this message. Condoms come in all colors, are designed and manufactured in forms that stimulate greater sexual sensation, and show sexually appealing scenes on outer packages. The point is that in both Zaire and the United States "sexual behavior is based on long-standing cultural traditions and social values and may be very difficult to change" (U.S. General Accounting Office 1987). This fact in turn makes it difficult for any person to manipulate the meaning of a condom without offending his or her sexual partner. Therefore many people resist using condoms even in high-risk situations (Giese 1987). The findings of an Urban Institute study highlight some of the reasons young men choose to use (or not use) condoms.

http://www.urban.org/periodcl/prr25_2c.htm
Document: Why Teenagers Do Not Use Condoms

 What attitudes and social characteristics are associated with consistent condom use among young men?

The dramaturgical model of social interaction is useful because it helps us see how the need to convey the right impressions can work to discourage people from behaving in ways that are not in their best (or others' best) interest. Goffman uses another theater analogy—staging behavior—to identify situations in which people are most likely to engage in impression management.

Staging Behavior

Just as the theater has a front stage and a back stage, so does everyday life. The **front stage** is the area visible to the audience, where people take care to create and maintain expected images and behavior. The **back stage** is the area out of the audience's sight, where individuals can "let their hair down" and do things that would be inappropriate or unexpected on the front stage. Because backstage behavior frequently contradicts frontstage behavior, we take great care to conceal it from the audience. Goffman uses the restaurant as an example. Restaurant employees do things in the kitchen and pantry (back stage) that they would not do in the dining areas (front stage), such as eating from customers' plates, dropping food on the floor and putting it back on a plate, and yelling at one another. Once they enter the dining area, however, such behavior stops. How often have you, as a customer, seen a server eat a scallop or a french fry from a customer's plate while in the dining area? But if you have ever worked in a restaurant, you know that this is fairly common backstage behavior.

The division between front stage and back stage is hardly unique to restaurants but is found in nearly every social setting. In relation to the AIDS crisis, a host of environments have a front stage and a back stage, including hospitals, doctors' offices, and blood banks. Much as restaurant personnel shield diners from backstage behavior, medical personnel shield patients from backstage behavior. For example, most people know little about the blood bank industry beyond what they see when they donate, sell, or receive blood. Although the public can research such industries and learn about their inner workings, most people do not have the time to study every industry or do not know what questions to ask about the industry that provides them with goods and services. For instance, most people probably give little thought to whom the blood industry collects blood from and would never think to ask the question of where the blood comes from. Thus they would be surprised to find out that Mexicans cross the border into the United States to give blood. In 1980, a few years before HIV was discovered, *Newsweek* magazine ran a story on the thousands of poor Mexicans who earn $10.00 by undergoing a procedure called *plasmapheresis* (technicians take blood, separate the red cells from the plasma, and reinject the donor with his or her own cells minus the plasma). Because red cells are returned to donors, they may undergo the procedure up to twice a week. The article highlighted the international nature of blood collection (Clark 1980).

The Back Stage of Blood Banks

Blood bank officials found themselves in a dilemma in 1981, when officials at the Centers for Disease Control made known their suspicion that HIV was being transmitted through blood products. This revelation meant that not only the United States' blood supply but also the world's blood supply was contaminated, because the United States supplies about 30 percent of the world's blood and blood products (U.S. Bureau of the Census 1992a; *The Economist* 1981, 1983).

Until spring of 1985 U.S. blood bank officials continued to affirm publicly their faith in the safety of the country's blood supply, insisting that there was no need to screen donors. Yet these officials never revealed to the public the many deficiencies in medical knowledge and in the level of technology that could jeopardize the safety of blood. Looking back on their policies, blood bank officials later argued that they practiced this concealment to prevent a worldwide panic. (In Goffman's terminology they did not want to lose their audience.) Such a panic would have brought chaos to the medical system, which depends on blood products. Still the delay in implementing screening exposed many people to infection.

Herein lies a potential connection to Zaire. Malaria is a common disease in Zaire, especially among children. The disease leaves its victims vulnerable to severe anemia, and blood transfusions are used to treat the anemia. As mentioned previously, the United States is a large exporter of blood products. Many countries that import this blood reexport it, and the Red Cross delivers blood products to countries in need. Therefore some of the blood used in Zaire is likely to have originated in the United States.

Although blood bank officials announced publicly that the risk of HIV infection from blood products was one in a million, knowledge of the backstage collection, production, and distribution of blood products leaves no doubt that the risks were in fact higher. Hemophiliacs especially were at risk because their plasma lacks the substance Factor VIII, which aids in clotting, or contains an excess of anticlotting material (U.S. Department of Health and Human Services 1990). In fact we now know that 50 percent of hemophiliacs were HIV-infected from Factor VIII treatments before the first case of AIDS appeared in this group ("Frontline" 1993).

Even after blood bank officials agreed to start screening blood for HIV infection in spring 1985 (after years of debate over whether to test), they still pronounced the blood supply safe. They did not, however, announce the shortcomings of the screening tests: (1) the antibodies that the test measures may not appear in the blood for 11 months after infection with the virus, and (2) a small but unknown percentage of HIV carriers never develops detectable antibodies (Kolata 1989).

In hindsight it is easy to criticize blood bank officials' response to the situation. Yet, in fairness to the blood industry, we must acknowledge the legitimacy of their wish not to induce worldwide panic, especially very early on when no tests were available to screen blood for the infection. As the head of the New York Blood Center argued, "You shouldn't yell fire in a crowded theater, even if there is a fire, because the resulting panic can cause more deaths than the threat" (Grmek 1990, p. 162). In addition we must recognize that it is impossible to eliminate every element of risk. Even so, one troubling fact suggests that their decision may have been motivated by profit. Although U.S. companies tested new blood for the domestic market, they did not test blood already stored in their inventories, and they continued to supply untested blood to foreign countries for at least six months after the tests were available ("Frontline" 1993; Hiatt 1988; Johnson and Murray 1988). Yukuo Yasuda, president of the Tokyo Hemophilia Fraternal Association, explains how imported concentrated blood products played a significant role in the transmission of HIV in Japan.

http://www.nmia.com/~mdibble/Japan2.html
Document: Japanese Hemophiliacs Suffering from HIV Infection

Q: How much blood plasma and concentrated blood products did Japan import from the U.S. in 1983, the year the threat of AIDS emerged as a major social problem? Who does Yasuda blame for the plight of HIV infection among Japanese hemophiliacs? How did the for-profit blood industry contribute to the problem in Japan?

Knowing about statuses, roles, impression management, and front/back stages helps sociologists predict much of the content of social interaction. However, when we interact, we do more than identify the statuses of the people involved and act to behave in ways consistent with role expectations. We also try to assign causes to our own and others' behaviors. That is, we posit explanations for behavior, and we may act differently depending on what explanations we come up with. A theoretical approach that helps us understand how we arrive at our everyday explanations of behavior is attribution theory.

Attribution Theory

Social life is complex. As we have seen, people need a great deal of historical, cultural, and biographical information if they are to understand the causes of even the most routine behaviors. Unfortunately it is nearly impossible for people to have this information at hand every time they want to understand the causes of behavior. For one thing "the real environment is altogether too big, too complex, and too fleeting for direct acquaintance" (Lippmann 1976, p. 178).

Yet, despite our limitations in understanding causes of behavior, most people do attempt to determine a cause anyway, even if they rarely stop to examine critically the accuracy of their explanations. As most of us well know, ill-defined, incorrect, and inaccurate perceptions of cause do not keep people from forming opinions and taking action. Such perceptions, however, result in actions that have real consequences. Sociologists William and Dorothy Thomas described this process very simply: "If [people] define situations as real they are real in their consequences" (Thomas and Thomas [1928] 1970, p. 572).

Attribution theory rests on the assumption that people assign (or attribute) a cause to behavior in order to make sense of it. People usually attribute cause to either dispositional traits or situational factors. **Dispositional traits** include personal or group traits such as motivation level, mood, and inherent ability. **Situational factors** include forces outside an individual's control such as environmental conditions or bad luck. When evaluating the causes of their own behavior, people tend to favor situational factors. When evaluating the causes of another's behavior, however, people tend to point to dispositional traits.

A memorable example of attributing cause to dispositional rather than situational characteristics appears in Colin Turnbull's *The Lonely African.* Turnbull describes how a European farm owner reacts to an African farmhand who wears a tie without a shirt: "'I've actually got a farmhand who wears a tie—but the stupid bastard doesn't realize you don't wear a tie without a shirt!'" (1962, p. 21). The owner attributes the behavior to a dispositional factor—the farmhand is simply stupid. If the farm owner had thought about the behavior in terms of situational factors, he would have found that the farmhand wears the tie "because it makes a bright splash of color, and is useful for tying up bundles, and refuses to wear the shirt that collects dirt and sweat and makes the Europeans smell so bad" (p. 21). Right or wrong, the attributions that people make shape the content of social interaction. In other words the way in which people construct reality affects how they treat and deal with different individuals and groups.

Throughout history, whenever medical professionals have lacked the knowledge or technology to combat a disease, especially one of epidemic proportions, the general population has tended to hold some groups of people within or outside of the society responsible for causing the disease (Swenson 1988). In the sixteenth and seventeenth centuries, for example, the English called syphilis "the French disease," and the French called it "the German disease." In 1918 the worldwide influenza epidemic, which infected more than a billion people and killed more than 25 million, was called "the Spanish flu" even though there was no evidence to support the idea that it originated in Spain.

In a similar vein U.S. medical researchers trying to map the geographical origin and spread of AIDS have hypothesized various interaction scenarios between specific groups of people who are inferred to behave in careless, irresponsible, or immoral ways to explain how AIDS has spread transcontinentally. The hypotheses usually assume that the disease originated in Zaire and then spread to the United States via Europe, Haiti, or Cuba. Yet there is no evidence to support the possibility that it did not spread transcontinentally from the United States. Two of the most prevalent hypotheses are as follows:

- The virus traveled from Zaire to Haiti to the United States. In the mid-1960s a large number of Haitians went to Zaire to fill middle-management positions in the newly independent state; in the 1970s Mobutu sent them home. American homosexuals vacationing in Port-au-Prince brought the virus back to New York and San Francisco.

- The virus traveled from Zaire to Cuba to the United States. Cubans brought it back from Angola, which shares a long border with Zaire. In the late 1970s the Cuban government purged the army of undesirables, including some homosexual veterans who had served in Angola. Many of these Cubans migrated to Miami.

For more theories on the origin of HIV/AIDS and critiques of those theories, see the document "What Are the Theories That HIV Is Man Made?"

ftp://ftp.cs.berkeley.edu/ucb/sprite/www/theories/index.html

Path: Follow any two of the paths.

Document: AIDS Theories

Q: Describe two of the theories covered in the site, and then study the critique posted for each theory you select. What are the strengths and weaknesses of each critique?

Attributing cause to dispositional factors seems to reduce uncertainty about the source and spread of the disease. The rules are clear: if we do not interact with members of this group or behave like members of that group, we can avoid the disease. Such rules provide us with the secure feeling that the disease cannot affect *us* because it affects *them* (Grover 1987).

In the United States many people believe that HIV and AIDS are confined to members of a few high-risk groups, notably hemophiliacs, male homosexuals, and intravenous (IV) drug users. Dispositional explanations for the high risk among members of these groups imply that these individuals "earned" their disease as a penalty for perverse, indulgent, and illegal behaviors (Sontag 1989). As late as 1985 medical and government officials referred to AIDS as the "gay plague," even though there was overwhelming evidence that the HIV infection was transmitted through heterosexual intercourse, needle sharing, and other exchanges of blood and blood products or other bodily fluids. In his book *And the Band Played On* journalist Randy Shilts (1987) argues that the public health and medical research response to AIDS was delayed several years because policymakers believed that casualties were limited to unpopular and socially powerless groups such as homosexuals, IV drug users, and Haitians.

Similarly dispositional thinkers might explain the high incidence of HIV infection and AIDS among both males and females in Zaire (or other African countries) in terms of excessive sexual promiscuity or polygynous marriages or in terms of rituals involving monkeys, female circumcision, and scarification. An article from the *Sacramento Bee*, "Myths of AIDS and Sex," gives examples of dispositional explanations that connect deviant sexual activity with the spread of AIDS in Africa.

ftp://ftp.cs.berkeley.edu/ucb/sprite/www/theories/africa2
Document: Myths of AIDS and Sex

Q: List two theories connecting AIDS in Africa with "bizarre sexual activities." What arguments might be advanced to contradict those theories?

Whereas Americans tend to regard HIV infection as originating in Africa and as being transmitted through bizarre and indulgent behaviors, Zaireans tend to view AIDS as an American disease:

> Westerners had brought AIDS to Africa with their "weird sexual propositions"—a view echoed by *La Gazette* in July 1987, which referred to Westerners coming to Africa with their "sexual perversions." "Many of the venereal diseases now found in Kenya," said an editorial in the *Kenyan Standard*, "were brought into the country by the same foreigners who are now waging a smear campaign against us."
> (THE PANOS INSTITUTE 1989, P. 75)

In Zaire and other African countries people often refer to the United States as the "United States of AIDS." Condoms are called "American socks," and aid in the form of condoms is known as "foreign AIDS" (Brooke 1987; Hilts 1988). People commonly believe that (1) AIDS arrived via rich American sports fans who came to Kinshasa to watch the Ali–Foreman boxing match in 1977, (2) the virus was manufactured in an American laboratory for military germ warfare and was unleashed deliberately on Africans, (3) the disease came from American canned goods sent as foreign aid, and (4) the AIDS epidemic can be traced to the way in which the polio vaccine was manufactured and administered to Africans 30 years ago. Return to the article "Myths of AIDS and Sex."

ftp://ftp.cs.berkeley.edu/ucb/sprite/www/theories/africa2
Document: Myths of AIDS and Sex

Q: According to the theory presented in the article, why does AIDS equally affect men and women in Africa? According to the information presented in the article, how would the West gain from an AIDS epidemic in Africa?

In each dispositional scenario one can infer that those giving the disease to Africans are morally suspect, evil plotters, profit-driven, or careless; these are all dispositional characteristics. Dispositional theories such as the ones just listed, whether American or African, are alike in that they define a clear culprit or scapegoat. A **scapegoat** is a person or a group that is assigned blame for conditions that cannot be controlled, that threaten a community's sense of well-being, or that shake the foundations of a trusted institution. Usually the scapegoat belongs to a group whose members are already vulnerable, hated, powerless, or viewed as different.

The public identification of scapegoats gives the appearance that something is being done to protect the so-called general public; at the same time, it diverts public attention from those who have the power to assign labels. In the United States identifying AIDS as the "gay plague" diverted attention from the blood banks and the risks associated with medical treatments involving blood products. Blood bank officials maintained that the supply was safe as long as homosexuals abstained from giving blood. In Zaire identifying AIDS as an American disease diverted attention from corrupt officials who were funneling money out of Zaire, leaving its people poor and malnourished and (by extension) vulnerable to HIV infection.

From a sociological perspective dispositional explanations that point to a group—or to characteristics supposedly inherent in members of that group—are simplistic and potentially destructive not only to the group but to the search for solutions. When the focus is a specific group and that group's behavior, the solution is framed in terms of controlling that group. In the meantime the problem can spread to members of other groups who believe they are not at risk because they do not share the problematic attribute of the groups identified as high-risk. This kind of misguided thinking about risk applies even to physicians and medical researchers.

Medical sociologist Michael Bloor (1991) and colleagues argue that the official statistics on the modes of transmission are influenced by researchers' beliefs about the relative riskiness of behaviors. For example, an HIV-positive male who has received a blood transfusion and who has had sexual relations with another male is placed in the transmission category "homosexual" rather than "blood recipient." This approach to classification ignores the possibility that a homosexual male can become HIV-infected through a blood transfusion or other means.

Similarly until 1993 the official definition of AIDS did not include HIV-related gynecological disorders such as cervical cancer as one of the conditions that constituted a diagnosis of AIDS. (Also not included under the official definition of AIDS until 1993 were HIV-related pulmonary tuberculosis and recurrent pneumonia.) Under the old definition HIV-positive persons were said to have AIDS only if they developed one of 23 illnesses, many of which were peculiar to the gay population with AIDS. Under the new, revised definition HIV-positive women with cervical cancer are officially diagnosed as having AIDS. Before this change physicians did not advise women with cervical cancer to be tested for HIV and did not give HIV-positive women with this condition the opportunity to be treated for AIDS (Barr 1990; Stolberg 1992). These two examples show that attributions about who "should" have AIDS affect the way in which the condition of AIDS is defined and influence the statistics about who has AIDS. They also show that such attributions affect the content of the physician–patient interaction.

In addition to acknowledging the shortcomings related to AIDS classification and diagnosis, we also have to acknowledge that we simply do not know how many people are HIV-infected worldwide or who is actually infected.

Determining Who Is HIV-Infected

To obtain information on who is actually infected with HIV, every country in the world would have to administer blood tests to a random sample of its population. Unfortunately (but perhaps not surprisingly) people resist being tested. Go to the site "AIDS in Africa?"

ftp://ftp.cs.berkeley.edu/ucb/sprite/www/theories/africa1

Document: AIDS in Africa?

Q: What are some of the shortcomings associated with AIDS tests in Africa? Who is classified as having HIV/AIDS in Africa? Explain.

A planned random sampling of the U.S. population sponsored by the Centers for Disease Control was aborted after 31 percent of the people in the pilot study refused to participate despite assurances of confidentiality (Johnson and Murray 1988). Two researchers involved with this project concluded that "it does not seem likely that studies using data on HIV risk or infection status, even with complete protection of individual identity, will be practical until the stigma of AIDS diminishes" (Hurley and Pinder 1992, p. 625).

The United States is not the only country whose people do not want to be tested. This resistance seems to be universal. For example, Zairean officials were reluctant to disclose the number of AIDS cases and infection rates to United Nations officials or to allow foreign medical researchers to test Zairean citizens because of their sensitivity to the unsubstantiated but widely held belief that Zaire was the cradle of AIDS (Noble 1989).

Random sampling, however, is the most dependable method we have of determining the number of HIV-infected persons. Until we have such information, we cannot know what factors cause a person with HIV to contract AIDS. Random blood samples would permit comparisons between the lifestyles of infected but symptom-free people and infected people who have developed AIDS or AIDS-related complex (ARC). From such comparisons we could learn which cofactors cause a healthy carrier to develop ARC or AIDS. Such cofactors might include diet, exposure to hazardous materials, or prolonged exposure to the sun or to tanning booth rays—anything that might compromise the immune system in such a way as to activate a dormant infection. For example, why do some people remain HIV infected for years without developing AIDS? Why do other people develop AIDS shortly after exposure to HIV? Why is HIV absent from the bodies of some patients with AIDS-like symptoms (Liversidge 1993)? Finally, how is it that at one time HIV may have been a harmless virus? How did the virus change to become the causative pathogen of AIDS? (Some scientists speculate that HIV has been around in an inactive state for at least a century.)

The point is that a person may develop AIDS as a result of other factors besides the behavior that causes a person to contract HIV infection. As long as there is no systematic and random sampling of populations, people will continue to speculate on these factors, either overestimating or underestimating the prevalence of infection and the projected numbers of AIDS cases worldwide.

This situation leaves people with a dilemma about how to deal with a complex health problem such as AIDS when there are so many unanswered questions. Most people do not have the time to inform themselves about all of the contextual forces underlying HIV and AIDS. Yet this does not stop people from acting or attributing cause on the basis of limited information. Even if people don't have the time to inform themselves, they do have the option of at least being critical of their information sources; for most people that source is television.

Public health officials believe that the media need to deliver at least one important message with regard to HIV: "It is not who you are; it is how you live and what you do" (Kramer 1988, p. 43). For the most part, however, this has not been the message transmitted to the American public. Because television news and information shows are an important source of information for most Americans (98 percent of American households have TV sets), we will examine how television producers present information in general and the AIDS phenomenon in particular.

Television: A Special Case of Reality Construction

When sociologists say that reality is constructed, they mean that people assign meaning to interaction or to some other event. When people assign meaning, they almost always emphasize some aspects of that event and ignore others. In this chapter we have looked at some of the strategies that people use to construct reality. Consider the following points:

- When people assign meaning, they tend to ignore the larger context.
- When people interact with others, they assign meaning first by determining their own social status in relation to the other parties and then by drawing upon learned expectations of how people in some social statuses are to behave.

- People create reality when they engage in impression management; that is, they manage the setting, their dress, their words, and their gestures to correspond to impressions they are trying to make.
- People control access to the back stage so that outsiders to the back stage form opinions on the basis of the front stage.
- People attribute cause to dispositional traits when evaluating others' behavior, and they attribute cause to situational factors when evaluating their own behavior.

People also assign meaning to events based on firsthand experiences. Television gives people access to events that they would never have the chance to experience if left to their own resources. Thus our analysis of how people come to construct the reality they do would not be complete if we ignored the format that television, especially television news, uses to present information about what is going on in the world.

Television conquers time and space: it allows us to see what is going on in the world as soon as it happens. However, consider the following features of national and local news—television at its most serious and most informative:

- The average length of a camera shot is 3.5 seconds.
- Every three or four news items, no matter how serious, are followed by three or four commercials.
- The news of the day is often presented as a series of sensationalized images.
- Approximately 15 news items are presented in a 30-minute news program with approximately 12 different commercials.
- Most news coverage of events focuses on the moment; each item is presented without a context.

In *Amusing Ourselves to Death* Neil Postman (1985) examines the format of news programs and the ways in which it affects how people think about the world. Overall, he believes, this format gives viewers the impression that the world is unmanageable and that events just seem to happen. More to the point, it gives the public only the most superficial facts about events in the world. For example, most people in the United States know that AIDS exists and they know basic facts about how the virus is transmitted and about how to reduce risk of transmission. However, a large percentage (71 percent) also admit that they do not know a lot about AIDS (U.S. Department of Health and Human Services 1992).

What characteristics of the news format produce this consequence? The brevity of camera shots is probably one of the main problems. Postman argues that with the average length of a shot being just 3.5 seconds, facts are pushed into and out of consciousness in rapid succession, so that viewers do not have sufficient time to reflect on what they have seen and to evaluate it properly. Furthermore commercials defuse the effects of any news event; they give the following message: I can't do anything about what is happening in the world, but I can do something to feel good about myself if I eat the right cereal, own the right car, and use the right hair spray.

Television is image-oriented; a picture represents a moment in time and by definition is removed from any context. This quality often causes television news to be sensationalistic. The highly publicized case of Ryan White (who died in April 1990 at age 18 from AIDS-related respiratory failure) illustrates just how sensationalistic news reports can be. As a child White had received HIV-infected Factor VIII while being treated for hemophilia and later was diagnosed as having AIDS. White was barred from attending Western Middle School in Kokomo, Indiana, in August 1985 after school officials learned that he had AIDS (Kerr 1990). When White returned to school in April 1986, reporters and camera crews covered the event in what the school principal termed a sensationalistic manner:

> "I understand that the media has a job to do, but I think there is a fine line between informing the public and creating controversy in order for a story to keep continuing.
>
> "It seems like the problems were brought out by those who jumped on the sensational. These were published and displayed on television and it created an excited atmosphere in what was really a pretty calm school situation. . . .
>
> "When Ryan did come back, there were 30 kids whose parents took them out of school. That's what made the news, but there were 365 other children who stayed in the school." (COLBY 1986, P. 19)

An image-oriented format tends to ignore those historical, social, cultural, and political contexts that would make the event more understandable. Viewers are left with vivid, sensationalistic images of enraged parents, an emaciated gay AIDS patient, a skid row drug addict, a prostitute, a family home destroyed by fearful neighbors, Africans walking to an AIDS clinic, or gays protesting discrimination. Rarely, however, do television producers present AIDS as a chronic but often manageable condition or portray those with AIDS as leading responsible lives. Imagine what people's initial attitudes toward AIDS might have been if the first discussion of the disease had centered around the life of someone like Dr. Rask instead of a small group of homosexual males. Trying to understand how Dr. Rask contracted HIV would certainly have told us more about the complex social origins of the disease than did focusing on homosexual practices.

This is not to deny that television is an important tool for informing large audiences. Despite the sensationalistic coverage most Americans know that AIDS exists and that it is related in some way to a virus, but some confusion remains about how HIV is transmitted. For example, about 29 percent of Americans believe that a person runs the risk of becoming infected from donating blood, and 10 percent don't know whether donating blood puts them at risk (U.S. Department of Health and Human Services 1992).

For more on the role of TV as a source of misinformation, see "Even My Own Mother Couldn't Recognize Me" by Jane Rhodes.

http://www.law.indiana.edu/fclj/v47/no1/jrhodes.html
Document: Even My Own Mother Couldn't Recognize Me

Q: With whom did the phrase "Even my own mother couldn't recognize me" originate? What circumstances led to the creation of this phrase? What criticisms of the media does Lonnie Guinier have?

To illustrate how little context television provides, you might list the things you have learned about AIDS from this chapter that you did not learn from the media. One might argue that there is not enough time to learn about AIDS from television because so many other important events are competing for attention in the news. Still this point does not eliminate the need for context in understanding events.

Another way to illustrate how little context television provides is to count how many events (other than sports and weather items) you remember from last night's news. If you can't remember many events, then the information may have been presented so quickly that you could not reflect long enough to absorb it. To remedy this problem, newscasters might reduce the number of stories, increase the time given to context, and cover stories in less reactive and more reflective ways. In the case of AIDS, for example, producers could show segments explaining how HIV differs from AIDS or how the television image of AIDS differs from the experience of AIDS. With regard to this last suggestion, television producers are quite good at covering and discussing presidential campaigns as media events. They certainly could cover AIDS as a media event as well.

Discussion Question

Based on the information presented in this chapter, what conclusions can you draw about the role of human activity in the origin, treatment, and perceptions of disease?

Additional Reading

For more on

- International travel and disease
  http://cdc2.cdc.gov/ncidod/EID/vol2no1/fritz.htm
  Document: Surveillance for Pneumonic Plague in the United States During an International Emergency: A Model for Control of Imported Emerging Diseases

- Social/human issues related to AIDS
  gopher://gopher.niaid.nih.gov/00/aids/nca/Working%20Group%20on%20Social-Human%20Issues
  Document: Working Group on Social-Human Issues: Executive Summary

- Zaire
  http://www.odci.gov/cia/publications/95fact/cg.html
  Document: Zaire

- HIV/AIDS
  http://www.nectec.or.th/users/craig/hiv-aids.htm
  Document: The Relationship Between HIV and AIDS

- Randy Shilts (author of *And the Band Played On: Politics, People and the AIDS Epidemic*)
  http://qrd.rdrop.com/qrd/aids/obits/shilts.obituary-SFChron
  Document: Randy Shilts Obituary

- Issues related to determining who is HIV-infected
  http://www.indiana.edu/~aids/news/news1.html#hiv
  Document: HIV Infection and AIDS in Rural America
  http://www.aoa.dhhs.gov/aoa/pages/agepages/aids.html
  Document: National Institute on Aging Age Page: HIV, AIDS, and Older Adults

- Blood and blood banks
  http://www.access.digex.net/%7Enpc/
  Document: Directory of News Sources

  http://www.aabb.org/docs/receive.html
  Document: Receiving a Blood Transfusion: What Every Patient Should Know

  http://www.webcom.com/~lef/texts/fda-lem-nov95.html
  Document: Offshore Drug Update

  http://www.web-depot.com/hemophilia/archives/iom_summary.html
  Document: Institute of Medicine Report: HIV and the Blood Supply: An Analysis of Crisis Decision Making

Social Organizations

Multinational Monitor is an online journal published by Essential Information, Inc. Founded by Ralph Nader in 1982, Essential Information, Inc., provides provocative information to the public on important topics neglected by the mainstream mass media and policymakers.

http://www.essential.org/monitor

Document: Multinational Monitor

Q: What is the purpose of the *Multinational Monitor?*

The December 1994 issue of the *Multinational Monitor* features the "Corporate Hall of Shame: The Ten Worst Corporations of 1994," by Russell Mokhiber.

http://www.essential.org/monitor/hyper/mm1294.html#topten

Document: The Ten Worst Corporations of 1994

Q: Mokhiber begins his article with the statement "those criminals who inflict the most damage on society are above average in the official measures of intelligence." What does he mean by this statement?

List the names of the 10 worst corporations of 1994, and in one sentence describe the actions that helped to place each corporation on the list. We will return to these cases throughout the chapter. As you read, note that there is one striking similarity across the 10 reports—no one person is assigned responsibility for the actions. The reports read as if the organizations themselves acted in the manner described. This suggests that the causes of the misconduct go beyond individual mistakes or technological failures. Rather something about the organizations themselves halts the spread of important information and prevents responsibility from falling on any one individual.

In this chapter we examine concepts that sociologists use to analyze **organizations,** coordinating mechanisms created by people to achieve stated objectives. In other words as coordinating mechanisms organizations channel the activities of people toward meeting a goal, whether that be to maintain order (for example, a police department); to challenge an established order (the Consumers Union of the United States); to keep track of people (the Census Bureau); to grow, harvest, or process food (PepsiCo); to produce goods (Sony); to make pesticides (Union Carbide); or to provide a service (a hospital) (Aldrich and Marsden 1988).

From a sociological perspective organizations can be studied apart from the people who staff them. This is because organizations have a life that extends to some degree beyond their members. This idea is supported by the simple fact that organizations continue on even as their members die, quit, or retire, or get fired, promoted, or transferred. Organizations are a taken-for-granted aspect of life:

> Consider, however, that much of an individual's biography could be written in terms of encounters with [organizations]: born in a hospital, educated in a school system, licensed to drive by a state agency, loaned money by a financial institution, employed by a corporation, cared for by a hospital and/or nursing home, and at death served by as many as five organizations—a law firm, a probate court, a religious organization, a mortician, and a florist. (ALDRICH AND MARSDEN 1988, P. 362)

Because organizations are so much a part of our lives, we rarely consider how they operate, how much power they have, and how much social responsibility they assume. The concepts that sociologists use to study organizations can help us understand how they can be powerful coordinating mechanisms that channel individual effort into achieving goals that benefit the lives of many people. At the same time, we can use these concepts to help us see how these coordinating mechanisms can contribute to ignorance, misinformation, and failure to take responsibility for known risks. These issues of accountability are particularly relevant when they involve technologies that have the potential to cause irreparable damage to people and to the environment. For this reason we will focus on one of the largest and most influential kinds of organizations—the multinational corporation.

A **multinational corporation** owns or controls production or service facilities in countries other than the one in which it is headquartered (U.S. General Accounting

Office 1978). The emphasis on multinational corporations allows us to consider several issues: (1) the cultural differences between workers who come from one country and members of management who come from another country, (2) the economic conditions of the host country, and (3) the fact that the total operations of a multinational corporation are not subject to the laws of one government. This last problem—regulating the multinational corporation—is a "terribly complex and thorny issue" that has yet to be resolved (Keller 1986, p. 12).

The Multinational Corporation: Agent of Colonialism or Progress?

The United Nations estimates that there are at least 35,000 multinational corporations (or just "multinationals") worldwide with 150,000 foreign affiliates (Clark 1993). Multinationals are headquartered disproportionately in the United States, Japan, and Western Europe. Multinationals compete against rival corporations for global market shares, and they plan, produce, and sell on a multicountry and even a global scale. In addition they recruit employees, extract resources, acquire capital, and borrow technology on a multicountry or worldwide scale (Kennedy 1993; Khan 1986; U.S. General Accounting Office 1978).

Multinationals establish operations in foreign countries in order to obtain raw materials or to make use of an inexpensive labor force (for example, a *maquila* assembly plant in Mexico, a lumber company in Brazil, a mining company in South Africa). They also establish subsidiary companies in foreign countries and employ their citizens to manufacture goods or provide services that are marketed to customers in those countries (for example, IBM Japan).

Hewlett-Packard is one example of an organization that epitomizes the far-flung nature of the multinational corporation that operates on a global scale. It has 95,000 employees working in 475 sales offices in 40 countries and at 55 factory sites in 15 countries. The offices and the factory sites are connected by computer and communications technology that integrates voice, video, and information sharing and generates a staggering 8 million pages of text each day (National Public Radio 1990).

Critics of multinational corporations, such as Essential Information, Inc., monitor them because of their potential to be engines of destruction. Such groups focus on the multinationals that exploit people and resources in order to manufacture products inexpensively and to take advantage of cheap and desperately poor labor forces, lenient environmental regulations, and sometimes nonexistent worker safety standards. Advocates of multinationals, in contrast, maintain that these corporations are agents of progress. They praise multinationals' ability to transcend political hostilities, to transfer technology, and to promote cultural understanding.

In reality no simple evaluation applies to all multinationals. Obviously at some level some multinationals "do spread goods, capital, and technology around the globe. Some do contribute to a rise in overall economic activity. Some do employ hundreds of thousands of workers around the world, often paying more than the prevailing wage" (Barnet and Müller 1974, p. 151). George Keller, chairman of the board and chief executive officer of Chevron Corporation, explains:

I don't want to sound like Pollyanna, but I'm proud that the U.S. companies have a well-deserved reputation for contributing to the communities where they operate overseas. They don't do it solely out of charity. They do it because . . . an improvement in the local economic and social infrastructure is essential if they are to operate effectively. In some developing nations, the multinationals may virtually create the basis of a modern economy.

Chevron faced that situation in Saudi Arabia in the mid-thirties. Our people had come to search for oil in what was then a sparsely inhabited desert.

To conduct our operations, we had to help create the necessary environment. We drilled water wells, built roads, and developed electrical power. As we made progress, the local communities grew and developed into prosperous cities. Schools and hospitals were built, and local industries emerged. (KELLER 1986, PP. 125–26)

Even so, the means employed to create the environments that multinationals need to achieve the maximum profit for owners and stockholders (the valued goal) are not necessarily those that alleviate a host country's problems of mass starvation, mass unemployment, and gross inequality. Critics argue that, if anything, multinationals can aggravate these problems because pursuit of profit outweighs most other considerations. This criticism is supported by *Fortune* magazine's list of the most admired corporations, which were rated on eight dimensions: (1) quality of management, (2) quality of products/services offered, (3) innovations, (4) long-term investment value, (5) financial soundness, (6) ability to attract/keep talented people, (7) community/environmental responsibility, and (8) use of corporate assets.

http://pathfinder.com/@@oz00hwuapotknqeg/fortune/magazine/specials/mostadmired/comebacks.html

Document: Fortune Magazine's Most Admired Corporations 1996

Q: What two characteristics correlate most strongly with a corporation's reputation? Which of the eight characteristics do directors, executives, and analysts name as most important to a company's reputation? Which are ranked as least important? The *Fortune* analysis points out that Digital's quality-of-management score improved the most over the past year largely because of the work of CEO Robert Palmer. What factors does *Fortune* name as reasons Palmer was such an effective CEO?

Features of Modern Organizations

Sociologist Max Weber gives us one framework for understanding the two faces of organizations—those capable of (1) efficiently managing people, information, goods, and services on a worldwide scale and (2) promoting inefficient, irresponsible, and destructive actions that can affect the well-being of the entire planet. Weber's ideas about organizations are built on an understanding of social actions and their significance in modern life.

Social Action

According to Weber the sociologist's main task is to analyze and explain the course and consequences of **social action,** actions influenced by other people, including the thoughts and feelings that lead to particular actions. Weber recognized the endless variety of social action and maintained that the sociologist's task is to make sense of it, not just to observe it (Lengermann 1974). In studying social action Weber emphasized the subjective meaning that the involved parties attach to their behavior. Certainly Weber's preoccupation with the forces that move people to action was influenced by the fact that he saw

> the thrust behind sociological curiosity as residing in the endless variety of societies. Everywhere one looks one sees variety. Everywhere one finds [people] behaving differently. . . . Scrutiny of the facts can show endless ways of dealing with the problems of survival, and an infinite wealth of ideas. The sociological problem is to make sense of this variety. (LENGERMANN 1974, P. 96)

In view of this variety Weber suggested that sociologists focus on the broad reasons people pursue goals, whatever those goals may be (for example, to make a profit, to earn a college degree, to be recognized by others). He believed that social actions can be classified as belonging to one of four important types: (1) traditional (a goal is pursued because it was pursued in the past), (2) affectional (a goal is pursued in response to some emotion), (3) value-rational (a goal is pursued because it is valued, with little or no consideration of foreseeable consequences or the appropriateness of the means chosen to achieve it), and (4) instrumental (a goal is pursued after it has been evaluated in relation to other goals and after thorough consideration of the various means to achieve it) (Abercrombie, Hill, and Turner 1988; Coser 1977; Freund 1968). Read the account of the Taj Mahal and the reason it was built.

http://www.lonelyplanet.com/dest/ind/nor.htm#taj
Document: Destination-Northern India and the Ganges

Q: Which one of Weber's four types of social action applies to the Taj Mahal? Explain.

Weber contended that ever since the onset of the Industrial Revolution, an individual's actions are less likely to be guided by tradition or emotion and more likely to be value-rational. Weber was particularly concerned about the value-rational action because the valued goal can become so all-important that people lose sight of the negative consequences resulting from the methods used to reach that goal.

According to Weber rationalization is a product of human technological and organizational ingenuity and proficiency, and it coincides with the specialization, the division of labor, and the mechanization that revolutionized the production process. Weber defined **rationalization** as a process whereby thought and action rooted in emotion (love, hatred, revenge, joy), in superstition, in respect for mysterious forces, and in tradition are replaced by thought and action grounded in the logical assessment of cause and effect or the means to achieve a particular end (Freund 1968).

Means-to-End (Value-Rational) Thought and Action

The thought and action guided by tradition, superstition, and emotion differ from the thought and action guided by the means-to-an-end logic of value-rational action. We can illustrate these differences by comparing two distinctly different meanings applied to trees. One meaning is assigned by science and industry; the other is held by the Bonda, a small tribe that lives in the Orissa Mountains in India.

The Bonda maintain a culture rooted in emotion, superstition, and respect for mysterious forces. The Bonda believe that spirits inhabit the earth, sky, and water; they believe that sickness, bad weather, and poor harvests are caused by evil spirits. They are particularly respectful of spirits who live in trees and plants. Bonda priests specify which trees can and cannot be cut down. The people do not touch those trees considered to be the homes of gods and genies. Some trees are left standing if the priests believe that their felling would displease phantoms or demons and would cause them to send poor harvests, sickness, and deaths (Chenevière 1987). In essence the belief that trees, plants, and animals possess souls leads people to feel reverence and respect for nature and to behave accordingly toward it.

Rationalization discredits the idea that plants and trees have spirits. Science and technology have enabled us to break down trees and plants into various components and assign them precise functions in the larger production process. In contemporary society, for example, trees are thought of as a means to an end—a source of food, wood, rubber, quinine (a drug used to combat malaria), turpentine (an ingredient of paint thinner and solvents), cellulose (used to produce paper, textiles, and explosives), and resins (used in lacquers, varnishes, inks, adhesives, plastics, and pharmaceuticals). From a value-rational viewpoint nature is something to use: "Rivers are something to dam; swamps are something to drain; oaks are something to cut; mountains are something to see and lakes are sewers to use for corporate waste" (Young 1975, p. 29). In this vein one can also argue that Monsanto Corporation has rationalized the cow and thereby refused to consider foreseeable consequences of this action.

http://www.essential.org/monitor/hyper/mm1294.html#topten
Document: The Ten Worst Corporations of 1994

 How did Monsanto rationalize the cow? What potential consequences did it ignore? What actions did it take to keep knowledge about the product from the consumer?

It is not that people in "rational" environments do not value nature on some level; rather they place greater value on the goals of profit, employment, convenience, and global competition. For example, in 1989 *The New York Times* reported on a study published in a leading scientific journal, *Nature*, that justified forest conservation because "revenues generated by harvesting edible fruits, rubber, oils, and cocoa from

2.5 acres of tropical rain forest are nearly two times greater than the return on timber or the value of the land if used for grazing cattle" (p. Y24). Presumably, if the return on timber were greater than that from harvesting the forest's product, there would be no support for conservation.

One can argue that rationalization has released people from the bondage of superstition and tradition and given them unprecedented control over nature. One major negative side effect of rationalization, however, is what Weber called the **disenchantment of the world**—a great spiritual void accompanied by a crisis of meaning. Disenchantment occurs when the very process of achieving a valued goal is such that it leaves people with a great spiritual void (as when a college student takes the easiest courses to achieve the goal of getting a diploma).

Rationalization

Weber made several important qualifications regarding rationalization. First, he used the term *rationalization* to refer to the way in which daily life is organized socially to accommodate large numbers of people, not necessarily to the way individuals actually think (Freund 1968). For example, the companies that make pesticides advertise them as a rational means for the quick and efficient killing of insects in the house and garden or on pets. Yet most people who buy and use these products have no idea how the chemicals work, where they come from, or what consequences they have other than the death of insects. Thus on a personal level people deal with pesticides as if they were magic.

Second, rationalization does not assume better understanding or greater knowledge. People who live in a value-rational environment typically know little about their surroundings (nature, technology, the economy). "The consumer buys any number of products in the grocery without knowing what substances they are made of. By contrast, 'primitive' man in the bush knows infinitely more about the conditions under which he lives, the tools he uses and the food he consumes" (Freund 1968, p. 20). Most people are not troubled by such ignorance but are content to let specialists or experts know how things work and how to make corrections when something goes wrong. People assume that if they ever need this information, they can consult an expert or go to the library and look it up.

Finally, instrumental action is rare. When people are determining a goal and deciding on the means (actions) to be employed, they seldom consider and evaluate competing goals or other, more appropriate but less expedient means of reaching the stated goal. For example, people often turn to technology as the means of solving problems that they define as important. Rarely does anyone evaluate the overall consequences of a technology for the quality of life on the planet. Scientist Klaus-Heinrich Standke (1986) offers four criteria by which to evaluate a prospective technology:

> Which technology is "appropriate"? A new school of thought attempts to optimize the parameters by which a prospective technology is to be judged. An "optimum technology" is accordingly that which
> 1. is directed towards the highest possible human goals;
> 2. uses mineral and energy resources most efficiently and preserves or enhances the environment;

3. preserves or enhances "good work" for the maximum number of human beings,
4. uses the very best scientific and technical information and combines them with the wisdom and highest values of the culture. (p. 66)

More often than not, people set valued goals without first considering the possible disruptive or destructive social consequences of the means or strategies used to reach them. In other words they do not consider **externality costs**—costs that are not figured into the price of a product but that are nevertheless a price we pay for using or creating the product. The cost of restoring contaminated and barren environments and assisting affected people to cope is one example of an externality cost (Lepkowski 1985). Return to "The Ten Worst Corporations of 1994" and focus on General Motors (GM) and Dole.

http://www.essential.org/monitor/hyper/mm1294.html#topten
Document: The Ten Worst Corporations of 1994

Q: What externality costs are associated with manufacturing pickup trucks with fuel tanks mounted outside the frame rails? What externality costs are associated with using U.S.-banned pesticides on bananas, pineapples, and other fruit grown in Central America?

Multinationals are not charities, of course, and one could argue that they are not responsible for overseeing how the people who purchase their products use them or for setting foreign policy. Nevertheless many people question whether the multinationals and other corporations should have the right to ignore the larger long-term effects of their products on people and on the environment. We can also ask whether executives and stockholders of multinationals can legitimately point to increases in a "poor" country's gross national product or exports as evidence that multinationals are agents of progress when by most social measures their presence has exacerbated world poverty, world unemployment, and world inequality (Barnet and Müller 1974).

In this regard consider the statement of Keith Richardson, the public affairs chief of B.A.T. Industries, the world's largest manufacturer of cigarettes. When asked

> whether the company didn't feel some obligation to put on their cigarette packs sold in the Third World the kind of health warnings they are required to carry in Britain, the United States, and other countries [he replied] "These are sovereign countries, and they are unenthusiastic about being told what to do by pressure groups in the UK and the United States. We fit in with what the government wants in each country. We let the marketplace decide. We do not try to impose."
> (MOSKOWITZ 1987, PP. 40–41)

This response is similar to the justification Nike's CEO gave for paying workers less than $1.00 per day.

http://www.essential.org/monitor/hyper/mm1294.html
Document: The Ten Worst Corporations of 1994

 How does Nike justify paying its workers low wages?

Closer to home, consider the role of automobile makers in creating American dependence on foreign oil. (Americans account for 5 percent of the world's population, yet they consume 40 percent of the gasoline in the world, two-thirds of which goes toward fuel for transportation.) Consider especially American consumers' preferences regarding the fuel efficiency of automobiles. Chevrolet offered two versions of its 1990 Corsica. From a standstill one model took 14 seconds to reach a speed of 60 miles per hour; the other took 10 seconds and used five more gallons of gas per mile (Wald 1990). Most consumers chose the faster, less fuel-efficient model. In view of the various world oil crises—most recently the Persian Gulf crisis—should corporations produce and market products that are not fuel-efficient, even if those product satisfy public wants? Regardless of our response to such questions, our views will be more realistic and better informed if we understand in general how organizations operate and in particular what systematic procedures they follow to produce and distribute goods and services in the most efficient (especially the most cost efficient) manner. A key to understanding how organizations function is the concept of bureaucracy.

The Concept of Bureaucracy

Weber defined **bureaucracy** as a completely rational organization—one that uses the most efficient means to achieve a valued goal, whether that be making money, recruiting soldiers, counting people, or collecting taxes. The following are some major characteristics of a bureaucracy that allow it to coordinate people so all of their actions center on achieving the goals of the organization:

- There is a clear-cut division of labor; each office or position is assigned a specific task toward accomplishing the organizational goals.

- Authority is hierarchical: each lower office is under the control and supervision of a higher office.

- Written rules and regulations specify the exact nature of relationships among personnel and describe the way in which tasks should be performed.

- Positions are filled on the basis of qualifications determined by objective criteria (academic degree, seniority, merit points, test results), not on the basis of personal considerations such as family ties or friendships.

- Administrative decisions, rules, regulations, procedures, and activities are recorded in a standardized format and are preserved in permanent files.

- Authority belongs to the position, not to the individual who fills the position or office. The implication is that one person can have authority over another on the job because he or she holds a higher position, but those in higher positions can have no authority over another's personal life away from the job.

- Organizational personnel treat clients as "cases [and] without hatred or passion, and hence without affection or enthusiasm" (Weber 1947, p. 340). This approach is necessary because emotions and special circumstances can interfere with the efficient delivery of goods and services.

Together these characteristics describe a bureaucracy as an **ideal type**—ideal not in the sense of being desirable but as a standard against which real cases can be compared. Anyone involved with an organization realizes that the actual operation departs from the ideal type. Thus one might ask, What is the use of listing essential characteristics if no organization exemplifies them? This list is a useful tool because it identifies important organizational features. (Note, however, that having these traits does not guarantee that things run perfectly; the rules and policies themselves can cause problems.) By comparing the actual operation to the ideal, one can determine the extent to which an organization departs from these traits or adheres to them too rigidly.

Factors That Influence Behavior in Organizations

On paper the job descriptions, the relationships among personnel, and the procedures for performing work-related tasks in organizations are well defined and predictable. However, the actual workings of organizations are not as predictable, because members of organizations vary in the extent to which they adhere to rules and regulations. Three factors that influence how people act are the informal relationships they form with others, the way they are trained to do their jobs, and the way their performances are evaluated.

Formal Versus Informal Dimensions

Sociologists distinguish between formal and informal aspects of organizations. The **formal dimension** consists of the official, written guidelines, rules, regulations, and policies that define the goals of the organization and its relationship to other organizations and with integral parties (for example, the government or the stockholders). This term also applies to the roles, the nature of the relationships among roles, and the way in which tasks associated with roles should be carried out to realize the goals. The **informal dimension** includes worker-generated norms that evade, bypass, do not correspond with, or are not systematically stated in official policies, rules, and regulations. In the most general sense this term applies to behavior that does not correspond to written plans (Sekulic 1978). Examples of worker-generated norms include unwritten rules about standards of interaction, the appropriate content of conversations between employees of different ranks, and the pace at which people should work.

Worker-generated norms are so much a part of daily life that people rarely think about them. Consequently they are best illustrated by situations in which members of an organization come from different cultures. *Twin Plant News*, a magazine that covers the *maquila* industry (*maquilas* are foreign-owned factories operating in Mexico), printed the details of an encounter between an American manufacturing manager (Frank) and a Mexican personnel director (Pablo). It shows that misunderstandings of informal norms governing expectations about how people in particular positions should relate to one another can disrupt organizational harmony. In the following account Frank has just arrived at the Mexican plant and is meeting Pablo for the first time:

> Frank wants to impress upon Pablo that he will be an effective employer. He instinctively takes on his most self-assured, logical, and factual approach, as this has worked in the past. He has prepared to talk about turnover figures, salaries, current staffing problems and other concerns they had in Chicago as related to the Mexican operation.
>
> Desiring to anticipate what Frank would expect from a person in his position, Pablo has asked his two assistants to join them and told the entire staff he would introduce Frank. When together, Pablo tries to be gracious, maintain good eye contact and ask about Frank's family. He also refers most of Frank's questions to his assistants to give them recognition and an opportunity to establish a relationship with Frank.
>
> Departing an hour later, Frank is determined to speak to the plant manager about getting a Personnel Director he can trust with the facts. Pablo seemed uneasy with his questions and kept looking to his two assistants for the answers. He also expected a person in Pablo's position to have a better command of the English language. [Never mind that Frank doesn't speak Spanish.]
>
> Once Frank left, Pablo asked his assistants whether they thought anyone would get along with this "gringo." He noted that Frank appeared in a hurry, was impatient with introductions, seemed uncomfortable with all the handshaking among the department staff and even appeared angry. (MCINTOSH-FLETCHER 1990, P. 32)

One area of informal organization that sociologists have studied quite extensively is worker-generated norms that govern output or physical effort. These include informal norms against working too hard (those who do so are often called "rate busters"), working too slowing, or slacking off, as well as norms about the number and length of coffee breaks and the length of lunch breaks. Another, more drastic form of informal action employees can take against employers is sabotage.

http://iww.org/labor/sabotage/s2.html
Document: General Forms of Sabotage

Q: Give a general definition of sabotage. What is the purpose of sabotage? How do employer and working-class sabotage differ?

Both positive and negative consequences are associated with informal norms. Informal norms about bending the rules, cutting through red tape, and handling unusual cases or problems, for example, can increase organizational efficiency and effectiveness. Informal norms about after-work activities, friendships, and unofficial communication channels can promote loyalty and worker satisfaction. However, informal norms that put workers and the public at risk can have destructive consequences. One example of such negative consequences can be found in the case of the 1994 oil spill off the coast of Puerto Rico.

gopher://justice2.usdoj.gov/00/press-releases/previous/April96/192.enr

Document: Top Manager and Corporations Convicted in Puerto Rico Oil Spill Disaster

: What formal procedures did the Bunker Group Incorporated and England Marine Services violate? What informal action did Captain McMichael take that directly resulted in the oil spill?

Trained Incapacity

If an organization is to operate in a safe, credible, predictable, and efficient manner, its members need to follow rules, guidelines, regulations, and procedures. Organizations train workers to perform their jobs a certain way and reward them for good performances. However, when workers are trained to respond mechanically to the dictates of the job, they risk developing what economist and social critic Thorstein Veblen (1933) called **trained incapacity,** the inability to respond to new and unusual circumstances or to recognize when official rules and procedures are outmoded or no longer applicable. In other words workers are trained to do their jobs only under normal circumstances and in a certain way. They are not trained to respond in imaginative and creative ways or to anticipate what-if scenarios so that they can perform under a variety of changing circumstances.

In her 1988 book *In the Age of the Smart Machine,* social psychologist Shoshana Zuboff distinguishes between work environments that promote trained incapacity and those that promote empowering behavior. Zuboff's conclusions are the result of more than a decade of field research in various work environments including pulp mills, a telecommunications company, a dental insurance claims office, a large pharmaceutical company, and the Brazilian offices of a global bank—all of which were learning to use computers. We focus here on the experiences of pulp mill workers.

Zuboff found that the pulp mill employees who had worked in conventional mills all of their lives were overwhelmed at first by the new condition of having to run the mill while seated at computer consoles. These comments illustrate their reactions:

> "With computerization I am further away from my job than I have ever been before. I used to listen to the sounds the boiler makes and know just how it was running. I could look at the fire in the furnace and tell by its color how it was burning. I knew what kinds of adjustments were needed by the shades of color I saw. A lot of the men also said that there were smells that told you different things about how it was running. I feel uncomfortable being away from these sights and smells. Now I only have numbers to go by. I am scared of that boiler, and I feel that I should be closer to it in order to control it." (p. 63)

> "When I go out and touch something, I know what will happen. There is a fear of not being out on the floor watching things. It is like turning your back in a dark alley. You don't know what is behind you; you don't know what might be happening. It all becomes remote from you, and it makes you feel vulnerable. It was like

being a new operator all over again. Today I push buttons instead of opening valves on the digester. If I push the wrong button, will I screw up? Will anything happen?" (pp. 63–64)

"With the change to the computer it's like driving down the highway with your lights out and someone else pushing the accelerator." (p. 64)

"What strikes me as most strange, hardest to get used to, is the idea of touching a button and making a motor run. It's the remoteness. I can start it from up here, and that is hard to conceive. I can be up in the control room and touch the keyboard, and something very far away in that process will be affected. It takes a while to gain confidence that it will be OK, that what you do through the terminal actually will have the right effects. . . . It's hard to imagine that I am sitting down here in front of this terminal and running a whole piece of that plant outside. The buttons do all the work." (p. 82)

Zuboff believes that management can choose to use computers as automating tools or informating tools. To **automate** means to use the computer to increase workers' speed and consistency, to provide a source of surveillance (for example, by checking up on workers or keeping precise records of the number of keystrokes per minute), and to maintain divisions of knowledge and thus a hierarchical arrangement between management and workers. The pulp mill workers' comments show that this choice has resulted in trained incapacity:

"Currently, managers make all the decisions. . . . Operators don't want to hear about alternatives. They have been trained to do, not to think. There is a fear of being punished if you think. This translates into a fear of the new technology." (p. 74)

"Sometimes I am amazed when I realize that we stare at the screen even when it has gone down. You get in the habit and you just keep staring even if there is nothing there." (p. 66)

"We had another experience with the feedwater pumps, which supply water to the boiler to make steam. There was a power outage. Something in the computer canceled the alarm. The operator had a lot of trouble and did not look at the readout of the water level and never got an alarm. The tank ran empty, the pumps tripped. The pump finally tore up because there was no water feeding it." (p. 69)

"We have so much data from the computer. . . . Operators are tempted not to tour the plant. They just sit at the computer and watch for alarms. One weekend I found a tank overflowing in digesting. I went to the operator and told him, and he said, 'It can't be; the computer says my level is fine.' I am afraid of what happens if we trust the computer too much." (p. 69)

Alternatively management can choose to use computers as informating tools. To **informate** means to empower workers with knowledge of the overall production process, with the expectation that they will make critical and collaborative judgments about production tasks. The pulp mill workers who use the computer as an informating tool experience work very differently than those who use it as an automating tool. The following quotes illustrate this point:

"Each number is telling you about something, and you draw a picture in your own mind. Each number has a picture, and each number is connected to another number, et cetera, . . . and you get a map. You see the number, the equipment, and all the pieces of the equipment related to it." (p. 87)

"To do the job well now you need to understand this part of the mill and how it relates to the rest of the plant. You need a concept of what you are doing. Now you can't just look around you and know what is happening; you can't just see it. You have to check through the data on the computer to see your effects. And if you don't know what to look for in the data, you won't know what's happening." (p. 94)

"[Before automation we] never expected them to understand how the plant works, just to operate it. But now if they don't know the theory behind how the plant works, how can we expect them to understand all of the variables in the new computer system and how these variables interact?" (p. 95)

"If something is happening, if something is going wrong, you don't go down and fix it. Instead, you stay up here and think about the sequence. . . . You get it done through your thinking. But dealing with information instead of things is very . . . well, very intriguing. I am very aware of the need for my mental involvement now. I am always wondering: Where am I at? It all occurs in your mind now." (p. 75)

"Things occur to me now that never would have occurred to me before. With all of this information in front of me, I begin to think about how to do the job better. And, being freed from all that manual activity, you really have time to look at things, to think about them, and to anticipate." (p. 75)

"The computer makes your job easier . . . but it also makes things more complicated. You have to know how to read it and what it means. That is the biggest problem. What does that number actually mean? You have to know this if you want to really learn how to trust the technology." (p. 81)

In addition to decisions about whether to automate or informate, management also decides who should receive training. According to an Economic and Social Research Council study conducted in Europe, only certain kinds of employees get work-related training.

http://www1.ifs.org.uk/research/education/workrelatedtraining.htm
Document: Work-Related Training Provides Big Rewards for Those Who Get It

 Which groups of people in Europe are more likely to receive work-related training? What are the benefits associated with such training? Do you think the European findings have relevance for the United States?

Statistical Records of Performance

In large organizations supervisors often compile statistics on absenteeism, profits and losses, customer satisfaction, total sales, and production quotas as a way to measure individual, departmental, and overall organizational performance. Such measures can be convenient and useful management tools because they are considered to be objective and precise and because they permit systematic comparison of individuals across time and departments. On the basis of numbers management can reward good performances through salary increases and promotions and can take action to correct poor performances. Measures of performance, however, are problematic when they are not reliable or valid. Consider the problems encountered by the Occupational Safety and Health Administration (OSHA) in its attempt to keep track of employers (corporations and businesses) who are guilty of repeated safety violations. Because OSHA's system for keeping track of repeat offenders is not valid, many employers who are repeat offenders are not caught.

http://www.oshadata.com/fsdr.htm
Document: The Defeat of Repeat

: What are the shortcoming in OSHA's data collection and record-keeping system?

In his 1974 book *On the Nature of Organizations*, sociologist Peter Blau examines the problems that can occur when managers use faulty measures of employee performance without taking these shortcomings into consideration. One problem with statistical measures of performance is that a chosen measure may not be a valid indicator of what it is intended to measure, or it may measure performance by too narrow a criterion. For example, occupational safety is often measured by the number of accidents that occur on the job. On the basis of this indicator, the chemical industry has one of the lowest accident rates of all industries. This indicator, however, has been criticized as too narrow and as lacking validity: chemical workers may be less likely to suffer physical injury on the job than to suffer illnesses whose symptoms go unrecognized as related to chemical exposures. Furthermore exposure-related illnesses may take years to develop.

Another example of a narrow indicator is one commonly used to regulate workers' exposure to chemicals: this measure is the acceptable number of deaths per 1,000 workers. OSHA defines as acceptable 6.2 cases of cancer per every 1,000 people who work with formaldehyde. For people who work with arsenic and benzene, 8 cases per 1,000 and 152 cases per 1,000, respectively, are considered acceptable (Shabecoff 1985). This definition means, for example, that after 152 out of 1,000 people who work with benzene die, the risk becomes unacceptable and working conditions must be investigated. Such a measure is problematic because it requires that a certain number of deaths be reached before a situation is considered unusual and before corrective action can be taken.

A second problem with statistical measures of performance is that they encourage employees to concentrate on achieving good scores and to ignore problems generated by their drive to score well. If quarterly profit is used as an indicator of corporate performance, management may lay off employees and cut costs in such critical areas as worker safety, plant maintenance, and employee training (Wexler 1989).

A third problem with statistical measures of performance is that people tend to pay attention only to those areas that are being measured and to overlook those that are not being measured or those for which no measures exist. For example, ecological economists criticize traditional economists because the latter ignore resource depletion and loss of human skills (through death, injury, or lack of training) when they calculate profit and use it as a measure of performance. For instance, the profits associated with the construction of the vast Northeastern Pipeline would be substantially reduced if the environmental costs were subtracted.

gopher://justice2.usdoj.gov/00/press_releases/previous/May96/233.enr
Document: Builder of Vast Northeastern Gas Pipeline Pleads Guilty

Q: What were the costs of the Northeastern Pipeline in environmental terms?

As we have seen, many potential problems are associated with statistical measures of performance. To ensure that important conditions such as occupational safety are monitored, it is advisable to develop thoughtful and accurate indicators to measure them, to assign responsibility to a definite position, and to tie the measures to the evaluation of performance by persons occupying that position. When no such system is in place, responsibility for accident prevention never rests squarely with specific people. This diffusion of responsibility also can occur when decision makers rely on experts for advice or when the power to make decisions is concentrated in the hands of a few people at the top.

Obstacles to Good Decision Making

In his writings about bureaucracy, Weber emphasized that power was located not in the person, but in the position that a person occupied in the organizational hierarchy. The kind of power described by Weber is clear-cut and familiar: a superior gives orders to subordinates, who are required to carry out those orders. The superior's power is supported by the threat of sanctions: demotions, layoffs, firings. Sociologists Peter Blau and Richard Schoenherr (1973) recognize the importance of this form of power but identify a second, more ambiguous type—expert power—which they believe is "more dangerous than seems evident for democracy and . . . is not readily identifiable as power" (p. 19).

Expert Knowledge and Responsibility

According to Blau and Schoenherr (1973) expert power is connected to the fact that organizations are becoming increasingly professionalized. **Professionalization** is a trend whereby organizations hire experts (such as chemists, physicists, accountants, lawyers, engineers, psychologists, or sociologists) who have formal training in a particular field or activity that is essential to achieving organizational goals. Experts are not trained by the organization, however; they receive their training in colleges and universities. Theoretically they are self-directed and are not subjected to narrow job descriptions or direct supervision. Experts use the frameworks of their chosen profession to analyze situations, solve problems, or invent new technologies. From the experts' viewpoints the information, service, or innovation they provide to the organization is technical and neutral. They do not necessarily think about or have control over the application of that information, service, or invention. Consider the cases of GM and Halcon from "The Ten Worst Corporations of 1994."

http://www.essential.org/monitor/hyper/mm1294.html#topten

Document: The Ten Worst Corporations of 1994

 Did GM and Halcon have in-house information that warned of a problem? How do these two cases illustrate the dilemma of expert power?

Blau and Schoenherr regard this arrangement between experts and organizations as problematic because it leaves nobody accountable for the actions of powerful corporations and because it complicates attempts to find individuals "whose judgments [are] the ultimate source of a given action" (pp. 20–21). This situation is complicated for two reasons. First, the recommendations and judgments of experts rest on specialized knowledge and training. The experts may understand principles of accounting, physics, biology, chemistry, psychology, or sociology, but their training for the most part is compartmentalized; they know one subject very well, but they do not know other subjects. For example, a chemist may be able to design a pesticide, but he or she has not been trained to consider the limitations of the people who use it. Similarly a sociologist may understand the abilities and limitations of people who use a pesticide but may not understand the technology. Because the sociologist does not understand the chemistry, he or she cannot design the details of a program to educate consumers.

Second, decision making in large organizations is complex because no single person provides all of the input that goes into a decision. A decision is a joint product of information and judgments by a variety of experts. Often the decision maker does not understand the principles underlying an expert's recommendations or consider other factors such as profit when reviewing those recommended actions. The problem is that when something goes wrong, the experts claim that they only provided information, suggestions, and recommendations and that their position does not give them the power to see that the information is used wisely. In turn management claims that it cannot predict the consequences of an invention or a service that only the experts understand.

Blau and Schoenherr emphasize that the majority of men and women who give expert advice or make decisions based on expert advice are decent people but that their training and viewpoint make them unable to anticipate and plan for the unintended consequences. The chemists who created pesticides and fertilizers believed that their inventions would help feed the world. Little did they know that these chemicals would be misused and overused to the point of precipitating an ecological crisis. This brings us to another issue related to decision making—oligarchy.

The Problems with Oligarchy

Oligarchy is rule by the few, or the concentration of decision-making power in the hands of a few persons who hold the top positions in a hierarchy. According to social critic Charles Snow (1961)

> one of the most bizarre features of any advanced industrial society in our time is that the cardinal choices have to be made by a handful of men . . . who cannot have firsthand knowledge of what those choices depend upon or what their results may be. [And by] "cardinal choices," I mean those which determine in the crudest sense whether we live or die. For instance, the choice in England and the United States in 1940 and 1941, to go ahead with work on the fission bomb: the choice in 1945 to use that bomb when it was made. (P. 1)

Political analyst Robert Michels (1962) believed that large formal organizations tended inevitably to become oligarchical, for the following reasons. First, democratic participation is virtually impossible in large organizations. Size alone makes it "impossible for the collectivity to undertake the direct settlement of all the controversies that may arise" (p. 66). Consider the size (as measured by the number of employees) of the Fortune 500 Global Corporation. "It is obvious that such a gigantic number of persons belonging to a unitary organization cannot do any practical work upon a system of direct discussion" (p. 65).

http://pathfinder.com/@@GUCN3gUAXevkNqeG/fortune/magazine/1995/950807/global500/workers.html

Document: Companies Ranked by Number of Employees

Q: Who are the top five employers among the Fortune 500? How many people does each organization employ?

Second, as the world grows more interdependent and as technology grows increasingly complex, many organizational features become incomprehensible to workers. As a result many employees work toward achieving organizational goals that they did not define, cannot control, may not share, or may not understand. This lack of knowledge prevents workers from participating in or evaluating decisions made by executives.

A danger of oligarchy is that those who make decisions may not have the necessary background to understand the full implications of those decisions. For example,

Warren Anderson, the chairman of the Union Carbide Corporation, stated, "It never entered my mind that an accident such as Bhopal could happen" (Engler 1985, p. 495).

http://www.essential.org/monitor/hyper/mm1294.html#ed

Document: Remembering Bhopal

Q: Once in the site, space down to page 13. How does the outcome of the Bhopal crisis reinforce oligarchy? Explain.

In addition decision makers may not consider the greater good and may become preoccupied with preserving their own leadership. Guarding against these effects of oligarchy requires that the average worker and the general public be interested, attentive, and informed. As more and more people hold jobs that require them to deal with science and technology, and as their daily lives become increasingly dependent on technology, people need to understand what is going on around them. Otherwise technology decisions will be made by others on their behalf, and they will be forced to accept the consequences. One model of corporate–citizen partnership may be found within the Saskatchewan uranium industry.

http://www.uilondon.org/uiabs95/michel.html

Document: Corporate Citizenship and the Saskatchewan Uranium Industry

Q: How much of the world's total population of uranium in 1994 came from the Saskatchewan uranium mines? How does the company's community relations program function?

To this point we have discussed a number of important concepts that help us understand how some characteristics of organizations make them coordinating mechanisms with the potential for both constructive and destructive consequences. Now we turn to Karl Marx and his concept of alienation to understand how workers are dominated so strongly by the forces of production that they remain uninformed or uncritical about their role in the production process.

Alienation of Rank-and-File Workers

Human control over nature increased with the development of increasingly sophisticated instruments and tools and with the growth of bureaucracies to coordinate the efforts of humans and machines. Machines and bureaucratic organizations combined to extract raw materials from the earth more quickly and more efficiently and to increase the speed with which necessities such as food, clothing, and shelter could be produced and distributed.

Karl Marx believed that increased control over nature is accompanied by **alienation,** a state in which human life is dominated by the forces of human inventions. Chemical substances represent one such invention; they have reduced both the physical demands and the specialized knowledge involved in producing goods. Fertilizers, herbicides, pesticides, and chemically treated seeds give people control over nature because they eliminate the need to fight weeds with hoes, they prevent pests from destroying crops, and they help people produce unprecedented amounts of food. In the long run, however, people are dominated by the effects of this invention. Heavy reliance on chemical technologies causes the soil to erode and become less productive, and it causes insects and disease-causing agents to develop resistance to the chemicals. Chemical technologies also have altered the ways in which farmers plant crops: planting patterns have changed from many species of sustenance crops planted together to a single cash crop, planted in rows. As a result most farmers have lost knowledge of how to control insects and diseases without chemicals by interplanting a variety of flowers, herbs, and vegetables. Farmers are now economically dependent on a single crop and on the chemical industry.

Consequences of Worker Alienation

Although Marx discussed alienation in general, he wrote more specifically about alienation in the workplace. He believed that alienation resulted from dividing up the production process so that a single product was assembled by many workers, each performing a specialized task. Mechanization, specialization, and bureaucratic organization have given people new control over production because these technologies reduce physical effort and increase the pace of work. But people in turn are controlled by these same technologies, which have caused workers to lose self-direction. As a result of technology, some workers have become replaceable, or as interchangeable as machine parts. They are treated as economic components rather than as active, creative social beings (Young 1975). Marx believed that the conditions of work usually are such that they impair an individual's "capacity to become a multidimensional, authentic being with human qualities of compassion, reflection, judgment, and action" (Young 1975, p. 27).

Marx believed that workers are alienated on four levels: (1) from the process of production, (2) from the product, (3) from the family and the community of fellow workers, and (4) from the self. Workers are alienated from the process because they produce not for themselves or for known consumers but for an abstract, impersonal market. In addition they do not own the tools of production. Workers are alienated from the product because their roles are rote and limited; no person can claim a product as the unique result of his or her labor. Workers are alienated from their families because home and work environments are separate. In other ways households are uprooted because large-scale enterprises take over the land or force families to move to areas where work is available. Furthermore workers are alienated from one another because they compete for a limited number of jobs. As they compete, they fail to consider how they might unite as a force and control their working conditions. Finally workers are alienated from themselves because "one's genius, one's skills, one's talent is used or disused at the convenience of management in the quest of private profit. If private profit requires skill, then skill is permitted. If private profit requires subdivision of labor and elimination of craftsmanship, then skill is sacrificed" (Young 1975, p. 28).

Alienation on these four levels is particularly evident among workers in developing countries, which are trying to lure foreign business. The foreign companies own the factory buildings, the tools, the machines, and the labor of workers from the host country. The work is often repetitious, mind-numbing, and tedious. In the worst cases workers handle chemicals about which "they may understand little, other than that their eyes tear, they cough harshly and they suffer recurring rashes and headaches" (Engler 1985, p. 493). Although studies on working conditions in developing countries are imprecise and sketchy, the preliminary information suggests that workers are widely exploited without regard for health consequences. For example, two studies found that one-third of Indians working in a DDT plant and one-fourth of all employees at battery plants were sick (Engler 1985). In addition the World Health Organization found that of the thousands of industries in Delhi, many are "cramped, poorly lighted, ill-ventilated spaces with atmospheres full of dust, gas, vapors and fumes," and that they operate without health controls (Crossette 1989, p. Y24).

Another sign that workers in developing countries are exploited and treated as dispensable is that hazardous substances banned in the United States and other Western countries are now being produced in the developing countries. Asbestos production is a typical example. The adverse effects of asbestos are well known: chronic exposure is linked to lung disease and to a fatal form of cancer (mesothelioma). In the United States asbestos products are gradually being replaced or phased out; for example, asbestos no longer is used to insulate homes and other buildings. However, for some products, most notably automobile brakes to fit cars made before 1990, asbestos is still used. Environmental Protection Agency limits on workers' exposure to asbestos have caused many Western factories to move production offshore to Taiwan, Mexico, and India, where regulations governing workers' safety are decades behind those in the United States. Asbestos dust in one Bombay, India, plant is like the dust "behind a bus on a dirt road in the dry season" (Castleman 1986, p. 62).

The factory owners and shareholders make large profits in part because they can locate their plants in places where minimum wages are lower, environmental protection regulations are weaker or nonexistent, and workers' health and safety laws are not systematically enforced or are less stringent. When the company loses money, the management at corporate headquarters can neglect the plant, allowing equipment, employees' skills, and sometimes already inadequate safety standards to deteriorate. The workers are thankful for any job and are hesitant about organizing to improve working conditions when such an action could mean the loss of their jobs and source of livelihood. Again return to the 1994 Corporate Hall of Shame and examine the cases of Dole Foods and Nike.

http://www.essential.org/monitor/hyper/mm1294.html#topten

Document: The Ten Worst Corporations of 1994

: How might you apply Marx's concept of alienation to the problems at Dole Foods and Nike?

Criteria for Evaluating Multinationals

This point about workers' hesitancy to speak out has led economist and U.S. Secretary of Labor Robert Reich to conclude that people must evaluate corporations, particularly multinationals, in a new light. According to Reich we cannot assume that, if corporations headquartered in a given country are prosperous and profitable, the workers in that country also are prosperous. Nor can we assume that executives from foreign-owned corporations are more likely than native-born executives to make decisions that will harm the economic well-being of the host country. Why? Because "American corporations and their shareholders can now prosper by going wherever on the globe the costs of doing business are lowest—where wages, regulations, and taxes are minimal. Indeed, managers have a responsibility to their shareholders to seek out just such business climates" (Reich 1988, p. 79). Reich advises people to be leery of business executives who point to large trade deficits as evidence that they need wage concessions, trade protection, and other incentives if their corporations are to compete on a global scale. In the case of the United States, a large portion of the trade deficits with other countries is the result of American corporations manufacturing products in those countries that are then sold in the United States under their own brand name. For example, 40 percent of the trade that Mexico does with the United States originates from U.S. subsidiaries in Mexico (Clark 1993). As Reich (1988) notes, "One of the ironies of our age is that an American who buys a Ford automobile or an RCA television is likely to get less American workmanship than if he had bought a Honda or [a] Matsushita TV" (p. 78).

In view of this global context, Reich argues that people should not be overconcerned with the nationality of a company or with trade deficits. Instead they should focus on the quality of jobs that a corporation brings to a community. The mistake often made by a local chambers of commerce and state governors when drumming up business is that they focus only on the number of jobs, not the quality of jobs or the corporate record. Inducing Volkswagen or Honda to set up a facility nearby employing 5,000 routine assembly-line jobs at a cost of $5 million in tax abatements may not be a great accomplishment. The point is that learning more about a corporation's record can help citizens decide whether the presence of a corporation in their community will be beneficial over the long term.

Organizations such as E-law and Essential Information, Inc., are dedicated to informing the public about multinationals that behave in irresponsible ways. As one example see how the citizens of West Eugene, Oregon, informed themselves about Hyundai.

http://www.igc.apc.org/elaw/update_summer_95.html#hyundai

Document: Hyundai Plans Chip Factory for Eugene

Q: What information did West Eugene citizens gather to assess the potential impact of the manufacturing plant on their community?

Reich (1990) summarizes the dilemma:

> Ultimately for a British, Dutch, or Japanese company to come into town isn't necessarily better or worse than a California or Massachusetts company. If they bring good jobs, build up the ability of the work force, this type of "foreign" investment is probably a big plus. But if jobs are unskilled or routine, and fail to add new skills, the investment may ultimately cause small towns to suffer in the long run. [For most towns] another generation of low-skill, low-paid assembly-line workers is not particularly beneficial in the long run. (p. 84)

Discussion Question

Based on the information presented in this chapter, what organizational issues would you focus on if you were asked to investigate a disaster such as Bhopal?

Additional Reading

For more on

- Externality costs
  gopher://justice2.usdoj.gov/00/press_releases/previous/March96/132.txt
  Document: U.S. Sues Eight Mining Companies for Vast Environmental Damage to the Idaho Pan Handle

- Informal aspects of organizations
  http://www.oshadata.com/fsoihu.htm
  Document: It's Confirmed—OSHA Inspectors Are Human!

  http://www.oshadata.com/fssy.htm
  Document: Seek and Ye Shall Not Find

- The dark side of organizations
  http://epawww.ciesin.org/national/epacoop.html
  Document: Integration and Use of Information Regarding the Human Dimensions of Environmental Change

  http://www.essential.org/monitor/hyper/list.html
  Document: List of Back Issues of Multinational Monitor

- Multinationals
  http://pathfinder.com/@@oz00hwuapotknqeg/fortune/magazine/1995/950807/global500/workers.html
  Document: Companies Ranked by Number of Employees

  http://pathfinder.com/@@oz00hwuapotkneg/fortune/magazine/1995/950807/global500/equity.html
  Document: Companies Ranked by Stockholders' Equity

  http://pathfinder.com/@@oz00hwuapotkneg/fortune/magazine/1995/950807/global500/assets.html
  Document: Companies Ranked by Assets

http://pathfinder.com/@@oz00hwuapotkneg/fortune/magazine/1995/950807/global500/profits.html
Document: Companies Ranked by Profits

- Alienation
  http://www1.ifs.org.uk/research/personal/wagechanges.htm
  Document: The Changing Face of Wages

- Oligarchy
  http://www.peak.org/%7edanneng/decision/targets.html
  Document: Atomic Bomb: Decision—Target Committee, May 10–11, 1945

Deviance, Conformity, and Social Control

Deviance, conformity, and social control are among the most complex issues in sociology because, as we will see in this chapter, almost any behavior or appearance can qualify as deviant under the right circumstances. **Deviance** is any behavior or physical appearance that is socially challenged and/or condemned because it departs from the norms and expectations of a group. **Conformity,** on the other hand, may be defined as behavior and appearances that follow and maintain the standards of a group. All groups employ **mechanisms of social control**—the methods used to teach, persuade, or force their members to conform to, not to deviate from, shared norms and expectations. Depending on the cultural circumstances a deviant act or appearance may be something as seemingly minor as wearing eyeglasses or withdrawing money from a bank or being the wrong size.

When sociologist J. L. Simmons (1965) asked 180 men and women in the United States from all age, educational, occupational, and religious groups to "list those things or types of persons whom you regard as deviant," a total of 1,154 items were listed.

> Even with a certain amount of grouping and collapsing, these included no less than 252 different acts and persons as "deviant." The sheer range of responses included such expected items as homosexuals, prostitutes, drug addicts, beatniks, and murderers; it also included liars, democrats, reckless drivers, atheists, self-pitiers, the retired, career women, divorcees, movie stars, perpetual bridge players, prudes, pacifists, psychiatrists, priests, liberals, conservatives, junior executives, girls who wear makeup, and know-it-all professors. (PP. 223–24)

Although this study was done more than 30 years ago, Simmons's conclusions are still relevant: almost any behavior or appearance can qualify as deviant under the right circumstances. The only characteristic common to all forms of deviance is "the fact that some social audience regards them and treats them as deviant" (p. 225).

It is difficult to generate a precise list of deviant behaviors and appearances because something that some people consider deviant may not be considered deviant by others. Likewise something that is considered deviant at one time and place may not be considered deviant at another. For example, cocaine and other now-illegal drugs once were legal substances in the United States. In fact "we unwittingly acknowledge the previous legality of cocaine every time we ask for the world's most popular cola by brand name" (Gould 1990, p. 74). Originally Coca-Cola was marketed as a medicine that could cure various ailments. One of its ingredients came from the coca leaf, which is also used to produce cocaine (Henriques 1993). As a second example consider that prior to 1920 in the United States installment buying, the process of postponing the full payment of goods purchased until a future date, was a sign of moral degeneracy. By 1920 this method of purchase was widely accepted.

http://www.lib.virginia.edu/journals/EH/EH37/Murphy.html
Document: The Advertising of Installment Plans

Q: For now read only the material covered through footnote 27 of this article. What historical factors may have contributed to a change in attitude about installment buying?

Such wide variations in what is perceived as deviant alert us to the fact that deviance exists only in relation to norms in effect at a particular time. The sociological contribution to understanding deviant behavior is that it goes beyond studying the individual and instead emphasizes the context in which deviant behavior occurs. Such an emphasis raises at least two general but fundamental questions about the nature of deviance. One obvious question is, How is it that almost any behavior or appearance can qualify as deviant under certain circumstances? It follows that any answers to this question must go beyond the deviant individual's personality or genetic makeup.

A second important question is, Who defines what is deviant? That is, who decides that a particular group, behavior, appearance, or person is deviant? The fact that an activity or an appearance can be deviant at one time and place and not another suggests that it must be defined as deviant by some particular process.

Deviance: The Violation of Norms

Some norms are considered more important than others. In this regard sociologist William Graham Sumner identified two kinds of norms—folkways and mores.

Folkways and Mores

Folkways are customary ways of doing things that apply to the details of life or routine matters—how one should look, eat, greet another person, express affection toward same-sex and opposite-sex persons, and even go to the bathroom. When Francisco Martins Ramos, an anthropology professor who teaches in Portugal, visited the United States, he found that folkways governing how to behave in public restrooms differed from those in his country. Ramos (1993) writes that

> the Portuguese newcomer will certainly be quite surprised with a form of cultural behavior never dreamt of: The American who urinates in public initiates conversation with the partner at his side, even if he does not know the latter! Themes of these occasional dialogues are the weather, football, politics, and so on. We can guess at the forced pleasure of the Portuguese, who heretofore has regarded urination as a necessary physical function, not as a social occasion. The public restroom! Is it an extension of the bar room? (p. 5)

Mores are norms that people define as essential to the well-being of their group or nation. Obvious examples of mores are norms that prohibit cannibalism or the unjust and deliberate taking of another person's life. People who violate mores usually are punished severely; they are ostracized, institutionalized in prisons or mental hospitals, and sometimes executed. In comparison to folkways, however, people consider mores to be "the only way" and "the truth." Consequently people consider mores to be final and unchangeable, and they view the mechanisms of social control that encourage conformity to mores to be more important.

Mechanisms of Social Control

Ideally socialization brings about conformity, and ideally conformity is voluntary. But when conformity cannot be achieved voluntarily, mechanisms of social control are used to convey and enforce shared norms and expectations. Such mechanisms are known as **sanctions**—reactions of approval and disapproval to behavior and appearances. Sanctions can be positive or negative, formal or informal. A **positive sanction** is an expression of approval and a reward for compliance, such as applause, a smile, or a pat on the back. A **negative sanction** is an expression of disapproval for noncompliance, such as withdrawal of affection, ridicule, ostracism, banishment, physical harm, imprisonment, solitary confinement, or even death. This Web site provides a global perspective on the use of the death penalty as a formal sanction.

http://www.best.com/~mlacabe/amnesty/info/eng/dpfacts.html
Document: Facts and Figures on the Death Penalty

Q: How many countries use the death penalty as a mechanism for social control? Which countries use the death penalty against juveniles? What are the arguments for the death penalty? Is the death penalty an effective deterrent?

Informal sanctions are spontaneous and unofficial expressions of approval or disapproval; they are not backed by the force of law. The following incident involving the son of a well-known psychologist, Sandra Bem, shows how informal sanctions usually are applied. Bem and her husband are attempting to raise their children not to be limited by those "attributes and behaviors the culture may have stereotypically defined as inappropriate" for one sex or the other (Monkerud 1990, p. 83). Their efforts to go against norms and expectations are challenged through informal sanctions at every step of the way:

> At age four, Bem's son, Jeremy, wore barrettes to nursery school. One day a boy repeatedly told him that "only girls wear barrettes." Jeremy tried to explain that wearing barrettes didn't make one a boy or girl: only genitalia did. Finally, in frustration, he pulled down his pants to show the boy that having a penis made him a boy. The boy responded, "Everybody has a penis; only girls wear barrettes." (p. 83)

This incident also shows how early socialization to norms works. Even at the age of four, children use sanctions to enforce the norms that they have learned from adults, teachers, and other sources such as television.

In contrast **formal sanctions** are backed by the force of law: they are planned and official expressions of approval or disapproval. Definite and systematic laws, rules, regulations, and policies specify (usually in writing) the conditions under which people should be rewarded or punished and define the procedures for allocating rewards and imposing punishments. The Texas Computer Crime Statutes represent one example of how laws specify conditions under which people will be punished.

gopher://wiretap.spies.com/00/Gov/US-State/compcrime.tx

Document: Texas Computer Crime Statutes

Q: Under which conditions should computers not be used in Texas?

Examples of formal sanctions include medals, cash bonuses, and diplomas on the one hand, and fines, prison sentences, and the death penalty on the other.

Sociologists use the word **crime** to refer to deviance that breaks the laws of society and is punished by formal sanctions. People in every society have different views of what constitutes crime, what causes people to commit crimes, and how to handle offenders.

The Functions of Deviance

According to sociologist Randall Collins (1982) Emile Durkheim presented one of the most sophisticated sociological theories of deviance. Durkheim ([1901] 1982) argued that although deviance does not take the same form everywhere, it is present in all societies. He defined deviance as those acts that offend collective norms and expectations. The fact that always and everywhere there are people who offend collective sensibilities led him to conclude that deviance is normal as long as it is not excessive and that "it is completely impossible for any society entirely free of it to exist" (p. 99). According to Durkheim deviance will be present even in a "community of saints in an exemplary and perfect monastery" (p. 100). Even in seemingly perfect societies, acts that most persons would view as minor may offend, create a sense of scandal, or be treated as crimes.

Durkheim drew an analogy to the "perfect and upright" person. Just as such a person judges his or her smallest failings with a severity that others reserve for the most serious offenses, so do societies that are supposed to contain the most exemplary people. Even among such exemplary individuals, some act or some appearance will offend simply because "it is impossible for everyone to be alike if only because each of us cannot stand in the same spot. It is also inevitable that among these deviations some assume a criminal character" (p. 100). What makes an act or appearance criminal is not so much the character or the consequences of that act or appearance, but the fact that the group has defined it as something dangerous or threatening to its well-being.

Durkheim maintained that deviance is functional for society for at least two reasons. First, Durkheim argued that a group that went too long without noticing crime or doing something about it would lose its identity as a group. The popularity of the crime-stopper TV shows (for example, "America's Most Wanted," "Unsolved Mysteries," and "American Justice") is one example of this integrative function. These shows offer viewers background information and encourage them to be on the lookout for criminals and to help police solve crimes. The internet contains 22,659 sites related to crime-stopper issues. The integrative nature of crime-stopper programs is reflected by the Peel Regional Police vision of their program: "Crime Stoppers is a non profit organization which rallies the community, the news media and the police in a collective campaign against crime. Crime Stoppers' mandate is to fight crime." The following represents a small sample of crime-stopper sites on the internet.

http://www.nj.com/crimestoppers/morriscounty/index.html
Document: New Jersey Online: Morris County CrimeStoppers

http://www.C-S-I.org/
Document: Crimestoppers International

http://www.matsu-crimestoppers.org/
Document: Matsu Crime Stoppers Inc.

http://www.fbi.gov/
Document: Federal Bureau of Investigation Home Page

Q: After viewing these Web sites, do you agree that crime has an integrative function? Explain.

In addition to its integrative function, deviance has another function. Deviance is functional because it is useful in making necessary social changes and in preparing people for such change. It is the first step toward what will be. Nothing would change if someone did not step forward and introduce a new perspective or new ways of doing things. Almost every invention or behavior is rejected by some group when it first comes into existence.

Durkheim's theory offers an intriguing explanation for why almost anything can be defined as deviant. Yet Durkheim did not address an important question: Who decides that a particular activity or appearance is deviant? One answer can be found from labeling theory.

Labeling Theory

A number of sociologists—including Frank Tannenbaum (1938), Edwin Lemert (1951), John Kitsuse (1962), Kai Erikson (1966), and Howard Becker (1963)—are linked to the development of what is conventionally called labeling theory. The scholar most frequently associated with labeling theory, however, is Howard Becker. In his 1963 work entitled *Outsiders: Studies in the Sociology of Deviance,* Becker stated the central thesis of *labeling theory:* "All social groups make rules and attempt, at some times and under some circumstances, to enforce them. When a rule is enforced, the person who is supposed to have broken it may be seen as a special kind of person, one who cannot be trusted to live by the rules agreed on by the group. He is regarded as an outsider" (p. 1).

As Becker's statement suggests, labeling theorists assume that rules are socially constructed and that rules are not enforced uniformly or consistently. These assumptions are supported by the fact that the definitions of what is deviant vary across time and place and that some people break rules and escape detection while others are treated as offenders even though they have broken no rules. Consider citizen's arrest. In the United States private citizens have the right to make arrests under certain conditions. Theoretically private citizens can arrest someone if they are dealing with a felony (for example, murder, rape, burglary) or misdemeanor (for example, traffic violations) and they have witnessed the crime or have reasonable cause to believe that that person committed the crime. However, very few people who witness such crimes choose to exercise their right to make an arrest. One exception is the Guardian Angels and an offshoot organization, the Cyber Angels. The case of citizen's arrest as implemented by the Guardian Angels and Cyber Angels illustrates the labeling theory view that deviance per se does not exist until someone takes notice and applies sanctions.

http://www.educom.edu/educom.review/review.96/jan.feb/sliwa.html
Document: Cyber-Cops: Angels on the Net

Q: How do the Guardian Angels make a citizen's arrest? How do Cyber Angels plan to police the internet?

Labeling theorists maintain that whether an act is deviant depends on whether people notice it and, if they do notice, on whether they label it as a violation of a rule and subsequently apply sanctions. Such contingencies suggest that violating a rule in and of itself does not make a person deviant. That is, from a sociological viewpoint a rule breaker is not deviant (in the strict sense of the word) unless someone notices the violation and decides to take corrective action. Labeling theorists suggest that for every rule a social group creates, there are four categories of people: conformists, pure deviants, secret deviants, and the falsely accused. The category to which one belongs depends on whether a rule has been violated and on whether sanctions are applied.

Conformists and Deviants

Conformists are people who have not violated the rules of a group and are treated accordingly. **Pure deviants** are people who have broken the rules and are caught, punished, and labeled as outsiders. As a result the rule breaker takes on the master status of deviant, an identification that "proves to be more important than most others. One will be identified as a deviant first, before other identifications are made" (Becker 1963, p. 33). We must remember that although pure deviants undeniably violate rules, rule enforcers select the people they apprehend and punish. The California Molester Identification Line illustrates this selective quality. California state law requires that the Child Molester Identification Line must be in place so that individuals and organizations can determine if someone is one of the state's 35,000 registered molesters.

http://www.ca-probate.com/molest.htm
Document: Child Molester Hotline Number

Q: How does this information illustrate the selective character of rule enforcement?

Consider another example of the selective characteristics of rule enforcement: how highway patrol officers choose from among all of the cars speeding along a highway which drivers to pull over. A study of vehicles' and drivers' characteristics on a stretch of the New Jersey Turnpike showed that fewer than 5 percent of the vehicles observed were late-model cars with out-of-state license plates driven by black males, yet 80 percent of the arrestees on that stretch of highway fit this profile (Belkin 1990). The drivers who are pulled over for speeding assume the status of deviant; those not stopped assume the status of secret deviant. As a final example consider the different penalties leveled against persons possessing two forms of cocaine: powder and crack.

http://www.acsp.uic.edu/lib/ussc/chapter1.htm
Document: Background and Methodology

Q: What penalties and sentencing guidelines are associated with powder versus crack cocaine?

Secret deviants are people who have broken the rules but whose violation goes unnoticed or, if it is noticed, goes unpunished. Becker (1963) maintained that "no one really knows how much of this phenomenon exists," but he is convinced that the "amount is very sizable, much more so than we are apt to think" (p. 20). The most recent U.S. Bureau of Justice survey of crime victims shows that many victims do not report crimes. According to the findings about 36 percent of all crimes and 42 percent of all violent crimes were reported to law enforcement officials in 1994. For a summary of findings go to a Bureau of Justice Statistics press release.

http://www.ojp.usdoj.gov/bjs/cvict.htm

Document: Bureau of Justice Statistics Crime and Victims Statistics

Q: Does the information about victim characteristics, offender characteristics, and type of crime offer clues for why some victims do not report incidents to the police?

The Falsely Accused

The **falsely accused** are people who have not broken the rules but who are treated as suspects and/or as if they have done so. Like pure deviants the falsely accused are labeled as outsiders. As with the phenomenon of secret deviants, no one knows how often people are falsely suspected or accused, but it probably happens more often than we think. In any case the status of *suspect* or *accused* often lingers even if the person is not charged or is cleared of all charges. For example, a sampling of records kept by drug enforcement officials at three major airports (Houston, New York, and Miami) shows that on average 50 percent of all people detained and taken in for X-rays were found not to have swallowed drug-filled pouches as suspected. (The pouches are usually condoms sealed with tape, from which smugglers can retrieve drugs after the pouches pass through their system.) This figure, of course, varies according to the individual airport. At the Houston International Airport, for example, a sampling of records showed that 4 out of 60 persons X-rayed were guilty (Belkin 1990).

The falsely accused usually are persecuted for what they are or because they are visible, not for what they supposedly did. A substantial number of people who are stopped by drug enforcement agents are detained because they fit a profile of a drug courier and a drug swallower. The profile includes the following traits: having dark skin and/or a foreign appearance, being obviously in a hurry, having an exotic hairstyle or wearing brightly colored clothing, purchasing a one-way ticket, paying cash for a ticket, changing flights at the last minute, flying to or from Detroit, Miami, or another large city, and flying to or from the Caribbean, Colombia, Jamaica, or other known drug-supplying regions (Belkin 1990).

The Stuart murder that took place in Boston in October 1989 represents another such case. Charles Stuart murdered his pregnant wife, Carole, and then deliberately and severely wounded himself. At that point he made a "frantic and heart-wrenching call" (Fox and Levin 1990, p. 66) for help to the state police. Stuart told police that he and his wife had been robbed and shot by an African-American man. The police then proceeded to question and round up African Americans in the neighborhood, one of whom Stuart identified as the killer. Later Stuart committed suicide upon learning that investigators were onto the real story: Stuart had murdered his wife in an attempt to collect her life insurance and had inflicted injuries on himself to make his story more believable. Such events as the Stuart murder lead us to ask a larger question: Under which circumstances are people most likely to be falsely accused?

Sociologist Kai Erikson (1966) identified a particular situation in which people are likely to be falsely accused of a crime: when the well-being of a country or a group is threatened. The threat can take the form of an economic crisis (depression or recession), a moral crisis (family breakdown, for example), a health crisis (AIDS, for example), or a national security crisis (such as war). At such times people need to define a seemingly clear source of the threat, because identifying the threat gives an illusion of control. When a catastrophe occurs, people look for someone or some group to blame. The person or group blamed likely is at best indirectly responsible, is in the wrong place at the wrong time, and/or is viewed as different.

This defining activity can take the form of a **witch-hunt,** a campaign against allegedly subversive elements with the purpose of investigating and correcting behavior that undermines a group or a country. In actuality a witch-hunt may not accomplish this goal because the real cause of the problem is often complex and may lie beyond the purview of the targeted person or group. Often the people who are defined as the problem are not in fact responsible but are intentionally set up as scapegoats. And sometimes the pettiest and most insignificant acts are classified as crimes against a group or country in order to divert the public's attention from the shortcomings of those in power, to unite the public behind a cause, or to divert people's attention from a disruptive event such as massive job layoffs. With regard to the 1993 debate over whether gays should be allowed to serve in the U.S. military, we might ask, Is it a coincidence that gays became the focus of media attention precisely when the Clinton administration and Congress were planning such large cuts in the defense budget?

The internment of more than 110,000 people of Japanese descent (80 percent of whom were American citizens) living on the West Coast of the United States during World War II is another example of a situation in which a group was targeted in conjunction with a crisis. Japanese Americans were forced from their homes and shipped to desert prison camps surrounded by barbed wire and guarded with machine guns (Kometani 1987). There was no evidence of anti-American activity on the part of Japanese Americans. Yet the wartime hysteria, combined with long-standing prejudices, led to the confinement of men, women, and children in concentration camps. Ironically the people convicted of spying for Japan during the war were Caucasian. Explore the Japanese American internment page for background on this historical event (**http://www.mit.edu:8001/people/cyu/Public/internment/main.html**), but pay special attention to two documents, "Special Interests and the Internment" and "European American Internment."

http://www.freedom-server.co.uk/Japanese-Internment.html
Document: The Freedom Server-Japanese Internment

http://www.netzonc.com/~adjacobs/
Document: European American Internment

: What was the main reason for the Japanese internment during World War II? Explain your answer. Were the Japanese the only ethnic groups interned during World War II? Explain.

Another example of false accusation took place after the war. In the late 1940s and 1950s, hundreds of Americans were accused of being communists. They were placed under investigation and, in some cases, on trial without sufficient evidence to support the charges. The existence of the falsely accused underscores the fact that the study of deviance must go beyond looking at people identified or labeled as rule breakers. The roles of rule makers and rule enforcers must be examined as well.

Rule Makers and Rule Enforcers

Howard Becker (1973) recommends that researchers pay particular attention to who the rule makers and rule enforcers are and to how they achieve power and then use it to define how others "will be regarded, understood, and treated" (p. 204). This topic, of course, interests not only labeling theorists but conflict theorists as well. According to conflict theorists those with the most wealth, power, and authority have the capacity to create laws and crime-stopping and -monitoring institutions. Consequently we should not be surprised to learn that law enforcement efforts focus disproportionately on crimes committed by the poor and by other powerless groups rather than on those committed by the wealthy and other politically powerful groups. The work of Amnesty International reminds us that some organizations need to monitor governments, specifically government officials who have access to and power over a country's mechanisms of social control.

http://www.io.org/amnesty/
Document: Amnesty International On-Line

: What are the goals of Amnesty International?

This disproportionate law enforcement focus on the less powerful gives the widespread impression that the poor, the uneducated, and minority group members are more prone to criminal behavior than are people in the middle and upper classes, the educated, majority group members, and the politically connected. Crime exists in all social classes, but the types of crimes, the extent to which the laws are enforced, the degree of access to legal aid, and the power to shape laws to one's advantage vary across class lines (Chambliss 1974). In the United States, for example, police efforts are directed at controlling crimes against individual life and property (such as robbery, assault, homicide, and rape) rather than against white-collar and corporate crime.

White-collar crime consists of "crimes committed by persons of respectability and high social status in the course of their occupations" (Sutherland and Cressey 1978, p. 44). **Corporate crime** is crime committed by a corporation as it competes with other companies for market share and profits. Usually white-collar and corporate crimes—such as the manufacturing and marketing of unsafe products, unlawful disposal of hazardous waste, tax evasion, and money laundering—are handled not by the police, but by regulatory agencies that have little power and minimal staff. In "Traditional Policing and Environmental Enforcement" Wayne Brewer writes about a relatively new policing responsibility—environmental crime enforcement.

gopher://justice2.usdoj.gov/00/fbi/May95/may3.txt
Document: Traditional Policing and Environmental Enforcement

Q: In the area of environmental pollution, why has there been a shift from civil law enforcement to criminal law enforcement? Who are potential polluters?

It is easier for white-collar offenders to escape punishment because they are part of the system: they occupy positions in organizations that permit them to carry out illegal activities discreetly. In addition white-collar and corporate crime "is directed against impersonal—and often vaguely defined—entities such as the tax system, the physical environment, competitive conditions in the market economy, etc." (National Council for Crime Prevention in Sweden 1985, p. 13). These crimes are without victims in the usual sense because they are "seldom directed against a particular person who goes to the police and reports an offense" (p. 13).

The Constructionist Approach

The **constructionist approach** focuses on the process by which specific groups (for example, illegal immigrants or homosexuals), activities (child abuse or drug use), conditions (teenage pregnancy, infertility, pollution), or artifacts (song lyrics, guns, art, eyeglasses) become defined as problems. In particular constructionists examine the claims-making activities that underlie this process. Claims-making activities include "demanding services, filling out forms, lodging complaints, filing lawsuits, calling press conferences, writing letters of protest, passing resolutions, publishing exposes, placing ads in newspapers, supporting or opposing some governmental practice or policy, setting up picket lines or boycotts" (Spector and Kitsuse 1977, p. 79). Claims-making activities also include setting up sites on the internet. One example of such sites is the Washington Coalition to Abolish the Death Penalty. **Claims makers** are people who articulate and promote claims and who tend to gain in some way if the targeted audience accepts their claims as true. The gain may be a feeling that justice has been served, monetary rewards, notoriety, peace of mind, and so on. Return to the document "The Advertising of Installment Credit."

http://www.lib.virginia.edu/journals/EH/EH37/Murphy.html

Document: The Advertising of Installment Plans

Q: What strategies did advertisers use to convince people that purchasing products on installment credit was not a sign of moral degeneracy?

Claims Makers

Claims makers include but are not limited to victims of discrimination, government officials, professionals (medical doctors, scientists, professors), and pressure groups (any group that seeks to influence public opinion and government decision makers in order to advance or protect its interests). Many claims-making organizations use the internet in this way. Some examples are the AIDS Quilt Organization, Fathers' Rights and Equality Exchange, the American Medical Association, the National Rifle Association, and Greenpeace.

http://www.aidsquilt.org

http://www.vix.com/free/index.html

http://www.ama-assn.org/what_new/what_new.htm

http://www.NRA.org

http://www.greenpeace.org

Q: Select two organizations from this list. Give an example of a claims-making activity each employs.

The success of a claims-making campaign depends on a number of factors, including access to the media, available resources, and the claims maker's status and skills at fund-raising, promotion, and organization (Best 1989). According to sociologist Joel Best (1989), when constructionists study the process by which a group or behavior is defined as problematic to society, they focus on who makes the claims, whose claims are heard, and how audiences respond. Constructionists are guided by one or more of the following questions: What kinds of claims are made about the problem? Who makes the claims? Which claims are heard? Why is the claim made when it is? What are the responses to the claim? Is there evidence that the claims maker has misrepresented or inaccurately characterized the situation? In answering this last question, constructionists examine how claims makers characterize a condition. Specifically they pay attention to any labels that claims makers attach to a condition, the examples they use to illustrate the nature of the problem, and their orientation toward the problem (describing it as a medical, moral, genetic, educational, or character problem).

Labels, examples, and orientation are important because they tend to evoke a particular cause for and solution to a problem (Best 1989). For example, to call AIDS a moral problem is to locate its cause in the goodness or badness of human action and to suggest that the solution depends on changing evil ways. To call it a medical problem is to locate its cause in the biological workings of the body and to suggest that the solution rests with a drug, a vaccine, or surgery. Similarly a claims maker who uses the example of a promiscuous homosexual male to illustrate the nature of the AIDS problem sends a much different message about AIDS than does a claims maker who uses hemophiliacs or HIV-infected children as examples. As another example consider the claims making associated with the death penalty.

http://www.gbiz.com/odell/

Document: Innocent Man on Death Row

Q: This site presents the case of Joseph Roger O'Dell, who is portrayed as an innocent man on death row. Review the information presented. Which material might make you believe the claim is credible? What evidence makes you suspicious of the claim? How would you verify the information posted at the site? On the basis of this information, would you add your name to the support group? Write O'Dell? Send e-mail to the governor of Virginia?

To better understand the constructionist approach, let's examine the accuracy of claims that alcohol and cigarettes are legal because they are less dangerous substances than cocaine and other now-illegal drugs. This claim is one reason that law enforcement officials generate profiles of people likely to be drug carriers and drug swallowers but not people likely to be tobacco or alcohol vendors.

Claims About the Dangers of Legal and Illegal Substances

In his 1990 article "Taxonomy as Politics: The Harm of False Classification," biologist Stephen Jay Gould describes a war of words between two claims makers—former Surgeon General C. Everett Koop and Illinois Representative Terry Bruce—which began after Koop announced that nicotine was no less addicting than heroin or cocaine: "Representative Terry Bruce (D-IL) challenged [Koop's] assertion by arguing that smokers are not breaking into liquor stores late at night to get money to buy a pack of cigarettes." Koop properly replied that the only difference resides in social definition as legal or illegal: "You take cigarettes off the streets and people will be breaking into liquor stores. I think one of the things that many people confuse is the behavior of cocaine and heroin addicts when they are deprived of the drug. That's the difference between a licit and an illicit drug. Tobacco is perfectly legal. You can get it whenever you want to satisfy the craving" (Gould 1990, p. 75).

Koop's hypothesis that cigarette smokers deprived of their drug would behave like cocaine and heroin addicts is supported by events in Italy, where the government has a monopoly over the manufacture and distribution of cigarettes. In 1992 approximately 13 million smokers were prevented from purchasing cigarettes because of a strike by the workers who distribute them. The forced abstinence resulted in the following behaviors: (1) panic buying and hoarding of the cigarettes still on the shelves when the strike began, (2) a run on nicotine gum and patches, (3) a sevenfold increase in traffic into Switzerland, (4) an underground market in cigarettes, and (5) robbery of cigarette vendors at gunpoint (Cowell 1992).

Gould (1990) argues that illegal drugs such as cocaine should be legalized and controlled, in the same way as alcohol and tobacco, and their use strongly discouraged. He would even accept, albeit reluctantly, a policy in which alcohol and tobacco had the same illegal status as drugs such as cocaine. Gould cannot find any rationale, however, for "an absurd dichotomy (legal versus illegal) that encourages us to view one class of substances with ultimate horror as preeminent scourges of life . . . while the two most dangerous and life-destroying substances by far, alcohol and tobacco, form a second class advertised in neon on every street corner of urban America" (p. 74). Gould's claim is supported by data from the National Institute on Drug Abuse, an agency of the U.S. government: for every one cocaine-related death in 1987 there were 300 tobacco-related and 100 alcohol-related deaths (Reinarman and Levine 1989). Gould (1990) does not deny that all such drugs mentioned are dangerous, but he finds that the illegal status for one set of substances and the legal status for another "permit a majority of Americans to live well enough with one, while forcing a minority to murder and die for the other" (p. 75). Gould argues that illegal status causes more social and medical problems than it solves and that the illegal status itself causes crime. For example, the responses to the prohibition of alcohol in the United States between 1920 and 1933 are remarkably similar to those associated with cocaine's illegal status. These responses include murder, payoffs, bribes, and international intrigue. There is one argument, seldom articulated, in favor of keeping drugs like cocaine illegal: prohibition might prevent the cocaine problem from reaching the tragic dimensions of the alcohol and tobacco problems.

Go to Chapter 8 of the "Cocaine and Federal Sentencing Policy" document. Read the sections "Congressional Concerns Leading to Powder Cocaine/Crack Cocaine Differential" and "Cocaine and Addiction."

http://www.acsp.uic.edu/lib/ussc/chapter8.htm

Document: Findings, Discussions, and Recommendations

Q: What is the logic underlying the different penalty structure for crack versus powder cocaine? Is that logic justified by research findings? Explain

We turn now to the theory of structural strain, which gives us insights into how society creates deviance by virtue of its valued goals and the opportunities it offers people to achieve those goals.

Structural Strain Theory

Robert K. Merton's theory of structural strain takes into account three factors: (1) culturally valued goals defined as legitimate for all members of society, (2) norms that specify the legitimate means of achieving these goals, and (3) the actual number of legitimate opportunities available to people to achieve the goals. According to Merton (1957) **structural strain,** or **anomie,** is any situation in which (1) the valued goals have unclear limits (that is, people are unsure whether they have achieved them), (2) people are unsure whether the legitimate means that society provides will lead to the valued goals, and (3) legitimate opportunities for meeting the goals are closed to a significant portion of the population. The rate of deviance is likely to be high under any one of these conditions. Merton used the United States, a country where all three conditions exist, to show the relationship between structural strain and deviance.

In the United States most people place a high value on the goal of economic affluence and social mobility. In addition Americans tend to believe that all people, regardless of the circumstances into which they are born, can achieve monetary success. Such a viewpoint suggests that success or failure results from personal qualities and that persons who fail have only themselves to blame.

Merton did not believe that the same proportion of people in all social classes accept the cultural goal of monetary success, but he did think that a significant number of people across all classes do so. He argued that "Americans are bombarded on every side" with the message that this goal is achievable, "even in the face of repeated frustration" (p. 137).

According to Merton considerable structural strain exists in the United States because this culturally valued goal has no clear limits. He believed there is no point at which people can say they have achieved monetary success: "At each income level . . . Americans want just about twenty-five percent more (but of course this 'just a bit more' continues to operate once it is obtained)" (p. 136).

Merton also maintained that structural strain exists in the United States because the legitimate means of achieving affluence and mobility are not entirely clear; it is the individual's task to choose a path that leads to success. That path might involve education, hard work, or natural talent. The problem is that education, hard work, and talent do not guarantee success. With regard to education, for example, many Americans believe that the diploma (especially the college diploma) in itself entitles them to a high-paying job. In reality, however, the diploma is only one component of many needed to achieve success.

Finally structural strain exists in the United States because too few legitimate opportunities are available to achieve desired goals. From New York City arrest records, which give us information on the characteristics of crack cocaine distributors who have been caught, we see that the most economically vulnerable individuals are those with the least access to legitimate economic opportunities.

http://www.acsp.uic.edu/lib/ussc/chapter4.htm#sectg

Document: The Role of Youth and Women in Crack Cocaine Distribution

Q: Which groups are identified as particularly vulnerable to the allure of profits to be made from drug distribution? Explain. What role do they play in the structure of this illegal activity?

Although all Americans are supposed to seek financial success, they cannot all expect to achieve it legitimately; the opportunities to do so are closed to many people, especially those in the lower classes. For example, many African-American males living in poverty believe that one seemingly sure way to achieve success is through sports. The opportunities narrow rapidly, however, as an athlete advances: fewer than 2 percent of college basketball players, for instance, even have a chance at the professional ranks. Merton believed that people respond in identifiable ways to structural strain and that their response involves some combination of acceptance and rejection of the valued goals and means. Merton identified the following five responses:

- **Conformity** is not a deviant response; it is the acceptance of the cultural goals and the pursuit of these goals through legitimate means.
- **Innovation** involves the acceptance of the cultural goals but the rejection of legitimate means to obtain these goals. For the innovator success is equated with winning the game rather than playing by the rules of the game. After all, money may be used to purchase the same goods and services whether it was acquired legally or illegally. Merton argued that when the life circumstances of the middle and upper classes are compared with those of the lower classes, the latter clearly are under the greatest pressure to innovate, although no evidence suggests that they do so, especially in light of the widespread prevalence of white-collar crime.
- **Ritualism** involves the rejection of cultural goals but a rigid adherence to the legitimate means of attaining those goals. This response is the opposite of innovation; the game is played according to the rules despite defeat. Merton maintained that this response can be a reaction to the status anxiety that accompanies the ceaseless competitive struggle to stay on top to get ahead. It finds expression in the cliches "Don't aim high and you won't be disappointed" and "I'm not sticking my neck out." Ritualism can also be the response of people who have few employment opportunities open to them. If one wonders how it is possible for a ritualist to be defined as deviant, consider the case of a college graduate who can find only a job bagging groceries at $4.50 an hour. Most people react as if this person is a failure even though he or she may be working full-time.
- **Retreatism** involves the rejection of both cultural goals and the means of achieving these goals. The people who respond in this way have not succeeded by either legitimate or illegitimate means and thus have resigned from society. According to Merton retreatists are the true aliens or the socially disinherited—the outcasts, vagrants, vagabonds, tramps, drunks, and addicts, "in the society but not of it."
- **Rebellion** involves the full or partial rejection of both goals and means and the introduction of a new set of goals and means. When this response is confined to a small segment of society, it provides the potential for subgroups as diverse as street gangs and the Old Order Amish. When rebellion is the response of a large number of people who wish to reshape the entire structure of society, the potential for a revolution is great.

The deviant responses to structural strain leave us with the question of how people learn such responses. One answer comes from the work of sociologists Edwin H. Sutherland and Donald R. Cressey, who offer us the theory of differential association.

Differential Association Theory

Sutherland and Cressey advance a theory of socialization called **differential association** to explain how deviant behavior, especially delinquent behavior, is learned. This theory centers on the idea that "when persons become criminal, they do so because of contacts with criminal patterns and also because of isolation from anticriminal patterns" (1978, p. 78). The contacts take place within **deviant subcultures,** groups that are part of the larger society but whose members adhere to norms and values that favor violation of the larger society's laws. That is, people learn techniques of committing crime from close association and interaction with people who engage in and approve of criminal behaviors.

Sutherland and Cressey maintain that impersonal forms of communication such as television, movies, and newspapers play a relatively small role in the genesis of criminal behavior. If we accept the premise that criminal behavior is learned, then criminals constitute a special type of conformist. That is, they conform to the norms of the group with which they associate. The theory of differential association does not explain how a person makes contact with deviant subcultures in the first place (unless, of course, the person is born into such a subculture). Once contact is made, however, the individual learns the subculture's rules for behavior in the same way that all behavior is learned. One deviant subculture in the United States that receives considerable media attention is youth gangs. "Mothers Against Gangs" is one site on the internet that addresses this problem. This organization has defined a plan for rescuing youths from gangs. The character of this plan suggests that deviance is something that is learned and consequently can be unlearned.

http://www.winternet.com/~jannmart/nkcmag.html
Document: Mothers Against Gangs

Q: Is deviant behavior something youths learn by association with a deviant subculture alone? Explain. What is the socialization component of the Mothers Against Gangs program? How would you reevaluate Sutherland and Cressey's theory in light of the information presented?

Many people take issue with Sutherland and Cressey's hypothesis that impersonal forms of communication play only a limited role in the genesis of deviant behaviors. In fact many claim that exposure to certain kinds of music (heavy metal), scenes (violent sexual), and writings (on the internet) have antisocial effects. The National Television Violence Study, the product of the joint effort of media scholars from a wide range of American universities and the nonprofit organization Mediascope, represent one attempt to assess the effects of TV violence on those who view it.

http://www.mediascope.org/mediascope/
Document: Mediascope

 Q: What were the study's major findings? How would you reevaluate Sutherland and Cressey's ideas in light of these findings?

Discussion Question

Based on the information presented in this chapter, how would you describe the sociological contribution to the study of deviance?

Additional Reading

For more on

- The relative nature of deviance
  http://www.acsp.uic.edu/lib/ussc/chapter1.htm
  Document: United States Sentencing Commission—Cocaine Federal Sentencing Policy

- Examples of claims-making activities
  http://tt.dx.com/tobacco/Misc/ash.kids.clinton.contest.html
  Document: Hey, Kids, Help Make History and Maybe Even Win $1000

- The falsely accused (witch-hunts)
  http://liquid2-sun.mit.edu/FAQ.html
  Document: Witch Hunt Frequently Asked Questions

- Plagues and the falsely accused
  http://jefferson.village.virginia.edu/osheim/plaguein.html
  Document: Plagues and Public Health in Renaissance Europe

- Rule makers/rule enforcers
  http://www.ojp.usdoj.gov/bjs
  Document: Bureau of Justice Statistics

  http://gopher.usdoj.gov/bop/facts.html
  Document: U.S. Bureau of Prisons Fact Sheet

  http://www.law.cornell.edu/supct/supct.table.html
  Document: U.S. Supreme Court Decisions

- Social control in selected countries
  http://lcweb2.loc.gov/frd/country.html
  Document: Army Area Handbook Access

  Select:
  China/Ch.13.00, Criminal Justice and Public Security
  Egypt/Ch.5.08, Crime and Punishment
  Japan/Ch.8.03, Public Order and Internal Security
  Indonesia/Ch.5.07, Criminal Justice System
  Israel/Ch.5.09 and /Ch.5.11, The Israel Police/Criminal Justice
  Philippines/Ch.5.05, Public Order and Internal Security
  Singapore/Ch.5.02, Public Order and Internal Security
  Somalia/Ch.5.03, Internal Security Concerns
  South Korea/Ch.5.07, Internal Security Concerns
  Yugoslavia/Ch.5.08, Internal Security Concerns

10 Social Stratification

Social stratification is the systematic process by which people in societies are ranked on a scale of social worth. Sociologists are interested in how the categories in which people are placed affect their **life chances**, a critical set of potential social advantages including "everything from the chance to stay alive during the first year after birth to the chance to view fine art, the chance to remain healthy and grow tall, and if sick to get well again quickly, the chance to avoid becoming a juvenile delinquent—and very crucially, the chance to complete an intermediary or higher educational grade" (Gerth and Mills 1954, p. 313). In this chapter we focus on both the sources and the functions of social stratification. Sociologists who study social stratification are interested in what criteria are used to categorize people and whether these criteria are something that people can achieve or something over which they have no control.

Achieved and Ascribed Characteristics

Almost any criterion can be used (and at one time or another has been used) to categorize people: hair color and texture, eye color, physical attractiveness, weight, height, occupation, age, grades in school, test scores, and many others. Two major kinds of criteria are used to categorize people: ascribed and achieved characteristics. **Ascribed characteristics** are attributes that people (1) have at birth (such as skin color, sex, or hair color), (2) develop over time (baldness, gray hair, wrinkles, retirement, reproductive capacity), or (3) possess through no effort or fault of their own (national origin or religious affiliation that was "inherited" from parents). **Achieved characteristics** are acquired through some combination of choice, effort, and ability. In other words people must act in some way to acquire these attributes. Some examples of achieved characteristics include occupation, marital status, level of education, and income.

Ascribed and achieved characteristics seem clearly distinguishable, but such is not always the case. Debate continues, for example, over whether homosexuality is achieved or ascribed. Some people argue that homosexuality is genetically based behavior over which people have no control; others insist that it is learned and ultimately a matter of choice. Sociologists are interested in those ascribed and achieved characteristics that take on social significance and **status value:** when that occurs, persons who possess one feature of a characteristic (white skin versus brown skin, blond hair versus dark hair, physician versus garbage collector) are regarded and treated as more valuable and more worthy than persons who possess other features (Ridgeway 1991). Sociologists are particularly interested in such situations because ascribed characteristics are attributes over which people have no control.

As one example of the importance of ascribed characteristics to life chances, consider the findings of an Urban Institute study. This study found that compared to their white counterparts, African-American male, but especially Hispanic, job applicants encounter greater discrimination during the application process.

http://www.cdinet.com/Rockefeller/Briefs/brief8.html
Document: Racial Discrimination in Hiring

What factors contributed to the high rate of discrimination against Hispanic applicants? (*Note:* To find out how the study was conducted, go to the end of the document and highlight "About the research.")

A classic demonstration of how ascribed characteristics can affect people's access to valued resources involves a third-grade class in Riceville, Iowa. In 1970 teacher Jane Elliot conducted an experiment in which she divided her students into two groups according to a physical attribute—eye color—and rewarded them accordingly. She did this to show her class how easy it is for people (1) to assign social worth, (2) to explain behavior in terms of an ascribed characteristic such as eye color, and (3) to build a reward system around this seemingly insignificant physical attribute. The following excerpt from the transcript of the program "A Class Divided" ("Frontline" 1985) shows how Elliot established the ground rules for the classroom experiment:

Elliot: It might be interesting to judge people today by the color of their eyes . . . would you like to try this?

Children: Yeah!

Elliot: Sounds like fun, doesn't it? Since I'm the teacher and I have blue eyes, I think maybe the blue-eyed people should be on the top the first day . . . I mean the blue-eyed people are the better people in this room. Oh yes they are, the blue-eyed people are smarter than brown-eyed people.

Brian: My dad isn't that . . . stupid.

Elliot: Is your dad brown-eyed?

Brian: Yeah.

Elliot: One day you came to school and you told us that he kicked you.

Brian: He did.

Elliot: Do you think a blue-eyed father would kick his son? My dad's blue-eyed, he's never kicked me. Ray's dad is blue-eyed, he's never kicked him. This is a fact. Blue-eyed people are better than brown-eyed people. Are you brown-eyed or blue-eyed?

Brian: Blue.

Elliot: Why are you shaking your head?

Brian: I don't know.

Elliot: The blue-eyed people get five extra minutes of recess, while the brown-eyed people have to stay in. . . . The brown-eyed people do not get to use the drinking fountain. You'll have to use the paper cups. You brown-eyed people are not to play with the blue-eyed people on the playground, because you are not as good as blue-eyed people. The brown-eyed people in this room today are going to wear collars. So that we can tell from a distance what colors your eyes are. [Now] on page 127—one hundred twenty seven. Is everybody ready? Everyone but Laurie. Ready, Laurie?

Child: She's a brown-eye.

Elliot: She's a brown-eye. You'll begin to notice today that we spend a great deal of time waiting for brown-eyed people. (pp. 3–5)

Once Elliot set the rules, the blue-eyed children eagerly accepted and reinforced them. During recess the children took to calling each other by their eye colors, and some brown-eyed children got into fights with blue-eyed children who called them "brown-eye." The teacher observed that these "marvelous, cooperative, wonderful, thoughtful children [turned into] nasty, vicious, discriminating little third-graders in a space of fifteen minutes" (p. 7).

This experiment illustrates on a small scale how categories and their status value are reflected in the distribution of valued resources. Because category determines status and is related to life chances, it is essential that we examine the shortcomings of classification schemes.

One major shortcoming of any classification scheme is that not all people fit neatly into categories designated as important. For example, the third-grade teacher whose experiment we described divided her students into just two categories—the blue-eyed and the brown-eyed. Such a classification scheme, however, does not accommodate people with green eyes, hazel eyes, gray eyes, or mixed-color eyes (one blue and one brown). This shortcoming in the classification scheme leaves us unsure about what to do with someone who does not fit into any of the designated categories. Typically, when people do not fit a category, others find ways to make them fit. In the third-grade class brown-eyed people were required to wear collars to make it absolutely clear who belonged to that category.

Such a strategy is not unique to that third-grade classroom experiment. Strategies to make people fit into categories are a part of all classification schemes—even those for sex, age, and race. For example, there is no perfect dividing line to separate people into the categories of "male" and "female." A small but significant number of babies are born hermaphrodites; that is they have both male and female reproductive organs. In the United States parents must choose what sex to put on a hermaphrodite child's birth certificate. Given the importance of sex categories, physicians tell parents that it is in the child's best interest to undergo a sex-clarifying operation. Because most people do not question their society's category system, it is difficult to see those categories as social constructions that everyone is more or less made to fit in and/or that the majority of people work to fit into. In order to demonstrate the many shortcomings associated with classification schemes, we will critique one major form of categorizing people—that of race.

Classifying People into Racial Categories

Most people in the United States equate **race** with groups of people who possess certain distinctive and conspicuous physical traits. Moreover racial categories represent "natural, physical divisions among humans that are hereditary, reflected in morphology, and roughly but correctly captured by terms like Black, White, and Asian (or Caucasoid, Mongoloid, or Negroid)" (Lopez 1994, p. 6). This three-category classification scheme, however, has many shortcomings, which immediately becomes evident when we imagine using it to classify the more than 5.6 billion people in the world. If we attempt this task, we would soon learn that three categories are not enough. This shortage of categories has a historical and social basis:

> The idea that there exist three races and that these races are Caucasoid, Negroid, and Mongoloid is rooted in the European imaginations of the Middle Ages, which encompassed only Europe, Africa, and the Near East. . . . The peoples of the American continents, the Indian subcontinent, East Asia, Southeast Asia and Oceania—living outside the imagination of Europe—are excluded for social and political reasons, not for scientific ones. Nevertheless, the history of science has been the failed efforts to justify these social beliefs. (LOPEZ 1994, P. 15)

Adding more categories, however, would not ease the task of classifying the world's billions of people because racial classification rests on the fallacy that clear-cut racial categories exist. Why is this a fallacy? First, many people do not fit clearly into a racial category because no sharp dividing line distinguishes, for example, black skin from white skin or curly hair from wavy hair. This lack of a clear line, however, has not discouraged people from trying to devise ways to make the line seem clear-cut. A hundred years ago in the United States, for example, there were churches that "had a pinewood slab on the outside door . . . and a fine toothed comb hanging on a string" (Angelou 1987, p. 2). People would go into the church if they were no darker than the pinewood and if they could run the comb through their hair without it snagging. At one time in South Africa, the state board that oversaw racial classification used a pencil test to classify individuals as white or black. If a pencil placed in the person's hair fell out, the person was classified as white (Finnegan 1986).

A second problem with the idea of clear-cut racial categories is that many millions of people have mixed ancestry and possess the physical traits of more than one race. Thus boundaries between races can never be fixed and definite if only because males and females of any race can produce offspring together.

A third shortcoming in systems of racial classification is that racial categories and guidelines for placing people in racial categories are often vague, contradictory, unevenly applied, and thus subject to change. To illustrate, consider that for the 1990 census, coders were instructed to classify as "white" those who said they were "white-black" but to classify as "black" those who said they were "black-white." Likewise the U.S. National Center for Health Statistics has changed its guidelines for recording race and ethnicity on birth and death certificates. Before 1989, for example, a child was designated as white if both parents were white; if only one parent was white, the child was classified according to the race of the nonwhite parent; if the parents were of different nonwhite races, the child was assigned the father's race; and if one parent's race was unknown, the infant was assigned the race of the parent whose race was known. After 1989 the rules for classifying newborns changed: now an infant's race is the same as the mother's (Lock 1993)—as if identifying the mother's race would present no challenges.

Finally, in trying to classify people by race, we would find a tremendous amount of variation even among people who possess all the traits designated as belonging to a particular race. For example, people classified as Mongoloid include Chinese, Japanese, Mayalans, Siberians, Eskimos, and Native Americans.

For more examples of the pattern of lumping many groups into one racial category, go to the U.S. Census Bureau for the official statement about which national and ethnic groups belong in each of the four racial categories. Note that the information on race is part of a larger technical document. Stop reading when the document shifts to a description of the variable "reference week."

http://www.census.gov/td/stf3/append_b.html#RACE

Document: Definitions of Subject Characteristics: Race

Q: What are two examples of how the government transforms many groups into one racial, ethnic, or ancestral category?

Sociologists are particularly interested in any classification scheme that incorporates the belief that (1) certain important abilities (such as athletic talent or intelligence), social traits (criminal tendencies, aggressiveness), and cultural practices (dress, language) are passed on genetically or (2) some categories of people are inferior to others by virtue of genetic traits and therefore should receive less wealth, income, and other socially valued items. Sociologists are interested in classification schemes because they have enormous consequences for society and for the relationships among people. One important dimension of stratification systems is the extent to which people "are treated as members of a category, irrespective of their individual merits" (Wirth [1945]

1985, p. 310). In this vein we will examine how "open" or "closed" a stratification system is.

"Open" and "Closed" Stratification Systems

Despite wide variations among forms of stratification systems, each falls somewhere on a continuum between two extremes: a **caste system** (or "closed" system), in which people are ranked on the basis of traits over which they have no control, and a **class system** (or "open" system), in which people are ranked on the basis of merit, talent, ability, or past performance. There are three main characteristics that distinguish a caste system from a class system of stratification: (1) the rigidity of the system (how difficult it is for people to change their category), (2) the relative importance of ascribed and achieved characteristics in determining people's life chances, and (3) the extent to which there are restrictions on social interaction between people in different categories. Caste systems are considered closed because of how rigid they are—rigid in the sense that ascribed characteristics determine life chances and that social interaction among people in different categories is restricted. In comparison class systems are open in the sense that no barriers exist to social interaction among people in different categories.

Class and caste systems are ideal types. An ideal type is ideal not in the sense of having desirable characteristics, but as a standard against which "real" cases can be compared. Actual stratification systems depart in some ways from the ideal types.

Caste Systems

When people hear the term *caste*, what usually comes to mind is India and its caste system, especially as it existed before World War II. India's caste system, now outlawed, was a strict division of people into four basic categories with 1,000 subdivisions. The strict nature of the caste system is shown by some of the rules that specified the physical distances that were to be maintained between people of different castes: "a Nayar must keep 7 feet (2.13m) from a Nambudiri Brahmin, an Iravan must keep 32 feet (9.75m), a Cheruman 64 feet (19.5m), and a Nyadi from 74 to 124 feet (22.6 to 37.8m)" (Eiseley 1990, p. 896).

Most sociologists use the term *caste* not to refer to one specific system, but to designate any scheme of social stratification in which people are ranked on the basis of physical or cultural traits over which they have no control and that they usually cannot change. Whenever people are ranked on the basis of such traits, they are part of a caste system of stratification.

The rank of any given caste is reflected in the public esteem accorded those who belong to it, in the power wielded by members of that caste, and in the opportunities of members of that caste to acquire wealth and other valued items. People in lower castes are seen as innately inferior in intelligence, personality, morality, capability, ambition, and many other traits. Conversely people in higher castes consider themselves to be superior in such traits. Moreover caste distinctions are treated as if they were absolute: their significance is never doubted or questioned (especially by those who occupy higher castes), and they are viewed as unalterable and clear-cut (as if everyone is supposed to fit neatly into a category). Finally there are heavy restrictions on the social interaction between people in higher and lower castes. For example, marriage between people of different castes is forbidden.

Apartheid: A Caste System of Stratification

Apartheid (the Afrikaans word for "apartness") was practiced in South Africa for hundreds of years. But it did not become official policy until 1948, when the conservative white Nationalist party, led by D. F. Malan, won control of the government. Once in power, the Nationalists passed hundreds of laws and acts mandating racial separation in almost every area of life and severely restricting blacks' rights and opportunities. The result was legalized political, social, and economic domination by whites over nonwhites. Examples of apartheid laws include (1) the Separate Amenities Act of 1953 (repealed on October 16, 1990), which authorized the creation of separate and unequal (in quantity and quality) public facilities—parks, trains, swimming pools, libraries, hotels, restaurants, hospitals, waiting rooms, pay telephones, beaches, cemeteries, and so on—for whites and nonwhites; (2) the Land Acts of 1919 and 1936 (repealed on June 5, 1991), which reserved 85 percent of the land for 5 million whites (14 percent of the population) and the remaining 15 percent for the rest of the population; (3) the Group Areas Act of 1950 (repealed on June 5, 1991), which designated separate living areas for whites and nonwhites; and (4) the Population Registration Act of 1950 (repealed on June 18, 1991), which required that everyone in South Africa be classified by race and issued an identification card denoting his or her race.

It is virtually impossible to give an overview of the effects of apartheid, a system that has inflicted enduring misery, economic damage, and daily indignities on South Africa's nonwhite population (Wilson and Ramphele 1989). Although apartheid has been legally abolished, the challenge lies in dismantling its effects on relationships between whites and nonwhites and on life chances for nonwhites, as well as overcoming its economic legacy.

http://curry.edschool.virginia.edu/go/multicultural/edusa/Socioecon.html

Document: Socio-Economic Conditions of South African Children

Q: Look over the headlines of the documents in this site. Relative to the countries of the world, what general statement can we make about economic and social inequality in South Africa?

Class Systems

In class systems of stratification "people rise and fall on the strength of their abilities" (Yeutter 1992, p. A13). Class systems of stratification contain economic and occupational inequality, but that inequality is not systematic. In other words there is no connection between a person's sex, race, age, or ascribed characteristics and his or her life chances. A class system differs from a caste system in that life chances are connected to merit, talent, ability, and past performance and not to attributes over which people have no control, such as skin color. In class systems people assume they can achieve a desired level of education, income, and standard of living through hard work. Furthermore people can change their class position upward during their own lifetimes, and their children's class position can be different from (and ideally higher than) their own.

Movement from one class to another is termed *social mobility*. There are many kinds of social mobility in class systems. **Vertical mobility** exists if the change in class status corresponds to a gain or loss in rank or prestige, as when a college student becomes a physician or when a wage earner loses a job and goes on unemployment. A loss of rank is **downward mobility** while a gain in rank is **upward mobility. Intergenerational mobility** is a change in rank over two or more generations; **intragenerational mobility** is movement (upward or downward) during an individual's lifetime. For one rough indication of changes in social mobility, see data from the 1993 Annual Housing Survey, which asks people questions about why they move.

http://www.census.gov/ftp/pub/hhes/housing/ahs/tab2-11.html
Document: Why Move?

Q: Which statistics suggest a change in economic status, either upward or downward?

In contrast to caste systems, in class systems distinctions between classes are not always clear. In principle people can move from one class to another, but in reality it is often difficult to identify a person's class through observation alone. That is, a person may drive a BMW and wear designer clothes—symbols of solid middle-class status—but may work at a low-paying job, live in a low-rent apartment, and have exceeded his or her line of credit. In other words a person may have champagne tastes and a beer income. Or a person may change class position through marriage, graduation, inheritance, or job promotions but may not assume the lifestyle of that class immediately.

India-born novelist Bharti Mukherjee, who now lives in the United States, was asked, "What does America mean to you as an idea?" Her answer clarifies the differences between class and caste systems of stratification:

> What America offers me is romanticism and hope. I came out of a continent of cynicism and irony and despair. A traditional society where you are what you are, according to the family that you were born into, the caste, the class, the gender. Suddenly, I found myself in a country where—theoretically anyway—merit counts, where I could choose to discard that part of my history that I want, and invent a whole new history for myself....
>
> [America represents] that capacity to dream and then to try to pull it off, if you can. I think that the traditional societies in which people like me were born really do not allow the individual to dream. To dream big. (1990, PP. 3–4)

Is Mukherjee's vision of the United States accurate? Is the United States a class system?

Does the United States Have a Class System?

Before we answer this question, let us recall that in a true class system ascribed characteristics do not determine social class. Although considerable economic inequality may be present, equality of opportunity exists. On paper the United States has a class system. The Declaration of Independence, the Preamble to the Constitution, and the Bill of Rights assert that all human beings are created equal, not in innate qualities like intelligence or physical endowment, but in the human right to full equality: "the right to equitable access to justice and freedom, and opportunity, irrespective of [sex], race, or religion or ethnic origin" (Merton 1958, p. 189). If individuals differ in innate endowment, "they do so as individuals, not by virtue of their group memberships" (p. 190). The documents that support the "American creed," however, are ambiguous and subject to various legal interpretations. This feature makes them remarkable but frustrating documents of freedom—historically many injustices have been interpreted as agreeing with them in spirit.

However much Americans want to believe that they live in a true class system, evidence indicates otherwise. The chances of economic and occupational success are more often than not connected to forces over which people have little control—social background, ascribed characteristics, and massive restructuring of the economy. To put it bluntly, economic inequality follows a clear pattern in the United States, and some groups are affected more than others by major changes in the economy. Although the pattern of inequality in the United States is not anywhere near as systematic as that which existed under apartheid and which exists today in South Africa, it is striking nonetheless.

Income Inequality The U.S. Bureau of the Census classifies people according to 21 income categories (under $5,000, $5,000–9,999, and so on up to over $100,000 per year). One picture of inequality in the United States is the income distribution by household, which can be found at the following website:

http://www.census.gov/cdrom/lookup

Document: 1990 Census Lookup

 Follow these instructions: choose "STF3C-part 1" and hit return, stay with the default options, and then page back up and highlight "submit"; then hit return. Use the same default options, highlight "submit," and hit return. Next, page down to select Table P80 (from a list of more than 300 tables) and hit return. Now go up to the beginning of the document with the up arrow and highlight "submit." Stay with the default options and then "submit." Use this table to figure the percentage of households in each income category. When this survey was done, there were 91,993,582 households in the United States.

It is important to understand that income data has its flaws and that at best it gives us a rough picture of inequality.

http://www.census.gov/td/stf3/append_b.html#INCOME

Document: Income in 1989

Q: What are three limitations of the income data? Based on this information, how accurate do you think income data is?

The income distribution table shows that there is income inequality in the United States. The presence of inequality alone does not refute a claim that the United States possesses a class system of stratification. But when we examine income for whites, blacks, and Hispanics, it becomes clear that income is connected with race and ethnicity.

http://www.census.gov/cdrom/lookup

Document: 1990 Census Lookup

Q: Follow the instructions for creating Table P80 (on p. 195), but this time select Tables P8, P10, P82, and P83 and create income distribution tables for blacks, whites, and Hispanics. Which of the three kinds of households are disproportionately concentrated in lower-income categories? (*Note:* To answer this question, you will have to calculate the percentage of households in each income category.)

Such differences suggest that the United States is not a class system in the true sense of the word. If it were a true class system, the percentages of families in each income category would be the same across all three groups.

Income distribution is even more uneven when comparing types of households—couple-headed, male-headed (no wife), female-headed (no husband), and black and Hispanic female-headed households in particular. Furthermore occupational inequality is an inescapable feature of the American socioeconomic landscape.

Occupational Inequality Occupations are considered to be segregated according to ascribed characteristics when some occupations (librarian, secretary, physician, lawyer, chief executive officer) are filled primarily by people of a particular race, ethnicity, sex, or age. In the United States we find a large number of occupations in which men and women are overrepresented or underrepresented. Go to the document "Detailed Occupation by Sex." Here we find a list of job titles and the number of males and females employed in each occupation. The first column of figures represents males, and the second column females.

gopher://una.hh.lib.umich.edu/00/census/summaries/eeous

Document: Detailed Occupations by Sex

Q: Page down the list of occupations until you find the categories "health diagnosing occupations" and "health assessment and treatment occupations." Use the data to create a table that shows the percentage of men and women in each occupational category. Which occupations contain a disproportionate number of males? A disproportionate number of females?

Occupational segregation is a problem when certain groups are concentrated in the low-paying, low-ranking jobs. When we compare the median income of black men, black women, and white women who work year-round and full-time with that of their white male counterparts, the impact of occupational segregation becomes evident. These figures can be found on the final four pages of the document "20 Facts on Women Workers."

gopher://eng.hss.cmu.edu/ 00ftp%3aEnglish.Server%3aGender%3aFacts%20on%20Working%20Women

Document: 20 Facts on Women Workers

 Q: What percentage of the male median income do white women, black women, and black men earn?

The fact that income and occupation are connected to race and sex (ascribed characteristics) forces us to conclude that the United States is at best a mixture of class and caste systems of stratification.

Mixed Systems: Class and Caste

Systems of stratification are usually a combination of class and caste. In the United States virtually every occupation contains members of different ethnic, racial, age, and sex groups. At the same time, however, some groups such as women and blacks are severely overrepresented or underrepresented in some occupations. Moreover in the United States a person's ascribed characteristics can overshadow his or her achievements in such a way that "no amount of class mobility will exempt a person from the crucial implications of . . . birth" (Berreman 1972, p. 399). For example, even though a person holds a high-ranking occupation, others may question or overlook that person's talent, merit, and accomplishments based on skin color or sex. The interrelations between caste and class

> can be brought readily to mind by thinking of the relative advantages and disadvantages which accrue in Western class systems to persons who occupy such occupational statuses as judge, garbage man, stenographer, airline pilot, factory worker, priest, farmer, agricultural laborer, physician, nurse, big businessman, beggar, etc. The distinction between class and birth-ascribed stratification can be made clear if one imagines that he encounters two Americans, for example, in each of the above mentioned occupations, one of whom is white and one of whom is black. This

quite literally changes the complexion of the matter. . . . Obviously something significant has been added to the picture of stratification in these examples which is entirely missing in the first instance—something over which the individual generally has no control, which is determined at birth, which cannot be changed, which is shared by all those of like birth, which is crucial to social identity, and which virtually affects one's opportunities, rewards, and social roles. The new element is race (color), caste, ethnicity (religion, language, national origin), or sex. (BERREMAN 1972, PP. 385–86)

Filmmaker Spike Lee, historian John Hope Franklin, television broadcaster Carole Simpson, and former professional basketball player Isiah Thomas have made insightful comments on how their race affects people's definitions of what they do:

> I want to be known as a talented young filmmaker. That should be first. But the reality today is that no matter how successful you are, you're black first. (LEE 1989, P. 92)

> It's often assumed I'm a scholar of Afro-American history, but the fact is I haven't taught a course in Afro-American history in 30-some-odd years. They say I'm the author of 12 books on black history, when several of those books focus on whites. I'm called a leading black historian, never minding the fact that I've served as president of the American Historical Association, the Organization of American Historians, the Southern Historical Association, Phi Beta Kappa, and so on.
>
> The tragedy . . . is that black scholars so often have their specialties forced on them. My specialty is the history of the South, and that means I teach history of blacks and whites. (FRANKLIN 1990, P. 13)

> A radio station executive in Honolulu read my name tag and said, "Carole Simpson, ABC. What does that stand for? African Broadcast Company?" (HEARD ON NPR)

> When [Larry] Bird makes a great play, it's due to his thinking, and his work habits. It's all planned out by him. It's not the case for blacks. All we do is run and jump. We never practice or give a thought to how we play. It's like I came dribbling out of my mother's womb.
>
> Magic [Johnson] and Michael Jordan and me, for example, we're playing only on God-given talent, like we're animals, lions and tigers, who run around wild in a jungle, while Larry's success is due to intelligence and hard work. (THOMAS 1987, P. D27)

The experiences of Lee, Franklin, Simpson, and Thomas show how the element of race affects the ways in which people evaluate accomplishments.

Two less personal examples demonstrate how class and caste stratification can operate together in the United States. Consider how positions are filled and how players relate to one another on coed softball teams. Usually females are assigned to the least central positions (those requiring the least amount of involvement in completing a play). Also, when a ball is hit toward a female player, at least one male comes over to "help out." Rarely does a female charge over to aid a male teammate. Finally, when a female player comes up to bat, the four outfielders move in close to the infield; when a male player bats, they stay in the outfield. (A female might be a good singles hitter, but it is difficult to hit a single when there are in essence nine infielders.)

We could make the case that females play the least central positions and are helped by males because they lack the skills and experience to field, throw, and hit the ball adequately. When experience and skill are the only criteria for assigning positions, a class system of stratification is at work. And when there is a predictable relationship between an ascribed characteristic such as sex and an achievement such as playing shortstop, this is a clue for sociologists to investigate the source of this pattern. Often the reasons for the pattern can be traced to the fact that an undetermined, yet significant, amount of male–female difference in talent and experience is imposed externally—that is, they are by-products of the ways in which males and females are socialized.

Starting in infancy, parents elicit more active and more physical behavior from sons than daughters. They also channel their children toward sex-appropriate sports. Girls are guided into predominantly noncontact, often individual sports, such as gymnastics, tennis, and swimming, that require grace, flowing movements, and flexibility. In contrast boys are encouraged to participate in team sports that involve contact, lifting, throwing, catching, and running. Because boys and girls participate in different kinds of athletic activity, they develop different skills and styles. Finally the number of organized teams for each sex (from T-ball to professional) shows that more human and monetary resources are devoted to male than to female athletic development.

In the case of coed softball, a class system of stratification is operating on one level because those who have the most ability and experience are assigned to the most central positions. Yet on another level a caste system is also at work, because social practices contribute to differences between males' and females' talent and experience.

Caste and class systems of stratification operate together in professional sports as well. Sociologists have long noted that blacks are concentrated in positions that require strength, speed, and agility while whites are concentrated in central leadership, "thinking," and playmaking positions (Loy and Elvogue 1971; Medoff 1977). In professional baseball, for instance, whites tend to play the infield (including pitcher and catcher) while blacks tend to play in the outfield. Because no on-field position in professional baseball excludes blacks completely, some element of class stratification must be at work. At the same time, however, coaches, who are predominantly white, obviously assign blacks to noncentral positions and whites to leadership positions. This hypothesis is supported by the fact that most elementary and high school athletes play on predominantly white or black teams, which means that there are black athletes who have experience in playing *all* positions. Sports sociologists believe that black athletes are removed systematically from positions of leadership and are assigned to less central positions as they advance from the elementary to the high school to the college to the professional level. This practice continues after their on-field professional career is ended: in comparison to their white counterparts, few blacks become head coaches, general managers, or executives.

From a sociological point of view, any person who explains these differences as due to biological differences between males and females and across racial and ethnic groups is, in the succinct words of social psychologist E. A. Ross ([1908] 1929), "too lazy" to trace these differences to the social environment or historical conditions.

Clearly social stratification is an important feature of society, one with significant consequences for the life chances of the advantaged and disadvantaged alike. In the remainder of this chapter we will examine various theories as to why stratification occurs and what forms it takes. One theory—functionalism—seeks to explain why resources are distributed unequally in society. A second set of theories deals with identifying the various strata within society.

A Functionalist View of Stratification

In their classic article "Some Principles of Stratification," functionalist sociologists Kingsley Davis and Wilbert Moore (1945) asked how stratification—the unequal distribution of social rewards—contributes to maintaining order and stability in society. They claimed that social inequality is the device by which societies ensure that the most functionally important occupations, particularly those requiring great amounts of talent and costly and rigorous training, are filled by the best-qualified people. Consider the average hourly earnings of temporary workers by occupational category, as well as paid vacations, health insurance, and training opportunities.

ftp://stats.bls.gov/pub/news.release/occomp.txt

Document: New Survey Reports on Wages and Benefits for Temporary Help Services Workers

Q: Study the wages and benefits tables for the temporary workers. Davis and Moore would argue that differences in hourly wages and other benefits are a result of occupational categories' functional importance. Do you agree with their explanation? Explain your answer in light of this data.

The Functional Importance of Occupations

Davis and Moore (1945) conceded that it is difficult to document the functional importance of an occupation, but they suggest two somewhat vague indicators: (1) the degree to which the occupation is functionally unique (that is, there are few occupations that can perform the same function adequately) and (2) the degree to which other occupations depend on the one in question. In view of these indicators, maids, cashiers, and janitors need not be rewarded highly because little training and talent are required to do those jobs. Even though we depend on garbage collectors to maintain sanitary environments, many people are able to do the work.

Davis and Moore argued that society must offer extra incentives in order to induce the most talented individuals to undergo the long and arduous training needed to fill the most functionally important occupations. They specified that the incentives must be great enough to prevent the best-qualified and most capable people from finding less functionally important occupations as attractive as the most important occupations.

Davis and Moore also conceded that the efficiency of a stratification system in attracting the best-qualified people is weakened when capable individuals are overlooked or are not granted access to training, when elite groups control the avenues of training (as through admissions quotas), and when parents' influence and wealth (rather than the ability of their offspring) determine the status that their children attain. Yet Davis and Moore believed that the system adjusts to such inefficiencies. When there are shortages of personnel for functionally important occupations, society must increase people's opportunities to enter those occupations; if this is not done, the society as a whole will suffer and will be unable to compete with other societies.

Thus the functionalist argument that society will adjust and that all will work out in the end receives considerable criticism because it introduces a moral question: Is the social inequality the best way to ensure that the most important occupations are filled by the most qualified people?

Critique of the Functionalist Perspective

The publication of "Some Principles of Stratification" prompted a number of articles that took issue with the fundamental assumption underlying the Davis and Moore theory—that social inequality is a necessary and universal device that societies use to ensure that the most important occupations are filled by the best-qualified people. Two especially insightful critiques were Melvin M. Tumin's "Some Principles of Stratification: A Modification of the Functional Theory of Social Stratification" (1956) and Richard L. Simpson's "A Modification of the Functional Theory of Social Stratification" (1956).

Neither Tumin nor Simpson believed that a position commands great social rewards simply because it is functionally important or because the available personnel are scarce. Some positions command large salaries and bring other valued rewards even though their contribution to the society is questionable. For instance, for the 1993 season the average salary of a major league professional baseball player was $1,116,946; 40 percent of the 650 athletes who played for the 26 major league teams were paid $1 million or more (Chass 1992, 1993). Elementary and secondary teachers, on the other hand, were paid an average of $34,098 per year (*The World Almanac and Book of Facts 1994*, 1993). This difference in pay raises the question of whether professional athletes and entertainers are more essential to society than teachers—or whether there are other forces that are at least equally important in defining occupational rewards and status.

ftp://stats.bls.gov/pub/news.release/occomp.txt

Document: New Survey Reports on Wages and Benefits for Temporary Help Services Workers

Q: How do wages for blue-collar temporary workers compare with those for white-collar temp workers? Do you believe that the blue-collar worker positions are less functionally important than the white-collar positions? Explain.

Critics of functionalism also question why a worker should receive a lower salary for the same job just because the person is of a certain race, age, sex, or national origin. After all, the workers are performing the same job, so functional importance is not the issue. This latter question is at the center of the "comparable worth" debate. Advocates of comparable pay for comparable work ask whether women who work in predominantly female occupations (registered nurse, secretary, daycare worker) should receive salaries comparable to those of men who work in predominantly male occupations that are judged to be of comparable worth (vocational education teacher, housepainter, carpenter, automotive mechanic). In this vein consider the overall pattern of wages for temporary workers.

ftp://stats.bls.gov/pub/news.release/occomp.txt
Document: New Survey Reports on Wages and Benefits for Temporary Help Services Workers

Q: Of the positions listed, which ones do you think are filled disproportionately by males? By females?

If you are unsure about whether a particular occupational category is filled disproportionately by men or women, the document "Detailed Occupations by Sex" contains useful data.

gopher://una.hh.lib.umich.edu/00/census/summaries/eeous
Document: Detailed Occupations by Sex

Q: What kind of pattern do you see? Which female-dominated occupations pay the highest salaries? The lowest salaries? Which male-dominated occupations pay the highest salaries? The lowest? Compare the lowest salaries in male-dominated occupations with the lowest salaries in female-dominated occupations, and the highest salaries in female-dominated occupations with the highest salaries in male-dominated occupations. How do you account for the inequality?

Tumin and Simpson argued that it is very difficult to determine the functional importance of an occupation, especially in societies characterized by a complex division of labor and in which every individual contributes to the overall operation. In light of this interdependence one could argue that every individual makes an essential contribution: "Thus to judge that the engineers in a factory are functionally more important to the factory than the unskilled workmen involves a notion regarding the dispensability of the unskilled workmen, or their replaceability, relative to that of the engineers" (Tumin 1953, p. 388). Even if engineers, supervisors, and CEOs have the more functionally important positions, how much inequality in salary is necessary to ensure that people choose these positions over unskilled ones? In the United States, for example, the average annual salary of the CEO of a Fortune 500 corporation is $2,025,485—93 times the average annual salary of factory workers. Are such high salaries really necessary to ensure that someone chooses the job of CEO over that of factory worker? Probably not. But such high salaries have been justified as necessary to recruit the most able people to run corporations in the context of a global economy. It is unclear whether such salaries accurately reflect the CEOs' contributions to society relative to those of factory workers. Even though unskilled workers might be replaced more easily than engineers or CEOs, an industrialized society depends on motivated and qualified people in all positions.

Finally both Tumin and Simpson argued that the functional theory of stratification implies that a system of stratification evolves as it does in order to meet the needs of the society. In evaluating such a claim, one must look at whose needs are being met by the system.

Analyses of Social Class

Although sociologists use the term *class* to refer to one form of stratification, they also use it to denote a category that designates a person's overall status in society. In any case sociologists ask two basic questions: How many social classes are there? and What constitutes a social class? For answers, we turn to the works of Karl Marx and Max Weber.

Karl Marx and Social Class

Karl Marx viewed every historical period as characterized by a system of production that gave rise to specific types of confrontation between the exploiting and exploited classes in society. Consequently Marx was interested in relationships between various social classes that make up a society. He gave several answers to the question, How many social classes are there? In *The Communist Manifesto* (1848) he named two major social classes: the bourgeoisie and the proletariat. In *Capital: A Critique of Political Economy* (1909) he named three social classes: wage laborers, capitalists, and landlords. In *The Class Struggles in France 1848–1850* ([1895] 1976) he named at least six: the finance aristocracy, the bourgeoisie, the petty bourgeois, the proletariat, landlords, and peasants. According to French sociologists Raymond Boudon and François Bourricaud (1989), a careful reading of Marx's writings suggests that he believed "that the number of classes to be defined depends on the reason why we want to define them" (p. 341). The fact that Marx paid so much attention to class divisions in society underscores his belief that the most important engine of social change is class struggle. A brief overview of these works clarifies this point.

In *The Communist Manifesto*, written with Friedrich Engels, Marx described how class conflict between two distinct classes propels society from one historical epoch to another. Over time free men and slaves, nobles and commoners, barons and serfs, and guild masters and journeymen have confronted one another. Marx observed that "society as a whole is more and more splitting up into two great hostile camps, into two great classes directly facing each other: Bourgeoisie and Proletariat."

> gopher://wiretap.spies.com/00/Library/Classic/manifesto.txt
> **Document: Manifesto of the Communist Party**
>
> **Q:** How does Marx define the bourgeoisie and the proletariat? What historical forces gave rise to the bourgeoisie?

In light of this historical theme, it is appropriate that in *The Communist Manifesto* Marx focused on different social classes he believed would usher society out of capitalism and into another era.

In *Capital: A Critique of Political Economy* (1909) Marx named three classes (wage laborers, capitalists, landlords), each comprised of people whose revenues or income "flow from the same common sources" (p. 1032). For wage laborers the source is wages; for capitalists, profit; for landlords, ground rent. Marx acknowledged that the boundaries separating landlords from capitalists are not clear-cut. For example, Henry Ford, the founder of Ford Motor Company, was both a landlord (because he owned a rubber plantation in Brazil) and a capitalist (because he owned factories and machines and purchased labor). Marx also acknowledged that each of the three classes can be subdivided further. The category of landlords, for instance, can be divided into owners of vineyards, farms, forests, mines, fisheries, and so on. In any case, because Marx was interested in distinguishing people according to their sources of income, a three-category social class scheme made sense.

The Class Struggles in France 1848–1850 ([1895] 1976) is a historical study of an event in progress—the 1848 revolutions against several European governments (Germany, Austria, France, Italy, Belgium), with special emphasis on France. In this book Marx sought to describe and explain "a concrete situation in its complexity" (Boudon and Bourricaud 1989, p. 341). He described the 1848 revolution as a struggle for the necessities of life and as "a fight for the preservation or annihilation of the bourgeois order" (Marx [1895] 1976, p. 56). The latter consisted of a finance aristocracy, which lived in obvious luxury among masses of starving, low-paid or unemployed people.

Marx outlined the major factors that triggered the 1848 revolution and explained why he believed it failed. The widespread discontent was fueled by two world economic events. One was the potato blight and the bad harvests of 1845 and 1846, which raised the cost of living and led to bloody conflict in France and on the rest of the continent. The other event was a general commercial and industrial crisis, which resulted in an economic depression and a collapse of international credit. The revolutions were centered in the cities, where the Industrial Revolution had created a proletariat from persons who had migrated there in search of work. Generally the workers were paid very low wages, lived in squalor, and lacked life's basic necessities.

Although the faces of the French government changed as a result of the 1848 revolution, the exploitive structure remained, and the workers ultimately were put down by "unheard of brutality" (p. 57). Marx believed that the uprising failed because, even though the workers displayed unprecedented bravery and talent, they were "without chiefs, without a common plan, without means and for the most part, lacking weapons" (p. 56). The revolution also failed because the "other" classes did not support the proletariat when they moved against the finance aristocracy.

It is difficult to apply Marx's ideas about social class in a total way because, as he made clear in *The Class Struggles in France 1848–1850*, the reality of class is very complex. He left us, however, with some useful ideas with which to approach social class. First, conflict between two distinct classes propels us from one historical epoch to another. Second, it is illuminating to view social class in terms of sources of income, because our understanding of social class moves beyond the simple notion of occupation (or relationship to the means of production). To help you identify potential sources of income, the site "step by step guide to building and buying a home" provides potential home buyers with a worksheet to estimate, among other things, "sources of income."

http://www.propertyguide.com/steps.htm#Step3
Document: Step Three: Home Buying Power

Q: What are the various sources from which people can draw income?

Finally Marx's ideas remind us that the conditions that lead to a successful revolt by an exploited class against the exploiting class are multifaceted and complex. He recognized that exploitive conditions can trigger uprisings but observed that other factors such as well-thought-out plans, effective leadership, the support of other classes, and access to weapons determine the success or failure of the revolt.

Max Weber and Social Class

Although Karl Marx did not consistently specify an exact number of social classes in society, he clearly stated that a person's social class is based on his or her relationship to the means of production. Max Weber, like Marx, did not specify how many social classes exist. For Weber, though, the basis for a social class was not the means of production, but the marketplace. Class situation ultimately was market situation: it was based on the chances of acquiring goods and services, obtaining a well-paying job, and finding inner satisfaction.

Weber ([1947] 1985) believed that people's class standing depends on their marketable abilities (work experience and qualifications), their access to consumer goods and services, their control over the means of production, and their ability to invest in property and other sources of income. Weber viewed class as a continuum of rungs on a social ladder. Persons completely unskilled, lacking property, and dependent on seasonal or sporadic employment occupy the bottom rung of the ladder or system. They form the "negatively privileged" class. Those at the very top—the "positively privileged"—monopolize the purchase of the highest-priced consumer goods, have access to the most socially advantageous kinds of education, control the highest executive positions, own the means of production, and live on income from property and other investments. Between the top and the bottom of the ladder is a continuum of rungs.

Weber stated that class ranking is complicated by status groups and political parties, of which there are many different kinds. He defined **status group** as a plurality of persons held together by virtue of a common lifestyle, formal education, family background, or occupation and "by the level of social esteem and honor accorded to them by others" (Coser 1977, p. 229). This definition suggests that wealth, income, and position are not the only factors that determine an individual's status group: "The class position of an officer, a civil servant or a student may vary greatly according to their wealth and yet not lead to a different status since upbringing and education create a common style of life" (Weber [1947] 1982, p. 73).

Weber (1982) defined *political parties* as organizations "oriented toward the planned acquisition of social power [and] toward influencing social action no matter what its content may be" (p. 68). Parties are organized to represent status groups and other interests; they exist at all levels (within an organization, a city, a country). The means employed by political parties to obtain power include violence, canvasing for votes, bribery, donations, the force of speech, suggestion, and fraud.

Weber argued that a "*uniform* class situation prevails only when completely unskilled and propertyless persons are dependent on irregular employment" (p. 69). We cannot speak of a uniform situation with regard to other social classes because class standing is complicated by such elements as occupation, education, income, status group membership, differences in property holdings, consumption patterns, and so on.

Weber's idea of top and bottom rungs of the social system inspires us to compare the situation of the wealthiest persons against that of the poorest. One indicator of the "wealth gap" between those at the very top and very bottom can be found in the housing characteristics data.

http://www.census.gov/ftp/pub/hhes/housing/ahs/tab2-3.html

Document: Big Homes?

Q: Select "tab2-13.html," "tab2-3.html," "tab2-8.html," "tab2-7.html," and "tab3-14.html" to complete the accompanying chart. There were 94,724,000 occupied households in the United States at the time of this survey. Figure the percentage in the top and bottom categories of each housing characteristic by dividing the number in that category by the total number of occupied households in the United States.

Housing characteristic	Number of households	Percentage of households (n/94,724,000)
Monthly cost		
Less than $100	_____	_____
More than $1500	_____	_____
Rooms		
1 room	_____	_____
10 or more rooms	_____	_____

Square footage per person
    Less than 200   _____   _____
    More than 1,500   _____   _____
Opinion of neighborhood
    Worst neighborhood (1) _____   _____
    Best neighborhood (10) _____   _____
Worth of house
    Less than $10,000   _____   _____
    More than $300,000   _____   _____

Weber's ideas about social class also draw our attention to the negatively privileged property classes. The proportion of negatively privileged persons tells us something important about the extremes of inequality in our own society.

Occupational Structure and Poverty in the United States

The existence of a negatively privileged property class in the United States can also be traced to structural factors—in particular to changes in the occupational structure, not to issues related to worker motivation. A Bureau of Labor Statistics (BLS) study found that from January 1991 to December 1993, 4.5 million workers were displaced from jobs they had held for three years or more. However, if we also consider workers who had held their jobs less than three years, some 9 million workers were displaced over that same period—many because of changes in the occupational structure.

ftp://stats.bls.gov/pub/news.release/disp.txt

Document: Worker Displacement During the Early 1990s

 How does the BLS report define displaced workers? What are the top reasons displaced workers lose their jobs? What category of workers is at highest risk of job displacement? What geographic region of the United States has the highest number of displaced workers?

This information may come as a surprise to some Americans who attribute mobility, whether upward or downward, to individual effort and who fail to consider changes in the occupational structure as the cause. Many Americans may not recognize that some groups are affected by changes in the occupational structure more strongly than others. An example of a group affected by changes in the occupational structure is workers in the poultry processing industry. In "Why Did Employment Expand in Poultry

Processing Plants?" economist Ron Hetrick identifies structural factors that contributed to this job growth.

http://stats.bls.gov/ceschick.htm

Document: Why Did Employment Expand in Poultry Processing Plants?

Q: What structural factors contributed to job growth in the poultry processing industry? What other meat processing industries lost jobs as a result of these factors? What are the wages associated with job growth in the poultry industry? With job decline in the industries that lost jobs?

In *The Truly Disadvantaged* (1987) and in other related articles and books, sociologist William Julius Wilson describes how structural changes in the U.S. economy helped create what he termed, in his 1990 presidential address to the American Sociological Association, the "ghetto poor." A number of economic transformations have taken place, including (1) the restructuring of the American economy from a manufacturing-based economy to a service- and information-based economy; (2) a labor surplus that began in the 1970s, marked by the entry of women and the large baby boom segment into the labor market; (3) a massive exodus of jobs from cities to the suburbs; and (4) the transfer of low-skill manufacturing jobs out of the United States to offshore locations. These changes are major forces behind the emergence of the ghetto poor or **urban underclass,** a "heterogeneous grouping of families and individuals in the inner city that are outside the mainstream of the American occupational system and that consequently represent the very bottom of the economic hierarchy" (Wilson 1983, p. 80).

Wilson (in collaboration with sociologist Loic J. D. Wacquant) looks at Chicago as case in point. (Actually the point applies to every large city in the United States.) In 1954 Chicago was at the height of its industrial power. Between 1954 and 1982, however, the number of manufacturing establishments within the city limits dropped from more than 10,000 to 5,000, and the number of jobs from 616,000 to 277,000. This reduction, in conjunction with the out-migration of stably employed working-class and middle-class African-American families and fueled by new access to housing opportunities outside the inner city, had a profound impact on the daily life of the people left behind. The exodus of the stably employed resulted in the closing of hundreds of local businesses, service establishments, and stores. According to Wacquant (1989) the single most significant consequence of these historical and economic events was the "disruption of the networks of occupational contacts that are so crucial in moving individuals into and up job chains . . . [because] ghetto residents lack parents, friends, and acquaintances who are stably employed and can therefore function as diverse ties to firms . . . by telling them about a possible opening and assisting them in applying [for] and retaining a job" (pp. 515–16).

The ghetto poor are the most visible and most publicized underclass in the United States. In addition demographers William P. O'Hare and Brenda Curry-White (1992) estimate that approximately 736,000 rural residents can be classified as an underclass, albeit a less visible one. Like their urban counterparts the rural underclass is concentrated in geographic areas with high poverty rates. The rural poor have been affected by economic restructuring, which includes the decline of the farming, mining, and timber industries and the transfer of routine manufacturing out of the United States.

http://www.cdinet.com/Rockefeller/Briefs/brief28.html
Document: White Poverty in America

Q: Why is more attention devoted to poverty among people classified as black and Hispanic than to people classified as white? What percentage of poor are classified as white? Why are poor people classified as white more likely than those classified as black or Hispanic to be lifted out of poverty?

The rural and urban underclass represent two distinct segments of the population that live below the poverty line, set at about $12,675 for a family of four. The poverty line or threshold varies depending on family size and geography.

http://aspe.os.dhhs.gov/poverty/poverty.htm
Document: The HHS Poverty Guidelines: One Version of the Federal Poverty Measure

Q: Do you believe the poverty threshold is set too low or too high? Explain.

On the basis of this definition, almost 32 million Americans (or 13 percent of the population) live below the poverty line. Included in this 32 million are 12 million people whose income is less than half the amount officially defined as the poverty level. However, the definition of poverty encompasses diverse groups of people, some of whom might not be considered really poor (such as graduate students and retired people who have assets but who live on a low fixed income). Still it is important to note that two out of every three poverty-level households are headed by women. For many of these women, "their only 'behavioral deviancy' is that husbands or boyfriends left them" (Jencks 1990, p. 42); in the case of many older women, their husbands have died. Two factors in particular account for women's disproportionate poverty: (1) the economic burden of children and (2) women's disadvantaged position in the labor market. These issues will be explored in an upcoming chapter.

Discussion Question

Based on the information presented in this chapter, what social factors do you think have affected your opportunities or life chances?

Additional Reading

For more on

- Hispanic as a racial-ethnic category
  http://www.census.gov/td/stf3/append_b.html#hispanic
  Document: Hispanic

- Projected changes in the occupational structure
  ftp://stats.bls.gov/pub/news.release/ecopro.txt
  Document: BLS Press Release: New 1994–2005 Employment Projections

- Downward mobility attributed to structural changes
  ftp://stats.bls.gov/pub/news.release/demdat.txt
  Document: Bureau of Labor Statistics Release: 1992 Demographic Data Book for State and Large Metropolitan Areas

- Status groups as defined by neighborhood characteristics
  http://propertyguide.com/steps.html#Step2
  Document: Step 2: Selecting a New Neighborhood

- Marxist perspective on capitalism
  http://members.gnn.com/dgude/capital.htm
  Document: The Truth About American Capitalism

  http://members.gnn.com/dgude/capital.htm#Caste
  Document: The American Caste System

- The poverty line
  http://www.cdinet.com/Rockefeller/Briefs/brief23.html
  Document: Where the American Public Would Set the Poverty Line

- Gender and income inequality
  http://gopher.census.gov:70/ls/Bureau/Pr/Subject/Income/cb95-129.txt
  Document: Female Householder Families Most Likely to Stay Poor

- African National Congress
  gopher://gopher.anc.org.za/00/anc/history/75years.87
  Document: Advance to Power—75 Years of Struggle

- Income distribution by gender
  http://www.census.gov/ftp/pub/hhes/laborfor/dewb9193/tableb.html
  Document: Dynamics of Economic Well-Being: Labor Force, 1991–1993

- Earnings before and after job turnover
  http://www.census.gov/ftp/pub/hhes/laborfor/dewb9092/jobturntab.html
  Document: Distribution of Average Weekly Earnings Before and After Job Turnover

Race and Ethnicity

Every ten years since 1790 (the year of the first census), the U.S. government has attempted to count the number of people living under its jurisdiction and classify them according to race. On the surface determining race seems like a relatively simple task: obviously we think race is something that is easily observable and we assume that everyone knows his or her race. Census Bureau data suggests that virtually everyone in the United States belongs to one of four broad racial categories: (1) white, (2) black, (3) American Indian, Eskimo, Aleut, or (4) Pacific Islander. The document "USA Statistics in Brief: Part 1" contains information on the number of people in each racial category.

gopher://gopher.census.gov/00/Bureau/Stat-Abstract/USAbrief/part1.prn
Document: USA Statistics in Brief: Part 1

Q: What percentage of the U.S. population is classified as belonging to each racial category?

Recently I asked students in a race and gender class if they knew of someone who might not fit neatly into only one category. Of the 70 students in this class, 19 responded in the affirmative. Here are three of those responses:

"My friend Debra's parents are both of mixed ancestry: her mother is Native American, Portuguese, and black; her father is French and black. Debra has the darkest skin of anyone in her family, but that does not save her from being teased by her friends for trying to be a white girl. Debra has light skin and fine hair, which led many people to assume that she is really not 'black.'"

"My friend Alena's father, a white American, met her mother in Thailand when he was stationed there while in the military. Alena has many of her mother's physical characteristics: dark skin, dark hair, big eyes, and full lips. Many people try to classify her as Hawaiian or Filipino. When people ask, 'Where are you from?' or 'What are you?' she usually answers, 'American.' It is obvious that she has been raised in America. She dresses and talks like an American. When she was a child, her paternal grandmother took care of her a lot, and all of her friends are white. She knows very little Thai and is Catholic (her mother is Buddhist)."

"Three years ago my good friends Cathy and Sam found out they were going to have their first baby. They spent the next nine months preparing to make everything perfect for the new arrival. Cathy eventually gave birth to a blue-eyed, blond-haired baby boy whom they named Michael. The hospital nurses who prepared the birth certificate told the parents that Michael would be classified according to the race of his father, black. Cathy was outraged by this idea, afraid that her son would be an object of discrimination. How will people react when Michael checks 'black' as his race on the various forms and applications that will come his way in the future? Will he be viewed as an impostor if he claims his black heritage?"

Apparently these student-generated examples are not unusual. For more "voices" of those who do not fit clearly into one racial category, go to the Interracial Voice homepage on the internet and skim the entries posted in "Letters/Voices to the Editor."

http://www.webcom.com/intvoice/letters1.html
Document: Letters/Voices to the Editor

http://www.webcom.com/intvoice/letters3.html
Document: Letters/Voices to the Editor

Q: Which one or two letters do you believe best capture the shortcomings of the U.S. system of racial classification? Why?

This brings us to an important question: How is it that racial categories are treated as mutually exclusive when we can identify many cases in which people have complex racial histories? Maybe race is not a biological fact, an inherited trait like eye color or hair color. Perhaps race refers to that which is produced through racial classification (Webster 1992). In this chapter we focus on the concepts of race and ethnicity, especially as they relate to the stigmatization and stratification of minority groups.

Racial Classification and the Idea of Race

The fact that everyone seems to fit into a single racial category is the result of the system of racial classification used in the United States and not biological reality. When most people meet someone, however, they do not think to learn the facts of that person's ancestry; instead they search for the visible clues that have come to be associated with a race and proceed to classify that person accordingly. The Interracial Voice Website is devoted to helping people understand the problems with the U.S. system of racial classification.

http://www.webcom.com/intvoice/

Document: Interracial Voice

Q: Select a document posted on this Website that attracts your attention. Describe how it helps readers to understand problems with racial classification in the United States.

The flaws of the four-category racial classification scheme used in the United States are especially evident when we come across people who do not fit into a single racial category, who must choose between categories, or who do not meet others' conceptions of what someone in a particular category should be like. Such cases tell us that **race** is not an easily observed characteristic immediately evident on the basis of physical clues, but is a category defined and maintained by people through a complex array of formal and informal mechanisms.

In addition to a question about race, the Census Bureau asks U.S. residents several ethnicity-related questions, including questions about Hispanic origin, ancestry, place of birth, and language (including a self-rating of ability to speak English).

http://www.census.gov/td/stf3/append_b.html

Document: Definition of Subject Characteristics

Q: Read entries connected with the following subject characteristics: ancestry, place of birth, Hispanic origin, and language spoken at home/ability to speak English. What are at least two strategies the U.S. government employs to simplify complex answers people give to ethnicity questions?

Although the Census Bureau asks a variety of questions related to ethnicity, the only ethnic categories it recognizes officially are (1) Hispanic/Spanish and (2) non-Hispanic origin. Contrary to popular belief, "Hispanic" is not a race. People classified as Hispanic can be of any race, because the history of Latin America is intertwined with that of Asia, Europe, the Middle East, and Africa. As a result of this interconnected history, the countries of Latin America are populated not by a homogeneous group known as Hispanics, but by native- and foreign-born persons, immigrant and nonimmigrant residents, and persons from every conceivable ancestry (not just Spanish). The Census Bureau keeps statistics on the number of Hispanics in each of the four official racial categories.

http://www.census.gov/cdrom/lookup
Document: 1990 Census Lookup

Q: Choose "STF3C-part 1"; stay with the default options and choose "submit"; stay with the default options and again choose "submit." Next, page down to select Table P12; choose "submit" after you have selected P12; stay with the default options and choose "submit." What percentage of the Hispanic-origin population is white? Black? American Indian, Eskimo, or Aleut? Asian or Pacific Islander? Other race? Why do you think 9,470,850 Hispanics are classified as "other race"?

As with race, the Census Bureau codes and manipulates data on ethnicity such that the various categories seem to be independent of one another. But again, the reality is that millions of people have mixed ancestry and possess a blend of physical and cultural traits. The case of culinary artist and chef Jeannette Holley shows that ethnic categories can never be clear-cut.

> The 37-year-old chef is thin and tall, with taffy-colored skin and sculptured black hair. Her deep, almond eyes allow only momentary glimpses into her private world of innerconnections. Her father is African-American, her mother Japanese, and Holley has spent most of her life in Asia, first outside Tokyo and later in Seoul, Korea. "I was raised an American. I lived in Asia for 17 years. I spoke English at home, although it wasn't my first language. . . . Language itself becomes a property of who you are. The less Japanese I speak, the less Japanese I feel." (BURNS 1993, P. 36)

Depending on the criteria one chooses to use—skin color, national origin, place of residence, ancestry, eye structure, language—Jeannette Holley could be classified as a member of any number of ethnic groups.

Charles Hirschman (1993) points out that there is no such thing as mutually exclusive or clear-cut ethnic categories. There has always been **ethnic blending**—"interethnic unions (interbreeding) and shifts in ethnic affiliation [such that] most ethnic communities are either amalgams of different peoples or have absorbed significant numbers of other groups through conquest, the expansion of national boundaries, or acculturation" (pp. 549–50). Thus, practically everyone in any society belongs to multiple ethnic categories.

Even though racial and ethnic classification schemes are problematic, many people argue that they know a white, black, Asian, or Arabic person when they see one. If they meet someone who does not fit the image, they say, "But you don't look like someone of African/Asian/Arabic descent." That's because their vision of human variety is limited by the images of people portrayed on airline posters, magazine ads, and television sitcoms (Houston 1991). Sometimes people will go to extremes to create an ethnic and racial group that fits an ideal image.

Sociologist Paul D. Starr (1978) notes that in the absence of distinctive skin color and other physical characteristics, people determine **ethnicity** on the basis of any number of other imprecise attributes—language, dress, jewelry, tattoos and other body scars, modes of expression, and residence. In determining someone's race and ethnicity, people rarely look beyond the most visible characteristics. Most people fail to consider the details of another person's life.

If race and ethnicity are such vague categories, perhaps the most appropriate definition of a racial or ethnic group is that its members believe (or outsiders believe) that they share a common national origin, cultural traits, or distinctive physical features. It does not matter whether this belief is based on reality. The point is that membership in an ethnic or racial group is a matter of social definition, an interplay of self-definition and others' definitions.

Sociologists are interested in racial and ethnic classification schemes that incorporate the assumption that specific abilities (athletic talent, intelligence), social traits (criminal tendencies, aggressiveness), and cultural practices (dress, language) are transmitted genetically. This assumption underlies the belief that some categories of people are inferior to others because of their race and/or ethnicity and therefore can be denied equal access to scarce and valued resources such as citizenship, decent wages, wealth, income, health care, and education. In other words sociologists are interested in the experiences of minority groups.

Minority Groups

Minority groups are subgroups within a society that can be distinguished from members of the dominant groups by visible and identifying characteristics, including physical and cultural attributes. Members of such subgroups are regarded and treated as inherently different from members of **dominant groups,** the racial and ethnic groups at the top of the socioeconomic hierarchy. For these reasons they are systematically excluded, consciously or unconsciously, from full participation in society and are denied equal access to positions of power, prestige, and wealth. Thus members of minority groups tend to be concentrated in inferior political and economic positions and isolated socially and spatially from members of the dominant groups.

On the basis of these characteristics, many groups can be classified as minorities, including some racial, ethnic, and religious groups; women; the very old and very young; and the physically different (for example, visually impaired people or overweight people). Although we focus on ethnic and racial minorities in this chapter, the concepts that follow can be applied to any minority.

Sociologist Louis Wirth (1945) made a classic statement on minority groups, identifying a number of essential traits characteristic of all such groups. First, membership is involuntary: as long as people are free to join or leave a group, no matter how unpopular the group, they do not by virtue of that membership constitute a minority. This first trait is quite controversial because the meaning of "free to join or leave" is unclear. Second, minority status is not necessarily based on numbers. In fact, a minority group may be the majority of the people in a society; the key to minority status is not size, but access to and control over valued resources. Third, and most important, is nonparticipation by the minority group in the life of the larger society. That is, minorities do not enjoy the freedom or the privilege to move within the society in the same way as members of the dominant group. Finally, and most troublesome, people who belong to a minority group are "treated as members of a category, irrespective of their individual merits" (Wirth 1945, p. 349) and often irrespective of context. In other words people outside the minority group focus on the visible characteristics that identify someone as belonging to a minority. This visible characteristic becomes the focus of interaction.

The characteristics that Wirth identifies as associated with minority group status indicate that minorities stand apart from the dominant culture. Some people argue that minorities stand apart because they do not wish to assimilate into mainstream culture. To assess this claim, we turn to the work of sociologist Milton M. Gordon, who has written extensively on assimilation.

Types of Assimilation

Assimilation is a process by which ethnic and racial distinctions between groups disappear. There are two main types of assimilation: absorption assimilation and melting pot assimilation.

Absorption Assimilation

With **absorption assimilation** members of a minority ethnic or racial group adapt to the ways of the dominant group, which sets the standards to which they must adjust (Gordon 1978). According to Gordon an ethnic or racial group is completely "absorbed" into the dominant group when it goes through all seven of these levels:

- *Level 1:* The group abandons its culture (language, dress, food, religion, and so on) for that of the dominant group (an action known as *acculturation*).
- *Level 2:* The group enters into the dominant group's social networks and institutions (*structural assimilation*).
- *Level 3:* The group intermarries and procreates with those in the dominant group (*marital assimilation*).
- *Level 4:* The group identifies with the dominant group (*identification assimilation*).
- *Level 5:* The group encounters no widespread prejudice from members of the dominant group (*attitude receptional assimilation*).
- *Level 6:* The group encounters no widespread discrimination from members of the dominant group (*behavior receptional assimilation*).
- *Level 7:* The group has no value conflicts with members of the dominant group (*civic assimilation*).

Be careful not to misconstrue absorption assimilation as a process whereby people designated as minorities decide to abandon their culture, enter the so-called mainstream, and disappear as a distinct group. Such a viewpoint suggests that the decision to assimilate is left to minorities, when in fact absorption assimilation has been a goal of many government policies. H. L. Dawes's 1899 essay in *The Atlantic Monthly* entitled "Have We Failed with the Indian?" reflects on absorption assimilation policies.

> http://etext.lib.virginia.edu/etcbin/
> browse-mixed-new?id=DawHave&tag=public&images/
> modeng&data=/texts/english/modeng/parsed
> Document: Have We Failed with the Indian?

Q: Which statements in this essay suggest that a goal of absorption assimilation drove government policies on Indian affairs? What was Dawes's answer to the question "Have we failed with the Indian?" If yes, why? If no, what factor did Dawes believe was responsible for the failure of government programs?

Gordon (1978) advances a number of hypotheses about how the various levels of assimilation relate to one another. For instance, he maintains that acculturation is likely to take place before the other six levels of assimilation. Gordon also suggests, however, that even if acculturation is total, it does not always lead to the other levels of assimilation.

Gordon proposes that a clear connection exists between the structural and marital levels of assimilation. That is, if the dominant group permits people from ethnic and racial minority groups to join its social cliques, clubs, and institutions on a large enough scale, a substantial number of interracial or interethnic marriages are bound to occur:

> If children of different ethnic backgrounds belong to the same play group, later the same adolescent cliques, and at college the same fraternities and sororities; if the parents belong to the same country club and invite each other to their homes for dinner; it is completely unrealistic not to expect these children, now grown, to love and to marry each other, blithely oblivious to previous ethnic extraction. (pp. 177–78)

Of the seven levels of assimilation, Gordon believes that the structural level is the most important because if it is attained, the other levels of assimilation inevitably will follow. Yet structural assimilation is very difficult to achieve in practice, because members of ethnic or racial minorities traditionally are denied easy and comfortable access to the dominant group's networks and institutions. In fact all of our important and meaningful primary relationships are confined largely to individuals of the same racial or ethnic group:

> From the cradle in the sectarian hospital to the child's play group, the social clique in high school, the fraternity and religious center in college, the dating group within which he [or she] searches for a spouse, the marriage partner, the neighborhood of his [or her] residence, the church affiliation and the church clubs, the men's and the women's social and service organizations, the adult clique of "marrieds," the vacation resort, and then, as the age cycle nears completion, the rest

> home for the elderly and, finally, the sectarian cemetery—in all these activities and relationships which are close to the core of personality and selfhood—the member of the ethnic group may if he [or she] wishes follow a path which never takes him [or her] across the boundaries of his [or her] ethnic structural network. (p. 204)

This scenario especially characterizes the primary group relations of **involuntary minorities,** ethnic and racial groups that did not choose to be a part of a country, but that were forced to become part of a country by enslavement, conquest, or colonization. Native Americans, African Americans, Mexican Americans, and native Hawaiians are examples of involuntary minorities in the United States. Unlike **voluntary minorities,** whose members come to a country expecting to improve their way of life, members of involuntary minorities have no such expectations. Their forced incorporation involves a loss of freedom and status (Ogbu 1990).

Melting Pot Assimilation

Assimilation need not be a one-sided process in which a minority racial and ethnic group disappears or is absorbed into the dominant group. Ethnic and racial distinctions also can disappear through **melting pot assimilation** (Gordon 1978). In this process one group accepts many new behaviors and values from another group, which results in a new, blended cultural system. Melting pot assimilation is total when significant numbers of people from each ethnic and racial group take on cultural patterns of the other, enter each other's social network, intermarry and procreate, and identify with the blended culture.

The melting pot concept can be applied to the various African ethnic groups imported to the United States as slaves. They were "not one but many peoples" (Cornell 1990, p. 376), who spoke many languages and came from many cultures. Slave traders capitalized on this diversity: "Advertisements of new slave cargoes frequently referred to ethnic origins, while slave owners often purchased slaves on the basis of national identities and the characteristics they supposedly indicated" (Cornell 1990, p. 376; Rawley 1981). Although slave owners and traders acknowledged ethnic differences among Africans, they treated Africans from various ethnic groups as belonging to one category of people—slaves. Because slave traders sold and slave owners purchased individual human beings, not ethnic groups, this treatment had the effect of breaking down ethnic concentrations. In addition slave owners tended to mix together slaves of different ethnic origins in order to decrease the likelihood of the slaves' plotting a rebellion. To communicate with one another, the slaves invented pidgin and Creole languages. They also created a common and distinctive culture based on kinship, religion, food, songs, stories, and other features. The harsh conditions of slavery, in combination with the mixing together of people from many ethnic groups, encouraged slaves to borrow aspects of one another's cultures and to create a new, blended culture.

Few people in the United States view assimilation as a process of mutual exchange in which members of the dominant group form and identify with a blended culture (Opitz 1992b). This is not to say that the dominant culture has not been shaped and influenced by racial and ethnic minorities; rather members of the dominant group more often than not fail to acknowledge the contributions of others to the society in which they all live.

Stratification Theory and Assimilation

Stratification theory is one major approach to understanding forces that work against assimilation (absorption or melting pot) between dominant and minority groups. This theory is guided by two assumptions. First, ethnic and racial groups are ranked hierarchically. The group at the top is known as the dominant group; the groups at the bottom are called minorities. Second, racial and ethnic groups compete with one another for scarce and valued resources. The dominant group retains the advantage because its members are in a position to preserve the system that gives them their advantages. Stratification theorists focus on the mechanisms employed by people in the dominant group to preserve inequality (Alba 1992). These include racist ideologies, prejudice and stereotyping, discrimination, and institutional discrimination.

Racist Ideologies

An **ideology** is a set of beliefs that are not challenged or subjected to scrutiny by the people who hold them. Thus ideologies are taken to be accurate accounts and explanations of why things are the way they are. On closer analysis, however, ideologies are at best half-truths, based on misleading arguments, incomplete analysis, unsupported assertions, and implausible premises. They "cast a veil over clear thinking and allow inequalities to persist" (Carver 1987, pp. 89–90). One such ideology is **racism.** People who adhere to this ideology believe that something in the biological makeup of an ethnic or racial group explains and justifies its subordinate or superior status.

Racist ideologies are structured around three notions: (1) people can be divided into categories on the basis of physical characteristics; (2) a close correspondence exists between physical traits and characteristics such as language, dress, personality, intelligence, and athleticism; and (3) physical attributes such as skin color are so significant that they explain and determine social, economic, and political inequalities. Any racial or ethnic group may use racist ideologies to explain their own or another group's behavior. For one example of racist ideology, see Thomas Jefferson's writings on slavery.

http://grid.let.rug.nl/~welling/usa/documents/jefslav_note.html
Document: Thomas Jefferson on Slavery

: What are some examples of racist statements in Thomas Jefferson's writing (that is, instances where he makes a connection between physical traits and other characteristics)? Did Jefferson believe that conditions of slavery might have negative effects on the situation of slaves? Explain.

The doctrine of Manifest Destiny represents a second example of racist ideology.

gopher://gopher.etext.org/00/Politics/Fourth.World/Americas/manifest.txt
Document: Reflections on Race and Manifest Destiny

Q: What is Manifest Destiny? What is the racist logic underlying this ideology? What actions did Manifest Destiny justify? The essay ends with the statement "A philosophy such as Manifest Destiny once internalized in the culture, is never really abolished, it merely adapts to the present conditions and transforms itself into a suitable logic for the times." Do you agree with this statement? Explain.

The premise of racial superiority is also at the heart of other rationalizations used by one group to dominate another. Sociologist Larry T. Reynolds (1992) observes that race as a concept for classifying humans is a product of the 1700s, a time of widespread European exploration, conquest, and colonization that did not begin to subside until the end of World War II. Racist ideology also supported Japanese annexation and domination of Korea, Taiwan, Karafuto (the southern half of the former Soviet island Sakhalin), and the Pacific Islands prior to World War II. Both the Japanese and Europeans used racial schemes to classify the people they encountered; the idea of racial differences became the "cornerstone of self-righteous ideology," justifying their right by virtue of racial superiority to exploit, dominate, and even annihilate conquered peoples and their cultures (Lieberman 1968).

In *The Mismeasure of Man* Stephen Jay Gould (1981b) describes instances in which so-called scientists have used the scientific method to verify the accuracy of racist ideologies. He reveals how millions of lives have been ruined by spurious correlation, by the mystique of quantification, or simply by sloppy science. Gould's article entitled, "The Politics of Census" a particularly troubling account, explains how the U.S. census figures for 1840 were used to justify slavery. In 1840 the census takers counted the mentally ill for the first time in census history. When Dr. Edward Jarvis, an authority on medical statistics, examined census tables showing the number of mentally ill persons by race and by state, he found that "one in 162 blacks was insane in free states, but only one in 1,558 in slave states. Moreover, insanity among blacks seemed to decrease in even gradation" as one traveled from the northernmost part of the country to the deep south. Jarvis also found that location had no bearing on rates of insanity among whites. On the basis of these findings, he "drew the conclusion that so many other whites would advance": slavery must be beneficial to blacks. A slave actually gains from not having "hopes and responsibilities which the free, self-thinking and self-acting enjoy and sustain, for bondage saves him from some of the liabilities and dangers of active self-direction" (Gould 1981a, p. 20).

Dr. Jarvis, however, was troubled by these conclusions and wondered whether slavery could possibly exert such a significant effect on mental stability. He decided to investigate and found a great many errors in the gathering and tabulation of the data. For example, the census showed that in Worchester, Massachusetts, 133 blacks out of a total black population of 151 were insane. Jarvis discovered, however, that the 133 were actually white patients in the state mental hospital at Worchester. He attempted to have the census declared invalid, but his efforts met with considerable opposition, led by John C. Calhoun, head of the U.S. State Department. Consequently the data stood as fact, despite the errors.

Prejudice and Stereotyping

A **prejudice** is a rigid and usually unfavorable judgment about an outgroup that does not change in the face of contradictory evidence and that applies to anyone who shares the distinguishing characteristics of that group. Prejudices are based on **stereotypes**—exaggerated and inaccurate generalizations about people who are members of an outgroup. Stereotypes "give the illusion that one [group] knows the other [and] confirms the picture one has of oneself" (Crapanzano 1985, pp. 271–72).

Stereotypes are supported and reinforced in a number of ways. One way is through **selective perception**, whereby prejudiced persons notice only those behaviors or events that support their stereotypes about an outgroup. In other words people experience "these beliefs, not as prejudices, not as prejudgments, but as irresistible products of their own observation. The facts of the case permit them no other conclusion" (Merton 1957, p. 424). For example, the stereotype that young blacks are more violent than their white counterparts can be supported by FBI statistics showing that blacks are three times as likely as whites to be arrested for aggravated assault. However, National Crime Victimization Survey data, based on interviews with victims, shows that blacks and whites are equally likely to commit aggravated assaults (32 per 1,000 blacks and 31 per 1,000 whites). The discrepancy between FBI figures and National Crime Victimization Survey data suggests that blacks are more likely than whites to be caught, arrested, and convicted, not that they are more likely to engage in violent acts (Stark 1990).

Stereotypes also persist in another way: when prejudiced people encounter a minority group member who contradicts the stereotype, they see that person as an exception. The fact that they have encountered someone who is "different" only serves to reinforce their stereotypes.

Finally prejudiced individuals keep stereotypes alive when they evaluate the same behavior differently at different times, depending on the person who exhibits that behavior (Merton 1957). For example, incompetent behavior on the part of racial and ethnic minority group members often is attributed to innate flaws in their biological makeup, whereas incompetence exhibited by someone from the dominant group is almost always treated as an individual issue. Similarly prejudiced people treat certain ideas, when expressed by members of a minority group, as more threatening than when those ideas are expressed by members of a dominant group. Return to the Manifest Destiny Web site.

gopher://gopher.etext.org/00/Politics/Fourth.World/Americas/manifest.txt
Document: Reflections on Race and Manifest Destiny

 Q: What are some examples from the document of situations in which the same behavior is defined differently depending on who is the focus of attention?

In *Immigrant Workers and Class Structure in Western Europe,* Stephen Castles and Godula Kosack (1985) argue that the roots of racist ideologies and prejudice are not "based on real dislike for the physical appearance of the groups concerned. Rather, the physical appearance and the inferiority which is alleged to go with it is [sic] used as an excuse" to justify the subordinate position of racial and ethnic minorities in society (p. 456). "The exploited group is stigmatized as inherently inferior in order to justify [that position]" (p. 458). The subordinate position then is cited as proof of the group's inferiority.

Prejudice and racism directed toward international labor migrants serve another function: they divert attention from the deficiencies of the profit-driven capitalist system, which lowers production costs by hiring persons who will work for lower wages, by introducing labor-saving devices, and by moving production from high-wage zones. All of these actions lead to unemployment. However, the blame for unemployment is placed on labor migrants rather than on the economic system, which plays off workers against one another and creates the unemployment (Bustamante 1993). Because prejudice and racism against labor migrants direct attention away from the capitalist system, which treats labor as a commodity, they undermine the chances that workers will unify and make real improvements in their wages, work conditions, and job security.

Discrimination

In contrast to prejudice **discrimination** is not an attitude, but a behavior: intentional or unintentional unequal treatment of individuals or groups on the basis of attributes unrelated to merit, ability, or past performance. Discrimination is behavior aimed at denying members of minority groups equal opportunities to achieve valued social goals (education, health care, long life) and/or blocking their access to valued goods and services.

Robert K. Merton explores the relationship between prejudice (the attitude) and discrimination (the behavior). He distinguishes between two types of individuals: the nonprejudiced (those who believe in equal opportunity) and the prejudiced (those who do not). Merton further asserts, however, that people's beliefs about equal opportunity are not necessarily related to their conduct. That is, some prejudiced people do not discriminate and some unprejudiced people actually do discriminate. The accompanying chart shows Merton's typology for types of people. The plus sign (+) indicates an attitude/behavior that supports equal opportunity; the minus sign (−) indicates an attitude/behavior that rejects equal opportunity.

	Attitude Dimension: Prejudice and Nonprejudice	Behavior Dimension: Discrimination and Nondiscrimination
Type I: unprejudiced nondiscriminator	+	+
Type II: unprejudiced discriminator	+	−
Type III: prejudiced nondiscriminator	−	+
Type IV: prejudiced discriminator	−	−

In Merton's typology **nonprejudiced nondiscriminators** (all-weather liberals) accept the creed of equal opportunity, and their conduct conforms to that creed. They represent a "reservoir of culturally legitimized goodwill" (Merton 1976, p. 193). People who act to realize these goals would qualify as nonprejudiced nondiscriminators.

Unprejudiced discriminators (fair-weather liberals) believe in equal opportunity but engage in discriminatory behaviors because it is to their advantage to do so or because they fail to consider the discriminatory consequences of some of their actions. For example, unprejudiced discriminators decide to move out of their neighborhood after a black family moves in because they are afraid that property values might start to decrease. Or a white personnel officer tells friends and neighbors (who also likely are white) about a job opening. This word-of-mouth method of recruiting reduces the chances that a minority group candidate will learn of the opening.

Prejudiced nondiscriminators (timid bigots) do not accept the creed of equal opportunity, but they refrain from discriminatory actions primarily because they fear the potential sanctions if they are caught. Timid bigots do not often express their true opinions about racial and ethnic groups. One kind of prejudiced nondiscriminator is the employer who chooses not to discriminate simply because he or she fears a lawsuit. The employer takes note of the average jury award for job discrimination and decides that such discrimination may be too costly.

http://www.baclaw.com/#Jury

Document: Jury Award Statistics

What is the average jury award for the various forms of discrimination? Which type of discrimination received the largest monetary awards?

Prejudiced discriminators (active bigots) reject the notion of equal opportunity and profess a right, even a duty, to discriminate. They derive "large social and psychological gains from [the] conviction that any white man (including the village idiot) excels over anyone from the hated group" (Merton 1976, p. 198). The members of this group are most likely to believe that they "have the moral right" to destroy the people whom they see as threatening their values and way of life. Of the four categories in Merton's typology, prejudiced discriminators are the most likely to commit **hate crimes,** actions aimed at humiliating minority group members and destroying their property or lives. The Phoenix Police Department provides a more comprehensive definition of hate crimes.

http://www.getnet.com/silent/hatecrimes/

Document: Phoenix Police Department: Hate Crimes

How does this police department define hate crimes? What are the four most frequently committed hate crimes? What is the role of the Hate Crimes Advisory Committee?

Sociologists Jack Levin and Jack McDevitt (1993) studied more than 450 hate crimes committed in Boston between 1983 and 1987. They found that almost 60 percent were turf-related; that is, they were directed against people who were walking in, driving through, moving into, or working in areas where their attackers did not believe they belonged. Levin and McDevitt found that two-thirds of the attackers were under age 29 and that 108 of the crimes involved whites attacking blacks while 95 involved blacks attacking whites. Almost 30 percent of the victims were women. Sociologist Ivan Light believes that although hate crimes often are motivated by deep prejudices against a racial or ethnic group, "the roots of intergroup conflict are as much in economic competition as . . . in negative stereotypes" (1990, p. B7). During tight economic times, for example, some whites may blame their unemployment on affirmative action policies that they believe favor minorities, and blacks may blame white discrimination and prejudice as the source of their economic problems.

Merton's typology is useful for understanding the complex relationship between prejudice and discrimination. People do not fit neatly into one category, however, as is the case with ex-neo-Nazi Floyd Cochran.

http://www.alternet.org/an/niot/cochran.html

Document: Memoirs of a Professional Racist

Q: Which categories in Merton's typology can be applied to the various "chapters" in Cochran's life? Explain your answer.

Institutionalized Versus Individual Discrimination

Sociologists distinguish between individual and institutionalized discrimination. **Individual discrimination** is any overt action on the part of an individual that depreciates minority group members, denies them opportunities to participate, or does violence to their lives and property. A person who tells offensive jokes or posts them on the internet represents one example of individual discrimination (see **http://wiretap.spies.com/ftp.items/Library/Humor/Jokes/offensiv.jok** for examples of offensive jokes). Please note that by citing this URL address, I am taking the risk of putting these jokes in the hands of those who might use them to deprecate minorities. The intended purpose is to show that this kind of discrimination exists and that these kinds of posted jokes cannot be classified simply as harmless entertainment.

Institutionalized discrimination is the established and customary way of doing things in society—the unchallenged rules, policies, and day-to-day practices that impede or limit minority group members' achievements and keep them in a subordinate and disadvantaged position. It is the "systematic discrimination through the regular operations of societal institutions" (Davis 1978, p. 30). The Civil Rights Act of 1964 addressed institutional discrimination when it banned discrimination in voting, employment, public facilities and accommodations, and so on (see **http://www.law.cornell.edu/uscode/42/ch21.html** for an overview). The act defines various actions that qualify as institutional discrimination; here we focus on unlawful employment practices.

http://www.law.cornell.edu/uscode/42/2000e-2.html

Document: Unlawful Employment Practices

Look over the list of unlawful employment practices. Do you know or have you heard of a case of an employer who has failed to comply with Civil Rights Act employment directives? Describe the situation.

Of the two levels of discrimination, institutionalized discrimination is the more difficult to identify, condemn, hold in check, and punish, because it can exist in society even if most members are not prejudiced. Institutionalized discrimination cannot be traced to the motives and actions of particular individuals; discriminatory actions result from simply following established practices that seem on the surface to be impersonal and fair. After all, the same rules supposedly apply to everyone.

Institutionalized discrimination can also be due to laws or practices that penalize minorities for the consequences of past prejudice and discrimination. Such discrimination can be overt or subtle. It is overt when laws and practices are designed with the clear intention of keeping minorities in subordinate positions, as in North Carolina's 1831 Act Prohibiting the Teaching of Slaves to Read. It is subtle when the discriminatory consequences of a practice are neither planned nor intended. Consider Proposition 187, passed in November 1994 in California, which "denies welfare, health care (except in life or death cases), and public education to undocumented immigrants." Although Proposition 187 has not become law in California because its constitutionality has been challenged in the federal courts, many people act as if it is the law.

http://hcs.harvard.edu/~perspy/may95/187cons.html

Document: The Aftermath of Prop 187: Licensing Human Rights Abuses Against Racial Minorities

Which groups have been most affected by the Proposition 187 vote? How have they been affected?

One might even argue that the way in which the dominant group implements affirmative action programs is a subtle form of institutionalized discrimination. As essayist and English professor Shelby Steele (1990) observes, affirmative action programs, as they typically have been carried out, have some troubling side effects because

> the quality that earns minorities preferential treatment is an implied inferiority. However this inferiority is explained—and it is easily enough explained by the myriad deprivations that grew out of our oppression—it is still inferiority.
>
> The effect of preferential treatment—the lowering of normal standards to increase black representation—puts blacks at war with an expanded realm of debilitation doubt. (pp. 48–49)

One wonders whether this side effect would exist if blacks were recruited for jobs and college scholarships in the same way they are recruited for college sports. Few people seem to be disturbed when black athletes are recruited to play at predominantly white colleges, possibly because school officials, coaches, recruiters, students, and alumni believe that these athletes' presence on campus benefits the school. If the same energy and positive attitude were applied to recruiting black college students or job applicants, some of the troublesome features of the affirmative action programs might be alleviated. In many ways the on-the-field integration of major league baseball is a model (although not a perfect one) for affirmative action programs. Syndicated columnist William Raspberry pointed this out on the "MacNeil/Lehrer Newshour":

> "The Major Leagues [until 1948 had] been an exclusive preserve of white men, skilled white men, but white men. And it was also very clear then that there were a lot of black baseball players as good as many of those whites who were in the big leagues. But the rules said you couldn't be a major league ballplayer if you were black. Branch Rickey make the breakthrough in hiring Jackie Robinson to say we're going to bust this up. . . . But once the decision was made to bust it up, a couple of things happened. One had to change . . . the ways baseball looked for and recruited ballplayers. You couldn't just look to the minor leagues anymore because blacks weren't there either. They had to look in some places they weren't accustomed to looking and had to use some techniques they weren't accustomed to using to find prospects. That's affirmative action in one sense and I buy it absolutely. But there was another piece of it. Nobody supposed that pretty good black baseball players who because they had been denied opportunity to play in the big leagues before then should have been given special breaks and brought into the big leagues with lesser skills than their white counterparts. The assumption was, and we bought it, black and white, that if we were given the opportunity to use our skills, and develop our skills, you didn't have to cut us any breaks." ("MACNEIL/LEHRER NEWSHOUR" 1991)

To complicate matters further, most people do not understand what affirmative action means; they equate affirmative action with quotas and with special preferences, especially for blacks. A review of affirmative action laws would eliminate these common misperceptions. For the latest amendment to the 1964 Civil Rights Act, see the Civil Rights Act of 1991.

gopher://wiretap.spies.com/00/Gov/US-Docs/civil91.act
Document: The Civil Rights Act of 1991

Q: Does the Civil Rights Act of 1991 specify quotas as a means of achieving affirmative action goals? What kinds of evidence must a plaintiff supply to demonstrate discrimination? Is affirmative action a program designed to protect only blacks from discrimination? If no, identify other protected groups.

To this point we have looked at societal barriers that prevent racial and ethnic minorities from adapting and/or assimilating. These barriers include racist ideology, dependence on foreign labor, prejudice, and discrimination (individual and institutional). Although we have viewed these barriers in a general way, we have not examined how they operate in everyday interaction. In *Stigma: Notes on the Management of Spoiled Identity* sociologist Erving Goffman gave us such a framework.

Social Identity and Stigma

When people encounter a stranger, they make many assumptions about what that person ought to be. Based on an array of clues such as physical appearance, mannerisms, posture, behavior, and accent, people anticipate the stranger's **social identity**—the category (for example, male or female, black or white, professional or blue-collar, under 40 or over 40, American-born or foreign-born) to which he or she belongs and the qualities that they believe, rightly or wrongly, to be "ordinary and natural" (Goffman 1963, p. 2) for a member of that category. The important quality underlying all stigmas is that those who possess them are seen by others not as multidimensional, complex persons, but as one-dimensional beings.

Goffman was particularly interested in social encounters in which one of the parties possesses a stigma. A stigma is an attribute that is defined as deeply discrediting because it overshadows all other attributes that the person might possess. Goffman defined three broad varieties of stigma: (1) physical deformities, (2) character blemishes attributed to factors such as sexual orientation, mental hospitalization, or imprisonment, and (3) stigmas of race, ethnicity, nationality, or religion as reflected in skin color, dress, language, occupation, and so on.

According to Goffman it is important to remember that the discrediting trait itself is not the stigma. Rather the stigma consists of the set of beliefs about the trait. Furthermore, from a sociological point of view, the person who possesses the trait should not be the focus of attention; the focus should be on the reaction to the trait. Thus Goffman maintained that sociologists should focus not on the attribute that is defined as a stigma, but on interaction—specifically interaction between the stigmatized and **normals.** Goffman did not use the term *normal* in the literal sense of "well-adjusted" or "healthy." Instead he used it to refer to those people who are in the majority or who possess no discrediting attributes. Goffman's choice of this word is unfortunate because some readers forget how Goffman intended it to be used.

Mixed Contact Between the Stigmatized and the Majority Population

In keeping with this focus, Goffman (1963) wrote about **mixed contacts,** "the moments when stigmatized and normals are in the same 'social situation,' that is, in one another's immediate physical presence, whether in a conversation-like encounter or in the mere co-presence of an unfocused gathering" (p. 12). According to Goffman, when normals and the stigmatized interact, the stigma comes to dominate the interaction in several ways.

First, the very anticipation of contact can cause normals and stigmatized individuals to try to avoid one another. One reason is that interaction threatens their sense of racial and ethnic "purity" or loyalty. Sometimes the stigmatized and the normals avoid one another in order to escape mutual scrutiny. Persons of the same race may prefer to interact with one another so as to avoid the discomfort, rejection, and suspicion they encounter from people of another racial or ethnic group ("I have a safe space. I don't have to defend myself or hide anything, and I'm not judged on my physical appearance" [Atkins 1991, p. B8]). Another reason that stigmatized persons and normals make conscious efforts to avoid one another is that they believe widespread social disapproval will undermine any relationship. With regard to interracial relationships, some people believe that racist attitudes will destroy even the "most perfect and loving interracial relationships [because] racism waits like a cancer, ready to wake and consume the relationship at any, even the innocuous, time" (Walton 1989, p. 77).

Goffman observed a second pattern that characterizes mixed contacts: upon meeting each other, each party is unsure how the other views or will act toward him or her. Thus the two parties are self-conscious about what they say and how they act. For the stigmatized the source of the uncertainty is not that everyone they meet will view them in a negative light and treat them accordingly. Rather the chance that they might encounter prejudice and discrimination is great enough to give them reason to be cautious about all encounters.

A third pattern characteristic of mixed contacts is that normals often define accomplishments by the stigmatized, even minor accomplishments, "as signs of remarkable and noteworthy capacities" (Goffman 1963, p. 14) or as evidence that they have met someone from the minority group who is an exception to the rule. In *Two Nations: Black and White, Separate, Hostile, Unequal,* Andrew Hacker (1992) notes that whites attending professional meetings or panel discussions tended to applaud longer and more strenuously at the introduction of black participants and at the end of their remarks. Although Hacker acknowledges that in some instances the applause was well deserved, he also maintains that many whites do this to let the black speakers know that they are "in the company of friendly whites" (p. 56).

As well as defining the accomplishments of the stigmatized as something unusual, normals also tend to interpret the stigmatized person's failings, major and even minor (such as being late for a meeting or cashing a bad check), as related to the stigma. Thus a minority group member on welfare is perceived differently than a person classified as white on welfare. The minority group member is on welfare simply because being on welfare is a "natural" outcome for someone of that racial or ethnic group; the white person is on welfare due to individual failure, not skin color.

A fourth pattern characteristic of mixed contacts is that the stigmatized are likely to experience invasions of privacy, especially when people stare:

> I am tired of walking into restaurants, especially with a group of African-Americans, and having the patrons and proprietors act as if they were being visited by Martians.
>
> I'm tired of being out with my 8-year-old nephew and his best friend and never being completely at ease because I know they will act like little boys. Their curiosity will be perceived as criminal behavior. (SMOKES 1992, P. 14A)

If the stigmatized show their displeasure at such treatment, normals often dismiss such complaints as exaggerated, unreasonable, or much ado about nothing. The normals argue that everyone suffers discrimination in some way and that the stigmatized do not have a monopoly on oppression. They announce that they are tired of the complaining and that perhaps the stigmatized are not doing enough to help themselves

(Smokes 1992). Those who respond in this way dismiss the larger social context and historical forces that shape interaction and place all responsibility for race-related problems on the shoulders of those groups in question. Feedback (letters) to *Up Front*, a publication of the Canadian-based organization Heritage Front, illustrates the nature of this response.

http://www.pathcom.com/~freedom/hf/upfront19.html
Document: Up Front, Issue 19

 Q: Skim through some of the correspondence posted on this site. What are two or three dominant themes that appear in these letters?

The discussion thus far may lead you to believe that members of racial and ethnic minority groups are passive victims at the mercy of the dominant group. This is not the case, however; minorities respond in a variety of ways to being treated as members of a stigmatized group.

Responses to Stigmatization

In *Stigma: Notes on the Management of Spoiled Identity* Goffman (1963) described five ways in which the stigmatized respond to people who fail to accord them respect or to treat them as individuals. One way is to attempt to correct that which is defined as the failing, as when people change the visible cultural characteristics that they believe represent barriers to status and belonging. For example, a person may straighten his or her hair; undergo plastic surgery or do other things to alter the shape of the nose, eyes, or lips; or enroll in a school to change an accent. The comments of a young woman, unhappy with her physical features, and of an instructor from an accent training school illustrate:

> "When I was about thirteen I started to straighten my 'horse hair' so that it would be like white people's hair that I admired so much. I was convinced that with straight hair I would be less conspicuous. I would squeeze my lips together so that they appeared less 'puffy.' Everything, to make myself beautiful and less conspicuous." (EMDE 1992, P. 103)

> "They came here [to American Accent Training, a Berkeley, California, school for correcting accents] for 'corrective surgery.' . . . Correcting an accent problem requires that students be willing to separate their identities from the way they talk. They have to be in effect willing to talk white middle class." (NATIONAL PUBLIC RADIO 1990)

Alternatively, instead of taking direct action and changing the "visible" attributes that normals define as failings, the stigmatized may attempt an indirect response. That is, they may devote a great deal of time and effort to trying to overcome the stereotypes or appear as if they are in full control of everything around them. They may try to be perfect—to always be in a good mood, to outperform everyone else, or to master an activity ordinarily thought to be beyond the reach of or closed to people with such traits. The following statements reflect the second response to stigmatization:

> "[You have to be on guard all the time;] there is no way to get away from it. Because if you do something like close the door to your room, people will start saying, 'Is she being angry? Is she being militant?' You can't even afford to be moody. A white girl can look spacey and people will say, 'Oh, she's being creative.' But if you walk around campus with anything but a big smile on your face, they'll wonder, 'Why is she being hostile?'" (ANSON 1987, P. 92)

> "Don't ever do anything bad, because people are always looking for you to do something bad. You not only have to be good, you have to be perfect. If you do something bad, it's not a mark against yourself but a mark against the entire race." (ANSON 1987, P. 129)

> "As far as that's concerned, I have a mask, too. I give the appearance of going through life in full control. Sometimes I can actually feel myself, right when I'm stepping out of the house, pull my shoulders back and take on a perfectly erect bearing. I can't walk relaxed at all. Walking through the streets like that, I'm unapproachable." (WIEDENROTH 1992, P. 176)

> "For awhile I actually practiced walking erect. That was during my school days, when I felt I wasn't accepted or taken seriously on several different sides. I built up a facade of 'Nobody can do anything to me.'" (OPITZ 1992A, PP. 176–77)

In another type of indirect response, the stigmatized take issue with the way that normals define a particular situation, or they take action to change the way normals respond to them. For example, in 1955 Rosa Parks, a black seamstress from Montgomery, Alabama, refused to give her seat on the bus to a white person. Her actions, which challenged a law requiring her to do so, triggered a boycott of Montgomery buses and helped spark the civil rights movement. Although such actions to change the way normals respond are in one way very direct, they fall into Goffman's category of indirect responses because the stigmatized person does not try to change his or her traits, but rather attempts to change the way that normals respond to those traits.

Sometimes the stigmatized respond in a third way: they use their subordinate status for secondary gains, including personal profit or "an excuse for ill success that has come [their] way for other reasons" (Goffman 1963, p. 10). If a black person, for example, levels a charge of racism and threatens to file a lawsuit in a situation in which he or she is justly sanctioned for poor job, academic, or other performance, that person is using his or her status for secondary gains. A fourth response is to view discrimination as a blessing in disguise, especially for its ability to build character or to teach a person about life and humanity. Finally the stigmatized can condemn all of the normals and view them negatively: "You build up these perceptions of whites, that whites are mean and vile, never trust a white person" (Anson 1987, p. 127).

Discussion Question

Based on the information presented in this chapter, do you believe "race" is the product of racial classification? If no, why not? If yes, will knowledge of this fact change the interaction dynamics, attitudes, and behavior with regard to race? Explain.

Additional Reading

For more on

- Racial categories
  http://www.vrx.net/aar/educate5.html
  Document: **When Racial Categories Make No Sense**

- The history of the concept of "race"*
  http://www.grapevine-sys.com/~newworld/racehist.html
  Document: **The Islamic Solution to the Issue of "Race"**

- Statistical Policy Directive No. 15 (the official statement on racial classification in the United States)
  http://ftp.fedworld.gov/pub/omb/re.fr2
  Document: **Standards for the Classification of Federal Data on Race and Ethnicity**

- UN International Convention on the Elimination of All Forms of Racial Discrimination
  http://wiretap.spies.com/ftp.items/Gov/Treaties/Treaties/racial.un
  Document: **International Convention on the Elimination of All Forms of Racial Discrimination**

- Racial discrimination in hiring
  http://www.cdinet.com/Rockefeller/Briefs/brief8.html
  Document: **Racial Discrimination in Hiring**

  http://epn.org/sage/rstill.html
  Document: **"Soft" Skills and Race: An Investigation of Black Men's Employment**

- Employment discrimination against Hispanics
  http://www.cdinet.com/Rockefeller/Briefs/brief9.html
  Document: **Employment Discrimination Against Hispanics**

- Bill of Rights for racially mixed people
  http://www.webcom.com/intvoice/rights.html
  Document: **Bill of Rights for Racially Mixed People**

- Undocumented immigration with focus on U.S. dependence on racial/national/ethnic groups
  http://www.law.indiana.edu/glsj/vol2/calavita.html
  Document: **U.S. Immigration Policy: Contradictions and Projections for the Future**

- An extreme form of institutional discrimination (the Nazi Holocaust)
  http://www.bmj.com
  Path: /BMJ/Medicine and Global Survival/Medicine and Global Survival: Current Issue
  Document: **Whither Nuremberg?: Medicine's Continuing Nazi Heritage**

---

*The value of the site lies with the historical account; the author's development of the Islamic solution argument is weak.

- Definitions and concepts of race and ethnicity
  http://www.inform.umd.edu:8080/EdRes/Topic/Diversity/Reference/Diversity_Dictionary
  Document: Diversity Dictionary

- Book reviews (race and ethnicity)
  gopher://RS6000.cmp.ilstu.edu/11/depts/polisci/COURSES/POS302
  Document: Race and Ethnicity Seminar Book Reviews

Gender

Sociologists define **gender** as social distinctions based on culturally conceived and learned ideas about appropriate behavior and appearance for males and females. "Appropriate" male and female behavior varies according to time and place. In the 1950s, for example, men were not expected to witness their children's births. Today, however, it is taken for granted that the father will be present. As another example, women in the United States typically remove the hair on their faces (even very small amounts of hair), on their legs, and under their arms. In contrast people in many European countries and elsewhere do not expect women to be hairless in these areas.

If we simply think about the men and women we encounter every day, we quickly realize that people of the same sex vary in the extent to which they meet their society's gender expectations. Some people conform to gender expectations, others do not. This variability, however, does not stop most people from using their society's gender expectations to evaluate their own and others' behavior and appearances in "virtually every other aspect of human experience, including modes of dress, social roles, and even ways of expressing emotion and experiencing sexual desire" (Bem 1993, p. 192).

Sociologists find gender a useful concept, not because all people of the same sex look and behave in uniform ways, but because a society's gender expectations are central to people's lives whether they conform rigidly or resist. For many people failure to conform to gender expectations, even if they fail deliberately or conform only reluctantly, is a source of intense confusion, pain, and/or pleasure (Segal 1990).

In this chapter we explore the concepts used by sociologists to analyze the connection between gender and life chances. In outlining this connection sociologists distinguish between sex (a biologically based classification scheme) and gender (a socially constructed phenomenon). They also focus on the extent to which society is gender-polarized—that is, organized around a clear-cut male–female distinction. In addition sociologists seek to explain gender stratification and the mechanisms by which people learn and perpetuate their society's expectations about gender-appropriate behavior and appearances. Finally sociologists explore the interactions between gender and such variables as race and ethnicity. We turn first to a definition of feminism.

Feminism in the United States

In the broadest sense a **feminist** is a man or woman who actively opposes *gender scripts* (learned patterns of behavior expected of males and females) and believes that men's and women's self-image, aspirations, and life chances should not be constrained by those scripts (Bem 1993). For example, a man should be free to choose to stay home and take care of the children rather than pursuing a full-time career; a female athlete should be able to develop her physique beyond what is considered feminine.

Unfortunately the term *feminist* evokes in many people extremely negative images and stereotypes of mannish-looking women who hate men and who find vocations such as mother and wife oppressive and unrewarding. Remember, however, that people who call themselves "feminists" are not a homogeneous group. This Web site contains an excellent discussion of the term *feminism* and a list of suggestions about how to discuss feminism in a constructive way.

http://www.cis.ohio-state.edu/hypertext/faq/usenet/feminism/info/faq.html
Document: The alt.feminism FAQ (Monthly Posting)

Q: What is a feminist? What might be the purpose of a "feminist" discussion group that includes both feminists and antifeminists? What are three basic premises on which all feminists might agree? Why do so many men participate in discussions on feminist issues? Study the tips for constructive exchange among feminist groups and between feminists and antifeminists. Do you know of anyone who has violated these rules for constructive discussion? Give examples.

Feminist literature in the United States has a relatively long history. The feminist scholarship that has been published by men and women, especially since 1960, is vast and wide-ranging (Komarovsky 1991). A clear feminist movement has been in place, however, since at least the mid-1840s. In 1848 Elizabeth Cady Stanton prepared a Declaration of Sentiments, the Declaration of Independence revised to include women—the "other half" of the United States—and listing social and economic inequities women have endured.

http://www.rochester.edu/SBA/declare.html
Document: Declaration of Sentiments

Q: What specific social, political, and economic aspects of society were women denied equal access to? Did any men sign the 1848 document?

Distinguishing Sex and Gender

Although many people use the terms *sex* and *gender* interchangeably, they do not have the exact same meaning. Sex is a biological concept; gender is a social construct.

Sex as a Biological Concept

A person's sex is determined on the basis of **primary sex characteristics,** the anatomical traits essential to reproduction. Most cultures divide the populations into two categories—male and female—largely on the basis of what most people consider to be clear anatomical distinctions. Like race, however, even biological sex is not a clear-cut category, if only because a significant (but unknown) number of babies are born **intersexed.** This is a broad term used by the medical profession to classify people with some mixture of male and female biological characteristics.

The intersexed group includes three very broad categories: (1) true hermaphrodites, persons who possess one ovary and one testis; (2) male pseudo hermaphrodites, persons who possess testes and no ovaries but some elements of female genitalia; and (3) female pseudo hermaphrodites, persons who have ovaries and no testes but some elements of male genitalia. For more on intersexuality and the physical characteristics defined as such, see the FAQ page posted by the Intersex Society of North America. Skim the answers to the questions defining intersexuality and the various intersexual conditions. Pay particular attention to the answer to the FAQ document question "What are the frequencies of these conditions?"

http://www.isna.org/FAQ.html
Document: Frequently Asked Questions

Q: Do we know how many people are intersexed? Explain. What is the estimated percentage of babies born with an intersexed condition? What does the existence of an intersexed group tell us about the male–female dichotomy?

Why, then, is there no official intersexed category? Instead parents of intersexed children collaborate with physicians to assign their offspring to one of the two recognized sexes. Intersexed infants are treated with surgery and/or hormonal therapy. The rationale underlying medical intervention is the belief that the condition "is a tragic event which immediately conjures up visions of a hopeless psychological misfit doomed to live always as a sexual freak in loneliness and frustration" (Dewhurst and Gordon 1993, p. A15). The Intersex Society of North America provides some information on recommendations for treatment of intersex infants and children.

---

http://www.isna.org/Pamphlets/recommendations.html

Document: Recommendations for Treatment of Intersex Infants and Children

Q: What is the current model for treating intersexed infants and children? In what year was the model developed? What is the "conspiracy of silence"? What is the "new paradigm" for the treatment of intersexed children?

Even adding a third category would not do justice to the complexities of biological sex. French endocrinologists Paul Guinet and Jacques Descourt estimate that on the basis of variations in the appearance of external genitalia alone, the category "true hermaphrodite" may contain as many as 98 subcategories (Fausto-Sterling 1993). This variation within only one intersex category suggests that "no classification scheme could [do] more than suggest the variety of sexual anatomy encountered in clinical practice" (p. A15).

The picture becomes even more complicated when we consider that a person's primary sex characteristics may not match the sex chromosomes. Theoretically one's sex is determined by two chromosomes: X (female) and Y (male). Each parent contributes a sex chromosome: the mother contributes an X chromosome, and the father an X or a Y depending on which one is carried by the sperm that fertilizes the egg. If this chromosome is a Y, then the baby will be a male. Although we cannot possibly know how many people's sex chromosomes do not match their anatomy, the results of mandatory "sex tests" of female athletes over the past 25 years have shown us that such cases exist and that a few women are disqualified from each Olympic competition and from other major international competitions because they "fail" the tests (Grady 1992). Perhaps the most highly publicized after-the-fact case is that of Spanish hurdler Maria José Martinez Patino, who although "clearly a female anatomically, is, at a genetic level, just as clearly a man" (Lemonick 1992, p. 65). Upon giving her the test results, track officials advised her to warm up for the race but to fake an injury so as not to draw the media's attention to her situation (Grady 1992). Patino lost her right to compete in amateur and Olympic events but subsequently spent three years challenging the decision. The IAAF (International Amateur Athletic Federation) restored her status after deciding that her X and Y chromosomes gave her no advantage over female competitors with two X chromosomes (Kolata 1992; Lemonick 1992).

In addition to primary sex characteristics and chromosomal sex, **secondary sex characteristics** are used to distinguish one sex from another. These are physical traits not essential to reproduction (breast development, quality of voice, distribution of facial and body hair, and skeletal form) that result from the action of so-called male (androgen) and female (estrogen) hormones. We use the phrase "so-called" because, although testes produce androgen and the ovaries produce estrogen, the adrenal cortex produces androgen and estrogen in both sexes (Garb 1991). Like primary sex characteristics none of these physical traits has any clear dividing lines to separate males from females. For example, biological females have the potential for the same hair distribution as biological males—follicles for a mustache, a beard, and body hair. Moreover females produce not only estrogen but also androgen, a steroid hormone that triggers hair growth.

Gender as a Social Construct

Whereas sex is a biological distinction, gender is a social distinction based on culturally conceived and learned ideas about appropriate appearance, behavior, and mental or emotional characteristics for males and females (Tierney 1991). The terms **masculinity** and **femininity** signify the physical, behavioral, and mental or emotional traits believed to be characteristic of males and females, respectively (Morawski 1991).

To grasp the distinction between sex and gender, we must note that no fixed line separates maleness from femaleness. Often we attribute differences between males and females to biology when in fact they are more likely to be socially created. In the United States, for example, norms specify the amount and distribution of facial and body hair appropriate for females. It is acceptable for women to have eyelashes, well-shaped eyebrows, and a well-defined triangle of pubic hair, but not to have hair above their lips, under their arms, on their inner thighs (outside the bikini line), or on their chin, shoulders, back, chest, breasts, abdomen, legs, or toes. Most men, and even most women, do not realize that women work to achieve these cultural standards and that their compliance makes males and females appear more physically distinct on this trait than they are in reality. We lose sight of the fact that significant but perfectly normal biological events—puberty, pregnancy, menopause, stress—contribute to the balance between the two hormones androgen and estrogen. Changes in the proportions of these hormones trigger hair growth that departs from societal norms about the appropriate amount and texture of hair for females. When women grow hair as a result of these events, they tend to think something is wrong with them instead of seeing it as natural. A "female balance" between androgen and estrogen is one in which a woman's hair is consistent with these norms.

The extreme measures taken by some women to eliminate facial hair are reflected in reports of physicians who intervened after women were harmed by commercial treatments. These treatments caused adverse side effects including wrinkling, scarring, discoloration, and cancerous growths due to X-ray treatments, as well as paralysis caused by depilatories containing thallium acetate, a highly toxic substance (Ferrante 1988).

Just as women strive to meet norms for facial and body hair, they work to achieve the idealized standards of feminine beauty as portrayed in such places as magazines and television. "Body Image and 'Eating Disorders,'" an excerpt from the *Barnard/Columbia Women's Handbook*, focuses on the tension between valued images and reality.

gopher://gopher.cc.columbia.edu:71/00/publications/women/wh27
Document: Body Image and "Eating Disorders"

Q: What are the social and demographic characteristics of the ideal female body as portrayed in the media? How has the weight of the female model changed over the past 25 years relative to average weight of women in the United States? How is the feminine ideal of thinness related to basic American values?

To this point we have made a distinction between sex and gender. Although sociologists acknowledge that no clear-cut biological markers distinguish males from females, they would not argue that biological differences do not exist. Sociologists are interested in the extent to which differences are socially induced. To put it another way, they are interested in the actions men and women take to accentuate differences between them—actions that lead to gender polarization.

Gender Polarization

In *The Lenses of Gender*, social psychologist Sandra Lipsitz Bem (1993) defines **gender polarization** as "the organizing of social life around the male–female distinction," so that people's sex is connected to "virtually every other aspect of human experience, including modes of dress, social roles, and even ways of expressing emotion and experiencing sexual desire" (p. 192). To understand how just about every aspect of life is organized around this distinction, we consider research by Alice Baumgartner-Papageorgiou.

In a paper published by the Institute for Equality in Education, Baumgartner-Papageorgiou (1982) summarizes the results of a study of elementary and high school students in which she asked the students how their lives would differ if they were members of the opposite sex. Their responses reflect culturally conceived and learned ideas about gender-appropriate behaviors and appearances. The boys generally believed that as girls their lives would change in negative ways. Among other things they would become less active and more restricted in what they could do. In addition they would become more conscious about their appearance, about finding a husband, and about being alone and unprotected in the face of a violent attack:

- "I would start to look for a husband as soon as I got into high school." (p. 5)
- "I would play girl games and not have many things to do during the day." (p. 10)
- "I'd have to shave my whole body." (p. 5)
- "I would not be able to help my dad fix the car and truck and his two motorcycles." (p. 12)

The girls generally believed that if they were boys they would be less emotional, their lives would be more active and less restrictive, they would be closer to their fathers, and they would be treated as more than sex objects:

- "I would have to stay calm and cool whenever something happened." (p. 7)
- "People would take my decisions and beliefs more seriously." (p. 13)

- "I would not have to worry about being raped." (p. 6)
- "It would not take me long to get ready for school." (p. 5)

These beliefs about how the character of one's life depends on one's sex seem to hold even among the college students enrolled in my introductory sociology classes. In the fall 1993 semester I asked students to take a few minutes to write about how their lives would change as members of the opposite sex. The men in the class believed that they would be more emotional and more conscious of their physical appearance and that their career options would narrow considerably:

- "I wouldn't always have to appear like I am in control of every situation. I would be comforted instead of always being the comforter."
- "My career options would narrow. Now I have many career paths to choose from, but as a woman I would have fewer."

The women believed that as men they would have to worry about asking women out on dates and about whether their major was appropriate. They also believed, however, that they would make more money, be less emotional, and be taken more seriously:

- "I'd have to remain cool when under stress and not show my emotions."
- "I think that I would change my major from 'undecided' to a major in construction technology."

These comments by high school and college students show the extent to which life is organized around male–female distinctions. They also show that students' decisions about how early to get up in the morning, what subjects to study, whether to show emotion, how to sit, and whether to encourage a child's athletic development are **gender-schematic decisions.** Decisions and viewpoints about any aspect of life are gender-schematic if they are influenced by a society's polarized definitions of masculinity and femininity rather than on the basis of other criteria such as self-fulfillment, interest, ability, or personal comfort. For example, college students make gender-schematic decisions about possible majors if they ask, even subconsciously, what the "sex" of the major is and, if it matches their own sex, consider it a viable option or, if it does not match, reject it outright (Bem 1993).

The extent to which college students' choice of major is gender-schematic has changed over the past 20 years. The report "Condition of Education—1995" contains a table that shows the proportion of bachelor's degrees earned by women in various fields relative to men at three points in time—1972, 1982, and 1992. For example, in 1972 the proportion of bachelor's degrees earned by women in biological sciences was .54. This means that for every 100 degrees awarded to men, 54 were awarded to women— a two-to-one difference.

http://www.ed.gov/pubs/CondOfEd_95/ovw3.html
Document: Educational Progress of Women

Q: Scroll down several screens until you come to the table "Ratio of the proportion of bachelor's degrees earned by women in a specific field relative to the proportion earned by men." In which academic disciplines were women most concentrated in 1972? In 1982? In 1992? Which academic disciplines show the greatest change in the proportion of degrees awarded relative to men?

Even sexual desire between men and women is organized around male–female characteristics unrelated to reproduction. Bem (1993) argues that "neither women nor men in American society tend to like heterosexual relationships in which the woman is bigger, taller, stronger, older, smarter, higher in status, more experienced, more educated, more talented, more confident, or more highly paid than the man, [but] they do tend to like heterosexual relationships in which the man is bigger, taller, stronger, and so forth, than the woman" (p. 163).

The negative consequences of channeling sexual desire according to age differences is evident when we consider that the average woman outlives her spouse by about nine years and the average life expectancy for women is seven years longer than that of men. This practice, in combination with differences in life expectancy, means that women can expect to live a significant portion of their lives as widows.

Not only is sexual desire between men and women influenced strongly by gender-polarized ideas, but emotions toward persons of the same sex are also influenced. **Social emotions** are internal bodily sensations that we experience in relationships with other people; **feeling rules** are norms specifying appropriate ways to express those sensations. When I asked students to comment on social emotions or "internal bodily sensations" that they had felt and expressed toward someone of the same sex, most indicated that other people generally made them feel uncomfortable and defensive about such feelings.

A society's feeling rules are so powerful that they even affect how people solve problems. For example, when human evolutionists discovered a pair of petrified footprints believed to be 3.5 million years old, they inferred that the prints belonged to a man and a woman, not to two women, two men, or an adult and a child. Ian Tattersall (1993), curator of the American Museum of Natural History, explains the logic underlying this conclusion:

> We know that [the people who left the footprints] were walking side by side because even though the individuals are of different size, because their footprint sizes are different, their stride lengths are matched. They must have been walking together. And if they were walking together, the footprints are so close that they must have been in some kind of bodily contact with each other. What the nature of that contact was we don't know. We have chosen to put the arm of the, of the male around the shoulder of the female. It's a bit anthropomorphic, but we couldn't think of a less, a more non-committal, if you want, kind of a gesture. (p. 13)

The point is that Tattersall could not imagine someone putting his or her arm on the shoulder of a same-sex person. In the United States physical contact between same-sex persons is reserved for specific situations. Men can give full body hugs during an athletic contest but cannot hold hands or put their hands on each other's shoulders while walking down the street. As of 1993 servicemen and servicewomen holding hands with someone of the same sex were subject to investigation (Lewin 1993). These norms against touching someone of the same sex are so powerful that some museum curators assume they existed 3.5 million years ago and construct exhibits that reflect such norms.

The information on gender and gender polarization suggests that one's sex has a profound effect on life chances—determining how long one can expect to live, what major one chooses in college, and whether one dates a shorter or taller person. Thus a person's sex is an important variable in determining his or her position in a society's system of social stratification.

Sexual Stratification

As discussed in Chapter 10, social stratification is the system societies use to place people in categories. When sociologists study stratification, they study how the categories in which people are placed affect their perceived social worth and their life chances. They are particularly interested in how persons who possess one category of a characteristic (male versus female reproductive organs) are regarded and treated as more or less valuable than persons who possess the other category. Sociologist Randall Collins (1971) offers a theory of sexual stratification to analyze this phenomenon.

Collins's theory is based on three assumptions: (1) people use their economic, political, physical, and other resources to dominate others, (2) any change in the way that resources are distributed in a society changes the structure of domination, and (3) ideology is used to justify one group's domination over another. In the case of males and females, males in general are physically stronger than females. Collins argues that because of differences in strength between men and women, the potential for coercion by males exists in every encounter they have with females. He maintains that the ideology of **sexual property**—which he defines as the "relatively permanent claim to exclusive sexual rights over a particular person" (p. 7)—is at the heart of sexual stratification and that for the most part women historically have been viewed and treated as men's sexual property.

Collins believes that the extent to which women are viewed as sexual property and are subordinate to men historically has depended, and still depends, on two important and interdependent factors: (1) women's access to agents of violence control, such as the police, and (2) women's position relative to men in the labor market.

Economic Arrangements

On the basis of these factors, Collins identifies four historical economic arrangements: low-technology tribal societies, fortified households, private households, and advanced market economies. Note that these are ideal types; the economic reality is usually a mixture of two or more types. Each arrangement is characterized by distinct relationships between men and women.

Low-technology tribal societies include hunting-and-gathering societies with technologies that do not permit the creation of surplus wealth, or wealth beyond what is needed to meet basic needs. In such societies the sex-based division of labor is minimal because all members must contribute if the group is to survive. Some evidence, however, shows that in hunting-and-gathering societies women perform more menial tasks and work longer hours than men. Men hunt large animals, for example, while women gather most of the food and hunt smaller animals. Because there is almost no surplus wealth, marriage does little to increase a family's wealth or political power.

Consequently daughters are not treated as sexual property in the sense that they are not used as bargaining chips to achieve such aims.

Fortified households are characteristic of preindustrial societies, in which there is no police force, militia, national guard, or other peacekeeping organization. Therefore the household is an armed unit, and the head of the household is its military commander. Fortified households "may vary considerably in size, wealth, and power, from the court of a king or great lord . . . down to households of minor artisans and peasants" (Collins 1971, p. 11). All fortified households, however, have a common characteristic: the presence of a **nonhouseholder class** consisting of propertyless laborers and servants. In the fortified household "the honored male is he who is dominant over others, who protects and controls his own property, and who can conquer others' property" (p. 12). Men treat women as sexual property in every sense: daughters are bargaining chips for establishing economic and political alliances with other households; male heads of household have sexual access to female servants; and women (especially in poorer households) bear many children, who eventually become an important source of labor. In this system women's power depends on their relationship to the dominant men. Because of their position, the servants in the household have few opportunities to form stable marriages or family lives.

Private households emerge along with a market economy; a centralized, bureaucratic state; and the establishment of agencies of social control that alleviate the need for citizens to take the law into their own hands. Thus private households exist when the workplace is separate from the home. Men are still heads of households, but they assume the role of breadwinner (as opposed to military commander), while women remain responsible for housekeeping and child rearing. As heads of households men control the property; it is a relatively new practice in the United States to obtain a mortgage or credit in the names of both husband and wife. Moreover men monopolize the most desirable and important economic and political positions. Collins states that a decline in the number of fortified households, the separation of work from home, smaller family size, and the existence of a police force to which women can appeal in cases of domestic violence gave rise to the notion of romantic love as an important ingredient in a marriage. In the marriage market men offer women economic security because they dominate the important, high-paying positions. Women offer men companionship and emotional support and strive to be attractive—that is, to achieve the ideals of femininity, which may include possessing an 18-inch waist and removing most facial and body hair. At the same time, they try to act as sexually inaccessible as possible because sexual access is something they offer men in exchange for economic security.

Advanced market economies offer widespread employment opportunities for women. Although women are far from being men's economic equals, some women now can enter into relationships with men with more than an attractive appearance; they also can offer an income and other personal achievements. Having more to offer, women can demand that men be physically attractive and meet certain standards of masculinity. This situation may explain why more commercial attention has been given to males' appearance—body building, hairstyles, male skin and cosmetic products—in the past decade.

Collins's theory of sexual stratification suggests that men and women cannot be truly equal in a relationship until women are men's economic equals. To reach this goal—still unrealized—the earnings gap between men and women must be closed, and fathers must share equally in household and child-rearing responsibilities.

Household and Child-Rearing Responsibilities

The Bureau of Labor Statistics has reported on earning differences between men and women.

http://www.dol.gov/dol/wb/public/wb_pubs/wagegap2.htm
Document: Earnings Differences Between Women and Men

Q: How is the earnings gap calculated? Why is it calculated in this way? What percentage of men's wages do women earn? What are the corresponding earnings gap figures? What factors account for this earnings gap? Can you think of factors other than those named in this report that might account for the earnings gap?

Whatever the reasons for the earnings gap, the difference in income between men and women suggests that women have a disadvantaged position in the labor market relative to men. This disadvantage helps to explain why after a divorce women's standard of living declines dramatically while men's standard of living increases. This is especially the case when children are involved because more often than not women bear primary responsibility for their care. Consequently women who maintain families have the most serious economic problems.

http://www.dol.gov/dol/wb/public/wb_pubs/wwmf1.htm
Document: Women Who Maintain Families

Q: How is a female-headed family defined in this report? How many families are maintained by women in the United States? What percentage of women who maintain families with children under age 18 work full-time? With children over age 18? What are the median incomes of families headed by women? By married couples? By men? How many families maintained by women are classified as poor?

In the United States research shows consistently that even among couples in which the spouses earn the same income and share egalitarian ideals, the husbands spend considerably less time than the wives in preparing meals, taking care of children, shopping, and performing other household tasks (Almquist 1992).

In addition to having equality in the labor market and the home, women must also have access to agents of violence control.

Access to Agents of Violence Control

Collins (1971) argues that women must have access to agents of violence control if they are to be men's equals. Agents of violence control includes persons authorized to make and enforce laws for the state or other organizations. Examples of such agents include the police, FBI, private security forces, prison guards, government officials, and the military. In the case of violence control, these groups have the power to enact and enforce laws that protect women from physical and/or psychological harm. Female circumcision, sex screening and selection, and rape are specific examples of violence against women. For two news items on this topic, see "India Bans the Use of Sex Screening Tests" and "Egypt Against Female Circumcision."

http://edie.cprost.sfu.ca/gcnet/ISS4-05d.html
Document: India Bans the Use of Sex Screening Tests

http://edie.cprost.sfu.ca/gcnet/ISS4-05e.html
Document: Egypt Against Female Circumcision

Q: What steps are the Indian and Egyptian governments taking to protect women from physical violence?

As another example of a situation in which women's access to agents of violence control is important, consider the treatment of women who have been raped. Even today in the United States, when women are raped or otherwise physically abused, they must prove to police, medical personnel, prosecutors and judges, and sometimes even family members that they did not do something to provoke the attack.

In the United States rape laws vary according to state; the strictest laws require that the defense prove force on the part of the man, and the most liberal laws make it a crime for a man to have sexual intercourse without the woman's consent (Burns 1992). Although legal reforms over the past 20 years have made it easier to prosecute rape cases, juries "still tend to blame the victim, particularly if she used alcohol or drugs, kept late hours, frequented bars, had an active sex life, or—in short—was in their eyes [of] 'questionable moral character'" (Jones 1994, p. 14). Furthermore juries are swayed by the appearance of the accused: the man has to "look like a rapist" (p. 14). Of course these dynamics are not peculiar to the United States. Jennifer Morrow reports on judges' reaction to the case of an Indian woman who accused five upper-caste men of rape.

gopher://csf.Colorado.Edu/OR149488-153662-/feminist/Femisa/95/dec95
Document: Rural Indian Women Protest Rape Verdict

Q: Why were the five men acquitted of rape charges?

We have seen that there are a number of sources and expressions of inequalities based on gender. Even where there is a physical basis for these inequalities (as in the case of physical strength), the inequalities themselves are social rather than biological realities. Even in the case of wartime rape, soldiers are not acting out a biological imperative. Their behavior reflects, among other things, the way their social reality has been redefined so that women are considered suitable targets of aggression and rape is considered acceptable under the circumstances.

Hazards in the Workplace

Although our discussion has focused on the inequality of women relative to men, this does not mean that men suffer no disadvantages in the labor market. On the one hand, men as a group, especially white males, have an economic advantage relative to women in the labor market. On the other hand, men are concentrated in the most hazardous occupations. The Bureau of Labor Statistics publishes data on fatal and nonfatal occupation-related illnesses and injuries. (*Note:* Depending on the browser you are using, some tables may be difficult to read.)

ftp://stats.bls.gov/pub/news.release/osh2.txt

Document: Characteristics of Injuries and Illnesses Resulting in Absences from Work, 1994

ftp://stats.bls.gov/pub/news.release/osh.txt

Document: Workplace Injuries and Illnesses in 1994

ftp://stats.bls.gov/pub/news.release/cfoi.txt

Document: National Census of Fatal Occupational Injuries, 1995

Q: What type of injuries and illnesses are most likely to be underreported in the Bureau of Labor Statistics survey? Which eight industries accounted for more than 30 percent of the 6.3 million nonfatal illnesses and injuries? How many industries are male-dominated? How many are female-dominated? What proportion of nonfatal illnesses and injuries are specific to men? Review the tables in "Workplace Injuries and Illnesses in 1994." Are there some injuries and illnesses for which females are overrepresented? How do you account for this? Now review Table 4 in "National Census of Fatal Occupational Injuries." Of the 6,588 fatalities in 1994, how many were men? Why do you think men have such high rates of injury, illness, and fatality?

Mechanisms of Perpetuating Gender Expectations

People vary in the extent to which they conform to their society's gender expectations. This fact, however, does not prevent us from using gender expectations to evaluate our own and others' behavior. For many people failure to conform (whether that failure is deliberate or reluctant) is a source of intense confusion, pain, or pleasure. This leads sociologists to ask what mechanisms explain how we learn and perpetuate a society's gender expectations. To answer this question, we examine three important factors: socialization, situational constraints, and ideologies.

Socialization

Socialization is a learning process that begins immediately after birth and continues throughout life. Through this process "newcomers" develop their human capacities, acquire a unique personality and identity, and internalize (take as their wont and accept as binding) the norms, values, beliefs, and language they need to participate in the larger community. Socialization theorists argue that an undetermined but significant portion of male–female differences are products of the ways in which males and females are socialized.

Child development specialist Beverly Fagot and her colleagues observed how toddlers in a play group interacted and communicated with one another and how teachers responded to the children's attempts to communicate with them at age 12 months and at age 24 months (Fagot et al. 1985). Fagot found no real sex differences in the interaction styles of 12-month-old boys and girls: all of the children communicated by gestures, gentle touches, whining, crying, and screaming. The teachers, however, interacted with the toddlers in gender-polarized ways. They were more likely to respond to girls when the girls communicated in gentle, "feminine" ways and to boys when the boys communicated in assertive, "masculine" ways. That is, the teachers tended to ignore assertive acts by girls and to respond to assertive acts by boys. Thus, by the time these toddlers reached two years of age, the differences in their communication styles were quite dramatic.

Fagot's findings may help explain the differing norms governing body language for males and females. According to women's studies professor Janet Lee Mills (1985), norms governing male body language suggest power, dominance, and high status, whereas norms governing female body language suggest submissiveness, subordination, vulnerability, and low status. Mills argues that these norms are learned and that people give them little thought until someone breaks them, at which point everyone focuses on the rule breaker. Such norms can prevent women from conveying a sense of security and control when they are in positions that demand these qualities, such as lawyer, politician, or physician. Mills suggests that women face a dilemma: "To be successful in terms of femininity, a woman needs to be passive, accommodating, affiliative, subordinate, submissive, and vulnerable. To be successful in terms of the managerial or professional role, she needs to be active, dominant, aggressive, confident, competent, and tough" (p. 9).

In addition to the ways in which adults treat children, children's toys figure prominently in the socialization process.

http://www.toysrus.com/bestsell.html

Document: Toys at the Top

Q: Study the lists of the top five best-selling toys in nine categories. Make three lists: (1) toys for boys, (2) toys for girls, and (3) gender-neutral toys. Which of the top-selling toys belong in each category? Based on the descriptions, what qualities and behaviors do "girl toys" and "boy toys" emphasize as important?

As a prime example of how toys contribute to socialization, consider Barbie dolls. These dolls have been on the market for more than 30 years and currently are available in 67 countries. Executives at Mattel, the company that created Barbie, are studying Eastern Europe as a potential new market. The company considers Barbie to be an aspirational doll—that is, the doll is a role model for the child. Barbie accounts for approximately half of all toy sales by Mattel (Boroughs 1990; Cordes 1992; Morgenson 1991; Pion 1993). Of girls between ages 3 and 11, 95 percent have Barbie dolls, which come in several different skin colors. Market analysts attribute Mattel's success with Barbie to the fact that "they generally have correctly assessed what it means to a little girl to be grown-up" (Morgenson 1991, p. 66). Return to "Toys at the Top." Is Barbie among the top five sellers?

Structural or Situational Constraints

Situational theorists agree with socialization theorists that biological makeup does not account for the social and economic differences between men and women. Rather the causes of these differences are structural and situational. For example, a structural constraint is that occupations are segregated by sex, so that women tend more often than men to be concentrated in low-paying, low-ranking, dead-end jobs. Even in professional and management positions, women are concentrated in specialties and fields that handle children and young adults, that involve supervising other women, and/or that are otherwise considered more feminine (a social work professor versus a mathematics and computer sciences professor).

These structural differences affect expectations about gender. The different social and physical demands and skills required to perform the jobs held by men and women function toward "channeling their motivations and their abilities into either a stereotypically male or a stereotypically female direction" (Bem 1993, p. 135). This point does not preclude the fact that men and women may limit their job searches to positions that are considered "sex-appropriate." However, considerable evidence supports the hypothesis that once women are hired, management steers male and female employees into different assignments and offers them different training opportunities and chances to move into better-paying positions.

Women are no less highly motivated than men to seek advancement in the workplace. The implication is clear: if we remove structural barriers to advancement, women will seek to improve their position. The character of structural barriers vary from one society to another. The World Watch Institute describes structural barriers that women living in subsistence economies face. (Theoretically subsistence economies produce only enough food for internal consumption, but not enough for export.)

gopher://gopher.igc.apc.org/00/orgs/worldwatch/worldwatch.news/10
Document: Gender Bias: Roadblock to Sustainable Development

Q: The World Watch Institute report maintains that discrimination against women is one important reason per capita income has declined in 50 countries and that this discrimination is reinforced by structural barriers, especially conventional strategies of economic development. What are those structural barriers?

This report describes the important economic role of female labor, but it does not explain why the division of labor between men and women is so lopsided with regard to raising crops and meeting basic family needs. In other words the report does not tell us what the men are doing while the women work. For hints about what men may be doing, read about employment and labor relations in the Philippines.

gopher://gopher.umsl.edu/00/library/govdocs/armyahbs/aahb4/aahb0269
Document: Employment and Labor Relations

Q: What might men be doing while the women work?

Sociologist Renee R. Anspach's (1987) research illustrates vividly how one's position in a social structure can channel behavior in stereotypically male or female directions. Anspach spent 16 months conducting field research (observing and interviewing) in two neonatal intensive care units (NICUs). Among other things she found that nurses (almost all of whom were female) and physicians (usually male) used different criteria to answer the question "How can you tell if an infant is doing well or poorly?" Physicians tended to draw on so-called objective (technical or measurable) information and immediate perceptual cues (skin color, activity level) obtained during routine examination:

> "Well, we have our numbers. If the electrolyte balance is OK and if the baby is able to move one respirator setting a day, then you can say he's probably doing well. If the baby looks gray and isn't gaining weight and isn't moving, then you can say he probably isn't doing well.
>
> "The most important thing is the gestalt. In the NICU, you have central venous pressure, left atrial saturations, temperature stability, TC (transcutaneous) oxymeters, perfusions (oxygenation of the tissues)—all of this adds in. You get an idea, when the baby looks bad, of the baby's perfusion. The amount of activity is also important—a baby who is limp is doing worse than one who's active."
> (ANSPACH 1987, PP. 219–20)

Although immediate perceptual and measurable signs were important to the nurses as well, Anspach found that the nurses also considered interactional clues such as the baby's level of alertness, ability to make eye contact, and responsiveness to touch:

"I think if they're doing well they just respond to being human or being a baby. . . . Basically emotionally if you pick them up, the baby should cuddle to you rather than being stiff and withdrawing. Do they quiet when held or do they continue to cry when you hold them? Do they lay in bed or cry continuously or do they quiet after they've been picked up and held and fed. . . . Do they have a normal sleep pattern? Do they just lay awake all the time really interacting with nothing or do they interact with toys you put out, the mobile or things like that, do they interact with the voice when you speak?" (p. 222)

Anspach concluded that the differences between nurses' and physicians' responses to the question "How can you tell if an infant is doing well or poorly?" can be traced to their daily work experiences. In the division of hospital and health-care labor, nurses interact more with patients than do physicians. The "Equal Employment Opportunity File" gives a broad overview of the health-care division of labor, specifically the proportion of males and females in each occupational category.

gopher://una.hh.lib.umich.edu/00/census/summaries/eeous
Document: Equal Employment Opportunity File (EEO)

Q: Space down to "Health Diagnosing Occupations" and "Health Assessment and Treating Occupations." Note that the first column of numbers is the number of males in each occupational category while the second column applies to females. Is Renee Anspach correct in her assumption that the occupational category "physician" is male-dominated while the category "nurses" is female-dominated? Explain.

Doctors and nurses also have access to different types of knowledge about infants' condition, which correspond to our stereotypes of how females and males manage and view the world. Because physicians have only limited amounts of daily interaction and contact with infants, they tend to rely on perceptual and technological (measurable) cues. Nurses, in contrast, are in close contact with infants throughout the day; consequently they are more likely to consider interactional cues as well as perceptual and technological ones.

Anspach (1987) suggests that one's position in the health-care division of labor "serves as a sort of interpretive lens through which its members perceive their patients and predict their futures" (p. 217). Her findings suggest that when physicians make life-and-death decisions about whether to withdraw or continue medical care, they should collaborate with NICU nurses so that they can consider interactional as well as technological and immediate perceptual cues. Anspach's findings also suggest that if nurses' experiences and opinions counted more in medical diagnosis, we might see a corresponding increase in the prestige and salary associated with this largely female position.

Anspach's research on the relative weight assigned to physicians' and nurses' experiences points to another problem that affects men's and women's life chances: institutionalized discrimination. **Institutionalized discrimination** is the established and customary way of doing things in society—the collection of unchallenged rules, policies, and day-to-day practices that impede or limit people's achievements and keep them in subordinate and disadvantaged positions on the basis of ascribed characteristics. This situation is "systematic discrimination through the regular operations of social institutions" (Davis 1979, p. 30).

Sexist Ideologies

Ideologies are ideas that support the interests of a dominant group but that do not stand up to scientific investigation. They are taken to be accurate accounts and explanations of why things are the way they are. On closer analysis, however, we find that ideologies are at best half-truths, based on misleading arguments, incomplete analysis, unsupported assertions, and implausible premises.

Sexist ideologies are structured around three notions:

1. People can be classified into two categories, male and female.
2. There is a close correspondence between a person's primary sex characteristics and characteristics such as emotional activity, body language, personality, intelligence, the expression of sexual desire, and athletic capability.
3. Primary sex characteristics are so significant that they explain and determine behavior and the social, economic, and political inequalities that exist between the sexes.

Sexist ideologies are so powerful that "almost everyone has difficulty believing that behavior they have always associated with 'human nature' is not human nature at all but learned behavior of a particularly complex variety" (Hall 1959, p. 67). One example of a sexist ideology is the belief that men are prisoners of their hormones, making them powerless in the face of female nudity or sexually suggestive dress or behavior. Another example is the belief that men are not capable of forming relationships with other men that are as meaningful as those that women form. Since the 1980s dozens of books have been written by men in response to these stereotypes (Shweder 1994). Assuming that all divorced or single men with children are "deadbeat" dads is still another example of sexist ideology at work. F.R.E.E. (Fathers' Rights and Equality Exchange) is one organization that is addressing this stereotype. While it is true that all divorced and single men with children cannot be classified as "deadbeat" dads, we must recognize that a significant number of women do not receive child support payments or do not receive the full amount due.

http://info-sys.home.vix.com/free/index.html
Document: The Fathers' Rights and Equality Exchange

Select the document "F.R.E.E. Reports 'Children Now' Child Support Collection Numbers Are Not Accurate." Describe F.R.E.E.'s critique of child support numbers. According to F.R.E.E. what percentage of fathers pay child support? Read the document "solid numbers," a hypertext link in the F.R.E.E. document. Is there evidence to support F.R.E.E.'s claim without qualification?

We might also add a fourth characteristic of a sexist ideology: gender nonconformity. **Gender nonconformists** are individuals who behave in ways that depart from concepts of masculinity or femininity and who thus are considered deviant, in need of fixing, and subject to negative sanctions ranging from ridicule to physical violence. Gender nonconformists include the following:

- Persons whose primary sex characteristics are not clear-cut (the intersexed)
- Persons whose secondary sex characteristics depart from the ideal conceptions of masculinity and femininity
- Persons whose interests, feelings, sexual orientation, choice of occupation, or academic major do not adhere to gender-polarized scripts
- Persons "who actively oppose the gender scripts of the culture" (Bem 1993, p. 167)

Sexist ideologies are reflected in social institutions. A good example is the military. One ideology that has dominated U.S. military policy since World War II is stated in a 1990 U.S. Department of Defense directive:

> Homosexuality is incompatible with military service. The presence of such members adversely affects the ability of the Armed Forces to maintain discipline, good order, and morale; to foster mutual trust and confidence among the members; to ensure the integrity of the system of rank and command; to facilitate assignment and worldwide deployment of members who frequently must live and work under close conditions affording minimal privacy; to recruit and retain members of the military services; to maintain the public acceptability of military services; and, in certain circumstances, to prevent breaches of security. (p. 25)

There is no scientific evidence, however, to support the claims in this directive. In fact, whenever Pentagon researchers (with no links to the gay and lesbian communities and with no ax to grind) found evidence that ran contrary to this directive, high-ranking military officials refused to release the information or found the information unacceptable and directed researchers to rewrite the reports. For example, when researchers found that sexual orientation is unrelated to military performance and that men and women known to be gay or lesbian displayed military suitability as good as or better than that of men and women believed to be heterosexual, the U.S. deputy undersecretary of defense, Craig Alderman, Jr., wrote the researchers that the "basic work is fundamentally misdirected" (Alderman 1990, p. 108). He explained that the researchers were to determine whether there was a connection between being a homosexual and being a security risk, not whether homosexuals were suitable for military service. Although the researchers found no data to support a connection between sexual orientation and security risks, Alderman maintained that the findings were not relevant, useful, or timely. The research that Alderman dismissed would have gone unnoticed had not Congressman Gerry Studds and House Arms Subcommittee Chairwoman Patricia Schroeder insisted it be released.

This example shows the role that ideologies play in setting policy. In this case the ideologies are that homosexuality is incompatible with military service and that being homosexual represents a security risk to the United States. The case of the military also alerts us to the fact that other variables, such as a person's sexual orientation, race, ethnicity, or social class, interact with gender to make the experience of being male and female different. To illustrate this interaction, we turn to the work of sociologists Floya Anthias and Nira Yuval-Davis, who have written about the interconnection among gender, race and ethnicity, and country (the state).

Gender, Race and Ethnicity, and the State

Ethgender refers to people who share (or are believed by themselves or others to share) the same sex, race, and ethnicity. This concept acknowledges the combined (but not additive) effects of gender, race, and ethnicity on life chances. Ethgender merges two ascribed statuses into a single social category. In other words a person is not a Croat and a woman but a Croatian woman; a person is not an African American and a man but an African-American man (Geschwender 1992). To complicate matters, the country or state that people of a particular ethgender inhabit (and their legal relationship to the state—as citizen, refugee, or temporary worker) has a significant effect on their life chances. Here the term **state** means a governing body organized to manage and control specified activities of people living in a given territory.

Everyone has some legal relationship to the state, whether as a citizen by birth or naturalization, a refugee, a temporary worker, an immigrant, a permanent resident, or an illegal alien. Moreover at birth everyone is classified as male or female and as a member of one race. Sociologists Floya Anthias and Nira Yuval-Davis (1989) give special attention to women, their ethnicity, and the state. They argue that "women's link to the state is complex" and that women "are a special focus of state concerns as a social category with a specific role (particularly human reproduction)" (p. 6). Anthias and Yuval-Davis maintain that the state's policies and discourse reflect its concerns about the kinds of babies (that is, their ethnicity) to which women give birth and about the ways in which the babies are socialized. They identify five areas of women's and men's lives over which the state may choose to exercise control. One should not conclude, however, that women accept state policies and programs without resistance. In fact women often work to modify these policies.

Women as Biological Reproducers of Babies of a Particular Ethnicity or Race

As factors that can underlie a state's population control policies, Anthias and Yuval-Davis (1989) name (1) "fear of being 'swamped' by different racial and ethnic groups" and (2) fear of a "demographic holocaust" (p. 8) (that is, a particular racial or ethnic group will die out or become too small to hold its own against other ethnic groups). Such policies can range from physically limiting numbers of a particular racial or ethnic group deemed undesirable to actively encouraging the "right kind" of women to produce more children. Policies that limit numbers include immigration control (limiting or excluding members of certain ethnic groups from entering a country and subsequently producing children), physical expulsion (which includes ethnic cleansing), extermination, forced sterilization, and massive birth control campaigns. Policies that encourage the "right kind" of biological reproduction include ideological mobilization (appeals to a woman's duty to her country), tax incentives, maternity leave, and other benefits.

Women as Biological Reproducers of the Boundaries of Ethnic or National Groups

In addition to implementing policies intended to encourage women to have or discourage them from "having children who will become members of the various ethnic groups within the state" (Anthias and Yuval-Davis 1989, p. 9), states also implement policies that define the "proper ways" to reproduce offspring. Examples include laws prohibiting sexual relationships with men or women of another race or ethnicity, laws specifying legal marriage if the child is to be recognized as legitimate, and laws connecting the child's ethnic and legal status to the ethnicity of the mother and/or father.

Although these laws apply to both men and women, the woman generally pays a heavier social price when the law is broken. For example, in *Wake Up Little Susie: Single Pregnancy and Race Before Roe v. Wade*, Rickie Solinger (1992) documents the options open to unmarried girls and women who faced pregnancy between 1945 and 1965. They included

> futilely appealing to a hospital abortion committee, [which at that time were not concerned about the question of when human life begins but about punishing single mothers]; being diagnosed as neurotic, even psychotic by a mental health professional; [being] expelled from school (by law until 1972); [becoming] unemployed; [enrolling] in a Salvation Army or some other maternities home; [and being] poor, alone, ashamed, threatened by the law. (P. 4)

Solinger argues that the policies and programs encouraged white women to give up their babies for adoption but encouraged African-American women to keep their babies to prevent them from having more.

Today, in contrast, many school districts across the United States offer on-site daycare, private tutoring, and special classes for pregnant girls. Still, considerable political debate revolves around the questions of whether single mothers should receive welfare, especially after they have a second child, and whether it might be to society's benefit to restigmatize single motherhood (L. Williams 1993). For the most part these debates rarely focus on stigmatizing the fathers of these children. Although there are federal and state programs in place to collect child support from so-called deadbeat dads, only about 50 percent of the women who have filed claims for child support receive the full amount (Department of Commerce 1989).

Women as Transmitters of Social Values and Culture

The state can institute policies that either encourage women to be the main socializers of their offspring or that leave socialization in the hands of the state. Examples include tying welfare payments to nonemployment so that mothers are forced to stay home with the children, instituting liberal or restrictive maternity leave policies, subsidizing daycare centers and providing opportunities to enroll children in preschool. Sometimes state leaders become concerned that children of members of particular ethnic or racial groups are not learning the cultural values and/or language they need to succeed in the dominant culture. This concern motivates them to fund programs that expose children to the necessary personal, social, and learning skills.

Women as Signifiers of Ethnic and Racial Differences

Political leaders often use various images of women to symbolize the most urgent issues they believe the state faces. For example, in wartime the state is represented as "a loved woman in danger or as a mother who lost her sons in battle" (Anthias and Yuval-Davis 1989, pp. 9–10). Men are called to arms to protect the women and children. Often political leaders present the image of a woman who meets the culture's ideal of femininity and who belongs to the dominant ethnic group. Sometimes political leaders use veiled language to evoke images of women of a certain ethnic or racial group as the source of a country's problems (for example, Albanian women who produce many children or African-American welfare mothers with no economic incentives to practice birth control). Frequently there is no evidence to support such generalizations. In "Fertility Among Women on Welfare: Incidence and Determinants," sociologist Mark R. Rank (1989) maintains that "it is impossible to calculate with any precision the fertility rate of women on public assistance" (p. 296) because the data available has serious flaws. "There is no way of judging whether the fertility rate of women on welfare is high or low" (p. 296) relative to the fertility rate of other women.

Women as Participants in National, Economic, and Military Struggles

States implement policies governing the roles that women and men can assume in crises, notably in war. Historically women have played supportive and nurturing roles, even in situations in which they have been exposed to great risks. In most countries women are not drafted; they volunteer to serve. In either case the state defines acceptable military roles. And if women do fight, they often do so as special units or in an unofficial capacity. In February 1994, for instance, Bosnian-Serb leaders announced that "the entire able-bodied population will be mobilized, either into military or labor units, and special women's units will be formed" (Kifner 1994, p. A4).

Regardless of their official roles in the war, women are affected by war. Since World War II civilians have suffered an estimated 80 percent of the deaths and casualties in war (Shaller and Nightingale 1992). Although women are killed, taken prisoner, tortured, and raped, they often are not trained formally and systematically to fight. As a result women occupy a different position than men during war. The fact that women's combat roles are limited does not mean that women are incapable of combat or that they do not fight wars.

Through its military institutions the state even establishes policies that govern male soldiers' sexual access to women outside military bases, both in general and in times of war. The Serbs, for example, sent female prisoners to places resembling concentration camps, in which many were raped; they also kept other women in brothel-like houses and hotels. We know too that during World War II Japanese military authorities forcibly recruited between 60,000 and 200,000 women, mostly Korean but also Chinese, Taiwanese, Filipina, and Indonesian, to work as sex slaves in army brothels in the war zone (Doherty 1993; Hoon 1992). They referred to them as "comfort women."

In *Let the Good Times Roll: Prostitution and the U.S. Military in Asia*, Saundra Pollock Surdevant and Brenda Stoltzfus (1992) examine "the sale of women's sexual labor outside the U.S. military bases" (p. vii). They present evidence that the U.S. military helps regulate prostitution; that retired military officers own some of the clubs, massage parlors, brothels, discotheques, and hotels; and that the military provides the women with medical care so as to prevent the spread of sexually transmitted diseases. In 1993 thousands of Filipina women who live near the Subic Bay Naval Base in the Philippines filed a class action suit against the United States, arguing that the United States has moral and legal responsibilities to support the estimated 8,600 children fathered by U.S. servicemen stationed at Subic Bay. Their suit is further evidence of the U.S. military's involvement in the lives of women abroad.

In identifying areas over which the state controls men's, but especially women's lives, Anthias and Yuval-Davis (1989) highlight the effects of race and ethnicity in combination with gender on people's life chances. In exploring the origins, manifestations, and consequences of the social differences between men and women, a key theme has been that these differences are social constructions rather than biological facts. We now apply these ideas to an issue of great concern to many college-age men and women in the United States: date rape.

Gender and Date Rape

In 1987 Mary P. Koss, Christine A. Gidycz, and Nadine Wisniewski published their findings from a nationwide survey about college students' sexual experiences. The researchers found that 57 percent of female college students said they had experienced some form of sexual victimization (from fondling to forced anal or oral intercourse) during the previous academic year. Fifteen percent said they had been victims of rape, and 12 percent said they had been victims of attempted rape. The researchers also found that most of the women knew the perpetrators. Because slightly more than one-fourth of the female student population reported that they had experienced a rape or an attempted rape, the researchers concluded that rape takes place more frequently than we are led to believe by official statistics such as the FBI's Uniform Crime Reports or the National Crime Victimization Survey.

Koss and her colleagues used the term **hidden rape** to refer to rape that goes unreported. Media accounts of this research coined the catchier, if less precise, phrases *date rape, acquaintance rape,* and *campus rape.* The last of these terms is especially imprecise because Koss did not ask whether the rapists were also college students or attended the same institution as the victim. A number of articles appearing in popular magazines such as *Time* and *Vogue* suggest that an epidemic of "date rape" is occurring, especially on college campuses.

Sociologists G. David Johnson, Gloria J. Palileo, and Norma B. Gray (1992) observed that between 1987, when Koss's study was published, and 1991 the media changed its focus from discovering and labeling a social problem on campus to debating whether so-called date rape is not actually something women claim when they have a "bad sexual experience" or when they don't hear again from their partner. In January 1991 these researchers replicated the Koss study. To do so they surveyed a sample of 1,177 male and female college students attending a southern university. They hoped (among other things) to clarify a number of unanswered questions clouding this debate, including the following:

- Is there an epidemic of date rape on U.S. college campuses?
- Were the female respondents defining "forcible" sexual experiences as rape or were the researchers defining such experiences as rape? In Koss's study, for example, female students were asked, "Have you ever had sexual intercourse when you didn't want to because a man threatened you or used some degree of physical force (twisting your arm, holding you down, etc.) to make you?" Critics point out that women who responded affirmatively to this question were not asked whether they considered that sexual experience to be rape, even though the experience, as worded, constitutes the legal definition of rape.
- Is so-called date rape something we might attribute to miscommunication rather than sexual aggression? Critics of the Koss findings often focused on this question. "This criticism emphasizes the ambiguity of sexual speech and nonverbal communication: When does 'no' really mean no, and when does it mean maybe, or even yes?" (Johnson, Palileo, and Gray 1992, p. 38).

With regard to the first question, Johnson and his colleagues found that the prevalence of rape at the campus they studied was strikingly similar to the prevalence recorded in Koss's national study in 1987. This fact suggests that date rape is not an epidemic (that is, a rapid spread or increase), in the strict sense of the word, on the campus they studied.

With regard to the second question, approximately half (51 percent) of the 149 women in this study who said they had experienced forced sexual relations said the experience was not rape; 12 percent said it was rape; 37 percent said that "some people would think it was close to rape" or that "many would call it rape."

With regard to the third question, the researchers found considerable miscommunication between men and women. Slightly more than one-third of the women reported that they never say "no" when they mean yes, and 66 percent reported that they have said "no" when they meant yes. This finding shows

> a significant number of female university students miscommunicate their sexual intentions. We expected this outcome, but were surprised by the magnitude of the miscommunication. That one in six females at a college campus today always says "no" when she means "yes" indicates the presence of a significant problem. Clearly the ambiguity of sexual communication is a significant barrier for the achievement of nonexploitive sexual relationships between men and women.
> (JOHNSON, PALILEO, AND GRAY 1992, P. 41)

In interpreting the data, we must be careful about the conclusions we draw. For one thing Johnson and his colleagues did not ask women who said they had experienced force whether they had said "no" when they meant yes in those instances. Also we cannot conclude that the women were defining as rape those instances when they reported that they had said "no" but meant yes. Finally, on the basis of questions asked in the Koss and Johnson studies, we cannot conclude that women are crying "rape" the morning after having "bad sex."

Some social commentators such as Katie Roiphe (1993) argue that the media hype and so-called feminist focus on date rape have created a mass hysteria on college campuses and that women are confused about what is rape. Such conclusions go far beyond any data that we have on this subject. Linda A. Fairstein (1993) takes an entirely different perspective than Roiphe and argues that she does not see hysteria on college campuses and does not think young women are confused about what is and is not rape. Fairstein believes that unsubstantiated conclusions such as Roiphe's trivialize the experience of rape.

This debate aside, we are still left with the question of why female respondents at this university at one time or another miscommunicated their intentions to their dates. For one answer to this question, we turn to a classic ethnographic study of a U.S. high school and its students, conducted by Jules Henry (1963). Although the study is more than 30 years old, Henry's observations are still relevant today. Henry argued that sending mixed signals is one strategy women can use to communicate their sexual interest while maintaining their "reputation" and not appearing "easy." The woman sends double messages (saying "no" while indicating in other subtle ways that she means yes), because the feeling rules are such that she can't communicate her desires directly without being considered "loose" or a "slut."

The fact that many women admit to sending mixed signals to preserve their reputation can complicate rape investigations. Whether women send mixed messages or not, some men may rationalize sexual misconduct with the excuse that the victims really were willing. Consider Sergeant Major William F. Merrill's essay "The Art of Interrogating Rapists."

gopher://justice2.usdoj.gov/00/fbi/January95/jan3.txt
Document: The Art of Interrogating Rapists

Q: How do contact rapists and sexual aggressor rapists differ? Which category of rapist is likely to rationalize sexual misconduct with the excuse that the victim was really willing? What approach does Merrill recommend for interrogating contact rapists?

Discussion Question

Based on the information presented in this chapter, what conclusions can you draw about the meaning of gender and about the social mechanisms that lead to gender polarization? Do you think if people become more flexible in their expectations about appropriate behavior for males and females that "societal decay" is inevitable? To put it another way, what are the benefits and hazards if expectations become more flexible?

Additional Reading

For more on

- Gender inequality in income
  http://www.cdinet.com/Rockefeller/FemMasc/wpb6.html
  Document: Comparable Worth: Theories and Evidence

- The glass ceiling
  gopher://gopher.etext.org/11/Politics/Womens.Studies/GenderIssues/GlassCeiling
  Document: various documents

- Women's suffrage
  http://lcweb2.loc.gov/ammem/rbnawsahtml/nawstime.html
  Document: One Hundred Years Toward Suffrage: An Overview

- The fluidity of sexuality and gender identity
  http://ezinfo.ucs.indiana.edu/~mberz/ttt/articles/rights
  Document: The International Bill of Gender Rights

  http://garnet.berkeley.edu:4248/Fantasies_of.html
  Document: Fantasies of Straight Men About Gays in the Military

- Victimization, power, and ideology
  http://english-www.hss.cmu.edu/bs/23/newitz.html
  Document: Myth of the Million Man March

- Domestic violence
  gopher://www.ojp.usdoj.gov/00/bjs/press/spousfac.pr
  Document: Husbands Convicted More Often Than Wives for Spouse Murder

  gopher://www.ojp.usdoj.gov/00/bjs/press/femvied.pr
  Document: Women Usually Victimized by Offenders They Know

- Gender nonconformists
  http://ezinfo.ucs.indiana.edu/~mberz/ttt/articles
  Path: /crossing-line.html
  Document: Crossing the Line

- UN efforts to empower women
  gopher://gopher.undp.org/00/ungophers/popin/unfpa/speeches/1995/summngo.asc
  Document: 180 Days–180 Ways Women_s Action Campaign

- Youth-of-the-world discussions of gender issues
  http://ux641a12.unicef.org/voy/past/voyI/tindex.html
  Document: Messages from Youth Indexed by Topic

# Population and Family Life

On August 20 and September 5, 1977, the United States launched the *Voyager 1* and *Voyager 2* probes into space to explore and photograph Jupiter, Saturn, Uranus, and Neptune. In 1990 the spacecraft left the earth's solar system. Attached to the outside of each was a gold-coated copper phonograph that contained 118 photographs of the planet and its inhabitants, 90 minutes of music from countries around the world, a collection of the sounds of the earth, and greetings in 60 different languages. This "portfolio" of the planet was made to "send to any possible extraterrestrial auditors information about the Earth and its inhabitants" (Sagan 1978, p. 33). Among the 118 photographs was one entitled "Family Portrait."

Imagine for a moment the photograph you would have selected to represent all of the various family arrangements on earth. Choosing a single photograph to represent family life in the United States—not to mention the entire planet—would be an overwhelming challenge. The difficulty is rooted in the fact that, even though every person is a member of a family (if only in the biological sense), "there is no concrete group which can be universally identified as 'the family'" (Zelditch 1964, p. 681). There is an amazing variety of family arrangements worldwide—a variety reflected in the numerous norms that specify how two or more people can become a family. These include norms that govern the number of spouses a person can have, the way a person should select a spouse, the ideal number and spacing of children, the circumstances under which offspring are considered legitimate (or illegitimate), the ways in which people trace their descent, and the nature of the parent–child relationship over the child's and parent's lives. In light of this variability we should not be surprised to learn that it is difficult to construct a definition of family and that "no general theory or universal model of the family can be formulated" (Behnam 1990, p. 549).

Most official definitions of family emphasize blood ties, adoption, or marriage as criteria for membership, and procreation and socialization of offspring as the primary functions. The U.S. Bureau of the Census (1993) uses the term family to mean "a group of two or more persons related by birth, marriage, or adoption and residing together in a household" (p. 5). This definition has been in effect since 1950. Between 1930 and 1950, however, the definition of family revolved around a head of the household and reflected living arrangements in a more rural America:

> A private family comprises a family head and all other persons in the house who are related to the head by blood, marriage, or adoption, and who live together and share common housekeeping arrangements. The term "private household" is used to include the related family members (who constitute the private family) and the lodgers, servants, or hired hands, if any, who regularly live in the home. (U.S. BUREAU OF THE CENSUS 1947, P. 2)

Changes in the official definition reflect dramatic changes in the size and composition of families over the past 20–30 years. The changing definitions also show that focusing on membership criteria and specific functions excludes many primary groups that have familylike qualities, including child-free couples, couples whose children no longer live at home, elderly people who live with their adult offspring, and unmarried heterosexual and homosexual couples with children.

Today changing lifestyles have influenced family structures to deviate from what society would consider the typical family, one consisting exclusively of parents and biological children. The nontraditional family has replaced the traditional family as a normal way of life for many people. The U.S. Census Bureau document gives a definition of a traditional family.

http://www.census.gov/cgi-bin/print_hit_bold.pl/pub/Press-Release/cb94-121.txt

Document: Half of Nation's Children Live in Non-Traditional Families

 Q: As you read this report, make note of the kinds of arrangements that vary from the traditional family. How many U.S. children live in a family that is nontraditional?

One problem with definitions of families is that they can be used to deny family-related benefits to people who don't conform to the accepted definition. Such benefits include insurance coverage, housing, time off from work to care for another family member, inheritances (especially in the absence of a will), and custody of children. In this vein a number of recent court decisions in the United States have ruled that criteria other than the socially recognized bonds of blood, marriage, or adoption must be considered in deciding whether a person belongs to a family and thus is entitled to family-related benefits. Other factors that influence family relationships include exclusivity, longevity, emotional support, and financial commitment (Gutis 1989, 1989b). For a look at how "families" that do not qualify as traditional according to socially recognized criteria but that do have familylike relationships receive family-related benefits, skim through this document.

http://sunsite.unc.edu/gaylaw/files/coleman.pag

Document: Analysis of the Opinion of the Attorney General . . .

Q: What strategy have some nontraditional families adopted to be recognized as a unit deserving of family-related benefits?

No matter what kind of family people are born into, live with, or form later, their lives are shaped by the following key episodes:

- Birth (even if it is only their own birth), which includes the number of children born and the spacing (amount of time) between births
- Work, which includes how much and what kind of work each family member must do, and where family members must travel to find work to sustain the family's standard of living
- Death, which includes how and when (infancy through old age) a family member dies

The sociologists who focus most on these key episodes are those who study human populations. The study of population includes an interest in (1) the number of people in social groupings (those who live within the boundaries of country, state, province, or other geographical area), (2) the composition of these social groupings (the various ages, the ratio of males to females, the percentage of people in various racial and ethnic groups, and so on), and (3) the factors that lead to changes in each social grouping's size and composition (Pullman 1992).

Population and family life are paired together in this chapter because the factors that affect population size and composition represent the sum of the key episodes common to all family life. In this chapter we look at the concepts and theories that sociologists draw upon to explain childbearing experiences (number and spacing of children), the type and timing of death, and migration, especially movement related to the search for work.

Changing definitions of family suggest that it is not especially useful to think about the family in terms of specific memberships or according to a single function (such as procreation or socialization). Instead we would do better to take a long view and consider key episodes that affect family size and composition. One such event that transformed the character of family life in very general but fundamental ways is the Industrial Revolution.

The Industrial Revolution and the Family

Traditionally we think of the Industrial Revolution as an event that began in England in the late eighteenth century and spread to other countries in Western Europe and to the United States. Actually this event forced people from even the most remote regions of the globe into a worldwide division of labor. The extent to which the Industrial Revolution has been realized in other parts of the world varies according to country and, in most cases, according to regions within countries.

The Industrial Revolution cut people off from the family and from the clan-oriented ways of life in which they had existed for most of history (Riesman 1977). The phrase "family and clan-oriented" refers to an environment in which change is slow (although not completely absent), in which family influences almost all facets of life (where one works, whom one marries, what one does), and in which people follow patterns that have endured for centuries. In clan-oriented societies life expectancy at birth is low, and the relationship to the food supply is precarious. In addition the division of labor is simple: the majority of people perform labor-intensive and subsistence-oriented tasks that are allocated on the basis of age, sex, and physical condition. Under these circumstances adult roles change very little from one generation to the next. Parents can assume that their children's daily lives will be much like their own and that their children will face virtually the same challenges and problems. Consequently parents pass on to their children a set of roles and time-tested rules and rituals that promote the group's survival (see the discussion in Chapter 7 on organic solidarity). Because life expectancy is short, children assume these adult roles very early (Riesman 1977).

In general, industrialization in Western societies is associated with dramatic changes in women's childbearing experiences, in the timing and causes of death, and in the kinds of work the majority of people do. Specifically the following changes occur:

- The number of children that women bear and the proportion of their reproductive life that women give to childbearing decreases substantially.

- Degenerative diseases and diseases often caused by lifestyle choices (such as cancer and heart disease) displace infectious and parasitic diseases (smallpox, yellow fever, polio, influenza, typhoid, scarlet fever, measles) as the primary causes of death. The decline in infectious and parasitic diseases results in a dramatic increase in the number of infants who survive the first year of life and the number of women who survive childbirth.

- Natural sources of power (human muscle, animal strength, wind, and water) are replaced with created power (fueled by oil, natural gas, coal, nuclear power, and hydroelectric power) and programmed machines (Bell 1989). Consequently the proportion of the population engaged in manual labor (the extraction and transformation of raw materials) declines, and a greater proportion is engaged in human and professional services (social work, education, health care, entertainment, publishing, sales, marketing, and so on).

At first glance this list of changes may mean very little, but try to imagine the profound effects of such changes on family life. To appreciate what it means for women to have control over the number and spacing of children, consider the reproductive experiences of two women of comparable social status, who lived at different times—Alice Wandesworth Thorton, who lived in the seventeenth century, and Jane Metzroth, a twentieth-century woman and a vice president in investment banking:

> At last [Alice] was persuaded to accept the proposal of one William Thorton, Esq., a gentleman less well-to-do than various other suitors, but more pious. Alice became pregnant almost immediately, and over the next fifteen years bore nine children, three of whom lived—a typical survival rate for her time. She remembers her pregnancies in terms of fevers, sweatings, nosebleeds, faintings, and agues; her deliveries were inevitably perilous, and her recoveries of many months duration. The babies who lived were subject to diarrhea, convulsions, or being rolled upon by their wet nurses. (SOREL 1984, PP. 318–19)

> "There were a couple of years' worth of thinking that went into the decision to have another child. We already had a school-age son, and for a long time I thought we'd just have one child, because working and raising a family and trying to have a normal kind of life, too, was kind of hard work. But after Erik was in school, and I saw him developing into a really neat little person, I kept trying to decide in my own mind whether to do it again. . . . I felt if we were going to add to the family, this would be the time. There were also financial considerations, and child care arrangements; we've had the same lady for almost four years now, and she's really terrific. And that helped me make the decision. . . . So with all those things combined, we said, 'Now's the time.' Plus, I'm in my thirties now, so I didn't want to put it off and then find out it couldn't be possible." (SOREL 1984, PP. 60–61)

In fueling these kinds of changes, as reflected in the lives of these two women, industrialization was indeed a revolution: "Within a few decades a social order which had existed for centuries vanished, and a new one, familiar in its outline to us in the late twentieth century, appeared" (Lengermann 1974, p. 28). However, some countries reaped far more fruit from the Industrial Revolution than did others.

Mechanized Rich Versus Labor-Intensive Poor Countries

We can classify the countries of the world into two broad categories with regard to industrialization: the mechanized rich and the labor-intensive poor. Comparable but misleading dichotomies include developed and developing, industrialized and industrializing, and First World and Third World. They are misleading because they suggest that a country is either industrialized or is not. The dichotomy implies that a failure to industrialize is what makes a country poor, and it camouflages the fact that as Europe and North America plunged into industrialization, they took possession of Asia, Africa, and South America and then established economies oriented to their industrial needs. The point is that labor-intensive poor countries were part of the Industrial Revolution from the beginning.

The World Bank, the United Nations, and other international organizations use a number of indicators to distinguish between mechanized rich and labor-intensive poor countries. These indicators include (1) **doubling time** (the estimated number of years required for a country's population to double in size), (2) **infant mortality** (the number of deaths in the first year of life for every 1,000 live births), (3) **total fertility** (the average number of children women bear over their lifetime), (4) **per capita income** (the average income that each person in a country would receive if the country's gross national product were divided evenly), (5) the percentage of the population engaged in agriculture, and (6) the **annual per capita consumption of energy** (the average amount of energy each person consumes over a year). When per capita energy consumption is low, it suggests that for the vast majority of people work is labor-intensive rather than machine-intensive. With labor-intensive work considerable physical exertion is required to produce food and goods. When the term *labor-intensive poor* (or an equivalent term) is used to characterize a country, it means that the country differs markedly on these and other indicators from countries considered to be industrialized (Stockwell and Laidlaw 1981). According to these measures 163 countries are labor-intensive poor and 45 are mechanized rich (U.S. Bureau of the Census 1989).

The official statistics used to identify labor-intensive poor and mechanized rich countries are counts of how many people have experienced the kinds of events discussed previously. For example, a high infant mortality rate tells us that for many families the death of an infant is a common experience. Similarly a high total fertility rate tells us that the population is growing rapidly. The *World Factbook* provides annual data on a number of such indicators for the countries of the world.

http://www.odci.gov/cia/publications/95fact/index.html

Document: 1995 World Factbook

Q: Use the data in the *World Factbook* to complete the accompanying chart. What do the numbers say about labor-intensive poor countries relative to mechanized-rich countries?

Labor-Intensive Poor	Population Doubling Time (years)	Infant Mortality (per 1,000 births)	Total Fertility	Per Capita GDP Income (U.S. $)	Per Capita Consumption of Electricity (annual)
Nigeria	3.16	72.6	6.31	122.6	NA
Bangladesh	_____	_____	_____	_____	_____
Pakistan	_____	_____	_____	_____	_____
Brazil	_____	_____	_____	_____	_____
India	_____	_____	_____	_____	_____
Indonesia	_____	_____	_____	_____	_____
Ethiopia	_____	_____	_____	_____	_____

	Mechanized Rich					
United States	___	___	___	___	___	___
Italy	___	___	___	___	___	___
Japan	___	___	___	___	___	___

The Theory of Demographic Transition

Demography, the study of population trends, is an area of specialization within sociology. Demographers study birthrates, death rates, and migration rates and their contribution to changes in family structure and population size.

ftp://coombs.anu.edu.au/coombspapers/coombsarchives/demography/what-is-demography.txt

Document: What Is Demography?

 Q: What are two examples of topics demographers study? Why study demography?

In the 1920s and early 1930s, demographers observed birthrates and death rates in various countries and noticed a pattern: both birthrates and death rates were high in the countries of Africa, Asia, and South America; death rates were declining while birthrates remained high in the countries of Eastern and Southern Europe; and birthrates were declining and death rates were low in the countries of Western Europe and North America. Demographers observed that the countries of Western Europe and North America had experienced all three of the conditions mentioned previously according to the following sequence:

- *Stage 1:* Birthrates and death rates were high until about the middle of the eighteenth century, at which time death rates began to decline.
- *Stage 2:* As the death rates decreased, the population grew rapidly because there were more births than deaths. The birthrates began to decline around 1800.
- *Stage 3:* By 1920 both birthrates and death rates had dropped below 20 per 1,000.

On the basis of these observations, demographers put forth the **theory of the demographic transition:** a country's birthrates and death rates are linked to its level of industrial or economic development.

This model documents the general situation and should not be construed as a detailed description of the experiences of any single country. Even so, we can say that all countries have followed or are following the essential pattern of the demographic transition, although they differ with regard to the timing of the declines and the rate at which their populations increase after death rates begin to decline. The theory of demographic transition includes more than this pattern; it is also an explanation of the events that caused birthrates and death rates to drop in the mechanized rich countries (with the exception of Japan). The factors that underlie changes in birthrates and death rates in the labor-intensive poor countries, however, are fundamentally different from those that caused such changes in mechanized rich countries. After we examine each stage of the demographic transition, we will consider the ways in which the demographic transition in labor-intensive poor countries differs from the experience of mechanized rich countries.

Stage 1: High Birthrates and Death Rates

For most of human history—the first 2–5 million years—populations grew very slowly (if at all). World population remained below 1 billion until A.D. 1800, at which point it began to grow explosively. In 1804 the world's population reached a milestone: 1 billion humans on earth. "World Population Milestones" documents trends in world population.

gopher://gopher.undp.org/00/ungophers/popin/wdtrends/mileston
Document: World Population Milestones

Q: How many years did it take after 1804 to reach a world population of 2 billion? Of 3 billion? Of 4 billion? Is there a pattern to the amount of time that elapses between each billion people added to the world population?

Demographers speculate that growth until 1800 was slow because **mortality crises**—frequent and violent fluctuations in the death rate caused by wars, famines, and epidemics—were a regular feature of life (Watkins and Menken 1985).

Stage 1 is often referred to as the stage of high potential growth: if something happened to cause the death rate to decline (for example, improvements in agriculture, sanitation, or medical care) population would increase dramatically. In this stage life is short and brutal; the death rate is almost always above 50 per 1,000 population. When mortality crises occur, the death rate can skyrocket. Sometimes half of the population is affected, as when the Black Plague (a form of the bubonic plague) struck Europe, the Middle East, and Asia in the middle of the fourteenth century and recurred for approximately 300 years. Within 20 years of its onset, the Black Plague killed an estimated three-fourths of the people in the affected populations.

Consult this document for additional information on the Black Plague and how it affected the death rate.

http://jefferson.village.virginia.edu/osheim/intro.html
Document: Plagues and Public Health in Renaissance Europe

Q: Read the introduction and select an account of the Black Plague as it affected one of the following cities: Florence, Pistoia, or Lucca. Why is the Black Plague an example of a mortality crisis?

Another mortality crisis, one that has not received as much attention as the Black Plague, affected the indigenous populations of North America when Europeans began arriving in the fifteenth century. A large proportion of the native population died because they had no resistance to diseases such as smallpox, measles, tuberculosis, and influenza, which the colonists brought with them. Others simply were killed by colonists because they refused to work as slaves on plantations. Historians debate what proportion of the native population died as a result of this contact; estimates range between 50 and 90 percent.

The key point is that in stage 1 average life expectancy at birth remained short—perhaps between 20 and 35 years—with the most vulnerable groups being women of reproductive age, infants, and children under age 5. (The high infant and child mortality rate pulled down the average life expectancy at birth; many people managed to live well beyond age 30.) Women typically bore large numbers of children, often as many as 10 or 12. Families remained small, however, because one infant in three died before reaching age 1 and another died before reaching adulthood. If the birthrate had not remained high, society would have become extinct. Demographer Abdel R. Omran (1971) estimated that in societies in which life expectancy at birth is 30 years, each woman must have an average of seven live births in order to ensure that two children survive into adulthood. And she must bear six sons to ensure that at least one survives until the father reaches age 65 (if the father lives that long).

In Western Europe prior to 1650, high mortality rates were associated closely with food shortages. Even when people did not die directly from starvation, they died from diseases brought on by their weakened physical state. Thus Thomas Malthus, a British economist and an ordained Anglican minister, concluded that "the power of population is so superior to the power in the earth to produce subsistence for man, that premature death must in some shape or other visit the human race" (Malthus [1798] 1965, p. 140). According to Malthus positive checks served to keep the population size in line with the food supply. He defined **positive checks** as events that increase mortality, including epidemics of infectious and parasitic disease, war, and famine. Malthus believed that the only moral way to prevent population growth from outstripping the food supply was delayed marriage and celibacy. He regarded any other method—such as infanticide, homosexuality, or sterility caused by sexually transmitted diseases—as immoral.

In most cases, however, famines do not occur simply because there are too many people and not enough food. Human affairs also play a major role. For example, the country of Somalia has gained worldwide attention in recent years because of mass starvation. Somalia's economy relies heavily on agriculture, but agricultural output has been reduced due to civil war and political unrest, causing widespread famine. In addition Somalia's environment suffers from deforestation, overgrazing, and soil erosion, which in turn leads to a decrease in agricultural production.

http://www.odci.gov/cia/publications/95fact/index.html
Document: 1995 World Factbook

 Can you identify another country that has been affected by deforestation, overgrazing, and/or soil erosion? What specific events led to the environmental problems?

Stage 2: The Transition Stage

Around 1650 mortality crises became less frequent in Western Europe, and by 1750 the death rate began to decline slowly in that region. The decline was triggered by a complex array of factors associated with the onset of the Industrial Revolution. The two most important factors were (1) increases in the food supply, which led improved overall nutrition and thus ability to resist diseases, and (2) public health and sanitation measures, including the use of cotton to make clothing and new ways of preparing food. Sociologist Holger Stub (1982) elaborates:

> The development of winter fodder for cattle was important; fodder allowed the farmer to keep his cattle alive during the winter, thereby reducing the necessity of living on salted meats during half of the year. . . . [C]anning was discovered in the early nineteenth century. This method of food presentation laid the basis for new and improved diets throughout the industrialized world. Finally, the manufacture of cheap cotton clothes became a reality after mid-century. Before then, much of the clothes were seldom if ever washed, especially among the poor. A journeyman's or tradesman's wife might wear leather stays and a quilted petticoat until they virtually rotted away. The new cheap cotton garments could easily be washed, which increased cleanliness and fostered better health. (P. 33)

Contrary to popular belief, advances in medical technology had little effect on death rates until the turn of the century, well after improvements in nutrition and sanitation had caused dramatic decreases in deaths due to infectious diseases.

Over a 100-year period the death rate fell from 50 per 1,000 population to below 20 per 1,000, and life expectancy at birth increased to about 50 years of age. As death rates declined, fertility remained high. It may even have increased temporarily, because improvements in sanitation and nutrition enable women to carry more babies to term. With the decrease in the death rate, the **demographic gap**—the difference between birthrates and death rates—widened, and population size increased substantially. **Urbanization,** an increase in the number of cities and in the proportion of the population living in cities, accompanied the unprecedented increases in population size. (As recently as 1850, only 2 percent of the world's population lived in cities with 100,000 or more residents.)

Around 1880 fertility began to decline. The factors that caused birthrates to drop are unclear and are subject to debate among demographers. But one thing is certain: the decline was not caused by innovations in contraceptive technology, because the methods available in 1880 had been available throughout history. Instead the decline in fertility seems to be associated with several other factors. First, the economic value of children declined in industrial and urban settings as children were no longer a source of cheap labor but became an economic liability to their parents. Second, with the decline in infant and childhood mortality, women no longer had to bear a large number of children in order to ensure that a few survived. Third, a change in the status of women gave them greater control over their reproductive lives and made childbearing less central to their lives. Scholars disagree, however, about the specific conditions under which women are able to control their reproductive lives.

Stage 3: Low Death Rates and Declining Birthrates

In the industrialized West around 1930, both birthrates and death rates fell below 20 per 1,000 population, and the rate of population growth slowed considerably. Life expectancy at birth surpassed 70 years, an unprecedented age. The remarkable reduction infant, childhood, and maternal mortality rates were such that accidents, homicides, and suicide have become the leading causes of death among young people. Because the risk of dying from infectious diseases is reduced, people who would have died of infectious diseases in an earlier era survive into middle age and beyond, when they face the elevated risk of dying from degenerative and environmental diseases (heart disease, cancer, strokes, and so on). For the first time in history, persons age 50 and over account for more than 70 percent of the annual deaths. Before this stage infants, children, and young women accounted for the largest share of deaths (Olshansky and Ault 1986).

As death rates decline, disease prevention becomes an important issue. The goal is to live not only a long life but a quality life (Olshansky and Ault 1986; Omran 1971). As a result people become conscious of the link between their health and their lifestyle (sleep, nutrition, exercise, and drinking and smoking habits). In addition to low birthrates and death rates, stage 3 is distinguished by an unprecedented emphasis on consumption (made possible by advances in manufacturing and food production technologies).

Theoretically all of the mechanized rich countries are in stage 3 of the demographic transition. At one time some sociologists and demographers maintained that the so-called Third World countries would follow this model of development as they industrialized. However, the nature of industrialization in labor-intensive poor countries is so fundamentally different from the version that occurred in the mechanized rich countries that these countries are unlikely to follow the same path.

The Demographic Transition in Labor-Intensive Poor Countries

One reason that the nature of industrialization in labor-intensive poor countries was so fundamentally different from that of mechanized rich countries is that most labor-intensive poor countries were once colonies of mechanized rich countries. The mechanized rich countries established economies oriented to their own industrial needs, not the needs of the countries they colonized. This helps explain why the model of the demographic transition does not apply to labor-intensive poor countries. When compared to mechanized rich countries, labor-intensive poor countries differ on several characteristics: they have (1) a faster decline in death rates, (2) relatively high birthrates despite declines in the death rate, (3) a more rapid increase in population size, and (4) greater (and unprecedented) levels of rural-to-urban and rural-to-rural migration.

Death Rates

The decline in death rates in the labor-intensive poor countries occurred much faster than in the mechanized rich countries; only 20–25 years (rather than 100 years) elapsed before the death rate fell from 50 per 1,000 population to less than 10 per 1,000. Demographers attribute the relatively rapid decline to cultural diffusion (see Chapter 5). That is, the labor-intensive poor countries imported some Western technology—pesticides, fertilizers, immunizations, antibiotics, sanitation practices, and higher-yield crops—which caused an almost immediate decline in the death rates. In addition governments and other organizations have identified and implemented low-cost alternatives to high-tech approaches to disease prevention.

The swift decline in death rates has caused the populations in developing countries to grow very rapidly. Some demographers believe that developing countries may be caught in a **demographic trap**—the point at which population growth overwhelms the environment's carrying capacity:

> Once populations expand to the point where their demands begin to exceed the sustainable yield of local forest, grasslands, croplands, or aquifers, they begin directly or indirectly to consume the resource base itself. Forests and grasslands disappear, soil erodes, land productivity declines, water tables fall, or wells go dry. This in turn reduces food production and incomes, triggering a downward spiral. (BROWN 1987, P. 28)

The countries of Nepal (between China and India) and Costa Rica (in Central America) represent two cases of the demographic trap. Because of population pressures, the peasants of Nepal have been forced to cultivate steep and forested hillsides, and the women have no choice but to collect feed for livestock and wood for cooking and heating from these sites. As the forests recede, the daily journey for fodder and fuel grows longer, and cultivation becomes more difficult. As a result family incomes have dropped, and diets have deteriorated. In fact the malnutrition rates in the Nepal villages are correlated strongly with deforestation rates (Durning, 1990).

In Costa Rica, a country once covered by tropical forest, the demographic trap is fueled by population growth and by land policies of the past two decades, which favor a small number of wealthy cattle ranchers (perhaps 2,000) and disregard the needs of peasants, who have become landless under these policies. About half the nation's cultivatable land is used to raise cattle, an industry that requires little labor and hence provides few employment opportunities for landless peasants. "The rising tide of landlessness has spilled over into expanding cities, onto the fragile slopes, and into the forests, where families left with little choice accelerate the treadmill of deforestation" (Durning 1990, p. 146).

Birthrates

Although still high, birthrates are beginning to show signs of decline in most developing countries. The demographic gap remains wide, however. It is not clear exactly which factors have caused total fertility to decline in developing countries. Sociologist Bernard Berelson (1978) has identified some important "thresholds" associated with industrialization and with declines in fertility, including the following:

- Less than 50 percent of the labor force is employed in agriculture.
- At least 50 percent of persons between the ages of 5 and 19 are enrolled in school.

- Life expectancy is at least 60.
- Infant mortality is less than 65 per 1,000 live births.
- Eighty percent of females between the ages of 15 and 19 are unmarried.

A number of countries in two major geographical regions of the world, sub-Saharan Africa and South-Central Asia, are experiencing fertility declines.

gopher://gopher.undp.org/00/ungophers/popin/wdtrends/fertilit
Document: New Fertility Declines in Sub-Saharan Africa and South-Central Asia

Q: What are two reasons for the decline in fertility? Are all countries in sub-Saharan Africa and South-Central Asia experiencing fertility declines? Explain.

To address the problem of high total fertility, governments and family planning organizations are turning their attention to men. Programs aimed at men recognize that the sole responsibility for limiting pregnancies no longer falls on the female. For example, the Zimbabwean government reported a 29 percent increase in use of family planning for males participating in the Zimbabwean Male Motivation Project in the late 1980s.

Population Growth

A country's annual population growth is calculated as (1) the number of people born in that country during the year minus (2) the number who died plus (3) the number who moved into the country minus (4) the number who moved out of the country. Another way to calculate population growth is to add the natural increase in population and the net migration. *Natural increase* is the number of births minus the number of deaths. *Net migration* is the number of people who move out of the country minus the number of people who move into the country. Generally population size increases faster in countries with a disproportionate number of young adults (men and women of reproductive age) than in countries with a disproportionate number of middle-aged and older people.

A population's age and sex composition is commonly represented by a **population pyramid,** a series of horizontal bar graphs, each of which represents a different five-year age cohort. (A **cohort** is a group of people who share a common characteristic or life event; in this case it includes everybody born in a specific five-year period.) Two bar graphs are constructed for each cohort, one for males and another for females; the bars are placed end to end, separated by a line representing zero. Usually the left-hand side of the pyramid depicts the number or percentage of males that make up each age cohort and the right-hand side depicts the number or percentage of females. The graphs are stacked according to age; the age 0–4 cohort forms the base of the pyramid and the 80+ age cohort is at the apex. The population pyramid allows us to view the relative sizes of the age cohorts and to compare the relative number of males and females in each cohort.

Generally a country's population pyramid approximates one of three shapes—expansive, constrictive, or stationary. The key word here is "approximate" because no country's population structure is a perfect representation of one pyramid. **Expansive pyramids,** characteristic of labor-intensive poor countries, are triangular; they are broadest at the base, and each successive bar is smaller than the one below it. The relative sizes of the age cohorts in expansive pyramids show that the population is increasing in size and that it is composed disproportionately of young people. **Constrictive pyramids,** which characterize some European countries, are narrower at the base than in the middle. This shape shows that the population is composed disproportionately of middle-aged and older people. **Stationary pyramids,** which are characteristic of most developed countries, are similar to constrictive pyramids except that all of the age cohorts in the population are roughly the same size, and fertility is at replacement level.

http://www.odci.gov/cia/publications/95fact/index.html
Document: 1995 World Factbook

Q: Select the country of Rwanda, which has a larger proportion of individuals in the 0–14 age group than in any other age category. Which population pyramid corresponds to this fact? Now choose Denmark, which has more middle-aged people than children or elderly. Which population pyramid corresponds to Denmark's age distribution? Now look up the Ukraine, Belgium, Paraguay, and Canada. Which type of pyramid represents each of their populations?

Demographers study the age–sex composition in order to place birthrates and death rates in a broader context. For example, between 1990 and mid-1991 the world's population increased by an estimated 90 million people. A significant proportion of this increase can be attributed to two countries—the People's Republic of China and India. Even though 90 countries had higher birthrates than India and 134 countries had higher birthrates than China (U.S. Bureau of the Census 1991), these two countries accounted for more than one-third (33 million) of the increase. Such a large share can be attributed to the fact that both countries have large populations to begin with (1.15 billion Chinese and 870 million Indians) and to the fact that more than 50 percent of the Indian and the Chinese populations are of childbearing age—between 15 and 49 (U.S. Bureau of the Census 1991).

The document "Age-Specific Fertility" contains information on how North America compares to other countries with regard to age-specific fertility.

gopher://gopher.undp.org/00/ungophers/popin/wdtrends/agespec.tab
Document: Age-Specific Fertility

Q: Which geographic area has the highest total fertility rate in the world? The highest fertility rate among its age 15–19 female population? Among its age 25–29 population?

Age–sex composition helps us understand in part why death rates in labor-intensive poor countries are the same as those in mechanized rich countries (and sometimes even lower). Brazil has an official death rate of 7 per 1,000 population; the United States has an official death rate of 9 per 1,000. Thirteen percent of the U.S. population, however, is over 65, compared to 5 percent of Brazil's population. If the chances of survival were truly equal in the two countries, the death rate should be substantially higher in the mechanized rich countries because a greater percentage of the population is older and thus at higher risk of death.

http://www.odci.gov/cia/publications/95fact/index.html
Document: 1995 World Factbook

Q: What is one example of a labor-intensive poor country with a death rate lower than that of the United States? Compare the percentage of people under 14 and the percentage of people over 65 in the country you named with their U.S. counterparts. Does this information help you to see why that country has a lower death rate than the United States?

Migration

Migration is a product of two factors: **push factors,** the conditions that encourage people to move out of an area, and **pull factors,** the conditions that encourage people to move into an area. "On the simplest level [and in the absence of force] it can be said that people move because they believe that life will be better for them in a different area" (Stockwell and Groat 1984, p. 291). Some of the most common push factors include religious or political persecution, discrimination, depletion of natural resources, lack of employment opportunities, and natural disasters (droughts, floods, earthquakes). Some of the most common pull factors are employment opportunities, favorable climate, and tolerance. Migration falls into two broad categories: international and internal.

International Migration The movement of people between countries is called **international migration.** Demographers use the term **emigration** to denote the departure of individuals from a country and the term **immigration** to denote the entrance of individuals into a new country. The document "South-to-North Migration Flows" contains the table "south-to-north migration flows," which documents the average annual number of immigrants to North America, Oceania, and Western Europe from other geographic regions between 1960 and 1990.

gopher://gopher.undp.org/00/ungophers/popin/wdtrends/inttab
Document: South-to-North Migration Flows

: Note that an average of 23,814 people emigrated from East and Southeastern Asia each year between 1960 and 1964. Of this number 15,088 were immigrants to the United States. How many people emigrated from East and Southeastern Asia each year between 1985 and 1989? Of that number how many were immigrants to the United States? Select one time period and one geographic area from which people have emigrated. What historical events do you think might "push" people to emigrate from their home country?

Unless their countries are severely underpopulated, most governments restrict the numbers of foreigners who are allowed to enter. Despite restrictions immigrants continue to comprise a large percentage of the American population. According to the U.S. Census Bureau, 1 in 11 U.S. residents in 1994 were foreign-born. The document "Foreign-Born Report Highlights" contains more information on the characteristics of the foreign-born population in the United States.

http://blue.census.gov/population/www/socdemo/foreign/foreign_rpt.html
Document: The Foreign-Born Population: 1994

: Which characteristic of the foreign-born population do you believe is the most significant? Explain.

Currently foreigners are barred from immigrating to the United States by 38 different conditions, including criminal, immoral, or subversive activities; physical or mental handicaps; and economic factors suggesting that the individual would be a drain on the welfare system or would compete unfairly with American workers for jobs.

Three major flows of intercontinental (and by definition international) migration occurred between 1600 and the early part of the twentieth century. The first was the massive exodus of Europeans to North America, South America, Asia, and Africa to establish colonies and commercial ventures. In some cases these immigrants eventually displaced native peoples and established independent countries (for example, in the United States, Brazil, Argentina, Canada, New Zealand, Australia, and South Africa). The second was the smaller flow of Asian migrants to East Africa, the United States (including Hawaii, which did not become a state until 1959), and Brazil, where they provided cheap labor for major transportation and agricultural projects. The third was the forced migration of some 11 million Africans by Spanish, Portuguese, French, Dutch, and British slave traders to the United States, South America, the Caribbean, and the West Indies.

POPULATION AND FAMILY LIFE   275

Internal Migration   In contrast to international migration, **internal migration** is movement within the boundaries of a single country—from one state, region, or city to another. Demographers use the term **in-migration** to denote the movement of people into a designated area and the term **out-migration** to denote movement out of a designated area. If the net migration is a negative number, more people have moved out of than into the area. Conversely, if net migration is a positive number, more people have moved into than out of the area. This document provides data on internal migration in the United States.

http://www.census.gov/ftp/pub/population/socdemo/migration/net-mig.txt
Document: Immigrants, Outmigrants, and Net Migration . . .

Q: Which states in the United States have the largest and smallest numbers of out-migrants? The largest and smallest numbers of in-migrants? Why might each state have high or low numbers of out-migrants? Of in-migrants? (*Note:* Keep in mind that raw numbers of in-migrants and out-migrants do not tell us what percentage of a state's population has left or the percentage of new people entering the state.)

One major type of internal migration is the rural-to-urban movement (urbanization) that accompanies industrialization. However, urbanization in labor-intensive poor countries differs substantially from that which occurred in mechanized rich societies:

> The world has never seen such extremely rapid urban growth. It presents the cities, especially in the developing countries, with problems new to human experience, as well as old problems—urban infrastructure, food, housing, employment, health, education—in new and accentuated forms. Many developing countries will have to plan for cities of sizes never conceived of in currently developed countries. High population growth in developing countries, whatever other factors enter the process, is inseparable from this phenomenon. (RUSINOW 1986, P. 9)

According to a report on rural and urban population estimates and projections compiled by the United Nations, today "45 percent of the world population are urban dwellers." The report states that by 2005 half of the world's population will live in urban areas.

gopher://gopher.undp.org/00/ungophers/popin/wdtrends/urban
Document: World Urbanization Prospects-The 1994 Revision

Q: Which country had the largest percentage of its population in urban areas in 1994? Which country had the smallest percentage?

The unprecedented growth of urban areas is caused by a number of factors. First, the cities in the mechanized rich countries grew in close proportion to the number of jobs created during their industrialization process a century and a half ago. Many Europeans who were pushed off the land were able to emigrate to sparsely populated places like North America, South America, South Africa, New Zealand, and Australia. If those who fled to other countries in the eighteenth and nineteenth centuries had been forced to make their living in European cities, the conditions would have been much worse than they actually were:

> Ireland provides the most extreme example. The potato famine of 1846–1849 deprived millions of peasants of their staple crop. Ireland's population was reduced by 30 percent in the period 1845–1851 as a joint result of starvation and emigration. The immigrants fled to the industrial cities of Britain, but Britain did not absorb all the hungry Irish. North America and Australia also received Irish immigrants. Harsh as life was for these impoverished immigrants, the new continents nonetheless offered them a subsistence that Britain was unable to provide. . . . there are no longer any new worlds to siphon off population growth from the less industrialized countries. (LIGHT 1983, PP. 130–31)

Another type of massive internal migration—rural-to-rural—is taking place in many labor-intensive poor countries. We know very little about the extent of this migration (except that it may involve as many as 370 million people worldwide) or its impact on the communities that the migrants enter and leave. We do know, however, that in this kind of migration people move to and from increasingly more marginal or fragile lands. This group of migrants is known as "environmental refugees."

Industrialization and Family Life

Sociologists have identified how the Industrial Revolution affected the key aspects of family life in mechanized rich countries, but they have failed to do so in labor-intensive poor countries. The operating assumption has been that an "Industrial Revolution" never occurred in the so-called developing countries when in fact they were indeed part of the Industrial Revolution.

The Yanomami of Brazil represent one group who are the "forgotten" participants in the Industrial Revolution.

gopher://gopher.etext.org/00/Politics/Fourth.World/Americas/yanomami.txt

Document: Yanomami in Peril

Q: How have the Yanomami have been affected by industrialization?

One reason that it is difficult to generalize about how industrialization affects family life is that industrialization and the changes that are thought to accompany it may run up against some political, cultural, or historical stumbling blocks. For example, birthrates are expected to decline with economic development and urbanization. Yet sociologists and demographers are still trying to understand the forces underlying decisions by American couples to have relatively large families in the two decades after World War II, a period of dramatic economic growth and the years that produced the baby boomers.

In addition government policies may have a strong influence on birthrates. For example, after the Romanian government legalized abortion in 1957, there were approximately 1,000 abortions for every 1,000 live births (about 80 percent of pregnancies ended in abortion), and the birthrate was about 15 per 1,000 population. In 1967, in an effort to increase the population and the number of future workers, President Nicolai Ceausescu drastically restricted the sale of contraceptives and placed tight restrictions on abortions. Under Ceausescu's policies Romanian women were required to undergo regular pregnancy tests in order to ensure that all pregnancies were carried to term, and they were given access to contraceptives and abortions only if they were over 45 years of age or already had four living children. Under these policies the birthrate climbed to 27.4 per 1,000 population (Burke 1989; van de Kaa 1987). By 1983, however, the birthrate had tumbled to 14 per 1,000 population, and Ceausescu announced more restrictions and closer monitoring.

The United Nations lists seven reasons for which abortions are permitted or not permitted around the world: (1) to save the woman's life, (2) to preserve physical health, (3) to preserve mental health, (4) after rape or incest, (5) in cases of fetal impairment, (6) for economic or social reasons, or (7) on request. Today Romanian women are allowed to terminate a pregnancy for all of these reasons, something not allowed during Ceausescu's presidency.

gopher://gopher.undp.org/00/ungophers/popin/wdtrends/charts.asc
Document: World Abortion Policies, 1994

: Which countries do not permit a woman to legally have an abortion for any reason?

A final reason that it is difficult to generalize about industrialization's effects on family life is that every family is unique and is composed of diverse personalities. Uniqueness makes it difficult to predict the effects of migration (and other events associated with industrialization) on family life. For example, increased migration unquestionably accompanied industrialization, but any statement about the effects of migration must consider the kind of migration (rural-to-rural, urban-to-urban, international, and so on) and the reason people migrate. Even if we narrow our analysis to a specific kind of migration for a particular reason, we must acknowledge that families do not respond in uniform ways.

These mixed findings are characteristic of family research in general. Consequently in the sections that follow we will consider how the effects of geographic mobility and the changes in life expectancy, residence, and the nature of work put new pressures on family life and may lead to changes in family functions and composition, and in the status of children. We will also consider how these factors may lead to changes in the division of labor between men and women.

The Effects of Geographic Mobility on Family Life

Everywhere industrialization has brought increased geographic mobility, which in turn affects family life. Some sociologists have found that geographic separation virtually guarantees that family members will meet less frequently than if they live close to one another and that over time this spatial distance results in emotional distance as well (Bernardo 1967; Parsons 1966). Other sociologists have found that although geographic separation hinders face-to-face interaction, it does not necessarily disrupt kin relationships because family members can keep in touch by phone or mail. In fact this latter group of sociologists found substantial interaction despite distance and concluded that separation can even enhance relationships. That is, separation teaches family members to not take one another for granted and to enjoy and appreciate the limited time available for interaction (Adams 1968; Allan 1977; Leigh 1982; Litwak 1960).

Some researchers, however, found that substantial interaction is more likely among persons recently separated than among those who have been separated for an extended period. Other researchers maintain that it is important to consider the variables that may confound the effects of distance—in particular the reasons people move away from family members and the extent of geographic dispersion of family members. In some cases lack of contact cannot be totally attributed to distance if the move liberates a family member from an already problematic family situation. As for geographic dispersion, if the majority of family members live in one geographic area and one or two members live outside that area, the modest degree of dispersion makes it easier for those who are visiting to see everyone in the family. But if all family members live in different cities, it is difficult for one person to visit them all. By definition some family members will not be contacted as often as others (Allan 1977).

The Consequences of Long Life

Since the turn of the century, the average life expectancy at birth has increased by 28 years in the mechanized rich countries and by 20 years (or more) in the labor-intensive poor countries. In *The Social Consequences of Long Life* sociologist Holger R. Stub (1982) describes at least four ways in which increases in life expectancy have altered the composition of the family since 1900. First, the chances that children will lose one or both parents before they reach age 16 have decreased sharply. In 1900 there was a 24 percent chance of such an occurrence; today the probability is less than 1 percent. At the same time, parents can expect that their children will survive infancy and early childhood. In 1900, 250 of every 1,000 children born in the United States died before age 1; 33 percent did not live to age 18. Today only 10 of every 1,000 children die before they reach age 1; fewer than 5 percent die before reaching age 18.

Second, the potential length of marriages has increased. Given the mortality patterns in 1900, newly married couples could expect their marriage to last an average of 23 years before one partner died (assuming they did not divorce). Today, assuming that they do not divorce, newly married couples can expect to be married for 53 years before one partner dies. This structural change may be one of several factors underlying the high divorce rates today. At the turn of the century,

> death nearly always intervened before a typical marriage had run its natural course. Now, many marriages run out of steam with decades of life remaining for each spouse. When people could expect to live only a few more months or years in an unsatisfying relationship, they would usually resign themselves to it. But the thought of 20, 30 or even 50 more years in an unsatisfying relationship can cause decisive action at any age. (DYCHTWALD AND FLOWER 1989, P. 213)

According to Stub (1982) divorce dissolves today's marriages at the same rate that death did 100 years ago.

Third, people now have more time to choose and get to know a partner, settle on an occupation, attend school, and decide whether they want children. Moreover an initial decision made in any one of these areas need not be final. The additional living time enables individuals to change partners, careers, or educational and family plans, a luxury not shared by their turn-of-the-century counterparts. Stub (1982) argues that the mid-life crisis is related to long life because many people "perceive that there yet may be time to make changes and accordingly plan second careers or other changes in lifestyle" (p. 12).

Finally the number of people surviving to old age has increased. (In countries where fertility is low or declining, the proportion of old people in the population—not merely the number—is also increasing.) In 1970 about 25 percent of people in their late fifties had at least one surviving parent; in 1980, 40 percent had a surviving parent. Even more astonishing, in 1990, 20 percent of people in their early seventies had at least one surviving parent (Lewin 1990). Although it has always been the case that a small number of people have lived to age 80 or 90, "there is no historical precedent for the aging of our population. We are in the midst of a new phenomenon" (Soldo and Agree 1988, p. 5).

Much has been written about the growing numbers of people over age 65 in the world, especially about issues related to caring for the disabled and the frail elderly. Yet the emphasis on this segment of the older population should not obscure the fact that the elderly are a rapidly changing and heterogeneous group. They differ according to sex, age (a 30-year difference separates persons age 65 from those in their nineties), social class, and health status. Most older people today are in relatively good health; in the United States only 5 percent of persons age 65 and older live in nursing homes (Eckholm 1990).

According to the U.S. Census Bureau the United States is experiencing an increase in the number of persons living past age 65. Women comprise more of the elderly population due to a higher death rate for men. The report states that "three out of four noninstitutionalized persons aged 65 to 74 and two out of three persons aged 75 or older considered their health to be good or better."

http://www.census.gov/cgi-bin/print_hit_bold.pl/pub/Press-Release/cb95-90.txt

Document: The Aging of the U.S. Population

Q: What is the most rapidly growing age group among the elderly? How long can people expect to live once they reach age 65? What are the leading causes of death among the elderly?

In most countries, including the United States, the disabled and frail elderly typically are cared for by female relatives (Stone, Cafferata, and Sangl 1987; Targ 1989). In the United States approximately 72 percent of these caregivers are women. Of this 72 percent 22.7 percent are spouses, 28.9 percent are daughters, and 19.9 percent are daughters-in-law, sisters, grandmothers, or some other female relative or nonrelative (Stone, Cafferata, and Sangl 1987). The major issue faced by even the closest of families is how to meet the needs of the disabled and frail elderly so as not to "constrain investments in children, impair the health and nutrition of younger generations, impede mobility," or impose too great a psychological, physical, and/or time-demand stress on those who care for them. This problem may be even more intense in developing countries, where the presence of older dependents "may alter the level of subsistence of other members in the family" (United Nations 1983, p. 570). In subsistence cultures, even in those that display considerable respect toward the elderly, "the decrepit elderly may be seen as too much of a burden to be supported during periods of deprivation and may be killed, abandoned, or forsaken" (Glascock 1982, p. 55). Although many programs exist and are being developed to support caregivers and to address the issues associated with caring for the disabled and frail elderly, most of these programs are still in the pilot stages. "It is going to take as many as 15 or 20 years to sort out what works" (Lewin 1990, p. A11).

The Status of Children

The technological advances associated with the Industrial Revolution decreased the amount of physical exertion and time needed to produce food and other commodities. As human muscle and time became less important to the production process, children lost their economic value. In nonmechanized, extractive economies, however, children still are an important source of cheap and unskilled labor for the family.

Middle- and upper-class children in mechanized rich economies are likely to live in settings that strip them of whatever potential economic contribution they might make to the family economy (Johansson 1987). In these economies the family's energies shift away from production of food and other necessities and toward the consumption of goods and services.

The U.S. Department of Agriculture estimates that the average family earning an annual income of at least $50,000 will spend about $270,000 to house ($84,000), feed ($61,000), clothe ($19,000), transport ($40,000), and supply medical care ($18,000 not covered by insurance) to a child until he or she is 22 years old (Rock 1990). Moreover the department estimates that economies of scale are minimal—that is, two children cost $419,000, three children $569,000, and four children $759,000. These figures represent a basic sum; the costs grow even higher when extras (summer camps, private schools, sports, music lessons) are included. Even if children go to work when they reach their teens, usually they use that income to purchase items for themselves and do not contribute to household expenses. Demographer S. Ryan Johansson argues that in the mechanized rich countries, couples who choose to have children bring them into the world not for economic reasons, but to provide intangible, "emotional" services—love, companionship, an outlet for nurturing feelings, and enhancement of dimensions of adult identity (Johansson 1987).

In the mechanized rich countries parents or guardians must also pay to have someone take care of children while they work. This cost is especially burdensome for low-income families. The U.S. Census Bureau states that in 1993 poor families spent a greater percentage of their income—two and a half times—on childcare than better-off families.

http://www.census.gov/cgi-bin/print_hit_bold.pl/pub/Press-Release/cb95-182.txt

Document: Child Care Costs Greater Burden for the Poor

Q: What percentage of their income do poor families who pay for childcare spend on it? Which childcare arrangement was used most often for children under 5?

Urbanization and Family Life

Urbanization encompasses two phenomena: (1) the migration of people from rural areas to cities so that an increasing proportion of the population comes to live in cities and (2) a change in the ties that bind people to one another. In mechanized rich countries people who live in urban settings usually are not part of self-sufficient economic units composed of family members. They have more opportunities to make contact with people outside the family network, and they come to depend on people and institutions other than the family (the workplace, hospitals and clinics, counseling services, schools, daycare centers, the media) to meet various needs. Their relationships with most of the individuals they meet in the course of a day are transitory, limited, and impersonal. This does not mean that the family ceases to be important to people's lives, however. Instead it is important in a different way: it becomes less an economic unit and more a source of personal support and indirect economic support. Most sociological research on the American family shows that the great majority of people have some family members living near them and that they define these interactions as meaningful and important (Goldenberg 1987).

Women and Work

In 1984 sociologist Kingsley Davis published "Wives and Work: The Sex Role Revolution and Its Consequences." In this article Davis identifies "as clear and definite a social change as one can find" (p. 401): between 1890 (the first year for which reliable data exists) and 1980, the proportion of married women in the labor force rose from less than 5 percent to more than 50 percent. Davis found that this pattern holds true for virtually every industrialized country except Japan. He attributed this change to the Industrial Revolution and its effect on the division of labor between men and women.

The Division of Labor and the Breadwinner System   Before industrialization—that is, for most of human history—the workplace was the home and the surrounding land. The division of labor was based on sex. In nonindustrial societies (which include two major types, hunting-and-gathering and agrarian), men provided raw materials through hunting or agriculture, and women processed these materials. Women also worked in agriculture and provided some raw materials by gathering food and hunting small animals, and they took care of the young.

The Industrial Revolution separated the workplace from the home and altered the division of labor between men and women. It destroyed the household economy by removing economic production from the home and taking it out of women's hands:

> The man's work, instead of being directly integrated with that of wife and children in the home or on the surrounding land, was integrated with that of non-kin in factories, shops, and firms. The man's economic role became in one sense more important to the family, for he was the link between the family and the wider market economy, but at the same time his personal participation in the household diminished. His wife, relegated to the home as her sphere, still performed the parental and domestic duties that women had always performed. She bore and reared children, cooked meals, washed clothes, and cared for her husband's personal needs, but to an unprecedented degree her economic role became restricted. She could not produce what the family consumed, because production had been removed from the home. (DAVIS 1984, P. 403)

Davis called this new economic arrangement the "breadwinner system." From a historical point of view, this system is not typical. Rather it is peculiar to the middle and upper classes and is associated with a particular phase of industrialization—from the point at which agriculture loses its dominance to the point at which only 25 percent of the population still works in agriculture. In the United States "the heyday of the breadwinner system was from about 1860 to 1920" (p. 404). This system has been in decline for some time in the United States and other industrialized countries, but it tends to recur in countries undergoing a particular phase of development.

Davis asks, "Why did the separation of home and workplace lead to this system [in most mechanized rich countries]?" The major reason, he believes, is that women had too many children to engage in work outside the home. This answer is supported by the fact that family size, in the sense of the number of living members, reached its peak from the mid-1800s to the early 1900s. This occurred because infant and childhood mortality declined while the old norms favoring large families persisted.

The Decline of the Breadwinner System The breadwinner system did not last long because it placed too much strain on husbands and wives and because a number of demographic changes associated with industrialization worked to undermine it. The strains stemmed from several sources. Never before had the roles of husband and wife been so distinct. Never before had women played less than a direct, important role in producing what the family consumed. Never before had men been separated from the family for most of their waking hours. Never before had men had to bear the sole responsibility of supporting the entire family. Davis regards these events as structural weaknesses in the breadwinner system. In view of these weaknesses the system needed strong normative controls to survive: "The husband's obligation to support his family, even after his death, had to be enforced by law and public opinion; illegitimate sexual relations and reproduction had to be condemned, divorce had to be punished, and marriage had to be encouraged by making the lot of the 'spinster' a pitiful one" (p. 406).

Davis maintains that the normative controls collapsed because of the strains inherent in the breadwinner system and because of demographic and social changes that accompanied industrialization. These changes included decreases in total fertility, increases in life expectancy, increases in the divorce rate, and increases in opportunities for employment perceived as suitable for women.

Declines in Total Fertility   The decline in total fertility began before married women entered the labor force. This fact led Davis to conclude that the decline itself changed women's lives in such a way that they had the time to work outside the home, especially after the children entered school. During the 1880s total fertility among white women was approximately 5.0; in the 1930s it averaged 2.4; during the 1970s it averaged 1.8. Not only did the number of children decrease, but reproduction ended earlier in women's lives, and births were spaced closer together than in earlier years. (The median age of mothers at the last birth was 40 in 1850; by 1940 it had fallen to 27.3.) Davis attributes the changes in childbearing patterns to the forces of industrialization, which changed children from an economic asset to an economic liability, and to the "desire to retain or advance one's own and one's children's status in a rapidly evolving industrial society" (p. 408).

Increased Life Expectancy   In view of the relatively short life expectancy in 1860 and the age at which women had their last child (age 40), the average woman was dead by the time her last child left home. By 1980, given the changes in family size, spacing of children, and age at last pregnancy, the average woman could expect to live 33 years after her last child left home. As a result childcare came to occupy a smaller proportion of a woman's life. In addition, although life expectancy has increased for both men and women, on average women can expect to live longer than men. In 1900 women outlived men by about 1.6 years on the average; in 1980 they outlived men by approximately 7 years. Yet, because brides tend to be three years younger on average than their husbands, married women can expect to live an average of 10 years after their husbands' death. In addition the distorted sex ratio caused by males' earlier death decreases the probability of remarriage. Although few women think directly about their husbands' impending death as a reason for working, the difference in mortality remains a background consideration.

http://www.census.gov/population/projection-extract/nation/npas9600.asc
Document: Resident Population of the United States . . .

Compare the number of females with the number of males in each five-year age category. For each age category calculate the ratio of females to males by dividing the number of males into the number of females. For example, in the United States as of July 1, 1996, there were 10,721,000 females and 10,649,000 males in the 30–34 age category. Thus there is roughly one female for every male. Calculate the sex ratio for the 45–49 age category. Has the ratio changed? What factors might account for this change?

Increased Divorce Davis traces the rise in the divorce rate to the breadwinner system—specifically to the shift of economic production outside the home:

> With this shift, husband and wife, parents and children, were no longer bound together in a close face-to-face division of labor in a common enterprise. They were bound, rather, by a weaker and less direct mutuality—the husband's ability to draw income from the wider economy and the wife's willingness to make a home and raise children. The husband's work not only took him out of the home but also frequently put him into contact with other people, including young unmarried women [who have always worked] who were strangers to his family. Extramarital relationships inevitably flourished. Freed from rural and small-town social controls, many husbands sought divorce or, by their behavior, caused their wives to do so. (pp. 410–11)

Davis notes that an increase in the divorce rate preceded married women's entry into the labor market by several decades. He argues that once the divorce rate reaches a certain threshold (above a 20 percent chance of divorce), more married women seriously consider seeking employment to protect themselves in case of divorce. When both husband and wife are in the labor force, the chances of divorce increase even more. Now both partners interact with people who are strangers to the family. Moreover, they live in three different worlds, only one of which they share.

Increased Employment Opportunities for Women Davis believes married women are motivated to seek work by changes in childbearing experiences, increases in life expectancy, the rising divorce rate, and the inherent weakness of the breadwinner system. This motivation became reality as opportunities to work increased for women. With improvements in machine technology, productivity increased, the physical labor required to produce goods and services decreased, and wages rose in industrialized societies. As industrialization matures, there is a corresponding increase in the kinds of jobs perceived as suitable for women (traditionally, nursing, clerical and secretarial work, and teaching).

Keep in mind that Davis's theory is intended to explain how industrialization and accompanying changes in life expectancy and total fertility in the mechanized rich countries affected the division of labor between males and females. It does not systemically address how industrialization affected the division of labor between males and females in labor-intensive poor countries. Furthermore Davis emphasizes overriding structural changes and does not address how the structural changes are played out in family relationships. For example, Davis does not address how changes in women's participation in the labor force affect the quality of family life, men's lives, children's lives, or their own lives. One thing is clear, however: the effects cannot be found in simple either/or terms.

As one example of this complexity, some researchers have assumed that employment for married women would cause them to experience great stress associated with juggling multiple roles—wife, mother, employee. However, other researchers have found that multiple roles can have a positive effect on women by "[buffering them] from the stress within each role." Actually the effect of multiple roles on women's quality of life is contingent on a number of factors.

http://www.hec.ohio-state.edu/famlife/bulletin/volume.1/bull13a.htm
Document: Multiple Roles and Women's Mental Health

Q: Which factors contribute to a positive balance among multiple roles?

Similarly Davis does not address how divorce affects relationships between former spouses or impacts children. The document "Parental Rights vs. What Is Right" offers some insights on the negative effects of divorce and outlines the characteristics of healthy relationships between divorced parents and their children.

http://www.divorce-online.com/ther_art/tract1.html
Document: Parental Rights vs. What Is Right

Q: What rule should divorced parents follow in considering the needs of children?

Discussion Question

Based on the information presented in this chapter, how would you summarize the ways in which the demographic structure of a society (birthrates, death rates, sex ratio, migration) is reflected in the makeup of the family?

Additional Reading

For more on

- Nontraditional family arrangements
  http://www.census.gov/population/socdemo/hh-fam/his9.dat
  Document: Average Number of Own Children Under 18 per Family, by Type of Family: 1955 to Present

- Internal migration in the United States
  http://www.census.gov/population/socdemo/migration/net-mig.txt
  Document: Inmigration, Outmigration, and Net Migration Between 1985 & 1990

  http://www.census.gov/population/socdemo/migration/mig-94.html
  Document: Geographical Mobility: March 1993–1994

- Global population estimates
  gopher://gopher.undp.org/00/ungophers/popin/wdtrends/pop1994
  Document: Population Figures for Countries of the World

- Birthrates and death rates
  http://www.census.gov/ftp/pub/statab/ranks/pg05.txt
  Document: Birth and Infant Mortality Rates in the United States by State and Area in 1992

- Child mortality estimates from 1990 to 1995 by geographic region of the world
  gopher://gopher.undp.org/00/ungophers/popin/wdtrends/child
  Document: Child Mortality Estimates

- Population growth in metropolitan areas in the United States
  http://www.census.gov/population/socdemo/migration/tab-A-3.txt
  Document: Annual Inmigration, Outmigration, and Net Migration for Metropolitan Areas: 1985–1994

- Childcare
  http://www.census.gov/population/socdemo/child/file1.dat
  Document: Child Care Expenditures

- Mortality crises/plagues
  http://www.ento.vt.edu/IHS/plagueHistory.html
  Document: Plagues Since the Roman Empire

- Demographic characteristics specific to children and teenagers
  gopher://cyfer.esusda.gov/11/CYFER-net/statistics/Kids_Count/kidscnt94
  Document: Kids Count Data Book, 1994

- Current health issues
  http://www.cdc.gov/nchswww/releases/fs_mor93.htm
  Document: 1996 News Releases and Fact Sheets

- Population growth in Asian and Eastern European successor states of the former Soviet Union
  gopher://gopher.undp.org/00/ungophers/popin/wdtrends/transit
  Document: The Demography of Countries with Economies in Transition

- Thomas Malthus
  http://socserv2.socsci.mcmaster.ca/~econ/ugcm/3ll3/malthus/popu.txt
  Document: An Essay on the Principles of Population

Appendix

How Demographers Measure Change

In order to discuss change, demographers must specify a time period (a year, a decade, a century) over which they keep track of how many times an event (a birth, a death, a move) occurs. The accompanying table shows the number of births and deaths that occurred in Brazil and the United States between July 1, 1991, and June 30, 1992.

The simplest way to express change is in absolute terms—that is, to state the number of times an event occurred. (In 1991 there were an estimated 3,955,000 births in Brazil and 3,553,000 births in the United States.) Expressing change in absolute terms is not very useful, however, for making comparisons between countries that have different population sizes. Consequently for comparative purposes demographers calculate rates of births, deaths, and migrations, usually per 1,000 people in the population. Rates are calculated by dividing the number of times an event occurs by the size of the population at the onset of the year and then multiplying that figure by 1,000. The 1991 death rate for Brazil is calculated as follows:

$$\frac{1,107,000}{158,202,000} \times 1,000 = 7$$

When the total number of people in the population is used as the denominator, the rates are called crude rates. Sometimes demographers wish to know how many births, deaths, or moves occur within a specific segment of the population (among males or among females ages 15–44). In these cases the denominator is the number of people in that segment of the population. For example, there are 35,356,000 women between the ages of 15 and 44 in Brazil. The age-specific birthrate for these women thus is calculated as follows:

$$\frac{3,955,000}{35,356,000} \times 1,000 = 112$$

	1995 Population	Births	Deaths	Births per 1,000 Population	Deaths per 1,000 Population
Brazil	158,202,000	3,955,000	1,107,000	25	7
United States	254,521,000	3,553,000	2,291,000	14	9

SOURCES: Text by Joan Ferrante, Northern Kentucky University (1991). Other data adapted from the *World Factbook, 1992*, pp. 47 and 358 (1992).

Education

In 1981 the Reagan administration formed the National Commission on Education to study the state of American education. Two years later the commission reported its conclusions in *A Nation at Risk*. The most famous lines in this report are these:

> The educational foundations of our society are presently being eroded by a rising tide of mediocrity that threatens our very future as a Nation and a people. . . .
> If an unfriendly foreign power had attempted to impose on America the mediocre educational performance that exists today, we might well have viewed it as an act of war. . . . We have, in effect, been committing an act of unthinking, unilateral educational disarmament. (THE NATIONAL COMMISSION ON EXCELLENCE IN EDUCATION 1983, P. 5)

The dramatic charges in this report set off a tidal wave of reforms across the 50 states and the more than 1,500 school districts. These included lengthening the school year, installing no pass/no play rules for sports, fining parents who failed to comply with attendance laws, extending the school day, enforcing competency testing of teachers and students, and increasing the number of required mathematics, English, foreign language, and science courses (Gisi 1985; White 1989).

In spite of the many reforms, however, one phrase captures their basic intent: more of the same. The reforms stipulated more tests, more class time, more homework, and more courses, but they failed to address the content and the quality of the educational experience. Critics argue that there was a gross disparity between the urgent tone of

A Nation at Risk and the superficial nature of the reforms that were enacted (Danner 1986).

Apparently the reforms proved ineffective, because at the September 1989 Governors Conference President Bush announced new educational goals to be achieved by the year 2000. And in April 1994 President Clinton signed the Goals 2000 Act, which sets aside $700 million in federal funds for states and school districts that implement programs aimed at meeting guidelines related to these goals; these include a 90 percent high school graduation rate and 100 percent literacy rate.

gopher//gopher.ed.gov:10001/00/initiatives/goals/legislation/g2k-1
Document: National Education Goals

Q: Read over the goals. Which one do you think is the most realistic? Why? Which one do you think is the least realistic? Why?

The belief that the inadequacies of the schools threaten the country's well-being is not unique to the 1980s and 1990s. Neither is the call for a restructuring of the American educational system. Indeed the recommendation that the educational system be more inclusive dates back to the mid-1800s, when educational leaders debated whether to establish universal public education. Although the problems and recommendations have remained essentially unchanged over the past 200 years, various events throughout this period have placed them in a different context and have given them a seemingly new sense of urgency. Between 1880 and 1920, for example, the public was concerned about whether the schools were producing qualified workers for the growing number of factories and businesses and whether they were instilling a sense of national identity (that is, patriotism) in the ethnically diverse student population. When the United States was involved in major wars—World War I, World War II, Korea, and Vietnam—the public was concerned about whether the schools were turning out recruits physically and mentally capable of defending the American way of life. When the Soviet *Sputnik* satellites were launched in the mid-1950s, Americans were forced to consider the possibility that the public schools were not educating their students as well as the Soviets were in mathematics and the sciences. In the 1960s civil rights events forced Americans to question whether the public schools were offering children from less advantaged ethnic groups and social classes the necessary knowledge and skills to compete economically with children from more advantaged groups. And beginning in the late 1970s, Americans confronted the possibility that their schools are educating an inferior work force, one that will be unable to compete in a global labor market and in an information-oriented economy.

For some insights into issues that are shaping the current education "crisis," see the Organization for Economic Cooperation and Development homepage for a summary of its most recent publications and reports related to education. Select "OECD Recent Publications" and then "April–May 1996," "January–March 1996," and any more recent listings. For each listing select the topic "education."

http://www.oecdwash.org
Document: Washington Center Home Page

Q: Review summaries of OECD publications related to education. Based on your reading, what three issues are shaping the "crisis" in education today?

The ongoing nature of the so-called education crisis in the United States and the corresponding criticisms aimed at the American system of education suggest that the schools are very visible and highly vulnerable targets, that "they are the stage on which a lot of cultural crises get played out" (Lightfoot 1988, p. 3). Inequality, poverty, chronic boredom, family breakdown, unemployment, illiteracy, drug abuse, child abuse, and ethnocentrism are crises that transcend the school environment. Yet we confront them whenever we go into the schools (Lightfoot 1988). Consequently the schools seem to be, in some way, both a source of problems and a solution for our problems.

In this chapter we give special emphasis to public education, the system in which 89 percent of students are enrolled in the United States, for two reasons. First, many critics, at home and abroad, maintain that the U.S. system of education is not adequate for meeting the challenges associated with global interdependence. Many employees claim that they are unable to find enough workers with a level of reading, writing, mathematical, and critical thinking skills needed to function adequately in the workplace. Such a human capital deficit weakens U.S. competitiveness in the global marketplace (U.S. Department of Education 1993a). Second, the research that compares the performance of American students with that of students in other countries (particularly in Asian and European countries) has disturbing implications: "compared with their peers in Asian and European countries, American students stand out for how little they work . . . [and] for how poorly they do" (Barrett 1990, p. 80).

What Is Education?

In the broadest sense **education** includes those experiences that stimulate thought and interpretation or that train, discipline, and develop the mental and physical potentials of the maturing person. An experience that educates may be as commonplace as reading a sweater label and noticing that it was made in Taiwan or as intentional as performing a scientific experiment to learn how genetic makeup can be altered deliberately through the use of viruses. In view of this definition and the wide range of experiences it encompasses, we can say that education begins when people are born and ends when they die.

Sociologists make a distinction, however, between formal and informal education. **Informal education** occurs in a spontaneous, unplanned way. Experiences that educate informally are not designed by someone to stimulate specific thoughts or interpretations or to impart specific skills. Informal education takes place when a child puts her hand inside a puppet and then works to perfect the timing between the words she speaks for the puppet and the movement of the puppet's mouth. Formal education is a purposeful, planned effort intended to impart specific skills and modes of thought. **Formal education,** then, is a systematic process (for example, military boot camp, on-the-job training, programs to stop smoking, classes to overcome fear of flying) in which someone designs the educating experiences. We tend to think of formal education as consisting of enriching, liberating, or positive experiences, but it can include impoverishing and narrowing occurrences (such as indoctrination or brainwashing) as well. In any case formal education is considered a success when the people instructed internalize (or take as their own) the skills and modes of thought that those who design the experiences seek to impart. This chapter is concerned with a specific kind of formal education—schooling.

Schooling is a program of formal and systematic instruction that takes place primarily in classrooms but also includes extracurricular activities and out-of-classroom assignments. In its ideal sense "education must make the child cover in a few years the enormous distance traveled by mankind in many centuries" (Durkheim 1961, p. 862). More realistically schooling is the means by which those who design and implement programs of instruction seek to pass on the values, knowledge, and skills they define as important for success in the world. This latter conception implies that what is taught in schools is only a part of the knowledge accumulated and stored throughout human history. This point, of course, raises questions about who has the power to select from the vast amounts of material available what students should study. Moreover what constitutes an ideal education—the goals that should be achieved, the material that should be covered, the techniques of instruction that should be used—is elusive and debatable. Conceptions vary according to time and place; they differ according to whether schools are viewed primarily as mechanisms by which the needs of a society are met or the means by which students learn to think independently, thus becoming free from the constraints on thought imposed by family, culture, and nation.

Social Functions of Education

Sociologist Emile Durkheim believed that education functions to serve the needs of society. In particular schools function to teach children the things they need to adapt to their environment. To ensure this end, the state (or other collectivity) reminds teachers "constantly of the ideas, the sentiments that must be impressed" upon children if they are to adjust to the milieu in which they must live. Otherwise "the whole nation would be divided and would break down into an incoherent multitude of little fragments in conflict with one another." Educators must achieve a sufficient "community of ideas and sentiments without which there is no society" (Durkheim 1968, pp. 79, 81). Such logic underscores efforts to use the schools as mechanisms for meeting the needs of society, whether they be to strip away ethnic and cultural identities in order to implant a common national identity, to transmit values, to take care of children while their parents work, to teach young people to drive, or to train a labor force. Consider the educational goal of training a labor force and some recommended strategies for incorporating employability skills into the instructional process as described in "Employability—The Fifth Basic Skill."

gopher://INET.ed.gov:12002

Path: ERIC.src; enter ED325659

Document: Employability—The Fifth Basic Skill

Q: Read the entire article and then look over Lankard's list of work/maturity skills. On a scale of 1–5, with 1 being "high emphasis on this skill" and 5 being "almost no emphasis on this skill," how would you rate the extent to which your high school emphasized each skill?

According to another quite different conception, education is a liberating experience that releases students from the blinders imposed by the accident of birth into a particular family, culture, religion, society, and time in history. Schools therefore should be designed to broaden students' horizons so that they will become aware of the conditioning influences around them and will learn to think independently of any authority. When schools are designed to achieve these goals, they can function as agents of change and progress.

These aims are not necessarily contradictory if what benefits the group also liberates the individual. For example, democracies and the free market system require an informed public that is capable of independent thought. Most sociological research, however, suggests that schools are more likely to be designed to meet the perceived needs of society than to liberate minds. (This point raises a question: Who defines the needs of society?) In either case a significant percentage of the population in every country seems to be **functionally illiterate**—that is, they do not possess the level of reading, writing, and calculating skills needed to adapt to the society in which they live. In fact, to many critics of the U.S. educational system, illiteracy in America has reached crisis proportions.

Illiteracy in the United States

In the most general and basic sense, **illiteracy** is the inability to understand and use a symbol system, whether it is based on sounds, letters, numbers, pictographs, or some other type of character. Although the term *illiteracy* is used traditionally in reference to the inability to understand letters and their use in reading and writing, there are as many kinds of illiteracy as there are symbol systems—computer illiteracy, mathematical illiteracy (or innumeracy), scientific illiteracy, cultural illiteracy, and so on.

If we confine our attention merely to languages, of which there are thought to be between 6,000 and 9,000 (including dialects), we can see that the potential number of literacies is overwhelming (Ouane 1990) and that people cannot possibly be literate in every symbol system. If a person speaks, writes, and reads in only one language, by definition he or she is illiterate in perhaps as many as 8,999 languages. Yet such a profound level of illiteracy rarely presents a problem because usually people need to know and understand only the language of the environment in which they live.

This point suggests that illiteracy is a product of one's environment—that is, people are considered illiterate when they cannot understand or use the symbol system of the surrounding environment in which they wish to function. Examples include not being able to use a computer, to access information, to read a map in order to find a destination, to make change for a customer, to read traffic signs, to follow the instructions to assemble an appliance, and to fill out a job application.

The contextual nature of illiteracy suggests that it is not "some sort of disease . . . like a viral infection that debilitates an otherwise healthy organism. . . . Illiteracy is a social phenomenon, not a natural one" in that it changes form whenever an environment changes to the point at which old literacy skills are no longer sufficient (Csikszentmihalyi 1990, p. 119).

In the United States (and in all countries for that matter) some degree of illiteracy has always existed, but conceptions of what people needed to know to be considered literate have varied over time. At one time people were considered literate if they could sign their names and read the Bible. At other times a person who had completed the fourth grade was considered literate. The National Literacy Act of 1991 defines literacy as "an individual's ability to read, write, and speak English and compute and solve problems at levels of proficiency necessary to function on the job and in society, to achieve one's goals, and to develop one's knowledge and potential" (U.S. Department of Education 1993b, p. 3). Today there are various estimates of the number of functionally illiterate adults in the United States. The U.S. Bureau of the Census (1982) estimates that 13 percent of the adult population (or 26 million adults) are illiterate. The National Alliance of Business estimates that 30 percent of high school students cannot write a letter seeking employment or information and that one 17-year-old in eight cannot read beyond a fifth-grade level (Remlinger and Vance 1989).

In 1988 the U.S. Congress requested that the Department of Education define literacy in the context of the new economic order and attempt to estimate how many Americans are illiterate. The project involved a representative sample of some 26,000 adults (13,600 were interviewed and 1,000 were surveyed in each of 11 states; in addition 1,100 federal and state prison inmates were interviewed). The researchers found that 21–23 percent of those contacted "demonstrated skills in the lowest level of prose, document, and qualitative proficiencies (level 1)" (U.S. Department of Education 1993b, p. xiv). Approximately 25–28 percent of those contacted performed at the next higher level of literacy proficiency. Terrance G. Wiley argues that before we draw conclusions about these numbers we must consider how illiteracy is actually assessed.

gopher://INET.ed.gov:12002

Path: ERIC.src; enter ED372664

Document: Estimating Literacy in the Multilingual U.S.

Q: What are the three approaches to literacy assessment? Which group is typically listed in the lowest levels of literacy ability? What are four problems with national literacy surveys? Does the information presented in this document change the way you think about illiteracy estimates? Explain.

Illiteracy and Schools

The fact that almost a quarter of the adult population could function at the lowest literacy level in a society with mandatory school attendance policies leads social critics, most notably government officials and business leaders, to point to the schools as one source of the problem. Upon close analysis we can see that the literacy problem cannot be solved by schools alone. Most Americans seem to believe that if "schools just did their jobs more skillfully and resolutely, the literacy problem would be solved" (L. Resnick 1990, p. 169). Not so, says Lauren B. Resnick, director of the Learning Research and Development Center at the University of Pittsburgh. She argues that policies designed to end illiteracy must consider situations in which people use or value written materials.

The complexity of estimating the literacy problem notwithstanding, most people are bothered by the fact that earning a high school diploma is no guarantee that one has acquired the skills needed to function in today's economy. People wonder how so many students could attend school for at least 12 years without acquiring sufficient reading, writing, and problem-solving skills to deal effectively with the work-related problems encountered in the new kinds of entry-level jobs. This question is complicated further by findings that American schoolchildren lag behind their Asian and European counterparts in nearly every subject, especially science and mathematics (Lapointe, Mead, and Phillips 1989). This finding holds true even for American students who score in the top 5–15 percent on achievement tests (Cetron 1988; Thomson 1989). "Generally the 'best students' in the United States do less well on the international surveys when compared with the 'best students' from other countries" (U.S. Department of Education 1992b, p. viii). These findings have prompted many critics to examine the ways in which schools in European and Pacific Rim countries differ from schools in the United States.

Insights from Foreign Education Systems

Previously we noted that in comparison with their counterparts in Pacific Rim and European countries American students—even the most able ones—do little school-related work and perform poorly in academic subjects.

> For example, only 13 percent of a select group of American 17-year-old students achieved algebra scores equal to [those of] 50 percent of 17-year-old Hungarians . . . 25 percent of Canadian 18-year-old students knew as much chemistry as a very select 1 percent of American high school seniors who had taken an advanced, second-year chemistry course . . . [and] 30 percent of South Korea's 13-year-old students were able to apply "advanced scientific knowledge" compared to 10 percent of American students of the same age. (THOMSON 1989, PP. 52–53)

A 1993 Department of Education study on gifted children (those who score in the top 3–5 percent of achievement and IQ tests) provides insights into why even the best American students perform poorly in the context of the international arena: most report that they study less than one hour a day and that they are bored in class because much of what is taught is a rehash of what they already know (*Los Angeles Times* 1993). Such findings are not meant to imply that Pacific Rim and European schools are operated perfectly or that American schools should emulate their systems. They do suggest, however, that it is important to learn why Americans as a group perform so poorly in the international arena.

Amount of Time Spent on Schooling In "The Case for More School Days" Michael Barrett (1990) describes how "American children receive hundreds of hours less schooling than many of their European or Asian mates and the resulting harm promises to be cumulative and lasting" (p. 87). This generalization seems to hold no matter how time spent on schooling is measured: by the length of the school year, week, or day; by the amount of time spent doing homework; by the amount of time that parents spend helping their children with homework; by the number of minutes that teachers spend on instruction (as opposed to disciplining or otherwise managing students); by the rate of absenteeism; or by the dropout rate. Barrett maintains that most Americans dismiss suggestions that the amount of time devoted to school-related learning, especially the number of school days, should be increased. Instead most Americans argue that we should learn to use more efficiently the time already allotted. Although Barrett recognizes that increases in time alone cannot improve our ability to compete internationally, he does suggest that Americans need to equalize the time they commit to learning. Barrett also notes the arrogance of thinking that we can accomplish in 180 days what the Europeans and Asians are accomplishing in 200–235 days, especially when that 180-day school year includes field trips, schoolwide assemblies, snow days, and teacher in-service days (in-service days count as official school days, but only teachers are required to attend school).

One of the most systematic and well-designed studies comparing the time spent on academic activities in three countries was done by Harold W. Stevenson, Shin-ying Lee, and James W. Stigler (1986). They compared mathematics achievement and the classroom and home environments of kindergartners, first-graders, and fifth-graders in three cities: Minneapolis (United States), Sendai (Japan), and Taipei (Taiwan). They found that across all three grades the Taiwanese, but especially the Japanese, consistently outperformed their American counterparts. At the fifth-grade level, however, the differences in test scores were the most striking:

> The highest average score of an American fifth-grade classroom was below that of the Japanese fifth-grade classroom with the lowest average score. In addition, only one Chinese [Taiwanese] classroom showed an average score lower than the American classroom with the highest average score. Equally remarkable is the fact that the lowest average score for a fifth-grade American classroom was only slightly higher than the average score for the best first-grade Chinese classroom. (p. 694)

After thousands of hours of classroom observation and after interviews with both mothers and teachers, Stevenson and his colleagues concluded that Americans devote significantly less time to academic activities either in school or at home and that American parents help their children less with homework. American parents, however, are more likely than their Taiwanese and Japanese counterparts to rate the quality of education at the schools their children attend as good or excellent: 91 percent of American parents, 42 percent of Chinese parents, and 39 percent of Japanese parents rate the quality as good or excellent. Stevenson (1992) maintains that one reason U.S. parents rate the quality as high is that their school system gives them no clear guidelines about what academic skills children in various grades should possess. Thus they have no baseline by which to judge their child's performance.

Cultural and Economic Incentives Scott Thomson (1989), executive director of the National Association of Secondary School Principals, argues that the poor international showing by Americans reflects a lack of cultural and economic incentives to do well. He offers the cases of South Korea and Germany as examples. With regard to cultural incentives American students are less likely than South Korean and German students to receive parental assistance with homework and to come from homes in which the family value structure is supportive of education. Moreover in the United States fewer television programs are aimed at educating youth, and the content of the programs and accompanying advertisements aired on American networks encourages students to consume, not to develop the intellect. When school is the backdrop to a television program or commercial, "more often than not, the school principals are portrayed as grumbling misfits, the teachers are blithering incompetents. Students who show the slightest interest in their studies are invariably depicted as wimps. The heroes are those who can best foul up the system" (O'Connor 1990, p. B1).

Economic incentives are another factor in the quality of education. Compared to South Koreans and Germans, Americans devote a smaller percentage of their gross national product to education, they pay their teachers lower salaries, and they make less of a direct connection between academic achievement in high school and the quality of future employment opportunities. Thomson believes that "this tendency to ignore classroom achievement in high school appears to be a peculiarly American phenomenon" (1989, p. 56), and it may help to explain U.S. Secretary of Labor Robert Reich's claim that "America may have the worst school-to-work transition system of any advanced industrial country. Short of a college degree, there is no way someone can signal to an employer that he or she possesses world-class skills" (1993, p. E1). European and Pacific Rim employers, even when hiring clerical and blue-collar workers, express significantly more interest in job applicants' academic achievements. In fact in many European countries grades and test scores are part of employment resumes (U.S. Department of Education 1992b). In South Korea "school is considered too important for students to work . . . students simply do not hold jobs while attending school; their time and energies are directed toward learning" (Thomson 1989, p. 57). In Japan "educational credentials and educated skills are central to employment, to promotion, and to social status in general" (Rohlen 1986, pp. 29–30).

As discouraging as these cross-national findings are for the country as a whole, it is important to acknowledge that students in states such as Iowa, North Dakota, and Minnesota score as high in mathematics, for example, as do students in Japan and Switzerland, countries with the best math scores (Sanchez 1993). In "What's Right with Schools?" Jayne Freeman points out that American students may compare more favorably with their foreign counterparts than data suggests.

gopher://INET.ed.gov:12002

Path: ERIC.src; enter ED378665

Document: What's Right With Schools?

Q: Read the section "How Do Our Students Compare with Students in Other Countries?" What do you think about Freeman's argument?

Such individual state and age-group accomplishments aside, things look bleak when we consider that within the United States the average elementary and secondary student misses approximately 20 days of school per year (Schlack 1992). Moreover the less economically advantaged social classes and minority groups have higher dropout rates, higher rates of absenteeism, and lower scores on standardized tests than do those from more economically advantaged groups (Horn 1987). Consider the following facts related to racial classification and dropout rates.

http://www.census.gov/cgi-bin/print_hit_bold.pl/pub/Press-Release/cb94-177.txt

Document: Changes in School Enrollment Levels of Nation's Race and Hispanic Origin Groups

Q: What is the dropout rate for people classified as African American? As Hispanic? As white?

Go to the Bureau of Labor Statistics document "College Enrollment and Work Activity of 1995 High School Graduates" to see how many students graduated from high school and dropped out of high school in the 1994–95 academic year.

ftp://stats.bls.gov/pub/news.release/hsgec.txt

Document: College Enrollment and Work Activity of 1995 High School Graduates

Q: How many students dropped out of high school between October 1994 and October 1995? How many graduated from high school in 1995? How many students drop out each year for every 100 who graduate?

The fact that American students spend little time on academic activities compared to their foreign counterparts, in conjunction with the high dropout rate, suggests that something about the American system of education turns many people away from academic pursuits. In the next sections we first examine the historical background of contemporary American education and then consider some of the most general and distinguishing characteristics that may contribute to Americans' ambivalence about education and learning.

The Development of Mass Education in the United States

The United States was the first country in the world to embrace the concept of mass education. In doing so, it broke with the European view that education should be limited to an elite few (for example, the top 5 percent or those who could afford it). In 1852 Massachusetts legislators passed a law making elementary school mandatory for all children, and within 60 years all of the states had passed compulsory school attendance laws. A number of factors other than the legal mandate encouraged parents to comply with attendance laws, however. First, as the pace of industrialization increased, jobs moved away from the home and out of neighborhoods into factories and office buildings. As mechanization increased, apprenticeship opportunities gradually disappeared. As family farms and businesses disappeared, parents could no longer train their children because familiar skills were becoming obsolete. Thus, with the home and work environments independent of each other, parents were no longer available to oversee their children. Second, a tremendous influx of immigrants to the United States between 1880 and 1920 created a large labor pool—a surplus—from which factory owners could draw workers, thereby eliminating the need for child laborers. This combination of events created an environment in which there was no place for children to go except to school.

At least two prominent features of early American education have endured to the present: (1) textbooks modeled after catechisms and (2) single-language instruction.

Textbooks

The most vocal early educational reformers such as Benjamin Rush, Thomas Jefferson, and Noah Webster believed that schools were an important mechanism by which a diverse population could acquire a common culture. They believed that the new "perfectly homogeneous" American was one who studied at home (not abroad) and who used American textbooks. To use Old World textbooks "would be to stamp the wrinkles of decrepit age upon the bloom of youth" (Webster 1966, p. 32).

The first textbooks in the United States were modeled after **catechisms,** short books covering religious principles written in question-and-answer format. Each question had one answer only, and in repeating the answer the question's wording was adhered to strictly (Potter and Sheard 1918). This format discouraged readers from behaving as active learners "who could frame questions, interpret materials, and reflect on the significance of what was presented. . . . No premium was placed on generating and inventing ideas or arguing about the truth or value of what others had written" (D. Resnick 1990, p. 18). The reader's job was to memorize the "right" answers to the questions. With this as the model, not surprisingly, textbooks tend to be written in such a way that the primary reason to read them is to find the "right" answers to the accompanying questions.

The influence of the catechisms on learning today is evident whenever students are assigned to read a chapter and answer the list of questions at the end. Many students discover that they do not need to read the material in order to answer the questions; they can simply skim the text until they find key words that correspond to those in the question and then copy the surrounding sentences.

In the past few years educators have questioned the value of textbooks as a learning tool. Although we know that some schools have abandoned textbooks, especially at the elementary level, we do not know the scope of this trend. Apparently educators in the field of elementary reading are leading the way as they increasingly abandon traditional textbooks modeled after catechisms in favor of children's literature books and daily writing projects (Richardson 1994). Historical novels and edited volumes that include essays, various perspectives, and so-called real or authentic literature are among the alternatives to textbooks.

Single-Language Instruction

The "peopling of America is one of the great dramas in all of human history" (Sowell 1981, p. 3). It involved the conquest of the native peoples, the annexation of Mexican territory along with many of its inhabitants (who lived in what is now New Mexico, Utah, Nevada, Arizona, California, and parts of Colorado and Texas), and an influx of millions of people from practically every country in the world. School reformers, primarily people of Protestant and British background, saw public education as the vehicle for "Americanizing" a culturally and linguistically diverse population, for instilling a sense of national unity and purpose, and for training a competent work force. As Benjamin Rush (1966) argued, "Let our pupil . . . be taught to love his family but let him be taught, at the same time, that he must forsake and even forget them, when the welfare of his country requires it" (p. 34). In order to meet these nation-building objectives, people had to learn to speak a common language. Consequently students were taught in English, the language of the established elite. Early reformers believed that the welfare of the country depended on a common culture that required people to forget their families' language.

Although the United States is hardly unique in pressuring its people to abandon their native tongues and learn to speak a common language, it is probably the only country in the world that places so little emphasis on learning at least one other language. Education critic Daniel Resnick (1990) believes that the absence of serious foreign language instruction contributes to the parochial nature of American schooling. The almost exclusive attention to a single language has deprived students of the opportunity to appreciate the connection between language and culture and to see that language is a thinking tool that enables them to conceptualize the world. Resnick states that the focus on a single language "has cut students off from the pluralism of world culture and denied them a sense of powerfulness in approaching societies very different from their own" (p. 25). It also denies those students from non-English-speaking heritages the means of reflecting on and fully appreciating their ancestors' lives.

In the United States, English became the language people needed to learn to speak if they were to enter the mainstream of society and have a chance at upward mobility. In this respect English was positioned against the language and even regional accents of parents and grandparents. English was the language "through which one's ethnic self was converted into one's acquired American personality . . . [and] as an instrument through which one could hide and mask ethnic (and regional) origins" (Botstein 1990, p. 63). Noah Webster, an influential nineteenth-century education reformer, wrote in the preface of his spelling book that the United States must promote a "uniformity and purity of language [and demolish] those odious distinctions of provincial dialects which are subject to reciprocal ridicule in different states" (1966, p. 33). The national memory of learning English, of breaking with the past, remains with subsequent generations in their ambivalent attitudes not just toward language acquisition but toward the meaning of learning (Botstein 1990).

Fundamental Characteristics of Contemporary American Education

A number of characteristics distinguish American education from other systems of education. They include the availability of college, the lack of a uniform curriculum, funding that varies by state and community, the belief that schools can be the vehicle for solving a variety of social problems, and ambiguity of purpose and value.

The Availability of College

One of the most distinctive features about the United States is that in theory anyone can attend college if he or she has graduated from high school or has received a GED. As a result the United States has the world's highest postsecondary enrollment ratio. Return to the Bureau of Labor Statistics document to see the percentage of high school graduates enrolled in college in the fall following graduation.

ftp://stats.bls.gov/pub/news.release/hsgec.txt

Document: College Enrollment and Work Activity of 1995 High School Graduates

Q: What percentage of 1995 high school graduates enrolled in college the following fall?

Another indicator of the widespread availability of education is that 20 percent of American four-year colleges and universities accept students regardless of what courses they took and grades they earned in high school or what scores they received on ACT or SAT tests. Seventy-four percent of colleges and universities offer remedial courses in reading, writing, and mathematics for those students who lack skills necessary to do college-level work. Thirty percent of all freshman students who entered college in 1989 took one or more remedial courses (U.S. Bureau of the Census 1993).

In most other countries education beyond high school is available only to a small minority. Consider the relationship between test scores and college enrollment in some European countries and Japan.

gopher://INET.ed.gov:12002

Path: ERIC.scr; enter ED355251

Document: National Assessments in Europe and Japan

Q: What proportion of students take entrance exams and attend college in France, the Federal Republic of Germany, England and Wales, and Japan?

From this perspective the educational opportunities in the United States are admirable. A college education is open to everyone regardless of previous educational failures; it is not reserved for a privileged segment. This policy reflects the American belief in equal opportunity to compete at whatever point in life one wishes to enter the competition: "We may not all hit home runs, the saying goes, but everyone should have a chance at bat" (Gardner 1984, p. 28).

At the same time, the unrestricted right to a college education seems to be connected with a decline in the value of a high school education and the effort put into achieving that level of education. As high school enrollments increased over time in the United States to include the middle class and eventually the poor and minority groups, the perceived value of the high school diploma declined and the perceived need for a college degree increased. As college enrollments increased and as a college degree came to be defined as important for success, the high school was no longer the last stop before young people entered the labor force. Consequently high schools were subject to less pressure to make sure that their graduates had acquired the skills necessary for literacy in the workplace. As a result the high school curriculum became less important and less rigorous (Cohen and Neufeld 1981).

Defenders of the American public schools argue that a less rigorous curriculum is a necessary consequence of trying to educate *all* citizens: if everyone is to pass through, the standards must be reduced. Even if we accept this logic, the consequences of such a policy are clear: the high school diploma loses its value, and the goal of achieving equality through compulsory and free education is undermined. Mass education accomplished through **social promotion** (passing students from one grade to another on the basis of age rather than academic competency), through awarding high school diplomas to people with fifth-grade reading abilities, and through issuing certificates of achievement to those who fail minimum competency tests is not the same as educating everyone. True equality in education is achieved only if everyone has an opportunity to earn a degree that is valued.

Differences in Curriculum

In the United States there is no uniform curriculum. Each of the 50 states sets broad curriculum requirements for kindergarten through high school; each school interprets and implements these requirements. Consequently, even with state guidelines, the textbooks, the assignments, the instructional methods, the staff qualifications, and the material covered vary across the schools within each state. In addition to curriculum differences across states, students in the same school are usually grouped or "tracked" on the basis of tests or past performance. For example, students enrolled in standard diploma (versus college preparatory, honors, or advanced) tracks often take fewer mathematics courses and different kinds of mathematics (general math instead of algebra) to meet a state's mathematics requirement. And they take English composition rather than creative writing to meet the state English requirement (*The Book of the States* 1992).

Although most countries also track students at some point in their school careers, all students are exposed to a core curriculum, even if not everyone can assimilate the material. The Japanese, for example, believe that a standardized curriculum is the only way to ensure that everyone has an equal chance at the rewards that education brings (Lynn 1988). "They put nearly their entire population through twelve tough years of basic training" (Rohlen 1986, p. 38). Although the Japanese have three tracks at the secondary level—academic, specialized vocational, and comprehensive—all Japanese students take classes in foreign language, social studies, mathematics, science, health and physical education, and the fine arts. The Japanese schools do not have gifted or self-paced learning programs (Rohlen 1986). They do not teach some students general mathematics and others algebra; everyone is taught algebra. In elementary school all Japanese children learn to read music and to play a wind instrument and a keyboard instrument (Rohlen 1986). Likewise in Germany all students take a foreign language, German, physics, chemistry, biology, mathematics, and physical education.

Another source of differences in curriculum in the United States is that some minority group members are more likely to take or be assigned to less demanding classes. For example, for 46 percent of white 17-year-olds, algebra II is the highest-level mathematics course they have taken. In contrast algebra II is the highest-level mathematics class taken by 41 percent of African-American and 32 percent of Hispanic 17-year-olds (National Science Board 1991).

E. D. Hirsch, Jr., author of a best-selling and controversial 1989 book *Cultural Literacy*, maintains that the nonuniform and diverse nature of the American curriculum has created

> one of the most unjust and inegalitarian school systems in the developed world. It happens that the most egalitarian elementary-school systems are also the best. . . . The countries that achieve these results tend to teach a standardized curriculum in early grades. Hungarian, Japanese, and Swedish children have a systematic grounding in shared knowledge. Until third graders learn what third graders are supposed to know, they do not pass on to fourth grade. No pupil is allowed to escape the knowledge net. (p. 32)

Surprisingly there is little evidence that tracking students into remedial or basic courses contributes to intellectual growth, corrects academic deficiencies, prepares students for success in higher tracks, or increases interest in learning. Instead the special curricula exaggerate and widen differences among students and perpetuate beliefs that intellectual ability varies according to social class and ethnic group (Oakes 1986a, 1986b).

Differences in Funding

A 1993 report on education in 24 countries sponsored by the Organization for Economic Cooperation and Development (OECD) concluded that no country has greater educational disparities between rich and poor than the United States (Sanchez 1993). In the United States schools differ not only with regard to curriculum requirements but also with regard to funding. Elementary and secondary schools receive approximately 6 percent of their funding from the federal government, 48 percent from the state government, and the rest from the local sources, primarily property taxes (U.S. Bureau of the Census 1993). Heavy reliance on state revenues is problematic because the less wealthy states generate less tax revenue than do the wealthier states. Of the 24 countries studied, the United States is third in spending per student at the secondary level and first at the primary level (Celis 1993a). However, there is considerable financial inequity across school districts.

gopher://INET.ed.gov:12002

Path: ERIC.src; enter ED350717

Document: Financial Equity in the Schools

 Q: What does the term *funding disparity* mean? Give an example. How have state legislatures responded to this inequity? What are the shortcomings of equalizing formulas? Is funding important to student achievement? Explain.

The heavy reliance on local revenue is also problematic because it causes funding disparities among schools within states. In this regard the courts in at least 28 states are in the process of evaluating claims that methods of financing have helped create unequal school systems within the state or have ruled that the methods of school financing are unconstitutional (Celis 1992, 1993b). A dramatic case in point is Kentucky. In response to a 1986 lawsuit filed by 66 mostly rural school districts challenging the state's system of funding education, the Kentucky Supreme Court declared on June 8, 1989, that it

> found the entire state system of public education deficient and unconstitutional. The court declared that every aspect of the public school system should be reconsidered and a new system created no later than April 15, 1990.
>
> The court concluded that a school system in which a significant number of children receive an inadequate education or ultimately fail is inherently inequitable and unconstitutional. (FOSTER 1991, P. 34)

To remedy this inequity, the court made the academic success of all students a constitutional obligation and required the state legislature to devise a system to ensure that every student was "learning at the highest level of which he or she is capable" (Foster 1991, p. 36). In 1990 the legislature passed the Kentucky Education and Reform Act (KERA), which set into motion the restructuring of education's rules, roles, and relationships. The sociological significance of KERA is that the state recognized that inequality could be corrected only by completely rethinking and overhauling the way education is delivered to students.

Education-Based Programs to Solve Social Problems

The United States uses education-based programs to address a variety of social problems including parents' absence from the home, racial inequality, drug and alcohol addictions, malnutrition, teenage pregnancy, sexually transmitted diseases, and illiteracy. Although all countries have education-based programs that address social problems, the United States is unique in that education is viewed as the primary solution to many of its problems. In the United States

> the process became familiar: discover a social problem, give it a name, and teach a course designed to remedy it. Alcoholism? Teach about temperance in every school. Venereal disease? Develop courses in social hygiene. Youth unemployment? Improve vocational training and guidance. Carnage on the highways? Give

driver education classes to youth. Too many rejects in the World War I draft? Set up programs in health and physical education.... In practice, turning real problems—death behind the wheel, the syphilitic body, the frustrated job seeker—into classroom issues [gives] ... concerned citizens the reassuring feeling that something [is] being done—however symbolically—about real problems. (TYACK AND HANSOT 1981, P. 13)

The importance of these programs notwithstanding, the schools alone cannot solve such complex problems. Other programs must be implemented in concert with education-based programs. For example, an unknown number of infants are born each year with alcohol-related problems. The National Institute on Alcohol Abuse and Alcoholism offers some advice on how to interpret fetal alcohol syndrome estimates.

http://www.niaaa.nih.gov/publications/aa13.htm
Document: Fetal Alcohol Syndrome

Q: What is fetal alcohol syndrome (FAS)? What are some of the symptoms associated with FAS that could interfere with learning? What are some of the methodological problems associated with various methods of estimating FAS? Which group in the United States has the highest estimated rate of FAS? Which group has the lowest? Can these estimates be treated as accurate indicators of group rates of FAS? What question will yield the most reliable response to questions about alcohol consumption? Why?

When these children reach school age, they will present special problems to the school system. In *The Broken Cord* Michael Dorris (1989) summarizes the concerns of teachers who work with children affected by alcohol before birth, including "difficulty staying on task, distracting other children, poor use of language, inability to structure their work time, and a constant need for monitoring and attention" (p. 241).

Ambiguity of Purpose and Value

As noted previously, in comparison to people in other countries, Americans tend to be ambivalent about the purpose and value of an education. In general the American public supports without question mass education and the right to a college education despite past academic history. Yet for many Americans elementary school, high school, and college are merely something to be endured; students count the days until they are "out." Ernest Boyer (1986) interviewed hundreds of students from public schools around the United States and found no one who could articulate why he or she was in school: "The most frequent response was, 'I have to be here.' They know it's the law. Or, 'If I finish this, I have a better chance at a job.' The 'this' remains a blank. Or, 'I need this in order to go to college.' Or, 'This is where I meet my friends.' Not once in all our conversations did students mention what they were learning or why they should learn it" (p. 43).

Most Americans tend to equate education with increased job opportunities even though since the late 1960s and early 1970s the country has produced college graduates faster than the economy could absorb them, at least into the kinds of jobs the college-educated expect (Guzzardi 1976). In 1990 (the last year for which figures are available) approximately 122.6 million workers were in the labor force. Of these workers 99.3 million held jobs that did not require a college degree. Thus slightly less than 20 percent of American workers were employed in jobs that required a college degree (Shelley 1992). Because approximately 29 million workers have had four or more years of college and only 23.2 million jobs require a college degree, about one college graduate in five is underemployed. This gap between the number of college-educated workers and the number of jobs requiring a college education is reflected in a 1989–90 survey of college degree recipients: 44 percent reported that they did not believe a degree was required for the job they had obtained in 1991. This figure is up from 37 percent in 1985 (U.S. Department of Education 1993a).

These findings do not mean that level of education is unrelated to occupation or income. Rather they indicate that a large proportion of college graduates are *under*employed, if only because there are not enough high-skill jobs available to absorb the increasing number of graduates. In view of this trend "job" seems a narrow criterion by which to evaluate an education. Yet high school and college students commonly evaluate their courses, especially general requirements, as useless because "I will never use it in the real world"—in particular on the job.

The American tendency to associate education almost exclusively with job advancement means that other benefits of education are underemphasized. Two such benefits are personal empowerment and civic engagement:

> Personal empowerment requires that people be able to think analytically and examine information critically; that they be able to think creatively—[to] go beyond the analysis and challenge assumptions, leap out of the present and imagine beyond where they are; and that they be able to act with a clear sense of integrity. Civic engagement requires that people learn how to use these skills while taking full part in the life of the larger community. (BOYER 1986, P. 43)

Open college enrollments, diverse and special curricula, unequal funding, the problem-solving burden, and a national ambivalence toward education explain in part the dropout rate, the high number of functional illiterates, and the United States's poor academic showing relative to its European and Pacific Rim counterparts. In the next section we look more closely at the role of the classroom environment.

A Close-Up View: The Classroom Environment

The classroom is where schooling takes place. In this section we examine what goes on in the classroom: the curriculum to which students are exposed, the practice of tracking, the ways in which students are tested, and the problems teachers face. The focus is on practices that lead to boredom and failure and that undermine the time and energy devoted to academic pursuits. Certainly there are schools in the United States with stimulating classroom environments. The problem is that there are not enough of them or that not enough students define any educational activity as stimulating.

The Curriculum

Teachers everywhere in the United States teach two curricula simultaneously—a formal curriculum and a hidden curriculum. The various academic subjects—mathematics, science, English, reading, and so on—make up the **formal curriculum.** Students do not learn in a vacuum, however. As teachers instruct students and as students complete their assignments, other activities are going on around them. Social anthropologist Jules Henry (1965) maintained that these other activities represent a **hidden curriculum.** The hidden curriculum, then, is all the things that students learn along with the subject matter. The teaching method, the types of assignments and tests, the tone of the teacher's voice, the attitudes of classmates, the number of students absent, the frequency of the teacher's absences, the number of interruptions during a lesson—all are examples of things going on as students learn the formal curriculum. These so-called extraneous events function to convey messages to students not only about the value of the subject but about the values of society, the place of learning in their lives, and their role in society.

The Case of Spelling Baseball Henry used typical classroom scenes (acquired from thousands of hours of participant observation) such as a session of "spelling baseball" to demonstrate the seemingly ordinary process by which a hidden curriculum is transmitted and to show how students are exposed simultaneously to the two curricula. Although Henry observed this scene in 1963, his observations hold more than 30 years later:

> The children form a line along the back of the room. They are to play "spelling baseball," and they have lined up to be chosen for the two teams. There is much noise, but the teacher quiets it. She has selected a boy and a girl and sent them to the front of the room as team captains to choose their teams. As the boy and girl pick the children to form their teams, each child chosen takes a seat in orderly succession around the room. Apparently they know the game well. Now Tom, who has not yet been chosen, tries to call attention to himself in order to be chosen. Dick shifts his position to be more in the direct line of vision of the choosers, so that he may not be overlooked. He seems quite anxious. Jane, Tom, Dick, and one girl whose name the observer does not know, are the last to be chosen. The teacher even has to remind the choosers that Dick and Jane have not been chosen.
>
> The teacher now gives out words for the children to spell, and they write them on the board. Each word is a pitched ball, and each correctly spelled word is a base hit. The children move around the room from base to base as their teammates spell the words correctly.
>
> The outs seem to increase in frequency as each side gets near the children chosen last. The children have great difficulty spelling "August." As they make mistakes, those in the seats say, "No!" The teacher says, "Man on third." As a child at the board stops and thinks, the teacher says, "There's a time limit; you can't take too long, honey." At last, after many children fail on "August," one child gets it right and returns, grinning with pleasure, to her seat. . . . The motivation level in this game seems terrific. All the children seem to watch the board, to know what's right and wrong, and seem quite keyed up. There is no lagging in moving from base to base. The child who is now writing "Thursday" stops to think

after the first letter, and the children snicker. He stops after another letter. More snickers. He gets the word wrong. There are frequent signs of joy from the children when their side is right. (HENRY 1963, PP. 297–98)

According to Henry learning to spell is not the most important lesson that students learn from this exercise. They are also learning important cultural values from the way in which spelling is being taught: they are learning to fear failure and to envy success. In exercises like spelling baseball, "failure is paraded before the class minute upon minute" (p. 300) and success is achieved after others fail. And "since all but the brightest children have the constant experience that others succeed at their expense they cannot but develop an inherent tendency to hate—to hate the success of others" (p. 296).

In an exercise such as spelling baseball, students also learn to be absurd. To "be absurd," as Henry defined it, means to make connections between unrelated things or events and not to care whether the connections are appropriate or inappropriate. From Henry's point of view spelling baseball teaches students to be absurd because there is no logical connection between learning to spell and playing baseball. "If we reflect that one could not settle a baseball game by converting it into a spelling lesson, we see that baseball is bizarrely *irrelevant* to spelling" (p. 300). Yet most students participate in classroom exercises like spelling baseball without questioning their purpose. Although some children may ask, "Why are we doing this? What is the point?" and may be told, "So you can learn to spell" or "To make spelling fun," few children challenge further the purpose of this activity. Students go along with the teacher's request and play the game as if spelling is related to baseball because, according to Henry, they are terrified of failure and because they want so badly to succeed.

Henry argued further that classroom activities such as spelling baseball prepare students to fit into a competitive and consumption-oriented culture. Because the American economy depends on consumption, the country benefits if its citizens purchase nonessential goods and services. The assignments that children do in school don't prepare them to question false or ambiguous statements made by advertisers; schools do not properly prepare demanding individuals to "insist that the world stand up and prove that it is real" (p. 49). Henry argued that this sort of training—this hidden curriculum—makes possible an enormous amount of selling that otherwise could not take place: "In order for our economy to continue in its present form people must learn to be fuzzy-minded and impulsive, for if they were clear-headed and deliberate, they would rarely put their hands in their pockets. . . . If we were all logicians the economy [as we know it] could not survive, and herein lies a terrifying paradox, for in order to exist economically as we are we must . . . remain stupid" (p. 48).

Reading Assignments   The way in which many children are taught to read provides another example of one way the hidden curriculum can function to convey more than the subject matter. Teachers, parents, and other adults tell children that reading is useful and important if they want to participate in society and to succeed in life. Yet children typically are assigned dull stories such as this one:

> Raccoon and Groundhog wanted to play a trick on Rabbit. They dashed into an old building to hide.
>
> "He'll never find us here," said Raccoon.

"What is that rope doing here?" Groundhog asked. Raccoon started to climb up the rope. Suddenly a bell rang. The bell rang and rang, until Raccoon jumped down. Rabbit poked his head through the door. "Did you ring for me?" he asked. (EARLY 1987, P. 97)

After reading this story, students are expected to answer painfully detailed questions such as the following: "'Did you ring for me?' he (cried/asked/shouted)" and "'He will never find us here,' said (Rabbit/Groundhog/Raccoon)."

Students are asked to accept the idea that books are an important source of information and knowledge, even though the assignments do not support this idea. Nobody in their right mind would want to learn to read just so they could answer questions like these (Bettelheim and Zeland 1981). In this case the hidden curriculum—the content of stories—conveys the message that reading is not a meaningful experience and that it adds little to life. In the long run a steady diet of these kinds of reading assignments and exercises teaches children to hate to read. It is no wonder that many students (even at the college level) report the following symptoms:

- Inability to feel pleasure while reading
- Inability to read at a normal tempo for 20 minutes or more
- Inability to make sense of a sentence even after reading it over and over again
- Inability to read without constantly wondering if they have attention-deficit syndrome (Clements 1992, p. A11)

The implications of dull reading assignments are far-reaching because "the ability to read is of such singular importance to a child's life in school that his [or her] experience in learning it more often than not seals the fate, once and for all, of his [or her] academic career" (Bettelheim and Zeland 1981, p. 5). The kind of dull reading assignments many students do may explain in part why according to a National Assessment of Educational Progress survey only about one-quarter to one-third of students are "solid readers who can handle reading assignments at their grade level."

http://www.ed.gov/bulletin/fall1993/fallread.html

Document: The Word on Reading

Q: What are the three levels of reading proficiency? What reading skills does one possess at each level? At what level of reading proficiency is the average fourth-grader in the United States classified? What factors help to predict reading proficiency?

When we compare the content of school-related reading that students do in the United States with that in other countries, as Bruno Bettelheim and Karen Zeland (1981) did in *On Learning to Read: The Child's Fascination with Meaning*, we understand more fully why many students report that they never read for fun or personal enjoyment. There is no point to the story "Around the City"; children are simply running around.

Around the City

All around the city,
All around the town,
Boys and girls run up the street,
Boys and girls run down.
Boys come out into the sun.
Boys come to play and run.
Girls come out to run and play,
Around the city, all the day.

All around the city,
All around the town,
Boys and girls run up the street,
Boys and girls run down. (P. 250)

In contrast "Mami, Please" speaks to questions most small children have when their mothers are busy. It also gives child readers insights into someone's life other than their own.

Mami, Please

"Mami, please, a piece of bread!" says the child.
"Yes," says the mother and cuts a piece of bread for the child.
"Mami, please, read me a story!" says the child.
"Later," says the mother.
"Why later?" asks the child.
"Listen!" says the mother. "Don't you hear anything?"
At that the child is very quiet and listens.
"Mami, please wash us!" call the dishes.
"Mami, please polish us!" call the shoes.
"Mami, please mend us!" call the stockings.
"Mami, please sweep me!" calls the floor.
"Mami, please fetch the milk!" calls the milk jug.
"Mami, please iron me!" calls the laundry in the basket.

Oh, what an awful noise! The child covers his ears.
At that the mother says: "That's how it goes all day."

Now the child says: "Jug, come, we'll help Mother. The two of us will go and fetch milk." (PP. 283–84)

Bettelheim and Zeland argue that Austrian children learn to read faster and better than American children, which they attribute to the content of the stories. The Austrian children read stories that address children's concerns and therefore are more likely to view reading as something that can add to their lives. In general the foreign primers reviewed by Bettelheim and Zeland "treat the beginning reader with respect for his [or her] intelligence, for his [or her] interest in the more serious aspects of life, and with the recognition that from the earliest age on he [or she] will respond positively to writings of true literary merit" (p. 303).

Jules Henry (1965) believed that students who have the intellectual strength to see through absurd assignments such as spelling baseball and who find it impossible to learn to accept such assignments as important may rebel against the system, refuse to do the work, drop out, or come to think of themselves as stupid. We cannot know how many students do poorly in school because they cannot accept the manner in which subjects are taught. The United States has diverse public school systems; different schools teach reading in different ways. Yet the vast majority of Americans, especially middle- and lower-class Americans, typically learn to read and to carry out these kinds of assignments in the manner described here. In addition students, especially young students, cannot articulate what they don't like about school; many come to believe that school is not for them and think they have failed rather than believing that school has failed them.

Tracking

Most schools in the United States arrange students in instructional groups according to similarities in past academic performance and/or on standardized test scores. In elementary school the practice is often known as **ability grouping;** in middle school and high school it is known as **streaming** or **tracking.** Under this sorting and allocation system, students may be assigned to separate instructional groups within a single classroom; they may be sorted with regard to selected subjects such as mathematics, science, and English; or they may be separated across the entire array of subjects.

The rationales that underlie ability grouping, streaming, or tracking (hereafter referred to as "tracking") include the following:

- Students learn better when they are grouped with those who learn at the same rate: the brighter students are not held back by the slower learners, and the slower learners receive the extra time and special attention needed to correct academic deficiencies.
- Slow learners develop more positive attitudes when they do not have to compete with the more academically capable.
- Groups of students with similar abilities are easier to teach.

In spite of these rationales, the evidence suggests that tracking does not lead to these benefits.

The Effects of Tracking Sociologist Jeannie Oakes (1985) investigated how tracking affected the academic experiences of 13,719 middle school and high school students in 297 classrooms and 25 schools across the United States.

> The schools themselves were different: some were large, some very small; some in the middle of cities; some in nearly uninhabited farm country; some in the far West, the South, the urban North, and the Midwest. But the differences in what students experienced each day in these schools stemmed not so much from where they happened to live and which of the schools they happened to attend but, rather, from differences within each of the schools. (1985, p. 2)

Oakes's findings were consistent with the findings of hundreds of other studies of tracking in terms of how students were assigned to groups, how they were treated, how they viewed themselves, and how well they did.

- *Placement:* Poor and minority students are placed disproportionately in the lower tracks.

- *Treatment:* The different tracks are not treated as equally valued instructional groups. There are clear differences with regard to the quality, content, and quantity of instruction and to classroom climate as reflected in the teachers' attitude and in student–student and teacher–student relationships. Low-track students consistently are exposed to inferior instruction—watered-down curriculum and endless repetition—and to a more rigid, more emotionally strained classroom climate.
- *Self-image:* Low-track students do not develop positive images of themselves because they are identified publicly and are treated as educational discards, damaged merchandise, or unteachable. Overall, among the average and the low-track groups, tracking seems to foster lower self-esteem and to promote misbehavior, higher dropout rates, and lower academic aspirations. Placement in a college preparatory track has positive effects on academic achievement, grades, standardized test scores, motivation, educational aspirations, and attainment—"and this positive relationship persists even after family background and ability differences are controlled" (Hallinan 1988, p. 260).
- *Achievement:* The brighter students tend to do well regardless of the academic achievements of the students with whom they learn.

These findings are reflected in the written answers that teachers and students gave to various questions asked by Oakes and her colleagues. For example, when teachers were asked about the classroom climate, high-track teachers tended to reply in positive terms: "There is a tremendous rapport between myself and the students. The class is designed to help the students in college freshman English composition. This makes them receptive. It's a very warm atmosphere. I think they have confidence in my ability to teach them well, yet because of the class size—32—there are times they feel they are not getting enough individualized attention" (Oakes 1985, p. 122). Low-track teachers replied in less positive terms: "This is my worst class. Kids [are] very slow—underachievers and they don't care. I have no discipline cases because I'm very strict with them and they are scared to cross me. They couldn't be called enthusiastic about math—or anything, for that matter" (p. 123).

There also were clear differences in the high-track and the low-track students' responses to the question, "What is the most important thing you have learned or done so far in this class?" The replies of high-track students centered around themes of critical thinking, self-direction, and independent thought:

- "The most important thing I have learned in this [English] class is to loosen up my mind when it comes to writing. I have learned to be more imaginative" (p. 87).
- "The most important thing I have learned in this [math] class is the benefit of logical and organized thinking; learning is made much easier when the simple processes of organizing thoughts have been grasped" (p. 88).

Low-track students were more likely to give answers that centered around themes of boredom and conformity:

- "I think the most important is coming into [math] class and getting out folders and going to work" (p. 89).
- "To be honest, nothing" (p. 71).

- "Nothing I'd use in my later life; it will take a better man than I to comprehend our world" (p. 71).

In addition to these effects, tracking can create self-fulfilling prophecies by affecting teachers' expectations of the academic potential and abilities of students placed in each track.

Teachers' Expectations and Self-Fulfilling Prophecies Tracking can become a **self-fulfilling prophecy,** a deceptively simple yet powerful concept that originated from an insight by William I. and Dorothy Swain Thomas: "If [people] define situations as real, they are real in their consequences" ([1928] 1970, p. 572). A self-fulfilling prophecy begins with a false definition of a situation. The false definition, however, is assumed to be accurate, and people behave as if the definition were true. In the end the misguided behavior produces responses that confirm the false definition (Merton 1957).

A self-fulfilling prophecy can occur if teachers and administrators assume that some children are "fast," "average," or "slow" and expose them to "fast," "average," and "slow" learning environments. Over time real differences in quantity, quality, and content of instruction cause many students to actually become (and believe that they are) "slow," "average," or "fast." In other words the prediction or prophecy of academic ability becomes an important factor in determining academic achievement: "The tragic, often vicious, cycle of self-fulfilling prophecies can be broken. The initial definition of the situation which has set the circle in motion must be abandoned. Only when the original assumption is questioned and a new definition of the situation is introduced, does the consequent flow of events [show the original assumption to be false]" (Merton 1957, p. 424).

In the tradition of symbolic interactionism, Robert Rosenthal and Lenore Jacobson (1968) designed an experiment to test the hypothesis that teachers' positive expectations about students' intellectual growth can become a self-fulfilling prophecy and lead to increases in students' intellectual competence. Rosenthal and Jacobson were influenced by animal experiments in which trainers' beliefs about the genetic quality of the animals affected the animals' performances. When trainers were told that an animal was genetically inferior, the animal performed poorly; when trainers were told that an animal was genetically superior, the animal's performance was superior. This happened despite the fact that there were no such genetic differences between the animals defined as dull or bright.

Rosenthal and Jacobson's experiment took place in an elementary school called Oak School, a name given the school to protect its identity. The student body was largely from lower-income families and predominantly white (84 percent); 16 percent of the students were Mexican Americans. Oak School sorted students into ability groups based on teachers' judgments and on reading achievement.

At the end of the school year, Rosenthal and Jacobson gave a test, purported to be a predictor of academic "blooming," to those students who were expected to return in the fall. Just before classes began in the fall, all full-time teachers were given the names of the white and Hispanic students from all three ability groups who had supposedly scored in the top 20 percent. The teachers were told that these students "*will* show a more significant inflection or spurt in their learning within the next year or less than will the remaining 80 percent of the children" (Rosenthal and Jacobson 1968, p. 66). Teachers were also told not to discuss the scores with the students or the students' parents. Actually the names given to teachers were chosen randomly; the differences between the children earmarked for intellectual growth and the other children were in the teachers' minds. The students were retested after one semester, at the end of the academic year, and after a second academic year.

Overall intellectual gains, as measured by the difference between successive test scores, were greater for those students who had been identified as "bloomers" than they were for those not so identified. Although "bloomers" benefited in general, some bloomers benefited more than others: first- and second-graders, Hispanic children, and children in the middle track showed the largest increases in test scores. Note that the "bloomers" received no special instruction or extra attention from teachers. The only difference between them and the unidentified students was the belief that the "bloomers" bore watching. Rosenthal and Jacobson speculated that this belief was communicated to "bloomers" in very subtle and complex ways, which they could not readily identify:

> To summarize our speculations, we may say that by what she said, by how and when she said it, by her facial expressions, postures, and perhaps by her touch, the teacher may have communicated to the ["bloomers"] that she expected improved intellectual performance.
>
> It is self-evident that further research is needed to narrow down the range of possible mechanisms whereby a teacher's expectations become translated into a pupil's intellectual growth. (p. 180)

North Central Regional Educational Laboratory, a nonprofit research, planning, and implementation organization, publishes *City Schools*, a research journal that focuses on the strengths of city children and schools. Such a positive focus requires that the deficit model, as a way of thinking about inner-city students, be replaced with a resilience model.

http://www.ncrel.org/ncrel/sdrs/cityschl.htm

Document: Who Are Today's City Kids? Beyond the "Deficit Model"

What is the deficit model? Why is the term "limited-English proficiency" an example of a deficit model "term"? How does the deficit model shape teacher expectations and educational policies? What is the asset model (also known as the resilience, hidden resource, or strengths model)?

The research on tracking and teachers' expectations shows that the learning environment affects academic achievement and that tracking and expectations are two mechanisms that contribute to the unequal distribution of knowledge to American citizens. Because teachers draw on test results to form their expectations about students' academic potential and because academic personnel place students in different ability groups on the basis of test results (along with teacher evaluations), tests represent another mechanism that contributes to the unequal distribution of knowledge and skills.

Tests

Tests are the primary tools used by teachers to measure academic achievement. The United States and Japan are the only two mechanized rich countries that regularly use multiple-choice examinations. Most other countries use essay tests, oral presentations, or demonstrations of skills. The United States is the only country in which commercial test publishers develop assessment instruments (SAT, ACT, and various achievement tests) and define test content (National Endowment for the Humanities 1991).

Almost everyone agrees that tests (especially the multiple-choice or true-false variety), when used as the primary measure of students' performance, reduce the motivations for learning and encourage rote memorization. For these reasons alone it is important to consider what tests actually measure. Frederick Erickson (1984) argues that tests, as currently designed and administered, do not measure overall cognitive competencies. Instead tests measure the test-taker's ability to determine what the test-maker is looking for. Moreover test-makers devise questions that require a single correct answer and that do not permit complex answers. Because tests are usually timed, they measure the student's ability not to get bogged down with the meaning of the questions. Erickson argues that cultural differences in interpreting test questions can lead to answers that make the child appear unintelligent. Erickson notes that teachers often take questions from test banks supplied by a textbook's publisher and that even the teachers have difficulty determining what the test-maker is asking. Erickson was struck by "their frustration at not being able to explain simple confusions their students may have about particular items . . . the child's reasoning is on the right track, it's just that the child is having trouble 'reading' the task cues of the item" (p. 534). In such cases we can say that the test measures a student's ability to answer confusing questions.

An example provided by neurologist A. R. Luria (1979) illustrates the kinds of problems that arise if students cannot "see" what the test-maker wants or if what the test-maker wants does not make sense. This example involves a seemingly simple exercise that most American first-graders are exposed to: determine which one of four objects does not belong. When Luria gave such a test to adult Russian peasants with no formal schooling, he could not get them to understand the exercise. From the peasants' perspective the answers he was looking for did not make sense.

> Rakmat, a thirty-year-old illiterate peasant from an outlying district, was shown drawings of a hammer, a saw, a log, and a hatchet. "They're all alike," he said. "I think all of them have to be here. See, if you're going to saw, you need a saw, and if you have to split something, you need a hatchet. So they're all needed here."
>
> We tried to explain the task by saying, "Look, here you have three adults and one child. Now clearly the child doesn't belong in this group."
>
> Rakmat replied, "Oh, but the boy must stay with the others! All three of them are working, you see, and if they have to keep running out to fetch things, they'll never get the job done, but the boy can do the running for them. . . . The boy will learn; that'll be better, then they'll all be able to work well together."
>
> "Look," we said, "here you have three wheels and a pair of pliers. Surely, the pliers and the wheels aren't alike in any way, are they?"
>
> "No, they all fit together. I know the pliers don't look like the wheels, but you'll need them if you have to tighten something in the wheels."

"But you can use one word for the wheels that you can't for the pliers—isn't that so?"

"Yes, I know that, but you've got to have the pliers. You can lift iron with them and it's heavy, you know."

"Still, isn't it true that you can't use the same word for both the wheels and the pliers?"

"Of course you can't."

We returned to the original group, including hammer, saw, and hatchet. "Which of these could you call by one word?"

"How's that? If you call all three of them a 'hammer,' that won't be right either."

"But one fellow picked three things—the hammer, saw, and hatchet—and said they were alike."

"A saw, a hammer, and a hatchet all have to work together. But the log has to be here, too!"

"Why do you think [the fellow] picked these three things and not the log?"

"Probably he's got a lot of firewood, but if we'll be left without firewood, we won't be able to do anything."

"True, but a hammer, a saw, and a hatchet are all tools?"

"Yes, but even if we have tools, we still need wood. Otherwise we can't build anything." (pp. 69–70)

Note that this "matching-like-things" assignment is one that most kindergarten, first-grade, and second-grade students do on a regular basis. Because students rarely have a chance to explain to teachers the logic underlying their answers, teachers are unlikely to learn why some students miss what seems to be an obvious match. Consequently teachers are likely to label students who perform poorly on tests as "slow learners."

Joy Hakim, author of the 10-volume series *A History for Us* (1993) for fifth-graders, found that adult authors are unable to anticipate which words 10-year-olds might become confused over. She recalled an incident in which a child, upon reading the sentence "Ulysses S. Grant sends a wire to President Lincoln asking him to join him in his ship on the James River," asked, "Why would Grant send Lincoln a piece of wire?" When students repeatedly fail tests and assignments because the directions do not make sense or because the students have not acquired an ability to see what the testmaker wants, they are likely to give up. These findings do not suggest that we should stop testing students, nor that we should never use multiple-choice tests. They do suggest, however, that teachers should pay more attention than they do to the kinds of questions they ask, the wording of questions, and the directions they give.

We have examined various practices within the schools that contribute to the unequal distribution of knowledge and skills. We cannot blame teachers and other school personnel entirely for this outcome, however. Teachers do not have exclusive control over the classroom environment and they cannot single-handedly create students who are interested in learning. For many teachers the environment in which they work can make teaching problematic.

The Problems That Teachers Face

Teachers' jobs are complex; teachers are expected to undo learning disadvantages generated by larger inequalities in the society and to handle an array of discipline problems. A Met Life Survey of U.S. teachers identifies some of the problems they face.

http://www.utopia.com/mailings/reportcard/
DAILY.REPORT.CARD170.html#Index4

Document: Something Old, Something New: The American Teacher Survey

 What percentage of teachers named drug use and drinking as a serious problem? Is violence in and around school unique to inner-city schools? Explain. Which problem did teachers name more frequently in 1995 than they did in 1985? Less frequently in 1995 than 1985?

In addition to facing discipline problems, American teachers work in environments that discourage systematic learning outside the classroom and collaboration with other teachers. Psychologist Harold Stevenson (1992) notes that Asian schools plan extracurricular activities after school hours. During this time they teach students computer skills and do not have to use classroom time to do so. In addition Stevenson notes that Asian teachers work together very closely in preparing lesson plans. The level of collaboration is equivalent to that needed to mount a theatrical production. In contrast teachers in the United States prepare lesson plans on their own. Asian teachers have more time to collaborate because they teach about 60 percent of the school day; in the remaining time they discuss ideas with other teachers. American teachers, however, are in the classroom at least 85 percent of the time.

The job of teaching is further complicated by the social context of education. Teachers in the United States must deal with students from diverse family and ethnic backgrounds and with a student subculture that values and rewards athletic achievement, popularity, social activities, jobs, cars, and appearance at the expense of academic achievement. We examine these aspects of social context in the next section.

The Social Context of Education

We turn to the work of sociologist James S. Coleman (the 1991 president of the American Sociological Association), who has studied both family background and the adolescent student subculture, two factors that affect the classroom atmosphere and the learning experience.

Family Background

James S. Coleman (1966) was the principal investigator of *Equality of Educational Opportunity*, popularly known as the Coleman Report. The project was supported by the U.S. government under the directive of the 1964 Civil Rights Act, which (1) prohibited discrimination for reasons of color, race, religion, or national origin in public places (restaurants, hotels, motels, and theaters); (2) mandated that the desegregation of public schools be addressed; and (3) forbade discrimination in employment (racial segregation in public schools was ruled unconstitutional by the Supreme Court in 1954). Coleman's intent was to examine the degree to which public education is segregated and to explore inequalities of educational opportunity in the United States. Coleman and his six colleagues surveyed 570,000 students and 60,000 teachers, principals, and school superintendents in 4,000 schools across the United States. Students filled out questionnaires about their home background and educational aspirations and took standardized achievement tests of verbal ability, nonverbal ability, reading comprehension, mathematical ability, and general knowledge. Teachers, principals, and superintendents answered questionnaires about their backgrounds, training, attitudes, school facilities, and curricula.

Coleman found that a decade after the Supreme Court's famous 1954 desegregation decision—*Brown* v. *Board of Education*—the schools were still largely segregated: 80 percent of white children attended schools that were 90–100 percent white, and 65 percent of African-American students attended schools that were more than 90 percent African American. Almost all students in the South and the Southwest attended schools that were 100 percent segregated. Although Mexican Americans, Native Americans, Puerto Ricans, and Asian Americans also attended primarily segregated schools, they were not segregated from whites to the same degree as were African Americans. The Coleman Report also found that white teachers taught African-American children but that African-American teachers did not teach whites: approximately 60 percent of the teachers who taught African-American students were African American, whereas 97 percent of the teachers who taught white students were white. When the characteristics of teachers of the average white student were compared with those of teachers of the average African-American student, the study found no significant differences in professional qualifications (as measured by degree, major, and teaching experience).

Coleman found sharp differences among ethnic groups with regard to verbal ability, nonverbal ability, reading comprehension, mathematical achievement, and general information as measured by the standardized tests. The white students scored highest, followed by Asian Americans, Native Americans, Mexican Americans, Puerto Ricans, and African Americans.

Contrary to what Coleman expected to find, there were on average no significant differences in quality between schools attended predominantly by the various ethnic groups and schools attended by whites. (Quality was measured by age of buildings, library facilities, laboratory facilities, number of books, class size, expenditures per pupil, extracurricular programs, and the characteristics of teachers, principals, and superintendents.) Surprisingly variations in the quality of a school did not have much effect on the students' test scores.

Test scores were affected, however, by family background and by the attributes of other students. The average minority group member was likely to come from an economically and educationally disadvantaged household and to attend school with students from similar backgrounds. Fewer of his or her classmates would complete high school, maintain high grade point averages, enroll in college preparatory programs, or be optimistic about their future. Coleman also found some support for the idea that "the higher achievement of all racial and ethnic groups in schools with greater proportions of white students is largely, perhaps wholly, related to effects associated with the student body's educational background and aspirations" (1966, pp. 307, 310). This finding does not mean that there is something magical about a white environment. The

Coleman Report examined the progress of African Americans who had participated in school integration programs and found that their scores were higher than those of their counterparts who attended schools with members of the same social class. The important variable is the social class of one's classmates, not ethnicity:

> Taking all these results together, one implication stands out above all: That schools bring little influence to bear on a child's achievement that is independent of his background and general social context; and that this very lack of an independent effect means that the inequalities imposed on children by their home, neighborhood, and peer environment are carried along to become the inequalities with which they confront adult life at the end of school. For equality of educational opportunity through the schools must imply a strong effect of schools that is independent of the child's immediate social environment, and that strong independent effect is not present in American schools. (p. 325)

Coleman's finding that school expenditures are not an accurate predictor of educational achievement (as measured by standardized tests) was used to support arguments against allocating additional funds to the public school system. Yet the finding that schools *do not* make a difference does not mean that schools *cannot* make a difference. A more accurate interpretation of this finding is that schools, as currently structured, have no significant effect on test scores; this conclusion implies that the educational system needs restructuring.

Coleman's findings about the composition of the student body and the higher test scores earned by economically disadvantaged African Americans in predominantly middle-class schools were used to support busing as a means of achieving educational equality. Although Coleman initially supported this policy, he later retracted his endorsement because busing hastened "white flight," or the migration of middle-class white Americans from the cities to the suburbs. This migration only intensified the racial segregation in city and suburban schools. As the ratio of white to African-American students dropped sharply, the positive effects of desegregation proved to be short-lived. As a result economically and educationally disadvantaged African Americans were sent from their deficient schools into equally deficient lower-class and lower-middle-class white neighborhoods. Coleman (1977) adamantly maintained that court-ordered busing alone could not achieve integration:

> With families sorting themselves out residentially along economic and racial lines, and with schools tied to residence, the end result is the demise of the common school attended by children from all economic levels. In its place is the elite suburban school . . . the middle-income suburban school, the low-income suburban school, and the central-city schools of several types—low-income white schools, middle-income white schools, and low or middle-income black schools. (pp. 3–4)

The findings of this Coleman study do not imply that a person is trapped by family background. Coleman never claimed that family background explains all of the variation in test scores. He did claim, however, that it was the single most important factor in his study. This does not mean that other factors do not play significant roles in academic achievement. When the U.S. Department of Education National Center for Educational Statistics (NCES) surveyed students, principals, and administrators to learn about student performance in mathematics, it considered a wide range of factors related to student, school, instructional, classroom, and community characteristics.

gopher://gopher.ed.gov:10000/00/tab/assess/naep/math/readme
Document: National Center for Education Statistics 1992 Mathematics Almanac

Q: What are some of the factors the NCES considered important to student performance?

The fact of school segregation has changed very little over the past three decades. In 1968 the federal government reported that 76 percent of African-American students and 55 percent of Hispanic students attended predominantly minority schools (schools in which 50 percent or more of the students are African American, Asian, Native American, and/or Hispanic). In 1991 the Harvard Project on School Desegregation found that figure to be 66 percent for African Americans and 74.3 percent for Hispanics. In some states such as Illinois, Michigan, New York, and New Jersey, more than 50 percent of the schools are 90–100 percent minority. As in the 1960s African-American and other minority students are significantly more likely to find themselves in schools where overall academic achievement is undervalued and low. The Harvard Project recommended that busing, the most widely used strategy for integrating schools, be supplemented by other strategies, such as finding ways to integrate neighborhoods and enforcing desegregation laws (Celis 1993b).

Many subsequent studies support the importance of family background to educational achievement (Hallinan 1988). For example, the International Association for the Evaluation of Educational Achievement tested students in 22 countries on six subjects. The association found that the "home environment is a most powerful factor in determining the level of school achievement of students, student interest in school learning, and the number of years of schooling the children will receive" (Bloom 1981, p. 89; Ramirez and Meyer 1980). Yet in this international study, in the Coleman study, and in other studies, home background (as measured by parents' ethnicity, income, education, and occupation) explains only about 30 percent of the variation in students' achievement. This finding suggests that factors other than socioeconomic status affect academic performance:

> In most if not all societies, children and youth learn more of the behavior important for constructive participation in the society outside of school than within. This fact does not diminish the importance of school but underlines the nation's dependence on the home, the working place, the community institutions, the peer group and other informal experiences to furnish a major part of the education required for a child to be successfully inducted into society. Only by clear recognition of the school's special responsibilities can it be highly effective in educating its students. (TYLER 1974, P. C74)

In view of these findings, the special responsibility of the schools is to not duplicate the inequalities outside the school. Over the past three decades researchers have found that "schools exert some influence on an individual's chances of success, depending on the extent to which they provide equal access to learning" (Hallinan 1988, pp. 257–58). Unfortunately, as we have learned, several characteristics of American education and practices within the schools—the hidden curriculum, test biases, self-fulfilling prophecies, and tracking—work to perpetuate social and economic inequalities. Now we turn to another problematic phenomenon that teachers confront daily—a student value system that deemphasizes academic achievements.

Adolescent Subcultures

Around the turn of the century—the early decades of late industrialization—less than 10 percent of teenagers ages 14–18 attended high school in the United States. Young people attended elementary school to learn the three Rs, after which they learned from their parents or neighbors the skills needed to make a living. As the pace of industrialization increased, jobs moved away from the home and out of the neighborhood into factories and office buildings. Parents no longer trained their children because the skills they knew were becoming outdated and obsolete, so children came to expect that they would not make their livings as their parents did. In short, as the economic focus in the United States shifted from predominantly farm and small-town work environments to the factory and office, the family became less involved in the training of its children and, by extension, less involved in children's lives. The transfer of work away from the home and neighborhood removed opportunities for parents and children to work together. Under this new arrangement family occasions became events that were consciously arranged to fit everyone's work schedule.

Coleman argued that this shift in training from the family to the school cut adolescents off from the rest of society and forced them to spend most of the day with individuals of their own age. Adolescents came "to constitute a small society, one that has most of its important interactions *within* itself, and maintains only a few threads of connection with the outside adult society" (Coleman, Johnstone, and Jonassohn 1961, p. 3).

Coleman surveyed students from 10 high schools in the Midwest to learn about the adolescent subculture. He selected schools representative of a wide range of environments: five schools were located in small towns, one in a working-class suburb, one in a well-to-do suburb, and three in cities of varying sizes; one of the schools was an all-male Catholic school. Coleman was interested in the adolescent **status system,** a classification of achievements resulting in popularity, respect, peer acceptance, praise, awe, and support, as opposed to isolation, ridicule, peer exclusion, disdain, discouragement, and disrespect. To learn about this system, Coleman asked students questions similar to the following:

- How would you like to be remembered—as an athlete, as a brilliant student, as a leader in extracurricular activities, or as most popular?
- Who is the best athlete? The best student? The most popular? The boy the girls go for most? The girl the boys most go for?
- What person in the school would you like most to date? To have as a friend?
- What does it take to get in with the leading crowd in this school?

Based on the answers to these and other questions, Coleman was able to identify a clear pattern common to all 10 schools: "Athletics was extremely important for the boys, and social success with boys [accomplished through being a cheerleader or being good-looking] was extremely important for girls" (Coleman, Johnstone, and Jonassohn 1961, p. 314). Coleman (1960) found that girls in particular did not want to be considered good students, "for the girl in each grade in each of the schools who was most often named as best student has fewer friends and is less often in the leading crowd than is the boy most often named as best student" (p. 338). A boy could be a good student or dress well or have enough money to meet social expenses, but to really be admired he also had to be a good athlete. Coleman also found that the peer group had more influence over and exerted more pressure on adolescents than did teachers and that a significant number of adolescents were influenced more by the peer group than by their parents.

In comparison to athletic and other achievements, why does the adolescent subculture penalize academic achievement? Coleman maintained that the manner in which students are taught contributes to their lack of academic interest: "They are prescribed 'exercises,' 'assignments,' 'tests,' to be done and handed in at a teacher's command" (Coleman, Johnstone, and Jonassohn 1961, p. 315). The academic work they do requires not creativity, but conformity. Students show their discontent by choosing to become involved in and acquiring things they can call their own—athletics, dating, clothes, cars, and extracurricular activities. Coleman noted that this reaction is inevitable given the passive roles that students are asked to play in the classroom.

> [One] consequence of the passive, reactive role into which adolescents are cast is its encouragement of irresponsibility. If a group is given no authority to make decisions and take action on its own, the leaders need show no responsibility to the larger institution. Lack of authority carries with it lack of responsibility; demands for obedience generate disobedience as well. But when a person or group carries the authority for his own action, he carries responsibility for it. In politics, splinter parties which are never in power often show little responsibility to the political system; a party in power cannot show such irresponsibility.... An adolescent society is no different from these. (COLEMAN, JOHNSTONE, AND JONASSOHN 1961, P. 316)

Athletics is one of the major avenues open to adolescents, especially males, in which they can act "as a representative of others who surround [them]" (p. 319). Others support this effort, identify with the athletes' successes, and console athletes when they fail. Athletic competition between schools generates an internal cohesion among students that no other event can. "It is as a consequence of this that the athlete gains so much status: he is doing something for the school and the community" (p. 260).

Coleman argued that because athletic achievement is widely admired, everyone with some ability will try to develop this talent. With regard to the relatively unrewarded arena of academic life, "those who have most ability may not be motivated to compete" (p. 260). This reward structure may explain why top students in the United States have difficulty competing with top students in many other countries: the United States does not draw into the competition everyone who has academic potential.

Again Coleman's findings should deliver the message that the peer group is a powerful influence on learning, but they should not leave the impression that the peer group's world does not overlap with the family or the classroom. In fact it seems more appropriate to consider how the multiple contexts of students' lives—family, peer group, school environment—are interrelated.

To this point we have examined a number of factors that help explain why American students, even the most able, learn so little relative to their Pacific Rim and European counterparts. These factors also help explain how skills and knowledge are transmitted unevenly across different social classes and ethnic groups. The educational problems in the United States, however, go beyond the uneven transmission of skills and knowledge to certain disadvantaged groups. A large segment of the student population simply tunes out. It is not that they cannot learn, but that they do not want to learn (Csikszentmihalyi 1990). Education critic Mihaly Csikszentmihalyi (1990) argues that few students, even good ones, pay attention while being taught:

> In a series of studies teachers were given electronic pagers, and both they and their students were asked to fill out a short questionnaire whenever the pagers signaled (the signal was set to beep at random moments during the fifty-minute periods). In a typical high school history class, the pager went off as the teacher was describing how Genghis Khan had invaded China in 1234. At the same moment, of the twenty-seven students only two were thinking about something even remotely related to China. One of these two students was remembering a dinner she had had recently with her family at a Chinese restaurant; the other was wondering why Chinese men used to wear their hair in ponytails. (p. 134)

Discussion Question

Based on the information presented in this chapter, how do you think historical and cultural factors contribute to the structure of contemporary education?

Additional Reading

For more on

- International rankings of educational attainment
  gopher://INET.ed.gov:12002
  Path: ERIC.src; enter ED328604
  Document: The International Association for the Evaluation of Educational Achievement

- Rural schools
  gopher://INET.ed.gov:12002
  Path: ERIC.src; enter ED317332
  Document: Small Schools: An International Overview

- Education statistics
  http://chronicle.merit.edu/.almanac/.almdem2.html
  Document: Demographics

  http://chronicle.merit.edu/.almanac/.almmon6.html
  Document: Money

- The dropout rate
  gopher://INET.ed.gov:12002
  Path: ERIC.src; enter ED386515
  Document: School Dropouts: New Information About an Old Problem

- The relationship between education and income
  ftp://stat.bls.gov/pub/news.release/hsgec.txt
  Document: 1995 College Enrollment and Work Activity of High School Graduates

- Goals 2000
  http://www.utopia.com/mailings/reportcard/DAILY.REPORT.CARD113.html
  Path: National Education Goals
  Document: Goals 2000: A Politically Charged Battleground

- Misuses of tests
  gopher://INET.ed.gov:12002
  Path: ERIC.src; enter ED315429
  Document: Five Common Misuses of Tests

- Low-income preschoolers
  http://www.cdinet.com/Rockefeller/Briefs/brief35.html
  Document: Changing the Odds for Low-Income Preschoolers

- Immigrants' educational needs
  http://www.rand.org/publications/MR/MR103/MR103.html
  Document: Newcomers in American Schools

- Families and student performance
  http://www.rand.org/publications/RB/RB8009
  Document: Student Performance and the Changing American Family

- Resilience research (versus the deficit model)
  http://www.ncrel.org/ncrel/sdrs/cityschl/city1_1b.htm
  Document: Resilience Research: How Can It Help City Schools?

  http://www.ncrel.org/ncrel/sdrs/cityschl/city1_1c.htm
  Document: Funds of Knowledge: A Look at Luis Moll's Research into Hidden Family Resources

- Student writing
  http://www.ed.gov/bulletin/summer1994/writnaep.html
  Document: How Well Do Students Write? Can They Persuade?

- Segregation in the schools
  gopher://INET.ed.gov:12002
  Path: ERIC.src; enter ED316616
  Document: Hispanic Education in America: Separate and Unequal

- Education-related news
  http://www.utopia.com/mailings/reportcard/
  Document: Daily Report Card—Index

15 Religion

In this chapter we examine religion from a sociological perspective. Such a perspective is useful because it allows us to step back and view in a detached way a subject that is often charged with emotion. Detachment and objectivity are necessary if we wish to avoid making sweeping generalizations about the nature of religions that are unfamiliar to us.

The sociological perspective on religion is similar to the perspective of religious tolerance presented by the Ontario Centre for Religious Tolerance. This site identifies two levels of tolerance: tolerance in thought and tolerance in action. Sociologists who study religion strive to achieve both types of tolerance.

http://www.kosone.com/people/ocrt
Document: Ontario Centre for Religious Tolerance

Q: How are "tolerance in thought" and "tolerance in action" defined in this document?

When sociologists study religion, they do not investigate whether God or some other supernatural force exists, whether certain religious beliefs are valid, or whether one religion is better than another. Sociologists cannot study such questions because they adhere to the scientific method, which requires them to study only observable and verifiable phenomena. Rather sociologists investigate the social aspects of religion, focusing on the characteristics common to all religions, the types of religious organizations, the functions and dysfunctions of religion, the conflicts within and between religious groups, the way in which religion shapes people's behavior and their understanding of the world, and the way in which religion is intertwined with social, economic, and political issues. We begin with a definition of religion. Defining religion is a surprisingly difficult task and one with which sociologists have been greatly preoccupied.

What Is Religion? Weber's and Durkheim's Views

In the opening sentences of *The Sociology of Religion*, Max Weber (1922) stated: "To define 'religion,' to say what it is, is not possible at the start of a presentation such as this. Definition can be attempted if at all, only at the conclusion of the study" (p. 1). Despite Weber's keen interest in and extensive writings about religious activity, he could offer only the broadest of definitions: religion encompasses those human responses that give meaning to the ultimate and inescapable problems of existence—birth, death, illness, aging, injustice, tragedy, and suffering (Abercrombie and Turner 1978). To Weber the hundreds of thousands of religions, past and present, represented a rich and seemingly endless variety of responses to these problems. In view of this variety, he believed it was virtually impossible to capture the essence of religion in a single definition.

Like Weber, Emile Durkheim believed that there was nothing as vague and diffused as religion. In the first chapter of his book *The Elementary Forms of the Religious Life*, Durkheim ([1915] 1964) cautioned that when studying religions, sociologists must assume that "there are no religions which are false" (p. 3). Like Weber, Durkheim believed that all religions are true in their own fashion—all address in different ways the problems of human existence. Consequently, Durkheim said, those who study religion must first rid themselves of all preconceived notions of what religion should be. We cannot attribute to religion the characteristics that reflect only our own personal experiences and preferences.

In *The Spiritual Life of Children* psychiatrist Robert Coles (1990) recounts his conversation with a 10-year-old Hopi girl, which illustrates Durkheim's point. The conversation reminds us that if we approach the study of religion with preconceived notions, we will lose many insights about the nature of religion in general:

> "The sky watches us and listens to us. It talks to us, and it hopes we are ready to talk back. The sky is where the God of the Anglos lives, a teacher told us. She [the teacher] asked where our God lives. I said, 'I don't know.' I was telling the truth! Our God is the sky, and lives wherever the sky is. Our God is the sun and the moon, too; and our God is our [the Hopi] people, if we remember to stay here [on the consecrated land]. This is where we're supposed to be, and if we leave, we lose God." [The interviewer then asked the child if she had explained all of this to the teacher.]
> "No."

"Why?"

"Because—she thinks God is a person. If I'd told her, she'd give us that smile."

"What smile?"

"The smile that says to us, 'You kids are cute, but you're dumb; you're different—and you're all wrong!'"

"Perhaps you could have explained to her what you've just tried to explain to me."

"We tried that a long time ago; our people spoke to the Anglos and told them what we think, but they don't listen to hear us; they listen to hear themselves." (p. 25)

The conversation between Cole and the Hopi child shows that the teacher's preconceived notions of what constitutes religion closed her off other kinds of religious beliefs and experiences.

Indeed the nature of religion is elusive. And there are many varieties of religious experiences (see http://www.kosone.com/people/ocrt/var_rel.htm for descriptions of 37 such varieties). In spite of these obstacles Durkheim identified three essential features that he believed were common to all religions, past and present: (1) beliefs about the sacred and the profane, (2) rituals, and (3) a community of worshipers. Thus Durkheim defined **religion** as a system of shared rituals and beliefs about the sacred that bind together a community of worshipers.

Beliefs About the Sacred and the Profane

At the heart of all religious belief and activity stands a distinction between two separate and opposing domains—the sacred and the profane.

The Sacred The **sacred** includes everything that is regarded as extraordinary and that inspires in believers deep and absorbing sentiments of awe, respect, mystery, and reverence. These sentiments motivate people to safeguard what is sacred from contamination or defilement. In order to find, preserve, or guard that which they consider sacred, people have gone to war, sacrificed their lives, traveled thousands of miles, and performed other life-endangering acts (Turner 1978).

Definitions of what is sacred vary according to time and place. Sacred things include objects (chalices, sacred documents, books), living creatures (cows, ants, birds), elements of nature (rocks, mountains, trees, the sea, sun, moon, or sky), places (churches, mosques, synagogues, birthplaces of religious founders), days that commemorate holy events, abstract forces (spirits, good, evil), persons (Christ, Buddha, Moses, Muhammad, Zarathustra, Nanak), states of consciousness (wisdom, oneness with nature), past events (the Crucifixion, the Resurrection, the escape of the Jews from Egypt, the birth of Buddha), ceremonies (baptism, marriage, burial), and other activities (holy wars, just wars, confession, fasting, pilgrimages). The following site lists examples of one kind of sacred phenomenon—days of religious celebration or significance.

http://www.kosone.com/people/ocrt/holy_day.htm
Document: Holy Days, Seasonal Days of Religious Celebration, Etc.

Select one of the sacred days listed in the document. Why is this day designated as a sacred event?

Durkheim ([1915] 1964) maintained that the sacredness springs not from the item, ritual, or event itself, but from its symbolic power and from the emotions that people experience when they think about or are in the presence of the sacred thing. The emotions are so strong that believers feel part of something larger than themselves and are outraged when others behave inappropriately in the presence of the sacred.

Ideas about what is sacred are such an important element of religious activity that many researchers classify religions according to the type of phenomenon that their followers consider sacred. One such typology consists of three categories: sacramental, prophetic, and mystical religions (Alston 1972).

In **sacramental religions** the sacred is sought in places, objects, and actions believed to house a god or a spirit. These may include inanimate objects (relics, statues, crosses), animals, trees, plants, foods, drink (wine, water), places, and certain processes (such as the way in which people prepare for a hunt or perform a dance). One example of a sacramental religion is Native American spirituality.

http://www.kosone.com/people/ocrt/nataspir.htm
Document: Native American Spirituality

 Read the descriptions of the various Native American religions. What are three examples of beliefs about the character of sacred phenomena that suggest that Native American spirituality is a sacramental religion? Are there some characteristics of Native American spirituality that depart from the characteristics of sacramental religion?

In **prophetic religions** the sacred revolves around items that symbolize significant historical events or around the lives, teachings, and writings of great people. Sacred books such as the Christian Bible, the Muslim Koran, and the Jewish Torah hold the records of these events and revelations. In the case of historical events, God or some other higher being is believed to be directly involved in the course and the outcome of the event (a flood, the parting of the Red Sea, the rise and fall of an empire). In the case of great people, the lives and inspired words of prophets or messengers reveal a higher state of being, "the way," a set of ethical principles, or a code of conduct. Followers seek to imitate this life. Some of the best-known prophetic religions include Judaism as revealed to Abraham in Canaan and to Moses at Mount Sinai (see **http://www.kosone.com/people/ocrt/judaism.htm**), Confucianism (founded by Confucius; see **http://www.kosone.com/people/ocrt/confuciu.htm**), Christianity (founded by Jesus Christ; see **http://www.kosone.com/people/ocrt/christ.htm**), and Islam (founded by Muhammad; see **http://www.kosone.com/people/ocrt/islam.htm**).

In **mystical religions** the sacred is sought in states of being that at their peak can exclude all awareness of one's existence, sensations, thoughts, and surroundings. In such states the mystic is caught up so fully in the transcendental experience that all earthly concerns seem to vanish. Direct union with the divine forces of the universe is of utmost importance. Not surprisingly mystics tend to become involved in practices such as fasting or celibacy in order to separate themselves from worldly attachments. In addition mystics meditate to clear their minds of worldly concerns, "leaving the soul empty and receptive to influences from the divine" (Alston 1972, p. 144). Buddhism (see **http://www.kosone.com/people/ocrt/buddhism.htm**) and philosophical Hinduism (see **http://www.kosone.com/people/ocrt/hinduism.htm**) are two examples of religions that emphasize physical and spiritual discipline as a means of transcending the self and earthly concerns.

Keep in mind that the distinctions between sacramental, prophetic, and mystical religion are not clear-cut. Because most religions incorporate or combine elements of other religions, religions cannot be placed in a single category, although one category often predominates. For example, there are elements of mysticism in all the major world religions. As a case in point, read the guidelines Jewish, Christian, Muslim, Buddhist, Hindu, and Taoist religions offer to achieve a state of humility.

http://www.realtime.net/~rlp/dwp/mystic/index.html
Document: Mysticism in World Religions

Q: Once in the document, space down until you see the names of these five religions. Read what each religion has to say about humility. What mystical elements run through each description?

According to Durkheim ([1915] 1964) the sacred encompasses more than the forces of good: "There are gods [that is, satans] of theft and trickery, of lust and war, of sickness and of death" (p. 420). Evil and its various representations, however, are almost always portrayed as inferior and subordinate to the forces of good: "in the majority of cases we see the good victorious over evil, life over death, the powers of light over the powers of darkness" (p. 421). Even so, Durkheim considered evil phenomena to be sacred because they are endowed with special powers and are the object of rituals (confessions, baptisms, penance, fasting, exorcism) designed to overcome or resist their negative influences.

The Profane Religious beliefs, doctrines, legends, and myths detail the origins, virtues, and powers of sacred things and describe the consequences of mixing the sacred with the profane. The **profane** is everything that is not sacred, including things opposed to the sacred (the unholy, the irreverent, the contemptuous, the blasphemous) and things that stand apart from the sacred although not in opposition to it (the ordinary, the commonplace, the unconsecrated, the temporal, the bodily) (Ebersole 1967). Believers often view contact between the sacred and the profane as dangerous and sacrilegious, as threatening the very existence of the sacred, and as endangering the fate of the person who allowed the contact. Consequently people take action to safeguard those things they regard as sacred by separating them from the profane. For example, some believers refrain from speaking the name of God in frustration, and they believe that a woman must cover her hair or her face and that a man must remove his hat during worship.

The distinctions between the sacred and the profane do not mean that a person, object, or idea cannot pass from one domain to another or that something profane cannot ever come into contact with the sacred. Such transformations and contacts are authorized through rituals—the active and most observable side of religion.

Rituals

In the religious sense **rituals** are rules that govern how people must behave in the presence of the sacred. These rules may take the form of instructions detailing the appropriate context, the roles of various participants, acceptable attire, and the precise wording of chants, songs, and prayers. Participants must follow instructions closely if they want to achieve a specific goal, whether the goal is to purify each participant's body or soul (confession, immersion, fasting, seclusion), to commemorate an important person or event (pilgrimage to Mecca, the Passover, the Last Supper), or to transform profane items into sacred items (water to holy water, bones to sacred relics). During rituals behavior is "coordinated to an inner intention to make contact with, or to participate in, the invisible world or to achieve a desired state" (Smart 1976, p. 6).

Rituals can be as simple as closing the eyes to pray or having the forehead marked with ashes; they can be an elaborate combination and sequence of activities such as fasting for three days before entering a sacred place to chant, with head bowed, a particular prayer for forgiveness. Although rituals usually are enacted in sacred places, some rituals are codes of conduct aimed at governing the performance of everyday activities—sleeping, walking, eating, defecating, washing, dealing with members of the opposite sex. Durkheim maintained that the nature of the ritual is relatively unimportant. What's important is that the ritual be shared by a community of worshipers and evoke certain ideas and sentiments that help individuals feel part of something bigger than themselves.

Communities of Worshipers

Durkheim used the word **church** to designate a group whose members hold the same beliefs with regard to the sacred and the profane, who behave in the same way in the presence of the sacred, and who gather in body or spirit at agreed-upon times to reaffirm their commitment to those beliefs and practices. Obviously religious beliefs and practices cannot be unique to an individual; they must be shared by a group of people. If this were not the case, the beliefs and practices would cease to exist when the individual who held them died or chose to abandon them. In the social sense religion is inseparable from the idea of church. The gathering and the sharing create a moral community and give worshipers a common identity. The gathering, however, need not take place in a common setting. When people perform a ritual on a given day or at given times of day, the gathering is spiritual rather than physical.

Durkheim ([1915] 1964) used the term *church* loosely, acknowledging that it could assume many forms: "Sometimes it embraces an entire people . . . sometimes it embraces only a part of them . . . sometimes it is directed by a corps of priests, sometimes it is almost completely devoid of any official directing body" (p. 44). Sociologists have identified at least five broad types of religious organizations (communities of worshipers): ecclesiae, denominations, sects, established sects, and cults. As is the case with

330 CHAPTER 15

most classification schemes, the categories overlap on some characteristics because the criteria by which religions are classified are not always clear.

Ecclesiae   An **ecclesiae** is a professionally trained religious organization that is governed by a hierarchy of leaders and that claims as its members everyone in a society. Membership is not voluntary; it is the law. Consequently considerable political alignment exists between church and state officials, which makes the ecclesiae the official church of the state. Ecclesiae formerly existed in England (the Anglican church), France (the Roman Catholic church), and Sweden (the Lutheran church). Today Islam is the official religion of Bangladesh and Malaysia; Iran has been an Islamic republic since the Ayatollah Khomeini took power in 1979; and Saudi Arabia is a monarchy based on Islamic law (*The World Almanac and Book of Facts 1991* 1990). For a brief overview of Saudi Arabia's political system, go to the following site. This description of the political system is part of a very large document, so you will have to scroll down to the section "Synopsis of the Political System."

gopher://dosfan.lib.uic.edu/
0Q%3aSaudi%20Arabia%20Country%3a2%3a5089%3a-1402852575%3a182417
Document: Saudi Arabia Country Commercial Guide

Q: What role does religion play in Saudi government?

Individuals are born into ecclesiae; newcomers to the society are converted; dissenters often are persecuted. Those who do not accept the official view emigrate or occupy the most marginal status in the society. The ecclesiae claims to be the one true faith and often does not recognize other religions as valid. In its most extreme form it directly controls all facets of life.

Denominations   A **denomination** is a hierarchical organization in a society in which church and state are usually separate; it is led by a professionally trained clergy. In contrast to an ecclesiae, a denomination is one of many religious organizations in the society. For the most part denominations are tolerant of other religious organizations; they may even collaborate to address and solve some problems in the society. Membership is considered to be voluntary, but most people who belong to denominations did not choose to do so. Rather they were born to parents who are members. Denominational leaders generally make few demands on the laity, and most members participate in limited and specialized ways. For example, they choose to send their children to church-operated schools, attend church on Sundays and religious holidays, donate money, or attend church-sponsored functions. The leaders of a denomination do not oversee all aspects of the members lives. Yet, even though laypersons vary widely in lifestyle, denominations frequently attract people of particular races and social classes; that is, their members are drawn disproportionately from specific social and ethnic groups.

There are eight major religious denominations in the world—Buddhism, Christianity, Confucianism, Hinduism, Islam, Judaism, Taoism (see **http://www.kosone.com/people/ocrt/taoism.htm**), and Shinto (see **http://www.kosone.com/people/ocrt/shinto.htm**)—each dominant in different areas of the globe. For example, Christianity predominates in Europe, North and South America, New Zealand, Australia, and the Pacific Islands; Islam, in the Middle East and North Africa; and Hinduism, in India.

Sects A **sect** is a small community of believers led by a lay ministry, with no formal hierarchy or official governing body to oversee the various religious gatherings and activities. Sects typically are composed of people who broke away from a denomination because they came to view it as corrupt. Therefore they created the offshoot to reform the religion from which they separated.

All of the major religions encompass splinter groups that have sought at one time or another to preserve the integrity of their religion. In Islam, for example, the most pronounced split occurred 1,300 years ago, approximately 30 years after the death of the Prophet Muhammad, over the issue of Muhammad's successor. The Shia maintained that the successor should be a blood relative of Muhammad; the Sunni believed that the successor should be selected by the community of believers and need not be related by blood. When Muhammad died, the Sunni (the great majority of Muslims) accepted Abu-Bakr as the caliph (successor). The Shia supported Ali, Muhammad's first cousin and son-in-law, and they called for the overthrow of the existing order and a return to the pure form of Islam. The Druze are another religious group that began as a sect of Islam; over time its members developed such distinct rules and rituals that they are no longer regarded as Muslim.

Similarly several splits have occurred within the Christian churches. In 1054, for example, the Eastern Orthodox churches rejected the pope as the earthly deputy of Christ and questioned the papal claim of authority over all Catholic churches in the world. The Protestant religions owe their origins to Martin Luther, who also challenged the papal authority and protested many of the practices of the Roman Catholic church. This protest is known as the Reformation because it involved efforts to reform the Catholic church and to cleanse it of corruption, especially with regard to paying for indulgences (the forgiveness of sins upon saying specific prayers or performing specific good deeds at the order of a priest). Luther believed that a person is saved not by the intercession of priests or bishops, but by private and individual faith (Van Doren 1991). Divisions also exist within various Protestant sects and between Catholic sects or denominations. Lebanon, for example, contains several offshoots of the Roman Catholic church, including Maronites, Greek Catholics, Greek Orthodox, Jacobites, and Gregorians.

People are not born into sects as they are with denominations; theoretically they convert. Consequently newborns are not baptized; they choose membership later in life, when they are considered able to decide for themselves. Sects vary on many levels, including the degree to which they view society as religiously bankrupt or corrupt and the extent to which they take action to change people in society.

Established Sects In some ways **established sects** resemble both denominations and sects. They are renegades from denominations or ecclesiae but have existed long enough to acquire a significant membership and to achieve respectability. At this stage in their history it is appropriate to refer to groups such as the Druze, the Shia, and the Baptists as established sects.

Cults Generally **cults** are very small, loosely organized groups, usually founded by a charismatic leader who attracts people by virtue of his or her personal qualities. Because the charismatic leader plays such a central role in attracting members, cults often dissolve after the leader dies. For this reason few cults last long enough to become established religions. Even so, a few manage to survive, as evidenced by the fact that the major world religions began as cults. Because cults are formed around new and unconventional religious practices, outsiders view them with considerable suspicion.

These groups vary according to purpose and to the level of commitment that the cult leaders demand of converts. Cults may draw members on the basis of highly specific but eccentric interests such as astrology, UFOs, or transcendental meditation. Members may be attracted by the promise of companionship, of a cure from illness, of relief from suffering, or of enlightenment. A cult may meet infrequently and strictly voluntarily (such as at conventions or monthly meetings), or the cult leaders may require members to break all ties with family, friends, and jobs and thus come to rely exclusively on the cult to meet all their needs.

Critique of Durkheim's Definition of Religion

Durkheim's definition of religion revolves around the most outward, most visible characteristics of religion. Critics argue, however, that the three essential characteristics—beliefs about the sacred and the profane, rituals, and a community of worshipers—are not unique to religious activity. This combination of characteristics, they say, can be found at many gatherings (sporting events, graduation ceremonies, reunions, political rallies) and in many political systems (Marxism, Maoism, fascism). On the basis of these characteristics alone, it is difficult to distinguish between an assembly of Christians celebrating Christmas, a patriotic group supporting the initiation of a war against another country, and a group of fans eulogizing James Dean.

In other words religion is not the only unifying force in society to make use of these three elements. **Civil religion,** another such force, is "any set of beliefs and rituals, related to the past, present and/or future of a people (nation), which are understood in some transcendental fashion" (Hammond 1976, p. 171). A nation's beliefs (such as individual freedom or equal opportunity) and rituals (parades, fireworks, singing of the national anthem, 21-gun salutes, and so on) often assume a sacred quality. Even in the face of internal divisions based on race, ethnicity, region, religion, or sex, national beliefs and rituals can inspire awe, respect, and reverence for the country. These sentiments are most notable on national holidays that celebrate important events or people (such as Washington's Birthday, Martin Luther King, Jr., Day, or Independence Day), in the presence of national monuments or symbols (the flag, the Capitol, the Lincoln Memorial, the Vietnam Memorial), and at times of war or other national crises.

Often political leaders appeal to these sentiments in order to win an election, to legitimate their policies, to rally a nation around a cause that requires sacrifice, or to motivate a people to defend their country, as President George Bush (1991) attempted to do in his State of the Union Address on January 7, 1991:

> I come to this house of the people to speak to you and all Americans, certain that we stand at a defining hour.

Halfway around the world, we are engaged in a great struggle in the skies and on the seas and sands. We know why we're there. We are Americans—part of something larger than ourselves.

For two centuries, we've done the hard work of freedom. And tonight we lead the world in facing down a threat to decency and humanity.

For two centuries, America has served the world as an inspiring example of freedom and democracy. For generations, America has led the struggle to preserve and extend the blessings of liberty. And today, in a rapidly changing world, American leadership is indispensable. Americans know that leadership brings burdens, and requires sacrifice.

But we also know why the hopes of humanity turn to us. We are Americans; we have a unique responsibility to do the hard work of freedom. And when we do, freedom works.

Unifying forces other than civil religion can be blood ties, the village, or a powerful leader. The point is that the traits that Durkheim cites as characteristics of religion apply to other events, relationships, and forces within society.

The shortcomings of Durkheim's definition of religion are also reflected in other definitions of religion. Sociologist Herbert Spencer suggested that religion is the recognition that all things are derived from or are dependent on a power that need not be a deity; it could refer just as well to nature, magic, astrology, science, or even the internet. With regard to the internet, consider the World Wide Cemetery. This site offers interested parties an opportunity to create an online memorial for loved ones who have died.

http://www.cemetery.org/about.html
Document: About the World Wide Cemetery

: What features might make the internet a power or a technological deity?

In *Boundaries: Psychological Man in Revolution* psychiatrist Robert Jay Lifton (1969) described how people who witnessed nuclear explosions viewed the bombs as if they were technological deities "capable of both apocalyptic destruction and unlimited creation" (p. 64). Brigadier General Thomas Farrell wrote the following account after he witnessed the first atomic bomb test at Alamogordo:

The effects could well be called unprecedented, magnificent, beautiful, stupendous and terrifying. No man-made phenomenon of such tremendous power had ever occurred before. . . . The whole country was lighted by a searing light with the intensity many times that of the midday sun. It was golden, purple, violet, gray and blue. It lighted every peak, crevasse and mountain range with a clarity and beauty that cannot be described but must be seen to be imagined. It was the beauty the great poets dream about but describe most poorly and inadequately. Thirty seconds after the explosion came, first the air blast pressing hard against people

and things, to be followed almost immediately by the strong, sustained awesome roar which warned of doomsday and made us feel that we puny things were blasphemous to dare tamper with the forces heretofore reserved to The Almighty. Words are inadequate tools for the job of acquainting those not present with the physical, mental and psychological effects. It had to be witnessed to be realized.
(LIFTON AND HUMPHREY 1984, p. 65)

Other reactions at Alamogordo included, "The Sun can't hold a candle to it," "I am become death, the shatterer of worlds," and "One felt as though he had been privileged to witness the birth of the world" (Lifton 1969, pp. 27–28).

Narrower definitions of religion are also problematic. Suppose that religion were defined as the belief in an ever-living god. This definition would exclude polytheistic religions such as Hinduism, which has more than 640 million adherents. It would also exclude religions in which a deity plays little or no role, such as Buddhism, which has more than 300 million adherents. Thus narrow definitions of religion are no improvement over broad ones.

Despite its shortcomings Durkheim's definition of religion is one of the best and most widely used. No sociologist with any standing in the discipline can study religion without encountering and addressing Durkheim's definition. The question, What is religion? is not just a sociological question; it is also a question asked by the government. Besides proposing a definition of religion, Durkheim also wrote extensively about the functions of religion. This work laid the foundation for the functionalist perspective of religion.

Functionalist and Conflict Perspectives on Religion

In addition to identifying the characteristics common to all religions, sociologists also examine the various social functions that religion serves for the individual and the group. Among other things religion functions to comfort believers in the face of uncertainty, to promote group unity and solidarity, to bind the individual to a group, and to regulate society in the face of severe disturbances and abrupt change. History shows, however, that religion is not always a constructive force. Conflict theorists point out that religion also can have repressive, constraining, and exploitative qualities.

The Functionalist Perspective

As far as we know, some form of religion has existed as long as humans have been around (at least 2 million years). In view of this fact functionalists maintain that religion must serve some vital social functions for the individual and for the group. On the individual level people embrace religion in the face of uncertainty: they draw on religious doctrine and ritual in order to comprehend the meaning of life and death and to cope with misfortunes and injustices (war, drought, illness). Life would be intolerable without reasons for existing or without a higher purpose to justify the trials of existence (Durkheim 1951).

People also turn to religious beliefs and rituals to help them achieve a successful outcome (the birth of a healthy child, a job promotion) and to gain answers to questions about the meaning of life: How did we get here? Why are we here? What happens to us when we die?

According to Durkheim people who have communicated with their god or with other supernatural forces (however conceived) report that they gain the inner strength and the physical strength to endure and to conquer the trials of existence. "It is as though [they] were raised above the miseries of the world. . . . Whoever has really practiced a religion knows very well . . . these impressions of joy, of interior peace, of serenity, of enthusiasm which are, for the believer, an experimental proof of his beliefs" (Durkheim [1915] 1964, pp. 416–17).

Religion functions in several ways to promote group unity and solidarity. First, the shared doctrine and rituals help create emotional bonds among those who believe. Second, all religions strive to raise individuals above themselves—to help them live a life better than they would lead if they were left to their own impulses. In this sense religion offers ideas of proper conduct that carry over into everyday life. When believers violate this code of conduct, they feel guilt and remorse. Such feelings in turn motivate them to make amends. Third, although many religious rituals function to alleviate individual anxieties, uncertainties, and fears, they also establish, reinforce, or renew social relationships, binding individuals to a group. Finally, religion functions as a stabilizing force in times of severe social disturbances and abrupt change. During such times many regulative forces in society break down. When such regulative forces are absent, people are more likely to turn to religion in search of a force that will bind them to a group. This tie helps people to think less about themselves and more about some common goal (Durkheim 1951), whether the goal is to work for peace or to participate more fervently in armed conflicts.

The fact that religion functions to meet individual and societal needs, in combination with the fact that people create sacred objects and rituals, led Durkheim to reach a controversial but thought-provoking conclusion: the something out there that people worship is actually society.

Society as the Object of Worship   If we operate under the assumption that all religions are true in their own fashion and that the variety of religious responses is virtually endless, we find support for Durkheim's conclusion that everything encompassed by religion—gods, rites, sacred objects—is created by people. That is, people play a fundamental role in determining what is sacred and how people should act in the presence of the sacred. Consequently at some level people worship what they (or those before them) have created. This point led Durkheim to conclude that the rule object of worship is society itself—a conclusion that many critics cannot accept (Nottingham 1971).

Let us give Durkheim the benefit of the doubt, however, and ask, Is there anything about the nature of society that makes it deserving of such worship? In reply to this question, Durkheim gave what sociologist W. S. F. Pickering called a "virtual hymn to society, a social Gloria in Excelsis" (Pickering 1984, p. 252). Durkheim maintained that society transcends the individual life because it frees us from the bondage of nature (as in nature and nurture). How is this accomplished? We know from cases of extreme isolation, severe neglect, and limited social contact that "it is impossible for a person to develop without social interaction" (Mead 1940, p. 135). In addition studies of mature and even otherwise psychologically and socially sound persons who experience profound isolation—astronauts orbiting alone in space, prisoners of war placed in solitary confinement, volunteers in scientific experiments placed in deprivation tanks—show that when people are deprived of contact with others, they lose a sense of reality and personal identity (Zangwill 1987). The fact that we depend so strongly on society supports Durkheim's (1984) view that for the individual "it is a reality from which everything that matters to us flows" (p. 252). For an interesting discussion of the importance of the group to individual life, see the document "Part III Proper Attitude at a Jewish Funeral."

http://shamash.nysernet.org/pirchei/death.html#III
Document: Part III Proper Attitude at a Jewish Funeral

Q: How do Durkheim's ideas compare with the views offered at this site?

Durkheim, however, did not claim that society provides us with perfect social experiences: "Society has its pettiness and it has its grandeur. In order for us to love and respect it, it is not necessary to present *it other than it is*. If we were only able to love and respect that which is *ideally perfect*, . . . God Himself could not be the object of such a feeling, since the world derives from Him and the world is full of imperfection and ugliness" (p. 253). Durkheim observed that whenever a group of people have a strong conviction (no matter what kind of group it is), that conviction almost always takes on a religious character. Religious gatherings and affiliations become ways of affirming convictions and mobilizing the group to uphold them, especially when they are threatened. In this sense religion is used for many purposes.

Critique of the Functionalist View of Religion To claim that religion functions as a strictly integrative force is to ignore the long history of wars between different religious groups and the many internal struggles among factions within the some religious group. Therefore, in speaking of the integrative function of religion, it is important to specify the segments of society to which this idea applies.

Sociologist Robert K. Merton (1957) argued that if religion were truly integrative in every sense of the word, there would be no conflict or tensions among religious groups within the same society, and people who claimed religious affiliations would not become involved in violent conflicts. Moreover, if religion were entirely an integrative force, religious beliefs and sacred symbols would never be important to ingroup–outgroup distinctions. That is, religious symbols that unite a community of worshipers would not unite them so strongly that they would be willing to destroy persons who did not adhere to their religion.

The functionalist perspective, then, tends to overemphasize the constructive consequences associated with religion's unifying, bonding, and comforting functions. Strict functionalists who focus only on the consequences that lead to order and stability tend to overlook the fact that religion can also unify, bond, and comfort believers in such a way that it supports war and other forms of conflict between ingroups and outgroups. The conflict perspective, on the other hand, acknowledges the unifying, comforting functions of religion but, as we shall see, views such functions as ultimately problematic.

The Conflict Perspective

Scholars who view religion from the conflict perspective focus on how religion turns people's attention away from social and economic inequality. This view stems from the work of Karl Marx, who believed that religion was the most humane feature of an inhumane world and that it arose from the tragedies and injustices of human experience. He described religion as the "sigh of the oppressed creature, the sentiment of a heartless world, and the soul of soulless conditions. It is the *opium* of the people" (*The World Treasury of Modern Religious Thought* 1990, p. 80). People need the comfort of religion in order to make the world bearable and to justify their existence. In this sense, Marx said, religion is analogous to a sedative.

Religion as a Tool of Oppression Even while Marx acknowledged the comforting role of religion, he focused on its repressive, constraining, and exploitative qualities. In particular he conceptualized religion as an ideology that justifies the status quo, rationalizing existing inequities or downplaying their importance. This aspect of religion is especially relevant with regard to the politically and economically disadvantaged. For them, argued Marx, religion is a source of false consciousness. That is, religious teachings encourage the oppressed to accept the economic, political, and social arrangements that constrain their chances in this life because their suffering will be compensated in the next.

Material published in 62 languages by the Watchtower Bible and Tract Society (1987) and distributed worldwide describes life in God's Kingdom:

> God's Kingdom will bring earthly benefits beyond compare, accomplishing everything good that God originally purposed for his people to enjoy on earth. Hatreds and prejudices will cease to exist. . . . The whole earth will eventually be brought to a gardenlike [paradise]. . . . No longer will people be crammed into huge apartment buildings or run-down slums. . . . People will have productive, satisfying work. Life will not be boring. (PP. 3–4)

This kind of ideology led Marx to conclude that religion justifies social and economic inequities and that religious teachings inhibit protest and revolutionary change. He went so far as to claim that religion would not be needed in a **classless society**—a propertyless society providing equal access to the means of production. In the absence of material inequality, there would be no exploitation and no injustice—experiences that cause people to turn to religion.

In sum Marx believed that religious doctrines turn people's attention away from unjust political and economic arrangements and that they rationalize and defend the political and economic interests of the dominant social classes. For some contemporary scholars this legitimating function is reflected by the fact that most religions allow only a specific category of people—men—to be leaders and to handle sacred items. A letter written to the editor of *Christianity Today* in reaction to the article "Women in Seminary: Preparing for What?" show how the "facts" of the Bible can be used to explain and justify such inequalities:

> If the Lord meant for a woman to lead the church in such roles as preacher, elder, pastor, minister, prophet, priest, et cetera, why didn't he provide early Christians with a scriptural prototype? Where in Scripture can a woman priest be found? A woman (literary) prophet? A woman apostle? A woman elder or pastor? Could it be

the Lord didn't intend for a woman to serve in any of these positions? It makes me wonder. (CHRISTIANITY TODAY 1986, P. 6)

Sometimes religion can be twisted in ways that serve the interests of dominant groups. During the time of slavery, for example, some Christians prepared special catechisms for slaves to study. The following questions and answers were included in such catechisms:

Q: What did God make you for?

A: To make a crop.

Q: What is the meaning of "Thou shalt not commit adultery"?

A: To serve our heavenly Father, and our earthly Master, obey our overseer, and not steal anything. (Wilmore 1972, p. 34)

Often political leaders use religion to unite their country in war against another. Both Iraqi president Saddham Hussein and U.S. president George Bush invoked the name of God to rally people behind their causes in the Gulf War of 1991:

In the name of God, the merciful, the compassionate: Our armed forces have performed their holy war duty of refusing to comply with the logic of evil, imposition and aggression. They have been engaged in an epic, valiant battle that will be recorded by history in letters of light. (NEW YORK TIMES 1991, P. Y1)

This we do know: Our cause is just. Our cause is moral. Our cause is right. May God bless the United States of America. (BUSH 1991, P. A8)

Critique of the Conflict Perspective on Religion The major criticism of Marx and the conflict perspective on religion is that religion is not always the sigh of the oppressed creature. On the contrary the oppressed often have used religion as a vehicle for protesting or working to change social and economic inequities. **Liberation theology** represents such an approach. Liberation theologians maintain that they have a responsibility to demand social justice for the marginalized peoples of the world, especially landless peasants and the urban poor, and to take an active role at the grass-roots level to bring about political and economic justice. Ironically this doctrine is inspired by Marxist thought in that it advocates raising the consciousness of the poor and teaching them to work together to obtain land and employment and to preserve their cultural identity. As one example of liberation theology, see Abdul Aziz Said and Brady Tyson's tribute to the bishops of the Catholic church in Brazil for issuing a call for the worldwide study of the UN's Universal Declaration of Human Rights.

gopher://server.gdn.org/00/Miscellaneous_Items/World_HR
Document: The World Human Rights Movement

Q: What is the significance of the Brazilian bishops' action? What are the basic elements of the UN's Declaration of Human Rights? What are the goals for 1998, the fiftieth anniversary of the declaration?

Sociologist J. Milton Yinger (1971) identified at least two conditions under which religion can become a vehicle of protest or change. In the first condition a government or other organization fails to achieve clearly articulated ideals (such as equal opportunity, justice for all, or the right to bear arms). In the second condition a society is polarized along class, ethnic, or sectarian lines. In such cases disenfranchised or disadvantaged groups may form sects or cults and may use seemingly eccentric features of the new religion to symbolize "their sense of separation" (p. 111) and to rally their followers to fight against the establishment or the dominant group. The United States, for example, has not achieved the ideal of equal opportunity irrespective of race, sex, religion, or ethnic group. One religion that emerged in reaction to this failure is the Nation of Islam.

In the 1930s black nationalist W. D. Farad, who went by a variety of names including Farad Muhammad, founded the Nation of Islam and began preaching in the Temple of Islam in Detroit. (When Farad disappeared in 1934, his chosen successor, Elijah Muhammad, took his place.) Farad taught that the white man was the personification of evil and that black people were Muslims but that their religion had been taken from them after they were brought to America as slaves. Farad also taught that the way out was not through gaining the "devil's" (the white man's) approval but through self-help, discipline, and education. Members of the Nation of Islam received an X to replace their slave names (hence Malcolm X). In the social context of the 1930s, this message was very attractive:

> You're talking about Negroes. You're talking about niggers, who are the rejected and the despised, meeting in some little, filthy, dingy little [room] upstairs over some beer hall or something, some joint that nobody cares about. Nobody cares about these people. . . . You can pass them on the street and in 1930, if they don't get off the sidewalk, you could have them arrested. That's the level of what was going on. (NATIONAL PUBLIC RADIO 1984)

The case of the Nation of Islam shows that the larger social, economic, and political context must be incorporated into any discussion of the role of religion in shaping human affairs. This line of investigation was particularly interesting to Max Weber, who examined how religious beliefs direct and legitimate economic activity.

Max Weber: The Interplay Between Economics and Religion

Max Weber was interested in understanding the role of religious beliefs in the origins and development of **modern capitalism,** "a form of economic life which involved the careful calculation of costs and profits, the borrowing and lending of money, the accumulation of capital in the form of money and material assets, investments, private property, and the employment of laborers and employees in a more or less unrestricted labor market" (Robertson 1987, p. 6).

In *The Protestant Ethic and the Spirit of Capitalism* (1958) Weber asked why modern capitalism emerged and flourished in Europe but not in China or India (the two dominant world civilizations at the end of the sixteenth century). He also asked why business leaders and capitalists in Europe and the United States were overwhelmingly Protestant. To answer these questions, Weber studied the major world religions and some of the societies in which these religions were practiced. He focused on understanding how norms generated by different religious traditions influenced the adherents' economic orientations and motivations.

On the basis of his comparisons, Weber concluded that a branch of Protestant tradition—Calvinism—supplied a "spirit" or ethic that supported the motivations and orientations that capitalism required. Unlike other religions that Weber studied, Calvinism emphasized **this-worldly asceticism**—a belief that people are instruments of divine will and that their activities are determined and directed by God. Consequently people glorify God when they accept the tasks assigned to them and carry them out in exemplary and disciplined fashion and when they do not indulge in the fruits of their labor (that is, when they do not use money to eat or drink or otherwise relax to excess). In contrast Buddhism, a religion that Weber defined as the Eastern counterpart to Calvinism, "emphasized the basically illusory character of worldly life and regarded release from the contingencies of the everyday world as the highest religious aspiration" (Robertson 1987, p. 7).

The Calvinists conceptualized God as all-powerful and all-knowing; they also emphasized **predestination,** the belief that God has foreordained all things including the salvation or damnation of individual souls. According to this doctrine people could do nothing to change their fate. To compound matters, only relatively few people were destined to attain salvation.

Weber maintained that such beliefs created a crisis of meaning among adherents as they tried to determine how they were to behave in the face of their predetermined fate. Such pressures led them to look for concrete signs that they were among God's chosen people, destined for salvation. Consequently accumulated wealth became an important indicator of whether one was among the chosen. At the same time, this-worldly asceticism "acted powerfully against the spontaneous enjoyment of possessions; it restricted consumption, especially of luxuries" (Weber 1958, p. 171). Frugal behavior encouraged people to accumulate wealth and make investments, important actions for the success of capitalism.

This calculating orientation was not part of Calvinist doctrine per se. Rather it grew out of this-worldly asceticism and the doctrine of predestination. In view of this distinction, it is important that we do not misread the role that Weber attributed to the Protestant ethic in supporting the rise of a capitalistic economy. According to Weber the ethic was a significant ideological force; it was not the sole cause of capitalism but "*one* of the causes of *certain aspects* of capitalism" (Aron 1969, p. 204). Unfortunately many people who encounter Weber's ideas overestimate the importance he assigned to the Protestant ethic for achieving economic success, and they draw a conclusion that Weber himself never reached: the reason that some groups and societies were disadvantaged was simply that they lacked this ethic.

Finally let us remember that Weber was writing about the origins of industrial capitalism, not about the form of capitalism that exists today, which heavily emphasizes consumption and self-indulgence. Weber maintained that once capitalism was established, it would generate its own norms and would become a self-sustaining force. In fact Weber argued that "capitalism produces a society run along machine-like, rational procedures without inner meaning or value and in which men operate almost as mindless cogs" (Turner 1974, p. 155). In such circumstances religion becomes increasingly insignificant in maintaining the capitalist system. Some sociologists believe that industrialization and scientific advances cause society to undergo unrelenting secularization, a process in which religious influences become increasingly irrelevant not only to economic life but also to most aspects of social life. There are also those who argue that as religion becomes less relevant to economic and social life in general, a significant number of people become fundamentalist. That is, they seek to reexamine their religious principles in an effort to identify and return to the most basic principles (from which believers have departed) and to hold those principles up as the definitive and guiding blueprint for life.

Two Opposing Trends: Secularization and Fundamentalism

Secularization and fundamentalism are processes that have become increasingly popular in the recent past. Each has grown in spite of the other's growth, or possibly in opposition to it.

Secularization

In the most general sense **secularization** is a process by which religious influences on thought and behavior are reduced. It is difficult to generalize about the causes and consequences of secularization because they vary across contexts. Americans and Europeans associate secularization with an increase in scientific understanding and in technological solutions to everyday problems of living. In effect science and technology assume roles once filled by religious belief and practice.

We can speak of two kinds of secularization: subjective and objective (Berger 1967). **Subjective secularization** is a decrease in the number of people who view the world and their place in it from a religious perspective. They shift from a religious understanding of the world, grounded in faith, to an understanding grounded in observable evidence and the scientific method. In the face of uncertainty people often turn away from religion. They come to perceive the supernatural as a distant, impersonal, and even inactive phenomenon. Consequently they come to believe less in direct intervention by the supernatural (forces that can defy the laws of nature and can affect the outcome of an event) and to rely more strongly on human intervention or scientific explanation:

> Consider the case of the lightning rod. For centuries, the Christian church held that lightning was the palpable manifestation of divine wrath and that safety against lightning could be gained only by conforming to divine will. Because the bell towers of churches and cathedrals tended to be the only tall structures, they were the most common targets of lightning. Following damage or destruction of a bell tower by lightning, campaigns were launched to stamp out local wickedness and to raise funds to repair the tower. Ben Franklin's invention of the lightning rod caused a crisis for the church. The rod demonstrably worked. The laity began to demand its installation on church towers—backing their demands with a threat to withhold funds to restore the tower should lightning strike it. The church had to admit either that Ben Franklin had the power to thwart divine retribution or that lightning was merely a natural phenomenon. (STARK AND BAINBRIDGE 1985, PP. 432–33)

There is considerable debate over the extent to which subjective secularization is taking place. Data collected by the Gallup Organization over the past 20 years show little change in the degree of importance that Americans say they assign to religion.

More than 90 percent have a religious preference; almost 70 percent are members of a church, mosque, temple, or synagogue; 40 percent attend church weekly; and almost 60 percent state that religion is very important in their lives (Gallup and Castelli 1989). Recent polls also show that almost 80 percent of Americans are "sometimes very conscious of God's presence" and that more than 80 percent agree that "even today, miracles are performed by the power of God" (p. 50). For information about how people in the United States compare with people in 16 other countries with regard to the belief in life after death, see "You Never Have to Die!"

http://www.Trinity.Edu/~mkearl/never.html
Document: You Never Have to Die!

Q: After considering the findings, the Web master asks, "For what reasons do you believe that post-life beliefs are greater in the United States than in other European countries, from where many of its ideas originated?" How would you answer this question?

These findings, however, do not necessarily prove that religion influences behavior and thought as strongly today as in the past. To reach such a conclusion, we would have to determine whether people are as willing now as they were in the past to leave matters in God's hands. We might speculate, for example, that people were more willing to put their trust solely in God before the advent of sophisticated medical technology. Today people who take such a position may be viewed as ignorant, stubborn, or uncaring. Consider the publicity surrounding cases of children who have died of diabetes, meningitis, tumors, and obstructed bowels because their parents believed that physical ailments could be cured by spiritual means and did not seek medical treatment for the children. The publicity and the accompanying outrage suggest that although an overwhelming number of Americans believe in miracles, their faith is not such that they would leave medical matters in the hands of supernatural forces. At the same time, although science may have the edge over religion in the harder aspects of medicine, religion still seems to have the advantage in the "soft side" of emotional healing and long-term support. "The Forgotten Player in Health Care Reform: Organized Religion" addresses religion's advantage in this area.

http://www.interaccess.com/ihpnet/andrews
Document: The Forgotten Player in Health Care Reform: Organized Religion

Q: What has been/is organized religion's contribution to the soft side of medicine?

Objective secularization is the decline in the control of religion over education, medicine, law, and politics and the emergence of an environment in which people are free to choose from many equally valid religions. For several reasons it is difficult to make generalizations about this type of secularization and the extent to which it exists in a society. First, objective secularization is not an even, inevitable process that eventually results in the dominance of science and rationality and in the end of religion. No matter how strongly science comes to dominate human life, it cannot provide solace for the inescapable problems of existence—birth, death, illness, aging, injustice, tragedy, and suffering—nor can it "formulate a coherent plan for life" (Stark and Bainbridge 1985, p. 431).

Another reason that it is difficult to generalize about objective secularization is that no aspect of political life seems to be fully secularized. For example, the American currency includes the slogan "In God We Trust"; it does not say "In Federal Reserve We Trust" (Haddad 1991). In addition in the United States public schools are not fully secularized, because religious expression is permitted in the schools. The U.S. Department of Education has issued a policy statement on this matter.

http://www.kosone.com/people/ocrt/prayer.htm
Document: Prayer in the Public Schools

Q: What kinds of religious activities are permitted in public schools? What kinds of activities are not?

Sometimes religious affiliations mask more important factors such as economic and political inequalities between conflicting parties. At the same time, because religious differences are used to distinguish the opposing groups and to mobilize members to fight one another, we also must investigate the qualities of religion that make it a factor in such violent conflicts (Stavenhagen 1991). It is especially important to examine how religion influences people's behavior in a specific context and how people use it to legitimize actions. In other words religion itself is usually not the problem; more often than not the problem lies with the way people choose to use religion to justify actions. In the United States the role of religion in human affairs is addressed in three important documents: (1) the First Amendment to the Constitution, (2) Title VII of the Civil Rights Act of 1963, and (3) the Religious Freedom Restoration Act of 1993. To learn more about these documents, see L. Schoefield's article.

gopher://gopher.usdoj.gov/1/fbi/June95/7june.txt
Document: Freedom of Religion and Law Enforcement Employment

Q: One of the major challenges for the courts, which must interpret and enforce the intent of these legal documents, is distinguishing situations in which people use religion to justify actions that interfere with others' constitutional rights from situations in which people's freedom to worship and to practice their religion are violated. The line is not always clear. With this in mind, explain the court decisions Schoefield summarizes. Do you agree with each decision? What do the rulings say about objective secularization in the United States?

Fundamentalism

In "Popular Conceptions of Fundamentalism" anthropologist Lionel Caplan offers his readers one of the clearest overviews of a complex religious phenomenon—**fundamentalism,** a belief in the timeless nature of sacred writings and the applicability of such writings to all kinds of environments. The label "fundamentalist" is applied popularly to a wide array of religious groups in the United States and around the world, including the Moral Majority in the United States, Orthodox Jews in Israel, and various Arab groups in the Middle East.

Religious groups labeled as fundamentalist are usually portrayed as "fossilized relics . . . living perpetually in a bygone age" (Caplan 1987, p. 5). Americans frequently employ this simplistic analysis to explain events in the Middle East, especially the political turmoil that threatens the interests of the United States (including its need for oil). But such oversimplification misrepresents fundamentalism and cannot explain the widespread appeal of contemporary fundamentalist movements within several of the world's religions.

The Complexity of Fundamentalism Fundamentalism is a more complex phenomenon than popular conceptions lead us to believe. First, it is impossible to define a fundamentalist in terms of age, ethnicity, social class, or political ideology because fundamentalism appeals to a wide range of people. Second, fundamentalist groups do not always position themselves against those in power; they are just as likely to be neutral or to fervently support existing regimes. Perhaps the most important characteristic of a fundamentalist is the belief that a relationship with God, Allah, or some other supernatural force provides answers to personal and social problems. In addition fundamentalists often wish to "bring the wider culture back to its religious roots" (Lechner 1989, p. 51).

Caplan suggests a number of other traits that seem to characterize fundamentalists. First, fundamentalists emphasize the authority, infallibility, and timeless truth of sacred writings as a "definitive blueprint" for life (Caplan 1987, p. 19). This characteristic should not be taken to mean that a definitive interpretation of sacred writings actually exists. Any sacred text has as many interpretations as there are groups that claim it as their blueprint. For example, even members of the same fundamentalist organization disagree about the true meaning of the texts they follow.

Second, fundamentalists usually conceive of history as a "process of decline from an original ideal state, [and] hardly more than a catalog of the betrayal of fundamental principles" (p. 18). They conceptualize human history as a "cosmic struggle between good and evil": the good is a result of dedication to principles outlined in sacred scriptures, and the evil is an outcome of countless digressions from sacred principles. To fundamentalists truth is not a relative phenomenon: it does not vary across time and place. Truth is unchanging and is knowable through the sacred texts.

Third, fundamentalists do not distinguish between the sacred and the profane in their day-to-day lives. All areas of life, including family, business, and leisure, are governed by religious principles, so that religious activity is not confined only to the church, mosque, or temple.

Fourth, fundamentalist religious groups emerge for a reason, usually in reaction to a perceived threat or crisis, real or imagined. Consequently any discussion of a particular fundamentalist group must include some reference to an adversary.

Fifth, one obvious concern of fundamentalists is to reverse the trend toward gender equality, which they believe is symptomatic of a declining moral order. In fundamentalist religions women's rights often are subordinated to ideals that the group considers more important to the well-being of the society, such as the traditional family or the right to life. Such a priority of ideals is regarded as the correct order of things.

Islamic Fundamentalism In *The Islamic Threat: Myth or Reality?* professor of religious studies John L. Esposito (1992) maintains that most Americans' understanding of fundamentalism does not apply very well to contemporary Islam. The term *fundamentalism* has its roots in American Protestantism and the twentieth-century movement that emphasizes the literal interpretation of the Bible. Fundamentalists are portrayed as static, literalist, retrogressive, and extremist. Just as we cannot apply the term *fundamentalism* to all Protestants in the United States, we cannot apply it to the entire Muslim world, especially when we consider that Muslims make up the majority of the population in as least 45 countries. The document "What Does Fundamentalism Really Mean?" offers one explanation for why the term *fundamentalism* is inappropriately applied to Muslim activity.

http://www.ais.org/~bsb/Herald/Previous/495/fundamentalism.html
Document: What Does Fundamentalism Really Mean?

Why is *fundamentalism* an inappropriate term for Muslim activity?

Esposito (1992) believes that a more fitting term is *Islamic revitalism* or *Islamic activism*. The form of Islamic revitalism varies from one country to another but seems to be characterized by the following themes: "A sense that existing political, economic, and social systems have failed; a disenchantment with, and at times a rejection of, the West; a quest for identity and greater authenticity; and the conviction that Islam provides a self-sufficient ideology for state and society, a valid alternative to secular nationalism, socialism, and capitalism" (p. 14).

In "Islam in the Politics of the Middle East" Esposito (1986) asks, "Why has religion [specifically Islam] become such a visible force in Middle East politics?" He believes that Islamic revitalism is a "response to the failures and crises of authority and legitimacy that have plagued most modern Muslim states" (p. 53). Recall that after World War I France and Britain carved up the Middle East into nation-states, with the boundaries drawn to meet the economic and political needs of Western powers.

Lebanon, for example, was created in part to establish a Christian state with ties to the West; Israel was created as a refuge for persecuted Jews when no country seemed to want them; the Kurds received no state; Iraq was virtually landlocked; and resource-rich territories were incorporated into states with very sparse populations (Kuwait, Saudi Arabia, the Emirates). Many of the leaders who took control of these foreign creations were viewed by their citizens "as autocratic heads of corrupt, authoritarian regimes that [were] propped up by Western governments and multinational corporations" (p. 54).

When Arab armies from six states lost "so quickly, completely, and publicly" to Israel in 1967, Arabs were forced to question the political and moral structure of their societies (Hourani 1991, p. 442). Had the leaders and the people deviated too far from or even abandoned Islamic principles? Could a return to an Islamic way of life restore confidence to the Middle East and give it an identity independent of the West? Questions of social justice also arose. Oil wealth and modernization policies had led to rapid increases in population and urbanization and to a vast chasm between the oil-rich countries such as Kuwait and Saudi Arabia and the poor, densely populated countries such as Egypt, Pakistan, and Bangladesh. Western capitalism, which was believed to be one of the primary forces behind these trends, seemed to be blind to social justice, promoting unbridled consumption and widespread poverty. Marxist socialism (a godless alternative) likewise had failed to produce social justice.

For many people Islam offers an alternative vision for society. According to Esposito (1986) Islamic activists (who are of many political persuasions, from conservative to militant) are guided by five beliefs: (1) Islam is a comprehensive way of life relevant to politics, state, law, and society; (2) Muslim societies fail when they depart from Islamic ways and follow the secular and materialistic ways of the West; (3) an Islamic social and political revolution is necessary for renewal; (4) Islamic law must replace Western-inspired or -imposed laws; and (5) science and technology must be used in ways that reflect Islamic values in order to guard against the infiltration of Western values. Muslim groups differ dramatically as to how quickly and by what methods these principles should be implemented. Most Muslims, however, are willing to work within existing political arrangements; they condemn violence as a method of bringing about political and social change.

The information presented in this section points to the interplay of religion with political, economic, historical, and other social forces. Fundamentalism cannot be viewed in simple terms, and any analysis must consider the broader context. A focus on context allows us to see that fundamentalism can be a reaction to many events and processes including secularization, foreign influence, failure or crisis in authority, the loss of a homeland, and rapid change.

Discussion Question

Based on the information presented in this chapter, why do you think religion is such a difficult phenomenon to define? What are the advantages and disadvantages associated with not being able to define the term? What roles does religion play in human activities?

Additional Reading

For more on

- Media coverage of religion
  http://www.missouri.edu/~c676747/religion/religion.html
  Document: Religion, Journalism, and the Internet

- World conflicts that involve religious factions
  http://www.emory.edu/CARTER_CENTER/demo.htm#conres
  Document: Conflict Resolution Program

- Religion versus science
  http://www.myna.com/~davidck/hawking.htm
  Document: David Cherniack Films: Transcripts—Stephen Hawking

- Descriptions of religions
  http://www.kosone.com/people/ocrt/var_rel.htm
  Document: Descriptions of 36 Religions, Faiths Groups and Ethical Systems

- Religious terms
  http://web.canlink.com/ocrt/glossary.htm
  Document: Glossary of Confusing Religious Terms

- Internet resources related to religion (general sites)
  http://www.pitts.emory.edu/ptl_rel-std.html
  Document: Pitts Theology Library: Religious Studies Resources on the Internet

  http://www.pitts.emory.edu/boblist.html
  Document: Internet Lists Related to Topics in Religion

- The Religious Freedom Restoration Act (RFRA)
  http://northshore.shore.net/rf/theact.html
  Document: Full Text of the Religious Freedom Restoration Act

  http://northshore.shore.net/rf/nowl.html
  Document: NOWL Panel Discussion on RFRA

- The Centers for Disease Control partnership with the faith community
  http://www.interaccess.com/ihpnet/alpha
  Document: Expanding the Public Health Envelope Through Faith Community

Social Change

Sociologists define **social change** as any significant alteration, modification, or transformation in the organization and operation of social life. Social change is an important topic in the discipline of sociology. In fact it is fair to say that sociology emerged as a discipline in an attempt to understand social change. Recall that the early sociologists were obsessed with understanding the nature and consequences of the Industrial Revolution—an event that triggered dramatic and seemingly endless changes in every area of social life.

When sociologists study change, they first must identify the aspect of social life they wish to study, which has changed or is undergoing change. The list of possible topics is virtually endless. Some examples include changes in the resources people use; changes in how people communicate with each other; changes in the amount of goods and services that people produce, sell to, or buy from others; changes in the average life span; and changes in how and where people work (Martel 1986). Often change is something that can be measured: that is, sociologists strive to identify the amount, range (how widespread), duration, pace (how fast), and/or direction of change. Upon identifying a topic, sociologists ask at least two key questions: What were the triggers of change? and What are the consequences of that change for social life? In this chapter we look at the concepts and theories that sociologists use to answer these complex questions. We give special focus to the triggers of change because the sociological concepts for analyzing the consequences of change and identifying the elements of society undergoing change have been covered in previous chapters.

Social Change: Causes and Consequences

When we think about a specific social change, we usually cannot identify a certain factor as the sole cause of the change. More often than not, change results from a sequence of events. An analogy may help clarify this point. Suppose that a wide receiver, after catching the football and running 50 yards, is finally tackled by a cornerback. One could argue that the cornerback caused the receiver to fall. Such an account, however, would not fully explain what actually happened. For one thing a tackle is not the act of one person; "there is present a simultaneous conflict of forces" (Mandelbaum 1977, p. 54) between the tackler (who is attempting to knock down the person with the ball) and the wide receiver (who is doing everything in his power to elude the tackler's grasp). To complicate matters, the wide receiver and the tackler are each members of a team, and their teammates' actions help determine how the play develops and ends.

Yet even though change is caused by a seemingly endless sequence of events, sociologists can identify some key events or factors that trigger changes in social life. These agents of change include the following:

- *Innovation:* the development of something new, whether an idea, a practice, or a tool
- *The actions of leaders:* people in positions of authority, including charismatic leaders and the power elite
- *Conflict:* clashes between groups over their shares of wealth, power, prestige, and other valued resources
- *Capitalism:* an economic system in which natural resources and the means of production and distribution are privately owned

An analysis of a social change is not complete until sociologists assess its consequences for the life of the society. As in identifying a cause, it is difficult to predict exactly how a specific change will affect society if only because we cannot pinpoint a time in the future when it will cease to have an impact. The effects of the 40-year arms race between the United States and former Soviet Union is a particularly dramatic example. Even as the weapons are now being dismantled, some materials will remain radioactive for at least 240,000 years—a time "so long that from the human standpoint, they might as well last forever" (Wald 1989, p. Y19).

Another reason sociologists find it difficult to predict the effects of a specific change is that people react to the change, and their reactions play a role in shaping the consequences. This does not mean that change is random or that it lacks a clear pattern. Rather we should view change as a "complex, unrepeatable, unpredictable historical result that is explainable after it happens." At the same time we must acknowledge that "if we wind the tape of life back and start over it may not happen again" (Gould 1990).

The unpredictable element of change—how people choose to react to it—should not be viewed as problematic for the study of change and its consequences. In fact this is a positive characteristic because it means that people are not passive agents; they create conditions that lead to change, and they react to those conditions as well. Sociologists, however, make sense of all these events by emphasizing "the more *general* properties of social experience, the everyday patterns of activity that appear again and again in the life of the social order" (Erikson 1971, p. 64). From a sociological perspective "every event has properties that can be subsumed under a more general heading" (Erikson 1971, p. 64), such as the creation of and reactions to innovation, the actions of leaders, conflict over scarce and valued resources, or the behaviors and policies shaped by capitalist principles. The remainder of this chapter discusses each of these agents of change, beginning with innovations.

Innovations

Innovation is the invention of something new—an idea, a process, a practice, a device, or a tool. Although all innovations build on existing knowledge and materials, they also go beyond that which already exists. They are syntheses, refinements, new applications, and reworkings of existing inventions.

Innovations also include **discoveries,** the uncovering of something that had existed before but had remained hidden, unnoticed, or undescribed. Innovations are sociologically significant because they change the ways in which people think and relate to one another.

Innovations can be classified broadly in one of two categories—basic or improving—although the distinction between the two is not always clear-cut. **Basic innovations** are revolutionary, unprecedented, or ground-breaking inventions that are the cornerstones for a wide range of applications. Examples of basic innovations include the invention of mass production, attributed to Henry Ford in 1904; the accidental discovery of radioactivity, attributed to French physicist Antoine Henri Becquerel around 1898; and the creation of the first sustained nuclear chain reaction by scientists working at the University of Chicago in 1942. Innovations are not always material devices, however; they can also be social inventions. Stuart Conger, the man responsible for setting up Saskatchewan NewStart, a Canadian Social Inventions Center, offers a definition of social inventions.

http://www.newciv.org/GIB/BOV/BV-2.HTML
Document: The History of Social Inventions

Read only enough of Conger's article to answer the following questions. What are social inventions? Give some examples. Conger maintains that social inventions include both organizations and procedures. How would you define each?

Improving innovations represent modifications of basic inventions in order to improve on them—that is, to make them smaller, faster, less complicated, or more efficient, attractive, durable, or profitable. Most present-day innovations are improving innovations. The "Inventor's Home Page" provides a look at some innovations posted on the internet.

http://ourworld.compuserve.com/homepages/invent/

Document: The Inventor's Home Page

 Q: Select two innovations from the list and read each description. Which qualities do the inventors claim are an improvement to an existing invention?

The sociological significance of any innovation is the effect it has on human relationships. Anthropologist Leslie White (1949) maintained that once a basic or an improving innovation has been invented, it becomes part of the cultural base, the size of which determines the rate of change.

Innovations and Rate of Change

White defined an invention as the synthesis of existing inventions. For example, the first airplane was a synthesis of many preexisting inventions, including the gasoline engine, the rudder, the glider, and the wheel. White suggested that the number of inventions in the cultural base increases geometrically—1, 2, 4, 8, 16, 32, 64, and so on. He argued that if a new invention is to come into being, the cultural base must be large enough to support it. If the Wright brothers had lived in the fourteenth century, for example, they never could have invented the airplane because the cultural base did not contain the ideas, materials, and inventions to support its creation. "It is a profound and necessary truth that the deep things in science are not found because they are useful; they are found because it was possible to find them" (Oppenheimer 1986, p. 11).

The seemingly runaway expansion, or increases in the volume, of new inventions prompted White (1949) to ask: Are people in control of their inventions, or do our inventions control us? For all practical purposes he believed that inventions control us. He supported this conclusion with two arguments. First, he suggested that the old adage "Necessity is the mother of invention" is naive, because in too many cases the opposite idea—that invention is the mother of necessity—is true. That is, an invention becomes a necessity because we find uses for that invention after it comes into being:

> We invent the automobile to get us between two points faster, and suddenly we find we have to build new roads. And that means we have to invent traffic regulations and put in stop lights [and build garages]. And then we have to create a whole new organization called the Highway Patrol—and all we thought we were doing was inventing cars. (NORMAN 1988, P. 483)

Second, White (1949) argued, when the cultural base is capable of supporting an invention, that invention will come into being whether people want it or not. White supported this conclusion by pointing to **simultaneous-independent inventions,** situations in which the same invention is created by two or more persons working independently of each other at about the same time (sometimes within a few days or months). He cited some 148 such simultaneous-independent inventions—including the telegraph, electric motor, microphone, telephone, microscope, steamboat, and airplane—as proof that someone will come along to make the necessary synthesis if the cultural base is ready to support a particular invention. In other words inventions such as the light bulb and the airplane would have come into being whether or not Thomas Edison and the Wright brothers (the people we traditionally associate with these inventions) had ever been born. According to White's conception inventors may be geniuses, but they also have to be born at the right place and the right time—that is, in a society with a cultural base sufficiently developed to support their invention.

White's theory suggests that if the parts are present, someone eventually will come along and put them together. The implications are that people have little control over whether an invention should come into being and that they adapt to inventions after the fact. Sociologist William F. Ogburn (1968) called this failure to adapt to a new invention **cultural lag.**

Cultural Lag

In his theory of cultural lag Ogburn distinguished between material and nonmaterial culture. As you will recall from Chapter 5, material culture includes tangible creations or objects—including resources (oil, trees, land), inventions (paper, guns), and systems (factories, sanitation facilities)—that people have created or, in the case of resources such as oil, have identified as having the properties to serve a particular purpose. Nonmaterial culture includes intangible creations such as beliefs, norms, values, roles, and language.

Although Ogburn maintained that both components are important agents of social change, his theory of cultural lag emphasized the material component, which he suggested is the more important of the two. Ogburn believed that one of the most urgent challenges facing people today is adapting to material innovations in thoughtful and constructive ways. He used the term **adaptive culture** for the portion of the nonmaterial culture (norms, values, and beliefs) that adjusts to material innovations. Such adjustments are not always immediate; sometimes they take decades, and sometimes they never are made. Futurist Thomas Doherty describes a key information technology that is responsible for social change.

http://www.magpage.com/~tdoherty/futures.html
Document: Futures Research: New Technologies for the 21st Century

Q: According to Doherty what is the key technological innovation causing social change today and into the next century? What is the first-order effect of this technology? Why does it take at least a generation to realize the changes caused by a major technological innovation? Who is likely to be the last to change? Why is the replacement of paper-and-ink communication with electronic communication slow in occurring?

Ogburn, however, was not a **technological determinist,** someone who believes that human beings have no free will and are controlled entirely by their material innovations. For one thing he noted that people do not adjust to new material innovations in predictable and unthinking ways; rather they choose to create them, and after they create them they choose how to use them. If people have the power to create material innovations, they also have the power to destroy them and to ban or modify their use.

The human ingenuity, effort, and planning that went into the construction of the atomic bomb and the many policies and initiatives aimed at controlling testing, production, and usage illuminate how people can exercise control over their inventions. The challenge lies, however, in convincing people that they need to address an invention's potential disruptive consequences (which are usually known in advance because someone points them out but often are ignored) before they have a chance to materialize.

In our discussion about innovations as a trigger of social change, we have emphasized material inventions (devices, tools, and equipment). Innovations also can be nonmaterial inventions such as a revolutionary idea.

Revolutionary Ideas

In *The Structure of Scientific Revolutions* Thomas Kuhn (1975) maintains that most people perceive science as an evolutionary enterprise; that is, over time scientists move closer to finding the solution to problems by building on their predecessors' achievements. Kuhn takes issue with this evolutionary view. He argues that some of the most significant scientific advances have been made when someone breaks away from the predecessors' achievements and challenges the prevailing paradigms. According to Kuhn **paradigms** are the dominant and widely accepted theories and concepts in a particular field of study. Paradigms gain their status not because they explain everything, but because they offer the best way of looking at the world for the time being. On the one hand, paradigms are important thinking tools; they bind a group of people with common interests into a scientific or national community. Such a community could not exist without agreed-upon paradigms. On the other hand, paradigms can be blinders, limiting the kinds of questions that people ask and the observations they make.

The explanatory value, and hence the status, of a paradigm is threatened by **anomaly,** an observation or observations that it cannot explain. The existence of an anomaly alone, however, is usually not enough to cause people to abandon a particular paradigm. According to Kuhn, before people abandon old paradigms, someone must articulate an alternative paradigm that accounts convincingly for the anomaly. Kuhn hypothesizes that the people most likely to put forth new paradigms are those who are least committed to the old paradigms—the young and those new to a field of study.

A **scientific revolution** occurs when enough people in the community break with the old paradigm and change the nature of their research or thinking in favor of the incompatible new paradigm. Kuhn considers a new paradigm incompatible with the one it replaces because it "changes some of the field's most elementary theoretical generalizations" (p. 85). The new paradigm causes converts to see the world in an entirely new light and to wonder how they could possibly have taken the old paradigm seriously. "When paradigms change, the world itself changes with them. Led by a new paradigm, scientists adopt new instruments and look in new places" (p. 111). The internet could be one such instrument that could lead to paradigm shift with regard to learning.

gopher://borg.lib.vt.edu/00/catalyst/v22n3/katz.v22n3
Document: Redefining Success: Public Education in the 21st Century

Q: How will new telecommunications technology lead to a paradigm shift with regard to conceptions of literacy?

The Actions of Leaders

The actions of leaders represent a second major trigger of social change. In the most general sense a *leader* is someone who has the power to influence others or who is in charge or in command of a social situation. Max Weber (1947) defined *power* as the probability that an individual can realize his or her will even against the resistance of others. The probability increases if that individual can force people to obey his or her commands or if the individual has authority over others. **Authority** is legitimate power such that people believe that the differences in power are just and proper—that is, people see an authority figure as entitled to give orders. Weber identified two types of authority—charismatic and legal-rational—that have important implications with regard to social change.

Charismatic Leaders as Agents of Change

Charismatic authority rests on the exceptional and exemplary qualities of the person issuing the commands. Charismatic leaders are obeyed because their followers believe in and are attracted irresistibly to the vision that the leaders articulate. Because the source of charismatic authority resides in the leader's exceptional qualities, not in tradition or established rules, the charismatic leader's actions and visions are not bound by rules or traditions. Consequently these leaders, by virtue of their special qualities, have the ability to unleash revolutionary changes; they can ask their followers to behave in ways that depart from rules and traditions.

Charismatic leaders often appear during times of profound crisis (such as economic depressions or wars), when people are most likely to be drawn to someone with exceptional personal qualities who offers them a vision of a new order different from the current, seriously flawed situation. The source of the charismatic leader's authority, however, does not rest with the ethical quality of the command or vision as evidenced by the fact that Adolf Hitler, Franklin D. Roosevelt, Mao Zedong, and Winston Churchill all were considered charismatic leaders. Most assume leadership during turbulent times and convey a powerful vision (right or wrong) of a group's destiny. Consider the case of Adolf Hitler.

http://www.scetv.org/scetv/over2.html
Document: South Carolina Voices: Lessons from the Holocaust

Q: What was the historical context that facilitated Hitler's rise to power?

A charismatic leader is more than popular, attractive, likable, or pleasant; a merely popular person, "even one who is continually in our thoughts" (Boudon and Bourricaud 1989, p. 70), is not someone for whom we would break all previous ties and give up our possessions. Charismatic leaders are so demanding as to insist that their followers make extraordinary personal sacrifices, cut themselves off from ordinary worldly connections, or devote their lives to achieving a vision that the leaders have outlined.

Charismatic authority is a product of the intense relationships between leaders and followers. From a relational point of view, then, charisma is a "highly asymmetric power-relationship between an inspired guide and a cohort of followers" (Boudon and Bourricaud 1989, p. 70) who believe in the promises and visions offered by the person with special qualities. For example, many Germans believed that Adolf Hitler's vision could help them recover from the humiliation and massive destruction they suffered at the hands of the Allies (Britain, France, Russia, Italy, and the United States) in World War I.

Charismatic leaders and their followers come to constitute an "emotional community" devoted to achieving a goal and sustained by a belief in the leader's special qualities. Weber argued, however, that at some point the followers must be able to return to a normal life and to develop relationships with one another on a basis other than their connections to the leader. Attraction and devotion cannot sustain a community indefinitely, if only because the object of these emotions—the charismatic leader—is mortal. Unless the charisma that unites a community is routinized, the community may disintegrate from exhaustion or from a void in leadership. **Routinized charisma** develops as the community establishes procedures, rules, and traditions to regulate the members' conduct, to recruit new members, and to ensure the orderly transfer of power. The case of M. C. Mehta, the most succesful environmental lawyer in Asia, illustrates routinized charisma.

http://www.igc.apc.org/elaw/asia/india/icela.html
Document: Goldman Prize Winner: M.C. Mehta

Q: Briefly describe M. C. Mehta's accomplishments. What has Mehta done to routinize charisma?

The point of this example is to show that charismatic authority must eventually come to rest on legal-rational grounds. That is, for a charismatic leader's cause to continue, it must be grounded in a position or organization, not in the personal qualities of one person.

The Power Elite: Legal-Rational Authority and Change

Legal-rational authority rests on a system of impersonal rules that formally specify the qualifications for occupying a powerful position. The rules also regulate the scope of power and the conduct appropriate to someone holding a particular position. In cases of legal-rational authority, people comply with commands, decisions, and directives because they believe that those who have issued them have earned the right to rule by virtue of the position they occupy.

Sociologist C. Wright Mills (1959, 1963, 1973) argued that the causes of **great changes**—events that profoundly affect the life chances of virtually every person in a society—can be traced to the decisions made by the **power elite,** those few people positioned so high in the social structure of leading institutions that their decisions have consequences affecting millions of people worldwide. For the most part the source of this power is legal-rational and resides not in the personal qualities of those in power, but in the positions that the power elite have come to occupy. "Were the person occupying the position the most important factor, the stock market would pay close attention to retirements, deaths, and replacements in the executive ranks" (Galbraith 1958, p. 146).

The amount of power wielded by the elite over the lives of others is related to the nature and the quality of instruments that they can use, by virtue of their position, to rule, control, and influence others. Instruments might include weapons, surveillance equipment, and specialized modes of communication. Mills argued that since World War II rapid advances in technology have allowed power to become concentrated in the hands of a few; those with access to such power can exercise an extraordinary influence not only over their immediate environment but over millions of people, tens of thousands of communities, entire countries, and the globe.

According to Mills (1963) the leading institutions are the military, corporations (especially the 200 or so largest American corporations), and government. "The power to make decisions of national and international consequence is now so clearly seated in political, military, and economic institutions that other areas of society seem off to the side and, on occasion, readily subordinated to these" (p. 27).

The origins of these institutions' power can be traced to World War II, when the political elite mobilized corporations to produce the supplies, weapons, and equipment needed to fight the war. For one measure of the extent to which the government, the military, and corporations worked together, consider this Bell Telephone System advertisement in the July 30, 1945, issue of *Life:*

> In the last five years the Bell System has furnished millions of telephones for war, including 1,325,000 head sets for air and ground forces and more than 1,500,000 microphones.... Also more than 1,000,000 airplane radio transmitters and receivers ... 4,000,000 miles of telephone wire in cables ... a vast quantity of switchboards, gun directors and secret combat equipment. That helps to explain why we are short of all kinds of telephone facilities here at home. (P. 3)

After the war, as Stalin moved to consolidate his power in Eastern Europe, Japan and the countries of Western Europe—their populations demoralized, their economies in ruins, their infrastructures devastated—had little choice but to accept help from the United States in the form of the Marshall Plan. U.S. corporations, unscathed by the war, were virtually the only companies in the world able to offer the services and products that the war-torn countries needed for rebuilding. The interests of the government, the military, and corporations became further entangled when the political elite decided that a permanent war industry was needed in order to contain the spread of communism. Thus over the past half-century these three institutions have become deeply and intricately interrelated in hundreds of ways, as the cases cited in the "Top Censored News Stories of 1994 and 1995" illustrate.

http://censored.sonoma.edu/ProjectCensored/Stories1994.html
Document: Top Censored News Stories of 1994

http://censored.sonoma.edu/ProjectCensored/Stories1995.html
Document: Top Censored News Stories of 1995

Q: Scan the list of the top censored news stories of 1994 and 1995. What are two or three examples of cases in which decisions made by the government benefited the military and industrial sectors? (*Note:* for more information on the process by which the top censored news stories are selected, see **http://zippy.sonoma.edu/ProjectCensored.**)

Because the three realms of institutions (military, government, and corporations) are interdependent and because decisions made by the elite of one realm affect the other two, Mills believed that it is in everyone's interest to cooperate. Shared interests cause those who occupy the highest positions in each realm to interact with one another. Out of necessity, then, a triangle of power has emerged. This is not to say that the alliance among the three is untroubled, that the powerful in each realm are of one mind, that they know the consequences of their decisions, or that they are joined in a conspiracy to shape the fate of a country or the globe.

Mills gave no detailed examples of the actual decision-making process at the power elite level. He was more concerned with understanding the consequences of this alliance than with tracing the decision-making process or assessing the consciousness or purity of motives. Mills acknowledged that members of the power elite are not totally free agents, subject to no controls. A chief executive officer of a major corporation is answerable to unions, OSHA, the FDA, or other regulatory bodies. Pentagon officials are subject to congressional investigations and budget constraints. Defense contractors are liable to the Federal False Claims Act, which gives a share in the settlement to any employee who can prove that the contractor has defrauded the government (Stevenson 1991). The president of the United States is constrained by bureaucratic red tape and by a sometimes slow-moving, politically oriented Congress. Mills questioned, however, whether these constraints on the power elite have "much significance when weighed against the areas of unrestricted action open to [them]" (Hacker 1971, p. 136).

To this point we have examined two important agents of change: innovations and the actions of people in positions of authority (charismatic leaders and the power elite). We turn now to a third agent of change—conflict—which is intertwined with these other two. In its most basic form conflict involves clashes between groups over their access to shares of wealth, prestige, and other valued resources. To see the connection between conflict and innovations as agents of change, recall that the introduction of innovations (discoveries, inventions, paradigms) disrupts the balance of power, causing conflict between those who stand to benefit and those who stand to lose from the widespread acceptance of an innovation. Likewise, to see the connection between conflict and power, recall that charismatic leaders and the power elite possess the authority to impose their will (for better or worse) despite resistance by others.

Conflict

Sociologist Lewis Coser points out in his 1973 essay "Social Conflict and the Theory of Social Change" that conflict will always exist, if only because there is never a perfect "concordance between what individuals and groups within a system consider their just due and the system of allocation" (p. 120). Conflict occurs whenever a group takes action to increase its share of or control over wealth, power, prestige, or some other valued resource and when these demands are resisted by those who benefit from the current distribution system.

Consequences of Conflict

Whether it involves violent clashes or public debate, conflict is both a consequence and a cause of change. In general we can say that any kind of change has the potential to trigger conflict between those who benefit from the change and those who stand to lose because of it. When the bicycle was invented in the 1840s, for example, horse dealers organized against it because it threatened their livelihoods. Some physicians declared that people who rode bicycles risked getting "cyclist sore throat" and "bicycle stoop." Church groups protested that bicycles would swell the ranks of "reckless" women (because bicycles could not be ridden sidesaddle).

Conflict can lead to positive change as well. It can be a constructive and invigorating force that prevents a social system from becoming stagnant, unresponsive, or inefficient. Conflict such as that resulting from the antinuclear, civil rights, and women's movements can create new norms, relationships, and ways of thinking. Conflict also can generate new and efficient technologies. The internet is one example of a technology whose origins are rooted in conflict.

http://scuba.uwsuper.edu/~rwhiffen/web-intro/history/welcome.html
Documents: Ten Minutes of Internet History

 Q: Read "The Cold War Beginnings" and "ARPANET." What historical event set the context for the development of the internet? What central question guided the basic features of the internet? What was the Rand Corporation's answer to that question? What is ARPANET? Who could belong to ARPANET?

Ironically the most destructive form of conflict—war—has generated advances in life-saving medical technologies. For example, during World War I many soldiers fighting on manure-covered farmlands contracted tetanus. In addition large numbers of soldiers were injured by machine gun shrapnel and bombshells. Physicians experimenting with antitoxins eventually found a cure for tetanus and also made considerable advances in reconstructive surgery. Similarly the kinds of injuries incurred during World War II motivated doctors to create a system of collecting and preserving blood plasma and to mass-produce an effective drug—penicillin—to treat wound infections (Colihan and Joy 1984). For over 35 years the Pentagon's Advanced Research Project Agency (ARPA) has funded civilian projects that it believes will have military applications. ARPA's primary mission is "to help maintain U.S. technological superiority and guard against unforeseen technological advances by potential adversaries." Press releases give an overview of the projects that the Defense Advanced Research Projects Agency (DARPA) funds.

http://www.arpa.mil/news.html
Document: DARPA News

Q: Browse the various news releases. What are three examples of technologies that DARPA has funded?

Whether conflict will lead to reform (improvements or alterations in current practices) or to revolution (complete and drastic change) depends on a broad range of factors and contingencies. Sociologist Ralf Dahrendorf (1973) has identified and described some of these factors in his essay "Toward a Theory of Social Conflict."

Structural Origins of Conflict

Dahrendorf (1973) asks two basic questions: (1) What is the structural source of conflict? and (2) What forms can conflict take? Dahrendorf's answers rest on the following assumptions. First, in every organization that has a formal authority structure (a state, a corporation, the military, the judicial system, a school system), clear dichotomies exist between those who control the formal system of rewards and punishment (and thus have the authority to issue commands) and those who must obey those commands or face the consequences (loss of job, jail, low grades, and so on). Second, a distinction between "us" and "them" arises naturally from the unequal distribution of power. In view of these assumptions the structural origins of conflict can be traced to the nature of authority relations. That conflict can assume many forms: it can be mild or severe; "it can even disappear for limited periods from the field of vision of a superficial observer" (p. 111). As long as an authority structure exists, however, conflict cannot be abolished.

Dahrendorf outlines a three-stage model of conflict in which progression from one stage to another depends on many things. He does not claim to give an exhaustive list with regard to the possible course of a conflict. In fact he reminds readers that the conflicts he names are some of the most obvious. The point is that conflict—its course and its resolution—is a complicated phenomenon in which many elements must be considered.

Conflict Model Stage 1 Every authority structure contains at least two groups with opposite and hidden interests. Those with power have an interest in preserving the system; those without power have an interest in changing it. These opposing interests, however, remain latent or below the surface until the groups (especially those without power) organize. Adam Michnik (1990), one of the founders of Solidarity (the labor movement in Poland led by Lech Walesa that pushed for major reforms and that played a major role in overthrowing communism), observes that "it is immeasurably difficult to trace the path on which a person . . . encounters other people just like himself, and at a certain point . . . [says] 'Let us join hands, friends, so that they will not pick us off one by one'" (p. 240). In the uranium mining town of Marysvale, Utah, for example, the miners who worked between 1948 and 1966 hauling the uranium ore that the Atomic Energy Commission purchased for nuclear weapons production started to become aware of their shared fate when the first miners began dying in the early 1960s (Schneider 1990).

Often a significant event makes powerless people aware that they share an interest in seeing the system changed. Vaclav Havel, ex-president of the former Czechoslovakia and now the president of the Czech Republic, believes that the Chernobyl accident of 1986 may have played an important role in bringing about the revolutions in Central and Eastern Europe. After Chernobyl people in the former Czechoslovakia dared to complain openly and loudly to one another (Ash 1989).

http://www.enn.com/feature/fe042296/feature1.htm
Document: Chernobyl 10 Years After: The Nightmare Accelerates

Q: What are some of the immediate environmental and health problems created by the Chernobyl accident that motivated people to "complain loudly and openly"? What effect did Chernobyl have on the nuclear power industry in the United States and around the world?

Sometimes people organize because they have nothing left to lose. As one East German scientist explained, "You don't need courage to speak out against a regime. You just need not to care anymore—not to care about being punished or beaten. I don't know why it all happened this year (in 1989). We finally reached the point where enough people didn't care anymore what would happen if they spoke out" (Reich 1989, p. 20).

Conflict Model Stage 2 If those without authority have opportunities to communicate with one another, the freedom to organize, the necessary resources, and a leader, then they will organize. Before *perestroika*, for example, when people in the Warsaw Pact countries demanded political freedoms, the Soviets sent in soldiers and tanks. But when Soviet leaders announced that they would not intervene in the internal affairs of other communist countries and condemned past interventions, people quickly mobilized to topple the pro-Soviet regimes. The speed of these revolutions suggests that the political strength of the communist governments rested on a monopoly on physical force, which, until 1989, they made clear they would use.

This example shows that people in positions of authority often use their positions to keep potentially damaging information from those who might use it to organize to change the system. For instance, the hallmarks of the nuclear weapons program in the United States (as well as the former Soviet Union) were an obsessive and unchallenged secrecy and an absence of outside oversight that allowed scientists, government leaders, and weapons contractors to emphasize production over the environmental and health concerns. As a result information about radioactivity and its health effects was kept from workers, soldiers, and residents in affected communities. U.S. government leaders censored, dismissed, and/or withdrew funds from researchers who sought to publicize such information. However, we know that thousands of people persisted in their effort to study the effects or to gain access to government records. Under the Clinton administration thousands of classified records have been declassified.

One technology that will certainly enhance opportunities for people to communicate with one another at the grass-roots level is the internet. Consider the communication dynamics illustrated by the document "Internet Day of Protest."

http://www.socool.com/socool/news/protest.html

Document: Internet Day of Protest

Q: What are some of the ways in which the internet facilitates communication and action?

Conflict Model Stage 3 Once organized, those without power enter a state of conflict with those in power. The speed and the depth of change depend on the capacity of those who rule to stay in power and on the kind and degree of pressure exerted from below. The intensity of the conflict can range from heated debate to violent civil war, but it is always contingent on many factors, including opportunities for mobility within the organization and the ability of those in power to control the conflict. If those who lack authority are confident that eventually they will achieve such a position, the conflict is unlikely to become violent or revolutionary. If those in power decide that they cannot afford to compromise and thus mobilize all their resources to thwart protests, two results are possible. First, the protesters may believe that the sacrifices are too great and then withdraw. Second, they might decide to meet the "enemy" directly, in which case the conflict becomes violent.

To this point we have discussed conflict in very general terms. The fourth important agent of change represents a specific kind of conflict, motivated by the pursuit of profit. This agent is capitalism, an economic system whose origins can be traced back 500 years.

Capitalism

Karl Marx believed that an economic system—capitalism—ultimately caused the explosion of technological innovation and the enormous and unprecedented increase in the amount of goods and services produced during the Industrial Revolution. In a **capitalist system** profit is the most important measure of success. To maximize profit, the successful entrepreneur reinvests profits in order to expand consumer markets and to obtain technologies that allow the manufacture or provision of the highest-quality and most cost-effective products and services.

The capitalist system is a vehicle of change in that it requires the instruments of production to be revolutionized constantly. Marx believed that capitalism was the first economic system capable of maximizing the immense productive potential of human labor and ingenuity. He also believed, however, that capitalism ignored too many human needs and that too many people could not afford to buy the products of their own labor. Marx ([1881] 1965) stated that capitalism already had unleashed "wonders far surpassing Egyptian pyramids, Roman aqueducts, and Gothic cathedrals . . . [and] expeditions that put in the shade all former Exoduses of nations and crusades" (p. 531). He argued that if this economic system were in the right hands—those of socially conscious people motivated not by a desire for profit or by self-interest, but by an interest in the greatest benefit to society—public wealth would be more than abundant and would be distributed according to need.

http://csf.colorado.edu/psn/marx/Archive/1849-WLC/wlc9.txt
Document: Karl Marx: Wage-Labor and Capital

Q: According to Marx what strategies do capitalists use to compete in the marketplace? How do these strategies affect workers?

Marx's theories influenced a group of contemporary sociologists—world system theorists—to write about capitalism as the agent of change underlying global interdependence.

World System Theory

Immanuel Wallerstein (1984) is the sociologist most frequently associated with *world system theory*, a modern theory concerned with capitalism. Since the early 1970s he has been writing a four-volume work (three volumes of which have been published) about the ceaseless expansion over the past 500 years of a single market force—capitalism. According to Wallerstein, although stagnant periods have occurred and some countries (the communist countries, for example) have tried to withdraw from the capitalist economy, no real contraction has occurred. "Hence, by the late nineteenth century, the capitalist world-economy included virtually the whole inhabited earth and it is presently striving to overcome the technological limits to cultivating the remaining corners; the deserts, the jungles, the seas, and indeed the other planets of the solar system" (p. 165).

Wallerstein distinguishes between the terms *world economy* and *world-economy*. People who use the term *world economy* (without the hyphen) envision the world as consisting of 160 or so national economies that have established trade relationships with one another. In this vision globalization is portrayed as a relatively new phenomenon and as a process by which the countries of the world have moved from relatively isolated, self-sufficient economies to economies that trade with one another to varying degrees. Although popular, this conception of global interdependence is not very accurate. The more accurate term (and conception) is **world-economy.** The world-economy is not recent; it has been evolving for at least 500 years and is still evolving. People who use the hyphenated term envision a world (encompassing hundreds of countries and thousands of cultures) interconnected by a single division of labor. In the world-economy economic transactions transcend national boundaries. Although each government seeks to shape the global market in ways that benefit its "own" corporations and national interests, no single political structure (world government) or national government has authority over the system of production and distribution.

Wallerstein argues that the world-economy is capitalist because "its economy has been dominated by those who operate on the primacy of endless accumulation, . . . driving from the arena those who seek to operate on other premises" (p. 15). Critics counter that this is an exaggeration, that many countries have economies that are not capitalist and that no country has an economy that runs on purely capitalist principles.

Wallerstein counters this criticism in part with the argument that the communist countries were the equivalent of huge state-owned capitalist corporations. All depended on the world-economy, and all traded on some levels with countries that were bitterly opposed to their political and economic system. Even before the collapse of the Soviet Union, for example, the United States exported corn and wheat there, and the Soviets exported chemicals, fuels, and minerals to the United States and natural and enriched uranium to several Western countries (Broad 1991). Another feature that makes the world-economy capitalist is the fact that profits from goods and services are distributed unevenly through the global market to a network of beneficiaries, most of whom live in the mechanized rich countries.

The Role of Capitalism in the Global Economy

How has capitalism come to dominate the global network of economic relationships? One answer lies in the ways in which capitalists respond to changes in the economy, especially to economic stagnation. Historically there have been five important responses, all designed to create economic growth:

1. Lower production costs by hiring employees who will work for lower wages (for example, by busting unions, buying out workers' contracts, or offering early retirement plans), by introducing labor-saving technologies (such as computerizing the production process), or by moving production facilities out of high-wage zones and into lower-wage zones inside or outside the country.
2. Create a new product that consumers "need" to buy, such as the videocassette recorder, the computer, or the fax machine.
3. Improve on an existing product and thus make previous versions obsolete. For example, Nike lists more 800 shoe models for 25 or more sports. Nike updates each shoe every six months, tempting "customers to lace on new pairs before last

year's wear out" (Calonius 1991, p. 26). Nike's "Air Timeline" gives an overview of improvements on an existing product (see **http://text.nike.com/background/ air_timeline**).

4. Expand the outer boundaries of the world economy and create new markets. Since the fall of the Berlin Wall in 1989, for example, American, Western European, and Japanese corporations have been expanding their markets into Eastern Europe, Russia, and the new states of Eurasia. Procter & Gamble, for instance, produces and markets detergent, toothpaste, shampoo, and diapers in the former Czechoslovakia, Hungary, and Poland (Rawe 1991). The Coca-Cola Company moved particularly quickly into Eastern Europe. Almost immediately after the Berlin Wall fell and East Germans started to visit West Germany and West Berlin, Coca-Cola was there, handing out free Coca-Cola. This event was very popular in West Berlin; people sought out the Coca-Cola vendor for samples. Within weeks Coca-Cola executives were in East Germany discussing the distribution of Coca-Cola there:

> Almost at the same time we reorganized: As soon as we saw the changes, we moved East Germany into the West German and E. C. group, and transferred the infrastructure, talent and technology of our West German operations into East Germany. Within weeks, we were shipping Coca-Cola in cans into East Germany to distributors with whom we had made agreements. Within a month or two, we were selling a million cases a month in East Germany. . . . By July, we had put our first East German production facility in place. Now we're no longer solely importing into East Germany, we're producing and distributing in the country. Next year we expect to sell 30 million cases and, by 1995, 100 million cases a year in East Germany. (GUTTMAN 1990, p. 16)

5. Redistribute wealth to enable more people to purchase products and services. Henry Ford was the first to do this on a large scale; in 1908 he came up with the revolutionary concept of paying workers a wage ($5.00 per day) large enough to allow them to purchase the products of their labor (Halberstam 1986). Because of the shortage of hard currency in Russia, the new states of Eurasia, and Eastern Europe, executives of American, Western European, and Japanese corporations have set up barter systems there (Holusha 1989). "Pepsico, for example, exports wooden chairs from Poland to its Pizza Hut franchises in the United States, and sells its soft drink to the Soviet Union in exchange for old submarines" (O'Sullivan 1990, p. 22).

As a result of these responses to economic stagnation, capitalism has spread steadily to encompass the globe. In addition every country of the world has come to play one of three different and unequal roles in the global economy: core, peripheral, and semi-peripheral.

The Roles of Core, Peripheral, and Semi-Peripheral Economies

Core economies include those of the mechanized rich countries—countries characterized by strong, stable governments. Core economies tend to be highly diversified. The G-7 countries (Japan, Germany, France, the United States, Canada, Great Britain, and Italy) are examples of core economies, absorbing nearly two-thirds of developing countries' exports. They import raw materials from labor-intensive poor countries and make use of free-trade zones around the world. The overwhelming majority of "the great global enterprises that make the key decisions—about what people eat and drink, what they read and hear, what sort of air they breathe and water they drink, and, ultimately, which societies will flourish and which city blocks will decay" (Barnet 1990, p. 59)—have their headquarters in countries with core economies.

Fortune magazine lists the top 500 global corporations, ranked by revenue.

http://pathfinder.com/@@5MeilwYAjvZJTTu0/fortune/magazine/1995/950807/global500/revenue.html

Document: Fortune Global 500

Q: Browse through the list and pick approximately 25 of the top 500. In which country is each company headquartered?

The revenue of these corporations exceeds the gross domestic products (GDP) of many countries (Currie and Skolnick 1988). For example, note that the 1995 revenue for Mitsubishi (no. 1) is $175.8 billion, for K mart (no. 36) is $34.3 billion, for American Express (no. 200) is $15.6 billion, and for ToyoSeikan (no. 500) is $7.8 billion. The *World Factbook* site lists the GDP for selected countries.

http://www.odci.gov/cia/publications/95fact/index.html

Document: 1995 World Factbook

Q: Randomly select 10 non-G-7 countries and look up their GDP. For each, is the GDP less than the revenues of the top 10 global corporations? Where does its GDP rank in relation to the revenues of the Global 500?

When economic activity weakens in the industrial world, the labor-intensive poor countries suffer because the amount of exports declines and price levels fall.

Labor-intensive poor countries have **peripheral economies,** which are not highly diversified; most of the jobs are low-paying and require few skills. Peripheral societies depend disproportionately on a single commodity such as coffee, peanuts, or tobacco or a single mineral resource such as tin, copper, or zinc. The aggregate GDP of all the peripheral economies in the world is less than that of the European Economic Community (mainly composed of the Western European countries) (Van Evera 1990). Peripheral economies have a dependent relationship with core economies that is rooted in colonialism, and they operate on the fringes of the world-economy. In the midst of widespread and chronic poverty in peripheral economies, however, there are islands of economic activity including off-shore manufacturing zones, highly vulnerable extractive and single-commodity economies, and tourist zones. Return to the *World Factbook*.

http://www.odci.gov/cia/publications/95fact/index.html
Document: 1995 World Factbook

Q: Based on the information presented in the "Economy" section, what are two countries that represent peripheral economies? Why?

Between the core and the periphery are the **semi-peripheral economies,** characterized by moderate wealth (but extreme inequality) and moderate diversification. Taiwan, Brazil, South Korea, and Mexico fall into this category. Semi-peripheral economies exploit peripheral economies and are exploited by core economies. By this definition Iraq and Kuwait were semi-peripheral economies before the Gulf War: the core economies relied on them for cheap oil, and they relied on peripheral economies for cheap labor. The extent of that reliance became clear after Iraq invaded Kuwait: at that point millions of migrant workers fled Kuwait, including Iranians (70,000), Iraqis (2.2 million), Yemenis (45,000), Sudanese (21,800), Egyptians (700,000), Jordanians (220,000), and Palestinians (30,000). Similarly Pakistanis (67,600), Indians (150,000), Bangladeshis (85,000), Vietnamese (16,000), and Filipinos (30,000) fled Iraq (Miller 1991). According to Wallerstein (1984) semi-peripheral economies play an important role in the world-economy because they are politically stable enough to provide useful places for capitalist investment if wage and benefit demands become too great in core economies.

World system theorists maintain that political upheavals are caused by uneven and unequal integration into the world-economy. Corporations take actions and governments make trade decisions that leave out some groups. Eventually oppressed groups may organize around one of two main themes: "class" (disadvantaged position in the labor force) and "nation" (common community, culture, language, territory, and ethnicity). According to this viewpoint it would be a mistake to view conflicts in Russia, the new states of Eurasia and central Europe, and Africa as simply ethnic in origin. Rather they center around deeply rooted, chronic, persistent inequalities and conflicts over scarce and valued resources (territory, minerals, and so on).

Discussion Question

Based on the information presented in this chapter, why do you think it is difficult to identify the causes and consequences of social change? How do sociologists approach the study of social change?

Additional Reading

For more on

- Capitalism
  http://www.ocf.berkeley.edu/~shadab/capit-2.html
  Document: Capitalism: Frequently Asked Questions (Theory)

http://www.ocf.berkeley.edu/~shadab/capit 3.html
  Document: Capitalism: Frequently Asked Questions (Practice)

- Karl Marx
  http://home.sol.no/hansom/papers/960506.htm
  Document: A Review of Reviews: Making Sense of Marx

- Social change
  http://www.hotwired.com/special/unabom/principles.html
  Document: Some Principles of History

  http://www.samizdat.com/global.html
  Document: Global Competition and the Long Road to General Prosperity

- Technological determinism
  http://www.aber.ac.uk/~dgc/tdet01.html
  Document: Technological or Media Determinism: Introduction

  http://www.aber.ac.uk/~dgc/tdet02.html
  Document: Technological or Media Determinism: Technology-Led Theories

- Strategies to realize a world economy
  gopher://gopher.undp.org/00/ungophers/unctad/efficiency/columbus
  Document: Columbus Ministerial Declaration on Trade Efficiency

- World system theory
  http://cil.andrew.cmu.edu/projects/World_History/Wall.html
  Document: The Development of a World Economic System

  http://www.worldbank.org/html/extpb/wdr95/WDRENG.html
  Document: Workers in an Integrating World

17 Internet Home Library

World Information

State of the World Indicators

http://www.igc.apc.org/millennium/inds/

This site provides links to various indicators for a quick overview of the state of the world's environment. Some indicators are water availability, species extinctions per day, and years until half of known crude oil is gone. There is information about each indicator as well.

World Demographics

http://www.prb.org/prb/media.htm#demonews

The Population Reference Bureau provides information related to population and population growth around the world.

World Factbook

http://www.odci.gov/cia/publications/95fact/xx.html

This CIA site contains information on the world as a unit. For example, it gives the unemployment rate, population size, total fertility, and so on for the world.

International Agencies and Information

http://www.lib.umich.edu/libhome/Documents.center/intl.html

This site is an alphabetical list of international agencies beginning with the Asian Development Bank and ending with the World Trade Organization. Web links to these agencies and the information each agency offers are at one's fingertips. There are also links to the full text of many international treaties.

Population Profile of the United States

National PopClock from the U.S. Bureau of the Census

http://www.census.gov/cgi-bin/popclock

The resident population of the United States is projected to the day, hour, minute, and second.

Population Profile of the U.S.: 1995

gopher://gopher.census.gov/00/Bureau/Pr/Subject/Pop/cb95-137.txt

This Census Bureau Web site contains a series of selected facts about the U.S. population. The site lists statistical facts (such as the percentage of 3- and 4-year-olds enrolled in nursery school, the share of households occupied by families, and so on) Individually the facts might seem trivial, yet collectively they offer an interesting profile of the U.S. population. Additional selected statistics on topics such as the percentage of the population with a high school diploma and the percentage of people living in urban areas are available at **gopher://gopher.census.gov/00/Bureau/Pr/Subject/Pop/cb94-34.txt**.

Statistical Abstract Frequently Requested Population Tables

http://www.census.gov/stat_abstract/pop.html

This Census Bureau site gives population totals for the resident U.S. population (1900–1994), as well as population counts for states, large metropolitan areas, and cities with 100,000 or more people. The population totals are subdivided according to sex, age, race, and household type.

Information on Specific Population Groups

Academe This Week

http://chronicle.merit.edu/.almanac/.almanac.html

This is a statistical portrait of higher education in the United States. It includes statistics on graduation rates at NCAA Division I institutions, the number of college degrees awarded, the average pay of full-time professors, and enrollment by age, sex, and race.

Data on the Elderly

http://aspe.os.dhhs.gov/GB/apena.txt

This compilation of statistics from government agencies such as the Bureau of Labor Statistics and the U.S. Bureau of the Census "presents historical and current data on the demographic and economic characteristics of the elderly, including information on population, life expectancy, labor force participation, marital status, living arrangements, poverty rates, and income." Data related to health care for the elderly can be found at **http://aspe.os.dhhs.gov/GB/apenb.txt.**

Data on Poverty

http://aspe.os.dhhs.gov/GB/apenh.txt

The tables in this site relate to poverty in the U.S. as calculated according to the official census definition of poverty. The tables show the population, the number of people living in poverty, and the poverty rate in 1992 by age, race, region, and family type. For information on the official definition of poverty, see **http://www.census.gov/cgi-bin/print_hit_bold.pl/pub/hhes/www/.povmea.html?poverty+threshold#first_hit.**

Education Attainment—Historical Tables

http://www.census.gov/population/socdemo/education/ext-table18.txt

This Census Bureau site presents data on "Educational Attainment of Persons 25 Years and Over by State: 1990, 1980, 1970," "Percent of Persons 25 and Over Who Have Completed High School or College: Selected Years 1940 to 1993," and "Mean Earnings of Workers 18 Years Old and Over, by Educational Attainment: 1975 to 1992."

Homeless Fact Sheets

http://nch.ari.net/facts.html

The Homeless Information Exchange and National Coalition for the Homeless present a series of fact sheets on homelessness, such as "How Many Homeless People Are There?" and "Homeless Families with Children." Each sheet answers questions and includes a list of recommended reading.

U.S. Immigration & Naturalization Home Page

http://www.usdoj.gov/ins/

This Justice Department report provides information on legal immigration to the United States during 1994. It includes the text of the Immigration Act of 1990, and it reports on the number of immigrants entering the United States under the act in 1992 and 1993. The report also gives information such as the age, sex, occupation, and place of intended residence for legal immigrants.

Infants, Children, and Teenagers

gopher://cyfer.esusda.gov/11/CYFER-net/statistics/Kids_Count/kidscnt94

This site is maintained by the U.S. Department of Agriculture and the National Library's Youth Development Information Center. It contains an extensive list of statistical tables related to infants, children, and teenagers. There are approximately 200 statistical tables covering such information as the number and percentage of all births to teenage mothers and the number and percentage of women in the labor force with children under age 6. Statistics on the number of youths participating in 4-H youth programs and on youth enrollment in specific 4-H projects and activities can be found at **gopher://cyfer.esusda.gov/11/CYFER-net/statistics/4hstats.**

Persons of Indian Ancestry

http://www.usgs.gov/doi/bia/ancestry/ancestry.html

This site describes how the government defines "Indian ancestry."

Prison Populations

http://149.101.22.3/bjs/crimoff.htm

The U.S. Bureau of Justice Statistics posts almost every kind of statistic on crime victims and criminal offenders but especially on the prison and jail population.

Race and Ethnicity Standards for the Classification of Federal Data on Race and Ethnicity

http://ftp.fedworld.gov/pub/omb/re.fr2

This file contains the text of *Statistical Policy Directive No 15: Race and Ethnic Standards for Federal Statistics and Administrative Reporting,* which has been in effect since 1977. It is the U.S. government's official policy on racial and ethnic classification. The file also contains an extensive critique of the directive with suggestions for changing the system of classification.

School Enrollment—Historical Tables

http://www.census.gov/cgi-bin/print_hit_bold.pl/pub/population/www/school.html?historical+tables#first_hit

This site contains seven tables related to school enrollment and dropout rates since 1947 for various age groups and for males and females.

Statistical Briefs

http://www.census.gov/ftp/pub/apsd/www/statbrief/

Statistical briefs are two- to four-page reports that summarize data from demographic surveys of the U.S. population and provide data on issues of public policy. These documents are in Adobe Acrobat's Portable Document Format (PDF). In order to view these files, you must have Netscape, and you will need Acrobat Reader, which is available for free from the Adobe Web site at **http://www.adobe.com/acrobat/.**

The Interace Database

http://www.compumedia.com/%7Emulato/InteraceDatabase.html

This site is a resource for those interested in the lives and treatment of mixed-race people. It lists articles, organizations, films, databases, journals, and other resources that consider the topic.

Top 25 American Indian Tribes for the United States

http://www.census.gov/ftp/pub/population/socdemo/race/indian/ailang1.txt

The racial statistics branch of the U.S. Census Bureau lists the top 25 American Indian tribes in the United States for 1980 and 1990. There is also a table that shows the percentage change in membership for each tribe between 1980 and 1990.

United States Census

http://www.census.gov/

This site allows you to search the United States Census by subject. Use the *search* option to find information on any population group in the United States.

Veterans (Male and Female)

http://www.va.gov/vafvet.htm

This is a statistical profile of the veteran population, including information on the number of veterans and the number of dependents and survivors of veterans. A historical profile of veteran services and benefits including statistics on the size of the veterans' health-care system is available at **http://www.va.gov/vafhis.htm.** Statistics on the female veteran population can be found at **http://www.va.gov/womenvet/ CenWomVet.htm.**

Country-Level Information

Background Notes on the Countries of the World

gopher://dosfan.lib.uic.edu/1D-%3A22525%3ABackground%20Notes%20Ser

This site contains statistical and general information on most of the countries of the world (but not the United States) and covers geography, people, education, economics, and membership in international organizations.

Country Destinations by Text Express

http://www.lonelyplanet.com/dest/text.htm#count

Designed for tourists, this site gives travel-related and background information on the countries of the world. It includes "Off the Beaten Path" and "Comments by Travelers" links. Travelers' comments are of sociological interest as they reflect the things tourists find important about their travel experience. Most comments focus on hotels and cuisine and indicate little interest in the lives and well-being of local peoples.

Country Health Profiles for the Americas Only

http://www.paho.org/english/country.htm

The Pan American Health Organization assesses the health situation in each country that is part of the Americas. For each country specific health problems (statistics and a brief overview), demographic characteristics (population size and distribution and age-specific population characteristics), and descriptions and statistics related to health services and resources are available.

Country List

http://www.et.byu.edu/%7Eeliasone/country.html

The International Business Directory provides country-specific links to Web sites housing information related to that country.

Country Reports on Economic Policy and Trade

gopher://dosfan.lib.uic.edu/1D-1%3A5843%3ACountry%20Rpt.%20Econ%20

This site provides access to the 1993 and 1994 *Country Reports on Economic Policy and Trade Practices.* You will find information on debt management policies, significant barriers to U.S. exports and investments, workers' rights, and much more.

Country Studies

http://lcweb2.loc.gov/frd/country.html

This site gives access to book-length information on Ethiopia, China, Egypt, Indonesia, Israel, Japan, Philippines, Singapore, Somalia, South Korea, and Yugoslavia. The site eventually will include over 60 country studies written by a multidisciplinary team of authors. It is a comprehensive source of information about all areas of life including politics, economics, culture, religion, population, history, and culture.

Heads of State and Heads of Government

http://www.geocities.com/Athens/1058/rulers.html

This site lists all heads of state and other top-ranking leaders (past and present), including term of office and year of birth and death (if applicable), of all currently existing countries and territories. It also lists leaders (past and present) of the Arab League, European Union, Organization of African Unity, the Organization of American States, and the United Nations. A special section chronicles the changes in leadership since January 1996.

Human Rights

gopher://dosfan.lib.uic.edu/1D-1%3A6071%3ACountry%20Rpts.%20Human

This site contains the 1993, 1994, and 1995 U.S. Department of State *Country Reports on Human Rights Practices.* For each country (including the United States) you will find the State Department's assessment of the political situation and human rights policies, as well as a report on specific human rights abuses.

International Demographic Data

http://www.census.gov/ftp/pub/ipc/www/idbsum.html

This Census Bureau site includes data on the population size of every country and territory in the world for 1950, 1960, 1970, 1980, 1990, and 1991–1995. Population size is also projected to the year 2000, as is the age-specific population size.

National Library of Australia Internet Site

http://www.nla.gov.au/gov/govinfo.html

This site contains information about and from the governments of Australia, New Zealand, and selected countries, territories, and states within Asia, Canada, the United Kingdom, the United States, Europe, Africa, and the Middle East. The information available on each country varies.

1995 CIA World Factbook

http://www.odci.gov/cia/publications/95fact/index.html

The *World Factbook* contains a wide range of information on all countries and bodies of water. There are brief summaries and statistics related to topics such as the unemployment rate, population size, and total fertility.

OneWorld Online Home Page

http://www.oneworld.org/index.html

This site is described by its creators as the largest collection of multimedia materials (text, graphics, audio, and video) in the world related to the issues of development, the environment, and human rights. It provides access to many resources including articles, journals, and guides. Fifty-seven groups, including Amnesty International, the Save the Children Fund, and UNICEF, contribute material to this site.

Population for the Countries of the World

gopher://gopher.undp.org/00/ungophers/popin/wdtrends/pop1994

This site gives population data for the countries of the world as of 1994.

Poverty Clock

http://www.undp.org/undp/poverty/clock.htm

This page defines poverty from a global perspective. It documents the number of people who are living on less than a dollar a day around the world and calculates the increase in poverty that occurs every minute. The "clock" can be viewed only on Netscape, but the information is available in a text format.

World Constitutions

gopher://wiretap.spies.com/11/Gov/World

This site provides links to many of the world's constitutions and other important historical documents in their entirety.

World List Servers

http://www.w3.org/hypertext/DataSources/WWW/Servers.html

This site provides links to homepages maintained by the different countries of the world. The amount of information available varies by country. For example, under "United Kingdom" there are links to government Web sites, specific city Web sites, Aberdeen University, and so on. Most of the homepages are in the country's official language.

World PopClock from the U.S. Bureau of the Census

http://www.census.gov/ipc-bin/popclockw

This site gives the total population of the world, projected to the day, hour, minute, and second.

State-Level Information

A Brief Guide to State Facts

http://phoenix.ans.se/freeweb/holly/state.htm

This site includes basic information about each of the United States, including state capital, nickname, motto, flower, bird, tree, song, date entered the Union, and so on.

State and Local Governments

http://www.nkn.net/dfwifma/govern.html#State

This site provides detailed information from the states about their governments. The quantity and quality of the information varies by state, but often there is a link to the state constitution and to state-sponsored publications. In addition there is usually information on government services, including the office of tourism, libraries, archives, and public records. Also there are links to local, national, and international Web sites covering the state.

State Profiles 1993—by the Small Business Association

gopher://gopher.umsl.edu/11/library/govdocs/states

The Small Business Association compiles data from 15 different private and government sources to give a statistical overview of business activities in the states, with emphasis on small business. This report includes general statistics about each state (top 5 industries, unemployment figures, number of business establishments, exports as a percentage of total U.S. exports). It also includes statistics related to small-business activity in each state (number of small businesses, small-business share of total employment, fastest-growing industries for small business, and so on).

State Rankings

http://www.census.gov/ftp/pub/statab/ranks/

Drawing on data from the 1994 U.S. *Statistical Abstract*, this site ranks states according to 24 characteristics, including value of exports, miles of motor vehicle travel, energy expenditures, and median household income. For more state rankings see **http://www.census.gov/stat_abstract/ranks.html.** This site contains rankings of the states according to things such as educational attributes, labor force composition, and number of motor vehicles.

U.S. State Fact Sheets

http://www.econ.ag.gov/pubs/usfact/

This site is maintained by the U.S. Department of Agriculture and posts the most recent farm and rural data for each of the 50 states along with other general statistics about each state. The data available includes farm characteristics, farm financial indicators, top 5 agricultural exports, top 5 agricultural commodities, and top 5 counties in agricultural sales. This data is also available for the nation as a whole. This site is updated three or four times a year.

City-Level Information (U.S. and World Cities)

City Destinations by Text Express

   http://www.lonelyplanet.com/dest/text.htm#city

This travel guide for selected cities around the world provides information about each city's history and hotel accommodations. It also includes links to other Web sites with information on each city's environment, attractions, activities and events, and transportation, as well as tips from other travelers. Those with Netscape can view a slide show for each of the cities.

City Net

   http://www.city.net/globals/city.map

This page is aimed at the traveler and offers information on 2,333 U.S. and international cities. It lists links to the most popular U.S. and international cities.

Top City Rankings

   http://www.census.gov/stat_abstract/ccdb.html

This Census Bureau site contains tables in which U.S. cities are ranked according to factors such as percentage of foreign-born residents, population size, percentage of workers using public transportation, and so on.

Travel Guides

   http://point.lycos.com/reviews/database/woci.html

This travel guide provides links to cities, states, and regions that have posted information about a city in the United States or abroad. There is no telling what kinds of information you might find about a city. For example, under "Alexandria, Egypt," there is information about Cleopatra, Egyptian history, and Alexandrians on the internet. Under "Sidney, Australia," there is weather information, a picture gallery, and a virtual tour of the city.

County-Level Information

County Business Patterns

   gopher://gopher.census.gov/11s/Bureau/Economic/County2

The Census Bureau gives employment numbers for states and counties according to economic sectors (agriculture, mining, construction, manufacturing, transportation, public utility, wholesale trade, finance, insurance, real estate, and services). The report also includes the number of establishments for each economic division and the total annual payroll. Read "General Explanation of CBP Series" for an explanation of how to interpret tables. Information is presented according to geographical region, state, and county. Choose "1992-1991 State Data by 2-Digit SIC, and by County" to get county data.

County Population Profiles

http://govinfo.kerr.orst.edu/usaco-stateis.html

This site is part of the U.S. Government Information Sharing Project. It allows you to create a "summary report" for any county in the United States. The summary report includes data on the total resident population (1992), percentage of the population under age 18, percentage of owner-occupied housing units, median income, unemployment rate, and elementary, high school, and college enrollment figures.

Population 1900–1990

gopher://gopher.census.gov/11s/Bureau/Population/Estimate/Cencount

This page gives the population of all counties or county equivalents in all 50 states from 1900 to 1990.

Electronic Books

Main On-Line Books Page

http://www.cs.cmu.edu/Web/books.html

This site is an index of over 1,800 online books that you can browse by author or title. Some foreign-language materials are available. Some books of particular interest to sociologists are listed here. Keep in mind that books can be listed only if someone has taken the time to put them online. Consequently this list represents only a very small proportion of books important to the discipline of sociology.

Charles Darwin

Origin of Species

http://www.literature.org/works/Charles-Darwin/origin/

Friedrich Engels

The Housing Question

http://csf.colorado.edu/psn/marx/Archive/1872-HQ/

Ludwig Feuerbach and the End of Classical German Philosophy

http://csf.colorado.edu/psn/marx/Archive/1886-ECGP/

The Peasant War in Germany

http://csf.Colorado.EDU/psn/marx/Archive/1850-PWG/

Gustave LeBon

The Crowd: A Study of the Popular Mind

http://etext.lib.virginia.edu/cgibin/toccer?id=BonCrow&tag=public&images=images/modeng&data=/lv1/Archive/eng-parsed&part=0

Thomas Malthus

An Essay on the Principle of Population

http://socserv2.socsci.mcmaster.ca/~econ/ugcm/3113/malthus/popu.txt

Karl Marx

The Class Struggles in France, 1848 to 1850
   http://csf.Colorado.EDU/psn/marx/Archive/1850-CSF/

The Communist Manifesto
   gopher://wiretap.spies.com/00/Library/Classic/manifesto.txt

Wage-Labor and Capital
   http://csf.Colorado.EDU/psn/marx/Archive/1849-WLC/

Alexis de Tocqueville

Democracy in America
   http://darwin.clas.virginia.edu/~tsawyer/DETOC/

Thorstein Veblen

The Theory of the Leisure Class
   http://socserv2.socsci.mcmaster.ca/~econ/ugcm/3ll3/veblen/leisure/index.html

Electronic Journals and Newsletters

CITYSCHOOLS
   http://www.ncrel.org/ncrel/sdrs/cityschl.htm

The first issue of the innovative journal *CITYSCHOOLS*, a research magazine about urban schools and communities, can be accessed at this site. *CITYSCHOOLS* rejects the "deficit model" as an approach to solving problems related to urban and inner-city schools and advocates a "resilience model" that emphasizes strengths.

Cover Stories from Previous Congressional Quarterly Weekly Report Stories
   gopher://gopher.cqalert.com/1D-1%3a40101%3a10PREVIOUS

From this site it's possible to read the cover stories from the *Congressional Quarterly Weekly* for any week in the previous year. The articles deal with political issues, most of which are in the news.

Education Policy Analysis Archives
   http://olam.ed.asu.ed/epaa/

This site contains an electronically published scholarly journal focusing on education policy at all levels and in all countries.

Environmental Protection Agency (EPA) Journal
   gopher://gopher.epa.gov/11/.data/epajrnal

The *EPA Journal* is published quarterly. The articles in each issue focus on an environmental theme such as Earth Day or environmental awareness.

Federal Bureau of Investigation (FBI) Law Enforcement Bulletin
   http://www.fbi.gov/leb/leb.htm

From this page it is possible to read past and present issues of the *Law Enforcement Bulletin*, a monthly publication by the FBI that centers around current trends and issues in law enforcement.

Interracial Voice

http://www.webcom.com/%7Eintvoice/

The *Interracial Voice* publishes articles that focus on the shortcomings of the U.S. racial classification system and that clarify the need for a new "racial" category for interracial individuals.

Global Stewardship Network

http://www.iisd.ca/linkages/gsn

This is a free news service covering issues related to global stewardship (that is, "caring for the earth and its current inhabitants, as well as a responsibility to leave to future generations a planet capable of sustaining life"). Past issues also are available at this site.

Hispanic Online

http://www.hisp.com

Hispanic Online is a monthly magazine for and about Hispanics, covering events, issues, and news of interest to the Hispanic community. From this site you can access the latest issue and selected articles from back issues. This site also posts the "Hispanic 100," a list of the top 100 U.S. corporations that provide the most opportunities for Hispanics.

Global Child Health News and Review Online Newspaper

http://www.gcnet.org/gcnet/gchnrhm.html

This news magazine covers issues related to children such as health and well-being, as well as laws that directly affect families and children. Only the current issue is available at this site.

Historical Newspapers

http://lcweb.loc.gov/global/ncp/extnewsp.html#hist

This private collection of historical newspapers focuses on the early American experience, including the Colonial period, the Revolution, and the presidencies of Washington and Jefferson.

Indiana Journal of Global Legal Studies

http://www.law.indiana.edu/glsj/glsj.html

From this site it is possible to read past and present issues of the *Indiana Journal of Global Legal Studies*. This interdisciplinary journal focuses on issues of global and local interest (the environment, AIDS, and so on), as well as markets, politics, technology, and culture.

Journal of Statistics Education Information Service

gopher://jse.stat.ncsu.edu/1

This site gives access to the *Journal of Statistics Education.* Many articles deal with subjects of interest to sociologists, and some articles represent especially good statistical analyses of social science data.

New Jour

http://gort.ucsd.edu/newjour/

This site lists new journals and newsletters that have become available on the internet in the past six months.

Populi

gopher://gopher.undp.org/00/ungophers/popin/unfpa/populi/9512populi

Populi, a United Nations Population Fund publication, emphasizes issues related to overpopulation and fertility.

Postmodern Culture

gopher://jefferson.village.Virginia.EDU/11/pubs/pmc

Postmodern Culture publishes interesting and creative work in the area of postmodernism.

Prison Legal News

gopher://gopher.etext.org/11/Politics/Prison.Legal.News

This site gives access to *Prison Legal News*, a monthly newsletter published by two prison inmates. The newsletter deals with court decisions and their effects on prisoners and families. The newsletters online are several years old, but they offer a unique perspective on prisoners' issues.

Register of Leading Social Sciences Electronic Journals

http://coombs.anu.edu.au/CoombswebPages/EJrnls-Register.html

This site compiles the leading online journals of value to researchers in the social sciences and humanities. Select the link "Alphabetical List of the Topics" and 19 screens of topics appear for you to choose from. Examples of topics are transportation, religion, and AIDS.

Society for the Study of Symbolic Interaction (SSSI): Papers of Interest

http://sun.soci.niu.edu/~sssi/papers/papers.html

This site provides links to papers posted by the Society for the Study of Symbolic Interaction that represent good examples of research from a symbolic interactionist perspective.

Sociological Research Online

http://www.soc.surrey.ac.uk/socresonline/

This page provides access to the latest issue of *Sociological Research Online,* an electronic journal that publishes "high quality applied sociology, focusing on theoretical, empirical, and methodological discussions which engage with current political, cultural, and intellectual topics and debates."

The Disability Rag

gopher://gopher.etext.org/11/Politics/Disability.Rag

This site gives access to *The Disability Rag*, an online magazine that deals with issues disabled people face, such as access to public accommodations and discrimination in hiring practices.

The Monster Magazine List

http://www.enews.com/monster

This is an extensive list of online magazines. If you know the title of the magazine you're looking for, use the "search" option. If you don't know the title, select "All Titles." If you need to know the names of magazines that cover a general topic, browse the subject headings. Not all the magazines represented on this site are free, and some require that you subscribe in order to read them.

U.S. Department of Education Publications

http://www.ed.gov/publications.html

This site is provided by the U.S. Department of Education and offers access to many of its publications, including newsletters, publications for parents, education statistics, and so on.

Resources of Interest to Sociologists

The following is a list of international, national, and regional sociological associations. Although each posts different information, you will frequently find information on student competitions, association activities and meetings, new publications of interest to sociologists, membership details, and links to sociological resources on the internet.

International Sociological Association

http://www.ucm.es/OTROS/isa/

American Sociological Association

http://www.asanet.org/

Mid South Sociological Association

http://www.uakron.edu/hefe/mssapage.html

North Central Sociological Association

http://www.miavxl.muohio.edu/~ajjipsonNCSA.HTMLX

Pacific Sociology Association

http://www.csus.edu/psa/psa.html

Society for Applied Sociology

http://www.indiana.edu/~appsoc/

Southern Sociological Association

http://www.MsState.Edu/Org/SSS/sss.html

Data on the Net

http://odwin.ucsd.edu/jj/idata

This site is maintained by the University of California, San Diego. It contains descriptions and links to 204 Web sites with numeric data ready to download, 63 data archives, and 109 social science gateways to social science resources.

Internet Resources for Sociology

http://library.byu.edu/crln/sociology.html

The Brigham Young University Library's Information Network has created links to internet resources of interest to sociologists. The resources available include sociology departments, net links for sociologists, the *Electronic Journal of Sociology,* and *Yahoo.*

Sociology Departments

http://www.shu.edu/~brownsam/vl/institut.html

This site provides links to sociology departments in the United States and around the world.

The Berkeley Sociology Center

gopher://infolib.lib.berkeley.edu/11/resdbs/soci

The Berkeley Sociology Center provides links to sociological journals and data archives, as well as the Emma Goldman Papers, which chronicle the life of Emma Goldman (1869–1940), a major figure in the history of American radicalism and feminism.

WWW Virtual Library

http://www.shu.edu/~brownsam/v1/Overview.html

Dr. Samuel R. Brown at Seton Hall Library maintains this library of resources of interest to sociologists. There are links to departments of sociology, research centers, discussion groups, electronic journals and newsletters, organizations, and other important resources.

Press Releases and Briefings

Where does the media gets its information? Much of the information and ideas for stories come in the form of press releases. Check out these sites for firsthand accounts of information released to the media.

Administration for Children and Families Press Releases

http://www.acf.dhhs.gov/ACFNews/press/

This site is maintained by the Department of Health and Human Services. These press releases offer statistics on children and information about welfare demonstrations and reform.

Amnesty International UK Press Releases

This site gives information about refugees, activists, and political prisoners in other countries; on tourist safety abroad; and on human rights abuses in the United States and around the world. Current press releases can be found at **http://www.oneworld.org/amnesty/ai_press.html.** Press releases from the past are located at **http://www.oneworld.org/amnesty/ai_press_archive.html.**

Bureau of Indian Affairs Press Releases

http://www.usgs.gov/doi/bia/press/index.html

This site gives access to recent press releases from the Bureau of Indian Affairs. Court decisions and their impact on Native American communities are the most common subjects covered.

Census Bureau Press Releases

http://www.census.gov/Press-Release/www

The press releases at this site can be searched by subject or by date. Many of the press releases announce new statistics related to a wide range of subjects, from aging to national population estimates.

Department of Veteran Affairs Press Releases

http://www.va.gov/pressrel/index.htm

This site provides links to press releases covering U.S. veterans, including veterans' health issues (such as Gulf War syndrome, Agent Orange exposure, and health-care improvements), celebrations (Memorial Day, Veteran's Wheelchair Games), new programs, activities, and publications.

International Labor Organization Press Releases

http://www.ilo.org/english/235press/pr/index.htm

The International Labor Organization, a United Nations agency, seeks to promote social justice and establish internationally recognized standards of human and labor rights. Its press releases focus on labor issues such as child labor, unemployment, underemployment, equality for women, and international labor standards.

International Monetary Fund (IMF) Press Releases

gopher://imfaix3s.imf.org/11/press

The 181 countries that belong to the International Monetary Fund have pledged to cooperate with one another to maintain a productive and stable world economic environment. Members make monetary contributions from which "all may borrow for a short time to tide them over periods of difficulty in meeting their international obligations." IMF press releases announce the credit and loans that it has approved.

NATO Press Releases

http://www.nato.int/docu/pr/pr96e.htm

The North Atlantic Treaty Organization (NATO) is an alliance of collective defense formed in 1949 and linking 14 European countries, the United States, and Canada. Since 1989 (the fall of the Berlin Wall and the symbolic fall of communism) NATO has worked to establish cooperation between the governments of Central and Eastern European countries and with the newly independent states of the former Soviet Union. Its press releases focus on activities related to this task.

Organization for Economic Cooperation and Development (OECD) Press Releases

http://www.oecdwash.org/PRESS/pr.htm

This Organization for Economic Cooperation and Development site provides access to press releases about OECD activities. The OECD is best known for its economic analyses and forecasts and for its advice to governments in the area of finance, investments, and job growth strategies. The *OECD Newsletter* is also posted on the Web site.

Press Releases by Country

http://www2.iadb.org/prensa/PCOUNTRY.HTM

This site is a compilation of press releases from many countries. This is not a comprehensive list of countries or press releases, however, and some of the press releases are several months old.

Population Reference Bureau Releases

http://www.prb.org/prb/pressrel.htm

The Population Reference Bureau's press releases focus on population-related issues (births, deaths, migrations) that affect life in the United States.

Today's Press Releases from the White House

http://library.whitehouse.gov/PressReleases-plain.cgi

If you want to know the official opinion of the White House on a subject in the news, this is the site for you. This site contains all of the press releases coming directly from the White House for the current day. The press releases typically are related to speeches made by the president, first lady, and members of the White House staff or to information about them. The site also provides access to yesterday's press releases. In addition to White House press releases, there are also daily press briefings from the White House press secretary. They can be found at **http://library.whitehouse.gov/Briefings-plain.cgi.**

U.S. International Trade Statistics Current and Past Press Release File

gopher://gopher.census.gov/11s/Org/econ/foreign_trade/Press-Release

These press releases deal with all aspects of trade (the movement of goods and services) between the United States and foreign countries.

World Bank Press Releases

http://www.worldbank.org/html/extdr/extme/press.htm

These press releases report on World Bank loans, the conditions under which the loans were issued, and the effect of the loans on some segment of society.

World Health Organization

http://www.who.ch/press/1996pres.htm

The World Health Organization acts as "the directing and coordinating authority on international health work." Submit a keyword such as "malaria" or "Zaire" to see a list of all the press releases related to that disease or to health-related issues in that country. You can also choose to scroll through the list of press releases.

World Resources Institute News Release

http://www.wri.org/wri/press/wr96-nr.html

"The mission of the World Resources Institute (WRI) is to move human society to live in ways that protect the Earth's environment and its capacity to provide for the needs and aspirations of current and future generations." The news releases for 1996–1997 focus on environmental imbalances associated with urbanization.

World Trade Organization (WTO) Press Releases

http://www.unicc.org/wto/Pressrel/press45.htm

The World Trade Organization, founded in 1995, resolves trade disputes, oversees trade policies, and facilitates trade negotiations. Its press releases report on WTO activities in these areas.

Social Issues

Documents in the News

http://www.lib.umich.edu/libhome/Documents.center/docnews.html

This frequently updated site contains government documents relating to current news events. The University of Michigan Documents Center reports on government actions that make the news and on official government responses to domestic and international events such as the Montana Freemen standoff and the bombing of the U.S. military compound in Saudi Arabia. Documents in the news are also available for 1995.

Government and the Information Superhighway

http://www.nla.gov.au/lis/govnii.html

This site is maintained by the National Library of Australia and provides links to policy statements by various governments regarding national and global networking.

Media Watchdog

http://www.ipl.org/cgi-bin/redirect?http://theory.lcs.mit.edu/~mernst/media

This site compiles online media-watch resources, including organizations, articles, censorship material, and other resources. The "Top Censored Stories" section is a good source of information on issues such as child labor, nuclear weapons, and the internet.

The Issues Page

http://www.igc.apc.org/igc/issues.html

This site is sponsored by the Institute for Global Communications, an organization dedicated to expanding and inspiring social action. It covers issues from acid rain to youth. Publications, organizations, conferences, and information sites are among the resources you will find at this site.

The State of the World's Children

http://www.unicef.org/sowc96/contents.htm

This site provides links to resources such as statistical tables for measuring children's well-being and statistics on children from regions around the world. There is also information about children affected by war and a report on the improvements that have taken place in the 50 years that UNICEF has been involved with children's needs.

THOMAS: Legislative Information on the Internet

http://thomas.loc.gov/

This site is an attempt to make federal legislative information freely available to the internet public. "Hot Topics" are those bills and amendments that are the subjects of floor action, debate, and hearings in Congress and that are frequently reported on by the popular media. You can search through topics such as foreign aid and urban affairs for relevant current legislation. From this site it is possible to read the full text of bills and find out who sponsored and cosponsored them.

United Nations High Commissioner for Refugees

http://www.unicc.org/unhcr/contents.htm

This site provides "reliable and current information and analysis on all aspects relating to refugees and displaced persons, including their countries of origin, legal instruments, human rights, minorities, situations of conflict, and conflict resolution."

Worldspeaker

http://www.tiac.net/users/worldspe/1/index1.htm

This site serves as a forum for international educational institutions to share ideas and news of their activities in order to promote awareness of global educational developments.

World Resources Institute

http://www.wri.org/wri/press/wr96-nr.html

This site contains parts of the report *World Resources 1996–1997*, which gives statistical projections about urbanization through the year 2025. If you are interested in a particular issue of urbanization, you may conduct a keyword search. For example, entering the keyword "urban poor" produced definitions of "urban," information on China and India, and facts about population, poverty, and land degradation.

General Sources of Statistics

Economic Statistics Briefing Room

http://www.whitehouse.gov/fsbr/esbr.html

This page provides access to current federal economic indicators such as disposable personal income, civilian labor force, and consumer price index. It also provides links to information produced by a number of federal agencies such as the Bureau of Economic Analysis, U.S. Census Bureau, Federal Reserve Board, and Bureau of Transportation Statistics.

Social Statistics Briefing Room

http://www.whitehouse.gov/fsbr/ssbr.html

This page provides access to current federal social statistics such as violent crime measures, demography (income by race, population of the U.S.), education (literacy rates, full-time graduate students), and health (leading causes of death, cases of measles). It also provides links to information produced by a number of federal agencies such as the Bureau of Justice Statistics, Federal Bureau of Investigation, U.S. Census Bureau, Health Care Financing Administration, and National Center for Education Statistics.

Statistical Abstract Frequently Requested Tables

http://www.census.gov/stat_abstract/

The Census Bureau has posted the tables for which there is the most demand, as measured by the number of requests and responses to user surveys. Topics range from crime and crime rates to U.S. exports and imports of merchandise.

Statistical Resources on the Web

http://www.lib.umich.edu/libhome/Documents.center/stats.html

This site provides links to statistics related to a large number of subjects ranging from agriculture to weather and including statistics of interest to sociologists.

U.S. National Debt Clock

http://www.brillig.com/debt_clock/

Debt clock maintainer Ed Hall posts this up-to-date report on the national debt. This page gives access to other sites concerned with the national debt and also answers some commonly asked questions about the national debt.

USA Statistics in Brief: Part 1

gopher://gopher.census.gov/00/Bureau/Stat-Abstract/USAbrief/part1

This site contains statistics on population, law enforcement, education, communications, transportation, and housing. The data is from 1992.

Daily News

CNN Interactive

http://www.cnn.com/

Ecola's Newsstand

http://www.ecola.com/news/

Ecola's Newsstand provides links to 1,811 (and counting) English-language newspapers, magazines, and computer-related publications from around the world.

Los Angeles Times

http://www.latimes.com/

News from Reuters Online

http://www.yahoo.com/headlines/

New York Times

http://www.nytimes.com/

In order to read *The New York Times* online, you must register the first time you visit this site. At the moment it is free.

TimeDaily

http://www.pathfinder.com/time/daily/

Today in History

http://www.historychannel.com/today/

This Week's Magazine

http://pathfinder.com/@@UwmuhQUAkt048K*w/time/magazine/domestic/toc/latest.html

USA Today

http://www.usatoday.com/

Washington Post

http://www.washingtonpost.com/

Foreign Newspapers

Library of Congress Foreign Newspapers

http://lcweb.loc.gov/global/ncp/oltitles.html#forn

On this site the newspapers are listed alphabetically. You'll do better if you know the name of the paper you're looking for or if you have time to look through the list for a particular country. Some of the newspapers are in English.

Newspaper Listing—Worldwide

http://www.dds.nl/~kidon/papers.html

This site provides links to numerous foreign newspapers. Most of the newspapers are in their respective languages, but some of them are in English.

U.S. Government Agencies

These are just a few of the government agencies represented on the internet. The sites listed here may give access to government documents, newsletters, schedules, historical documents, statistics, and so on.

Centers for Disease Control

http://www.cdc.gov/

Consumer Information Center

http://www.pueblo.gsa.gov/textonly.htm

This page is maintained by the Consumer Information Center of the U.S. General Services Administration. It allows you to read reports of consumer goods such as cars and toys.

Department of Defense

http://www.dtic.dla.mil/defenselink

Department of Education

http://www.ed.gov/

Department of Housing and Urban Development

http://www.hud.gov/

Department of Justice

http://www.usdoj.gov/

Department of Labor

http://www.dol.gov/

Environmental Protection Agency

http://www.epa.gov/

National Aeronautics and Space Administration

http://www.nasa.gov/

U.S. Bureau of the Census

http://www.census.gov/

Federal Government Agencies

http://www.lib.lsu.edu/gov/fedgov.html

This site gives access to hundreds of government agencies. Use it if you need to find an agency not listed previously.

Government Resources Via the Web—National, State, and Local

http://www.nkn.net/dfwifma/govern.html

This site provides access to 34 screens of links to government documents. Documents are available from the following agencies or sources: U.S. federal government, U.S. Congress, Supreme Court, executive branch, Department of Defense, Library of Congress, federal economy, governmental departments and agencies and state and local governments.

Historical Documents

Historical Letters, Documents, Essays, and Speeches

http://history.cc.ukans.edu/carrie/docs/docs_us.html

The University of Kansas has posted an extensive list of historical documents beginning with Christopher Columbus's 1494 letter to the queen and king of Spain and ending with the 1993 Freedom of Information Act.

Search the White House Virtual Library

http://www.whitehouse.gov/WH/html/library-plain.html

This site allows users to search the White House databases for White House documents, radio addresses of the president, executive orders, and White House photographs. It is also possible to browse some historical national documents.

U.S. Historical Documents

gopher://wiretap.spies.com/11/Gov/US-History

This site contains the text of a number of important documents beginning with the Declaration of Arms in 1775 and ending with the U.S. State Department's release of the text of the agreements reached at the 1944 Yalta Conference attended by Roosevelt, Churchill, and Stalin (see "World War II Documents").

Global Organizations

Amnesty International Home Page

http://www.oneworld.org/amnesty/index.html

Amnesty International is devoted to the cause of human rights. The group focuses on prisoners of conscience, abuse by opposition groups, asylum seekers, and those in exile.

International Labor Organization

http://ilo.org/

The International Labor Organization determines international labor regulations such as the minimum standards of labor rights, working conditions, occupational safety standards, and so on.

International Red Cross/Red Crescent

http://www.icrc.org/icrcnews/212e.htm

The International Red Cross cares for the sick and wounded in war and helps relieve suffering from pestilence, floods, fires, and other disasters. The hospitals, doctors, and nurses of the Red Cross remain neutral during war. The Red Crescent functions as the Red Cross in Turkey.

The Carter Center

http://www1.cc.emory.edu/CARTER_CENTER/

The Carter Center was founded by former President Jimmy Carter and is dedicated to fighting disease, hunger, poverty, conflict, and oppression by working for development, urban revitalization, and global health.

The International Monetary Fund

http://www.self-gov.org/freeman/8904ewer.htm

The International Monetary Fund (IMF) was established to maintain fixed exchange rates among the different currencies of the world. The IMF does this by making short-term loans to nations with temporary balance-of-payment deficits.

The World Bank

http://www.worldbank.org/html/extdr/about.html

The World Bank is composed of five organizations that lend money to developing nations.

The World Council of Churches

http://www.wcc.coe.org/oikumene.html

Nearly all Christian traditions are represented in the World Council of Churches, which is a fellowship of 330 churches from 120 countries in all continents.

United Nations Home Page

http://www.unicc.org/Welcome.html

The United Nations, consisting of 126 countries, was formed to promote international peace, security, and cooperation.

World Health Organization WWW Home Page

http://www.who.ch/

The World Health Organization strives to help people attain the highest level of health through technical projects and programs.

Dictionaries, Thesauruses, and Quotations

ARTFL Project: Roget's Thesaurus Search Form

http://humanities.uchicago.edu/forms_unrest/ROGET.html

A thesaurus is a dictionary that groups words with similar meanings together. This resource is helpful when you find yourself using the same word over and over.

Bartlett's Familiar Quotations

http://www.columbia.edu/~svl2/bartlett/

This is a great source of phrases for personal as well as academic situations. For example, if you need to say something insightful about love after a fight, Richard Edward's (circa 1523–1566) "the anger of lovers renews the strength of love" might help. If you need to write an essay on the value of learning, Aschylus' (525–456 B.C.) line "learning is ever in the freshness of its youth, even for the old" is a good starting point.

Easton's Bible Dictionary

http://ccel.wheaton.edu/easton/ebd/ebd.html

This site is helpful when you come across a biblical reference but are unsure of the meaning or when you are curious about how the Bible treats a subject such as adoption or war.

Online Dictionaries

http://www.twics.com/~vladimir/dic.html

This is an extensive list of foreign-language dictionaries, from Arabic vocabulary lists to a Welsh–German dictionary.

Quotations Home Page

http://www.lexmark.com/data/quote.html

Choose "quotations by topic" if you are looking for quotations related to a specific topic. Choose "really miscellaneous" if you are looking for the statements of a particular person. This site offers a number of other search categories including "recent quotes," "advice," "sarcasm," and so on. Look for sociological concepts such as conformity, family inequality, and labor in the list of topics. See how closely sociological ideas correspond with the ideas conveyed in the quotes.

Webster's English Dictionary

http://c.gp.cs.cmu.edu:5103/prog/webster

Simply enter the word for which you need a definition and select "Lookup definition."

Encyclopedias and General References

Encyclopedia Smithsonian

http://www.si.edu/welcome/faq/start.htm

Topics included in this encyclopedia are determined by public demand for information on a topic. As of June 1996, there were nine general topics: armed forces history, anthropology, mineral sciences, musical history, physical sciences, services, conservation of textiles, transportation history, and vertebrate zoology.

Finding Data on the Internet: Links to Potential Story Data

http://www.probe.net/~niles/links.html

This site is maintained by journalist Robert Niles with the intention of making it easy for other journalists to find statistics and data on the internet. The following topics are covered: basic reference data, agriculture, aviation, banks and businesses, crime, economy and population, education, energy, finding people, health, immigration, law, military, nonprofits, politics, and weather. For each of these topics links to facts and information are provided. This site also provides links to the *World Factbook*, the Library of Congress, and so on.

Free Internet Encyclopedia

http://www.cs.uh.edu/~clifton/encyclopedia.html

Creators of this site suggest that a more accurate name than "Free Internet Encyclopedia" is "Free Internet Encyclopedia Index" because this is an encyclopedia of information on the internet. This encyclopedia has two divisions: Macroreference (large general topics such as Africa, courts, and so on) and Microreference (short bits of information on specific topics such as asthma, Jane Austen, and so on).

Judith Bower's Law Lists

http://flair.law.ubc.ca/guests/bowers.html

The resources at this site include information on law in Canada, the United States, and selected foreign countries. It provides links to newspapers and journals, as well as English and foreign-language reference materials. It also has a general information section, which includes a currency converter, *World Factbook 1996*, and Canadian flight information, among other things.

Library of Congress Home Page

http://lcweb.loc.gov/

This is a source for links to publications, foreign and U.S. newspapers, the government, Congress, copyright laws and procedures, events and exhibits, and special collections.

Reference Center of the Internet Public Library

http://ipl.sils.umich.edu/ref/RR/

This site provides links to internet resources on the following subjects: general reference, arts and humanities, business and economics, computers and the internet, education, entertainment and leisure, health and medical sciences, law, government and political science, science and technology, and social sciences.

Usenet FAQs

http://www.cis.ohio-state.edu/hypertext/faq/usenet/

This document contains an extensive list of Usenet frequently asked questions (FAQs). The list is alphabetized by topic. You can search by newsgroup name, archive name, subject, or keyword.

Writing Resources

A Guide for Writing Research Papers Based on Modern Language Association Documentation

http://155.43.225.30/mla.htm

This guide covers the various stages of the research process with advice on how to gather material, keep track of sources, take notes, document and cite references, and so on. There is also an informative article on plagiarism and examples of how to cite print and electronic references.

Bibliographic Formats for Citing Electronic Information

http://www.uvm.edu/~xli/reference/estyles.html

Do you have questions about how to cite information from the World Wide Web? This site covers the MLA (Modern Language Association) and APA (American Psychological Association) citation formats for electronic information. It presents clear examples for almost any type of electronic resources. Both formats are commonly used across many academic disciplines.

Elements of Style

http://www.columbia.edu/acis/bartleby/strunk/

This is a guide to the proper use of the English language, with special focus on commonly misused words and expressions. It focuses on the rules of usage and writing principles most commonly violated.

English as a Second Language Home Page

http://www.lang.uiuc.edu/r-li5/esl/

Whether you are someone who is learning English as a second language, a native speaker who needs to brush up on vocabulary, or someone who is interested in how the English language is presented to foreign-language speakers, this site is beneficial. This site presents links related to "Listening and Speaking," "Reading," and "Writing." Your computer must have audio capabilities in order to use "Listening and Speaking."

Grammar Handbook

gopher://gopher.uiuc.edu/11/Libraries/writers/g1

Written by students at the University of Illinois, this handbook covers the parts of speech, phrases, clauses, sentences and sentence elements, and common usage problems. The section on common usage problems (when to use a colon versus a semicolon, where in a sentence to place modifiers) is especially useful for fine-tuning writing assignments.

On-Line Resources for Writers

http://www.ume.maine.edu/%7Ewcenter/resource.html

This site gives links to a wide range of writing resources on the internet, including grammar, quotes, dictionaries, thesauruses, foreign-language dictionaries, citation guides, English as a second language, composition and rhetoric, and much more.

Handy Reference Guides

Abbreviations for International Organizations and Groups

http://odci.gov/94fact/appendb.html

Do you need to know what a particular abbreviation stands for? Or would you like to know the abbreviation of an organization for a project or paper? This site lists accepted abbreviations for international organizations and groups starting with the "Arab Bank for Economic Development in Africa (ABEDA)" and ending with the "Zangger Committee (ZC)."

Calculators On-Line

http://www-sci.lib.uci.edu/HSG/RefCalculators.html

This site is a list of calculators for almost everything you can imagine. You can figure out the maximum hull speed on your sailboat, determine your financial net worth, calculate your body mass, do simple and complex math, and much more.

Calendar Generator

http://www.stud.unit.no/USERBIN/steffent/kalender.pl?+1996

Are you interested in knowing what day your birthday will be on in the year 2005? Or do you need to know what day of the week it was when Abraham Lincoln was born on February 12, 1809? You can submit a year between 1754 and 3000 to generate a 12-month calendar for that year.

Currency Converter

http://bin.gnn.com/cgi-bin/gnn/currency

Select the country that uses the currency you wish to have converted, and the value of all other countries' currencies will be calculated in relation to that currency.

Finding an Email Address

http://sunsite.oit.unc.edu/~masha/

Do you need to find someone's e-mail address or check to see if they have one? This UNC site uses a question-and-answer format to help you find an address. Examples of questions that help to narrow your search are "What region or country is he/she located in?" and "Is he/she on a network other than the internet?"

Language

http://www.travlang.com/languages/

Do you need to talk to someone who speaks a foreign language? Identify the language(s) you speak and the one that you want to learn. The computer will display common words and phrases such as "yes," "no," and "you're welcome" and words and phrases that will be useful when shopping, asking for directions, establishing a time and place to meet, and so on.

Mathematical Notation, Weights, and Measures

http://www.odci.gov/cia/publications/95fact/appendf.html

Do you need help converting liters into quarts or meters into yards? The CIA has posted conversion tables that list almost every metric measure and its English-system equivalent.

Search for an Area Code

gopher://coral.bucknell.edu:4320/7areacode

Do you need to know an area code? Submit the name of the city and state and the three-digit area code appears.

United States Postal Service Zip Code Lookup and Address Information

http://www.usps.gov/ncsc/

Submit a city name and its zip code appears. Submit a zip code and the name of the city to which that zip code has been assigned appears. The U.S. Postal Service answers commonly asked questions about zip codes and makes recommendations for addressing mail.

Study Abroad

Cultural Immersion

http://www.nrcsa.com/

The National Registration Center for Study Abroad gives information about immersion classes in 30 different countries. The online information covers program descriptions, dates, and fees.

Semester at Sea's Home Page on CampusNET

http://campus.net/educat/semester/

"The Semester at Sea is a floating university allowing students to experience diverse cultures while getting credit from the University of Pittsburgh." This page provides access to a general overview of the program, the mission statement, campus information, and information on the student body, enrollment, the academic program, the faculty, courses, and so on.

Study Abroad Home Page

http://www.studyabroad.com/

This site is a resource for students to learn about study-abroad programs in 65 countries. Read the document "Consumer Information" before making any decisions about a program.

Voluntary Service Overseas

http://www.oneworld.org/vso/

The Voluntary Service Overseas is an organization that recruits people ages 20–70 to work in developing countries. This page answers questions about the program's goals, discusses volunteering, describes job openings, and gives general information about the program.

Coping with College

Funding College

http://www.ed.gov/prog_info/SFA/FYE/index.html

This U.S. Department of Education publication provides information about federal financial aid programs for college students.

List of American Universities Home Pages

http://www.clas.ufl.edu/CLAS/american-universities.html

This site gives the homepages of American universities granting bachelor or advanced degrees. It provides links to international universities, Canadian universities, and community colleges as well.

Preparing Your Student for College

http://www.ed.gov/pubs/Prepare/

This U.S. Department of Education guidebook covers areas such as choosing a college, financing an education, and doing long-range planning. The guidebook also lists and defines important terms related to the college experience, such as B.A. and B.S.

Scholarships and Fellowships

http://web.studentservices.com/search/

Searching for money to fund education is as easy as entering your name, address, and major. A list of scholarships for which you might be eligible will appear on the screen, and information about other scholarships for which you might be eligible will be e-mailed to you as they become available.

ERIC Digests

ERIC is a clearinghouse for educational material. The following are selected ERIC publications related to study skills and the college experience. The URL for ERIC is **gopher://INET.ed.gov:12002**. Once you are in the site select "ERIC.src." Once in the ERIC Web site, you will be prompted to enter a keyword. Use the numbers to the left of the titles listed below as the keyword for immediate access. Occasionally the number will not work. When this happens, enter the title at the query prompt.

Study Skills

ED250694	Qualities of Effective Writing Programs
ED250696	Vocabulary
ED291205	Critical Presentation Skills—Research to Practice
ED296347	Audience Awareness: When and How Does It Develop?
ED300805	Note-Taking: What Do We Know About the Benefits?
ED301143	Learning Styles

ED302558 Improving Your Test-Taking Skills
ED318039 How to "Read" Television: Teaching Students to View TV Critically
ED327216 Information Skills for an Information Society: A Review
ED326304 How Can We Teach Critical Thinking?
ED372756 Information Literacy in an Information Society
ED385613 Making the A: How to Study for Tests

College

ED284510 Self-Study in Higher Education: The Path to Excellence
ED266339 Selecting a College: A Checklist Approach
ED284514 Student Stress: Effects and Solutions
ED284526 Reducing Stress Among Students
ED286938 Alternatives to Standardized Tests
ED351079 First-Generation College Students

Career Guides

Career Connections

http://www.netline.com/cgi/sendto?site=http://www.career.com/

From this site it is possible to learn about job openings from around the country and get in touch with employers. You can search for jobs by company, category, or location, and this site profiles jobs for new graduates. There is a link to "Hot Jobs," which lists immediate openings and the qualifications needed for them.

Career Magazine

http://www.netline.com/cgi/sendto?site=http://www.careermag.com/

This site provides access to over 14,000 current job listings from around the world. It also has information on salary resources and companies that recruit college graduates and informative news articles.

CareerPath

http://www.netline.com/cgi/sendto?site=http://www.careerpath.com/

From this site it's possible to search newspaper employment ads from 10 major cities. There is no fee charged for this service, but you must register to use it.

Career Shop's Resume, Job and Employment Site

http://www.careershop.com/cshop/

This site is "an on-line database of resume profiles and employment opportunities, designed to assist job seekers and hiring employers alike. Job seekers can post resume profiles in our resume database and perform job searches of our Job Openings database—on-line and free of charge!" There is also advice on preparing for an interview and promoting yourself to employers.

Jobs

http://www.netline.com/cgi/sendto?site=http://ageninfo.tamu.edu/jobs.html

This site contains a listing of universities and corporations with available jobs, and it tells you how to contact employment services, employment recruiters, and the like.

JobTrack

http://www.jobtrak.com/

The creators of this site call it "the premier site for recruiting college students and recent graduates. The company has formed partnerships with over 300 college and university career centers and is utilized by over 150,000 employers." There is a section on graduate schools that includes information on topics such as preparing graduate school application essays and grants in graduate study.

ERIC Digests

ERIC is a clearinghouse for educational material. The following are selected ERIC publications related to jobs/careers. The URL for ERIC is **gopher://INET.ed.gov:12002**. Once you are in the site, select "ERIC.src." Once in the ERIC Web site, you will be prompted to enter a keyword. Use the numbers to the left of the titles listed below as the keyword for immediate access. Occasionally the number will not work. When this happens, enter the title at the query prompt.

ED292974   Workplace Literacy Programs
ED346318   Job Search Methods
ED376274   Job Search Skills for the Current Economy

The Best of the World Wide Web

Cool Site Winners

http://cool.infi.net/

This page changes daily and highlights the favorite sites of celebrities. Each day a new celebrity is the "agent of cool," and a short biography is given along with access to his or her favorite sites. This is a good site for those interested in the sociology of popular culture.

New Sites on the Web

http://webcrawler.com/select/nunu.new.html

This is a list of new sites posted on the Web in the past week. The new sites are grouped under the headings "Arts and Entertainment," "Business," "Computers," "Education," "Politics," "Medicine," "Diseases," "Humanities," "Culture," "Hobbies," "Technology," "Sports," and "Travel."

The Best of the Best of the Web

http://198.105.232.4/powered/bestofbest.htm

This page showcases some of the most innovative sites on the internet, but it is best viewed with a graphics-capable internet browser.

Top Web Sites

http://point.lycos.com/categories/index.html

This page gives access to the top Web sites related to a variety of subjects such as education, government, and the world.

Key Concepts

Ability grouping the arranging of elementary school students into instructional groups according to similarities in past academic performance and/or on standardized test scores.

Absorption assimilation a process of cultural blending in which members of a minority group adjust to the ways of the majority group.

Achieved characteristics attributes acquired through some combination of choice, effort, and ability. In other words people must act in some way to acquire the attribute. Examples include occupation, marital status, level of education, and income.

Achieved status positions acquired through effort and ability.

Active adaptation a biologically based tendency to adjust to and resolve environmental challenges.

Adaptive culture the portion of the nonmaterial culture (norms, values, and beliefs) that adjusts to material innovations.

Advanced market economies an economic arrangement that offers widespread employment opportunities to women as well as to men.

Alienation a state in which human life is dominated by the forces of human inventions.

Annual per capita consumption of energy the average amount of energy each person in a nation consumes over a year.

Anomaly an observation or observations that a paradigm cannot explain and that threaten the paradigm's explanatory value and hence its status. (See also **paradigms.**)

Anomie see **structural strain.**

Apartheid a former policy in South Africa that was a rigid system of racial classification designed to promote and maintain white supremacy.

Ascribed characteristics attributes that people (1) have at birth (such as skin color, sex, or hair color), (2) develop over time (such as baldness, gray hair, wrinkles, retirement, or reproductive capacity), or (3) possess through no effort or fault of their own (such as national origin or religious affiliation that was "inherited" from parents).

Ascribed status positions that people are born into, grow into, or otherwise acquire through no fault or virtue of their own.

Assimilation a process by which ethnic and racial distinctions between groups disappear.

Authority legitimate power in which people believe that the differences in power are just and proper and that a leader is entitled to give orders.

Automate to use the computer to increase workers' speed and consistency, to provide a source of surveillance, and to maintain divisions of knowledge and thus a hierarchical arrangement between management and workers.

Back stage the region out of sight where individuals can do things that would be inappropriate or unexpected on the front stage.

Basic innovations revolutionary, unprecedented, or ground-breaking ideas, practices, and tools that are the cornerstones for a wide range of applications.

Beliefs conceptions that people accept as true about how the world operates and about the place of the individual in the world.

Bourgeoisie the owners of the means of production. (See **means of production**.)

Browser tool for accessing information on the World Wide Web; browsers can be character-based and graphical.

Bureaucracy in theory, a completely rational organization—one that uses the most efficient means to achieve a valued goal.

Buyers persons in a targeted territory to which commodities are marketed and sold.

Capitalism an economic system in which natural resources and the means of producing and distributing goods and services are privately owned.

Caste system any scheme of social stratification in which people are ranked on the basis of physical or cultural traits over which they have no control and that they usually cannot change.

Catechisms short books covering religious principles written in question-and-answer format.

Charismatic authority authority that rests on the exceptional and exemplary qualities of the person issuing the commands.

Church according to Durkheim a group whose members hold the same beliefs with regard to the sacred and the profane, who behave in the same way in the presence of the sacred, and who gather together in body and spirit at agreed-upon times to reaffirm their commitment to those beliefs and practices.

Civil religion "any set of beliefs and rituals, related to the past, present and/or future of a people (nation), which are understood in some transcendental fashion" (Hammond 1976, p. 171).

Claims makers people who articulate and promote claims and who tend to gain if the targeted audience accepts their claims as true.

Classless society a propertyless society providing equal access to the means of production.

Class system any scheme of social stratification in which people are ranked on the basis of merit, talent, ability, or past performance.

Cohort a group of people sharing a common characteristic or life event.

Collective memory experiences shared and recalled by significant numbers of people.

Concepts powerful thinking and communication tools that enable us to give and receive complex information in an efficient manner.

Conformists people who have not violated the rules of a group and are treated accordingly.

Conformity (1) behavior and appearance that follow and maintain standards set by a group; (2) the acceptance of the cultural goals and the pursuit of these goals through legitimate means.

Constrictive pyramids population pyramids that are characteristic of some European societies, most notably Switzerland and the former West Germany, and that are narrower at the base than in the middle. This shape shows that the population is composed disproportionately of middle-aged and older people. (See also **population pyramid**.)

Constructionist approach a sociological approach that focuses on the process by which some groups, activities, conditions, or artifacts become defined as social problems.

Content (of interaction) the cultural factors (norms, values, beliefs, material culture) that guide interpretations, behavior, and dialogue during interaction.

Context (of interaction) the larger historical circumstances that bring people together.

Control variables variables suspected of causing spurious correlations.

Core economies the highly diversified economies of mechanized rich nations characterized by strong, stable governments.

Corporate crime crime committed by a corporation as it competes with other companies for market share and profits.

Correlation a relationship between two variables such that a change in one variable is associated with a change in another.

Correlation coefficient a mathematical representation of the extent to which a change in one variable is associated with a change in another.

Counterculture a subculture that conspicuously challenges, rejects, or clashes with the central norms and values of the dominant culture.

Crime deviance that is punished by formal sanctions.

Cults generally very small, loosely organized groups, usually founded by a charismatic leader who attracts people by virtue of his or her personal qualities.

Cultural genocide the outcome of a situation in which people of one society define the culture of another society not only as offensive but as so intolerable that they attempt to destroy it.

Cultural lag the failure to adapt to a new invention; a situation in which adaptive culture fails to adjust in necessary ways to a material innovation. (See also **adaptive culture.**)

Cultural markers distinctive characteristics that are used to clearly classify people into distinct cultural units.

Cultural relativism a perspective in which elements of foreign cultures are viewed in their cultural context, not in isolation or by standards of a home culture.

Culture shock the physical and mental strain that people from one culture experience when they must reorient themselves to the ways of a new culture.

Data printed, visual, and spoken materials. Data becomes information after someone reads it, listens to it, or views it.

Dearth of feedback a factor in creating poor-quality data. Much of the data that is televised and published is not subject to honest, constructive feedback because there are too many messages and not enough critical readers and listeners to evaluate the data before it is released or picked up by the popular media. Without feedback the creators cannot correct their mistakes; thus the data they produce becomes poor in quality.

Demographic gap the difference between birthrates and death rates.

Demographic trap the point at which population growth overwhelms the environment's carrying capacity.

Demography a subdiscipline within sociology that studies population trends.

Denomination a formal, hierarchical, well-integrated organization in a society in which church and state are usually separate.

Dependent variable the variable affected by a change in the independent variable. (See also **variable.**)

Deviance any behavior or physical appearance that is socially challenged and condemned because it departs from the norms and expectations of a group.

Deviant subcultures groups that are part of the larger society but whose members adhere to norms and values that favor violation of the larger society's laws.

Differential association a theory of socialization that explains the origins of delinquent behavior. It refers to the idea that "when persons become criminal, they do so because of contacts with criminal patterns and also because of isolation from anti-criminal patterns" (Sutherland and Cressey 1978, p. 78).

Diffusion the process by which an idea, an invention, or some other item is borrowed from a foreign source.

Discoveries the uncovering of something that had existed before but had remained hidden, unnoticed, or undescribed.

Discrimination the intentionally or unintentionally unequal treatment of individuals or groups on the basis of attributes unrelated to merit, ability, or past performance. The treatment may be based on such attributes as skin color, weight, religion, ethnicity, or social class. Discrimination is behavior aimed at denying members of minority groups equal opportunities to achieve valued social goals (education, health care, long life) and/or blocking their access to valued goods and services.

Disenchantment of the world Max Weber's phrase for a great spiritual void accompanied by a crisis of meaning. It occurs when people focus so uncritically on the ways they go about achieving a valued goal that they lose sight of that goal.

Dispositional traits personal or group traits such as motivation level, mood, and inherent ability.

Division of labor work broken down into specialized tasks, with each task performed by a different set of persons.

Dominant group the ethnic and racial group at the top of the hierarchy.

Doubling time the estimated number of years required for a country's population to double in size.

Downward mobility a change in social class that corresponds to a loss in rank or prestige.

Dramaturgical model a model in which interaction is viewed as though it were theater, people as though they were actors, and roles as though they were performances presented before an audience in a particular setting.

Dysfunctions parts that have disruptive consequences to the system or to some segments of society.

Ecclesiae a formal and well-integrated religious organization led by a hierarchy of leaders that claims as its members everyone in a society.

Education those experiences that stimulate thought and interpretation or that train, discipline, and develop the mental and physical potentials of the maturing person.

Emigration the departure of individuals from a country.

Engram physical traces formed by chemicals produced in the brain that store the recollections of experiences.

Established sects religious organizations that share characteristics of denominations and sects. They are renegades from denominations or ecclesiae but have existed long enough to acquire a significant membership and to achieve respectability.

Ethgender related to people who share (or are believed by themselves or others to share) the same sex and race and ethnicity.

Ethnic blending the process by which ethnic barriers break down and contact between members of different ethnic groups increases, such that differences between ethnic groups decrease.

Ethnicity a shared (or belief in a shared) national origin, ancestry, distinctive and visible cultural traits (religious practice, dietary habits, style of dress, body ornaments, or language), and/or socially important physical characteristics.

Ethnocentrism a viewpoint in which one's home culture is used as the standard for judging the worth of foreign ways.

Everyday mingling people's routine talking, looking, and/or listening as they live their lives. The settings in which people mingle are endless; they include the workplace, the home, the neighborhood, and the school.

Expansive pyramids population pyramids that are characteristic of labor-intensive poor nations. They are triangular in shape, broadest at the base and with each successive bar smaller than the one below it. The relative sizes of the age cohorts in expansive pyramids show that the population is increasing in size and that it is composed disproportionately of young people. (See also **population pyramid**.)

Externality costs costs that are not figured into the price of a product but that are nevertheless a price we pay for using or creating a product. An example of externality costs is the cost of restoring contaminated and barren environments and of assisting people to cope.

Facade of legitimacy an explanation that members in dominant groups give to justify exploitive practices; a justifying ideology.

Falsely accused people who have not broken the rules but who are treated as if they have done so.

Family two or more people related to one another by blood, marriage, adoption, or some other socially recognized criteria.

Feeling rules norms specifying appropriate ways to express the bodily sensations that we experience in relationships with other people. (See **social emotions**.) These rules are so powerful that they affect how people solve problems.

Femininity feminine characteristics; the physical, behavioral, and mental or emotional traits believed to be characteristic of females.

Feminist a man or woman who actively opposes gender scripts and believes that men's and women's self-image, aspirations, and life chances should not be constrained by those scripts.

Folkways norms that apply to routine matters.

Formal curriculum the various academic subjects such as mathematics, science, English, reading, physical education, and so on. (See also **hidden curriculum**.)

Formal dimension (of an organization) the official written guidelines, rules, regulations, and policies that define the goals of the organization and its relationship to other organizations and integral parties. This term also applies to the roles, the nature of the relationships among roles, and the way in which tasks should be carried out to realize the goals.

Formal education a systematic, purposeful, and planned effort intended to impart specific skills and modes of thought. (See also **informal education** and **schooling**.)

Formal sanctions definite and systematic laws, rules, regulations, and policies that specify (usually in writing) the conditions under which people should be rewarded or punished and that define the procedures for allocating rewards and imposing punishments. Examples of formal sanctions include medals, cash bonuses, diplomas, fines, prison sentences, and the death penalty. (See also **informal sanctions**.)

Fortified households preindustrial arrangements in which there is no police force, militia, national guard, or other peacekeeping organization. The household is an armed unit, and the head of the household is its military commander.

Front stage the region where people take care to create and maintain expected images and behavior.

Function the contribution of a part to the larger system and its effect on other parts in the system.

Functionally illiterate the lack of reading, writing, and calculating skills needed to adapt to a given society.

Fundamentalism a complex religious phenomenon that involves a belief in the timeless nature of sacred writings and a belief that such writings are applicable to all kinds of environments including high-technology societies.

Games structured and organized activities that almost always involve more than one person.

Gender social distinctions based on culturally conceived and learned ideas about appropriate behavior, appearance, and mental or emotional characteristics for males and females.

Gender nonconformists (1) persons whose primary characteristics are not clear-cut (the intersexed); (2) those whose secondary characteristics depart from the ideal conceptions of masculinity and femininity; (3) those whose interests, feelings, sexual orientation, choice of occupation, or academic major do not match gender-polarized scripts; and (4) those "who actively oppose the gender scripts of the culture" (Bem 1993, p. 167).

Gender polarization "the organizing of social life around the male–female distinction, so that people's sex is connected to virtually every other aspect of human experience, including modes of dress, social roles, and even ways of expressing emotion and experiencing sexual desire" (Bem 1993, p. 192).

Gender-schematic decisions choices related to any aspect of life that are influenced by society's polarized definitions of masculinity and femininity rather than on the basis of other criteria such as self-fulfillment, interest, ability, or personal comfort.

Generalizability the extent to which the findings of a research project can be applied to the population from which the sample is drawn.

Generalized other a system of expected behaviors, meanings, and points of view.

Great changes events whose causes lie outside ordinary people's characters or their immediate environments but profoundly affect their life chances.

Group two to 20 people who interact with one another in meaningful ways.

Hate crimes actions aimed at humiliating members of a minority group and destroying their property or lives.

Hawthorne effect a phenomenon whereby observed persons alter their behavior when they learn they are being observed.

Hidden curriculum all the things that students learn along with academic subject matter.

Hidden rape rape that goes unreported.

Hypothesis a trial explanation put forward as the focus of research that predicts how the independent and dependent variables are related. (See also **independent variable** and **dependent variable**.)

Ideal type a standard against which real cases can be compared.

Ideologies fundamental ideas that support the interests of dominant groups.

Ideology a set of ideas that do not hold up under the rigors of scientific investigation and that support the interests of dominant groups.

Illiteracy the inability to understand and use a symbol system, whether it is based on sounds, letters, numbers, pictographs, or some other type of character.

Immigration the entrance of individuals into a new country.

Impression management the process by which people in social situations manage the setting, their dress, their words, and their gestures to correspond to the impressions they are trying to make or the image they are trying to project.

Improving innovations modifications of basic inventions in order to improve on them—that is, to make them smaller, faster, less complicated, or more efficient, attractive, durable, or profitable. (See also **basic innovations**.)

Independent variable the variable of cause; a change in this variable brings about a change in the dependent variable. (See also **variable**.)

Individual discrimination any overt action on the part of an individual that deprecates minority group persons, denies them opportunities to participate, or does violence to their lives and property.

Infant mortality the number of deaths in the first year of life for every 1,000 live births.

Informal dimension (of an organization) those dimensions of organizational life that include worker-generated norms that evade, bypass, do not correspond with, or are not systematically stated in official policies, rules, and regulations.

Informal education education that occurs in a spontaneous, unplanned way.

Informal sanctions spontaneous and unofficial expressions of approval or disapproval; they are not backed by the force of law. (See also **formal sanctions**.)

Informate to use the computer to empower workers with knowledge of the overall production process, with the expectation that they will make critical and collaborative judgments about production tasks.

Information data that someone has read, listened to, or viewed.

Information explosion an unprecedented rate of increase in the volume of information due to the revolution in voice, data, and image processing, transmission, and storage that resulted from the development of the computer and telecommunications. These two inventions combine to create an intricate maze of electronic pathways that can move new and old information around the world in seconds.

Ingroup those groups with which people identify and to which they feel closely attached, particularly when that attachment is founded on hatred for another group. (See also **outgroup**.)

In-migration the movement of people into a designated area. (See also **out-migration**.)

Innovation (1) as a response to structural strain, the acceptance of the cultural goals but the rejection of legitimate means to obtain these goals; for the innovator success means winning the game rather than playing by the rules of the game; (2) the development of something new—an idea, a practice, or a tool.

Institutionalized discrimination the established and customary ways of doing things in society—the rules, policies, and day-to-day practices that we take for granted and do not challenge, which impede or limit minority group members' achievements and keep them in a subordinate and disadvantaged position.

Institutionally complete a term that describes a subculture whose members do not interact with anyone outside the subculture.

Intergenerational mobility a change in social class over two or more generations.

Internalization the process of taking as one's own and accepting as binding the norms, values, beliefs, and language of one's culture.

Internal migration movement within the boundaries of a single nation—from one state, region, or city to another.

International migration the movement of people between countries.

Internet a vast network of computer networks linking businesses, libraries, government agencies, universities, and private organizations.

Intersexed a term used to classify people with some mixture of male and female biological characteristics.

Interviews face-to-face sessions or telephone conversations between an interviewer and a respondent in which the interviewer asks the respondent questions and records his or her answers.

Intragenerational mobility a change in social class during an individual's lifetime.

Involuntary minorities ethnic and racial groups that do not choose to be a part of a country.

Issues public matters that can be explained by factors outside an individual's control and immediate environment.

Latent dysfunctions the unintended, unanticipated negative consequences that a part causes in some segment of society.

Latent functions the unintended, unrecognized, and unanticipated or unpredicted consequences that contribute to the smooth operation of the system.

Legal-rational authority power that rests on a system of impersonal rules that formally specifies the qualifications for occupying a powerful position.

Liberation theology an approach to the role of religion in society maintaining that organized religions have a responsibility to demand social justice for the marginalized peoples of the world, especially landless peasants and the urban poor, and to take an active role at the grass-roots level in order to bring about political and economic justice.

Life chances opportunities that include "everything from the chance to stay alive during the first year after birth to the chance to view fine art, the chance to remain healthy and grow tall, and if sick to get well again quickly, the chance to avoid becoming a juvenile delinquent—and very crucially, the chance to complete an intermediary or higher educational grade" (Gerth and Mills 1954, p. 313).

Local internet access provider nearby host institutions that offer internet access to individuals who cannot afford direct connection.

Looking-glass self phrase coined by Charles Horton Cooley to describe the way in which a sense of self develops: we visualize how we appear to others, we imagine a judgment of that appearance, and we develop a feeling somewhere between pride and shame.

Low-technology tribal societies hunting-and-gathering societies with technologies that do not permit the creation of surplus wealth, or wealth beyond what is needed to meet basic needs (food and shelter).

Manifest dysfunctions the expected or anticipated disruptions that a part causes in some segment of the system.

Manifest functions the intended, recognized, expected, or predictable consequences of a part for society.

Market involves transactions between buyers and sellers. (See also **buyers** and **sellers**.)

Masculinity masculine characteristics; physical, behavioral, and mental or emotional traits believed to be characteristic of males.

Material culture all of the physical objects or substances people have borrowed, discovered, or invented and to which they have attached meaning. It includes natural resources such as plants, trees, and minerals or ores, as well as items people have converted from natural resources into other forms for a purpose.

Means of production the resources—land, tools, equipment, factories, transportation, and labor—essential to the production and distribution of goods and services.

Mechanical solidarity societal order based on a common conscience or uniform thinking; characteristic of preindustrial societies.

Mechanisms of social control all of the methods that people employ to teach, persuade, or force others to conform.

Mechanization the addition of external sources of power such as oil or steam to hand tools and to modes of transportation.

Melting pot assimilation a process of cultural blending in which the groups involved accept many new behaviors and values from one another.

Methods of data collection the research procedure used to gather relevant data to test hypotheses.

Migration the movement of people from one area to another.

Minority groups subgroups within a society that can be distinguished from members of the dominant groups by visible and identifying characteristics, including physical and cultural attributes.

Mixed contacts social situations in which the stigmatized and normals are in each other's company. (See also **normals**.)

Modern capitalism "a form of economic life which involved the careful calculation of costs and profits, the borrowing and lending of money, the accumulation of capital in the form of money and material assets, investment, private property, and the employment of laborers and employees in a more or less unrestricted labor market" (Robertson 1987, p. 6).

Mores norms that people consider pivotal to the well-being of the group.

Mortality crises frequent and violent fluctuations in the death rate caused by war, famine, and epidemics, during which time the death rate has no limit.

Multinational corporation an enterprise that owns or controls production or service facilities in countries other than the one in which it is headquartered.

Mystical religions religions in which the sacred is sought in states of being that at their peak can exclude all awareness of one's existence, sensations, thoughts, and surroundings.

Nature human genetic makeup or biological inheritance.

Negative sanction an expression of disapproval for noncompliance; the punishment may be withdrawal of affection, ridicule, ostracism, banishment, physical harm, imprisonment, solitary confinement, or even death. (See also **positive sanction**.)

Nonhouseholder class propertyless laborers and servants usually residing within fortified households. (See also **fortified households**.)

Nonmaterial culture intangible creations that cannot be identified directly through the senses but that exert considerable influence over people's behavior.

Nonparticipant observation a research procedure that involves detached watching and listening; the researcher does not interact or become involved with those being studied.

Nonprejudiced nondiscriminators all-weather liberals; persons who accept the creed of equal opportunity and whose conduct conforms to that creed.

Normals those people who are in the majority or who possess no discrediting attributes.

Norms the written and unwritten rules that specify the behavior appropriate to specific situations.

Nurture the environment or the interaction experiences that make up every individual's life.

Objective secularization a twofold process: the decline in the control of religion over education, medicine, law, and politics, and the emergence of an environment in which people are free to choose from many equally valid religions the one to which they wish to belong.

Objectivity a position taken by researchers in which they do not let personal and subjective views about the topic influence the outcome of the research.

Obligations the relationship and behavior that a person enacting a role must assume toward others in a particular status.

Observation method of data gathering in which the researcher not only watches and listens but remains open to other considerations. Success results from identifying what is worth observing. (See also **nonparticipant observation** and **participant observation**.)

Oligarchy rule by the few, or the concentration of decision-making power in the hands of a few persons who hold the top positions in an organization's hierarchy.

Operational definitions clear and precise definitions and instructions about how to observe and measure the variables being studied.

Organic solidarity order based on interdependence and cooperation among a wide range of diverse and specialized tasks; this type of solidarity characterizes industrial societies.

Organization a coordinating mechanism created by people to achieve stated objectives—whether to maintain order; to challenge an established order; to keep track of people; to grow, harvest, or process food; to produce goods; or to provide a service.

Outgroup a group of individuals toward which members of an ingroup feel separateness, opposition, or even hatred. (See also **ingroup**.)

Out-migration movement out of a designated area. (See also **in-migration**.)

Paradigms the dominant and widely accepted theories and concepts in a particular field of study.

Participant observation a research procedure in which a researcher does one or more of the following: joins a group and participates as a member, interacts directly with those whom he or she is studying, assumes a position critical to the outcome of the study, or lives in a community under study.

Per capita income the average share of income that each person in a country would receive if the country's gross national product were divided evenly.

Peripheral economies the economies of labor-intensive poor countries, where most of the jobs are low-paying and require few skills. Peripheral economies are not highly diversified.

Play a voluntary and often spontaneous activity, with few or no formal rules, that is not subject to constraints of time.

Population (1) the total number of individuals, groups, households, items, or entities that could be studied; (2) a specialty within sociology that focuses on the number of people in and composition of various social groupings that live within specified boundaries and the factors that lead to changes in that social grouping's size and composition.

Population pyramid a series of horizontal bar graphs, each of which represents a different five-year age cohort. (See **cohort**.) Two bar graphs are constructed for each cohort, one for males and another for females; the bars are placed end to end, separated by a line that represents zero. Usually the left-hand side of the pyramid depicts the number or percentage of males that make up each age cohort and the right-hand side depicts the number or percentage of females. The graphs are stacked according to age; the age 0–4 cohort forms the base of the pyramid and the 80+ cohort is at the apex. The population pyramid allows us to view the relative sizes of the age cohorts and to compare the relative numbers of males and females.

Positive checks events that increase mortality, including epidemics of infectious and parasitic disease, war, and famine.

Positive sanction an expression of approval and a reward for compliance. Such a sanction may take the form of applause, an approving smile, or a pat on the back. (See also **negative sanction**.)

Power elite those few people positioned so high in the social structure of leading institutions that their decisions have consequences that affect millions of people worldwide.

Predestination the belief that God has foreordained all things, including the salvation or damnation of individual souls.

Prejudice a rigid judgment about an outgroup, usually unfavorable, that does not change in the face of contradictory evidence and that applies to anyone who shares the distinguishing characteristics of that group.

Prejudiced discriminators active bigots; persons who reject the "American creed" and profess a right, even a duty, to discriminate.

Prejudiced nondiscriminators timid bigots; persons who do not accept the "American creed" but who refrain from discriminatory actions primarily because they fear the sanctions they may encounter if they are caught.

Primary groups major socializing agents, especially in the early years, because they give newcomers their first exposure to the "rules of life." These groups are characterized by face-to-face contact and strong ties among members.

Primary sex characteristics the anatomical traits essential to reproduction. Most cultures divide the population into two categories—male and female—largely on the basis of what most people consider to be clear anatomical distinctions.

Private households an arrangement that exists when the workplace is separate from the home, where men are heads of households and assume a breadwinner role, and where women remain responsible for housekeeping and childrearing.

Probabilistic model a model in which the hypothesized effect does not always result from a hypothesized cause.

Profane everything that is not sacred, including those things opposed to the sacred (the unholy, the irreverent, the contemptuous, the blasphemous) and those things that, although not opposed to the sacred, stand apart from it (the ordinary, the commonplace, the unconsecrated, the temporal, the bodily).

Professionalization (within organizations) a hiring trend in organizations in which experts are hired who have formal training in a particular subject or activity that is essential to achieving organizational goals.

Proletariat those who must sell their labor to the bourgeoisie.

Prophetic religions religions in which conceptions of the sacred revolve around items that symbolize significant historical events or around the lives, teachings, and writings of great people.

Pull factors the conditions that encourage people to move into a particular area. Some of the most common pull factors are employment opportunities, favorable climate, and the relative absence of discrimination.

Pure deviants people who have broken the rules and are caught, punished, and labeled as outsiders.

Push factors the conditions that encourage people to move out of an area. Some of the most common push factors include religious or political persecution, discrimination, depletion of natural resources, lack of employment opportunities, and natural disasters.

Race a group of people who possess certain distinctive and conspicuous physical characteristics.

Racism an ideology that maintains that something in the biological makeup of a specific racial or ethnic group explains its subordinate or superior status.

Random sample a sample drawn in such a way that every case in the population has an equal chance of being selected.

Rationalization as defined by Max Weber, a process whereby thought and action rooted in emotion (love, hatred, revenge, joy), in superstition, in respect for mysterious forces, and in tradition are replaced by thought and action grounded in the logical assessment of cause and effect or means and ends.

Rebellion the full or partial denunciation of both goals and means and the introduction of a new set of goals and means.

Reentry shock reverse culture shock.

Reflexive thinking stepping outside the self to observe and evaluate it from another's viewpoint.

Reliability the extent to which the operational definition gives consistent results.

Religion according to Emile Durkheim a system of shared beliefs and rituals about the sacred that bind together a community of worshipers. (See also **sacred**.)

Representative sample a sample with the same distribution of characteristics as the population from which it was selected.

Research a fact-gathering and fact-explaining enterprise governed by strict rules.

Research design a plan for gathering data to test hypotheses.

Research methods the procedure that sociologists and other investigators use to formulate meaningful research questions and to collect, analyze, and interpret facts in ways that other researchers can duplicate.

Research-methods literate the ability to know how to collect data that is worth putting into the computer and to interpret the data that comes out of it.

Resocialization the process of discarding values and behaviors unsuited to new circumstances and replacing them with new, more appropriate values and standards of behavior.

Retreatism the rejection of both cultural goals and the means of achieving these goals.

Reverse ethnocentrism the tendency to see the home culture as inferior to a foreign culture.

Rights the behaviors that a person assuming a role can demand or expect from others.

Ritualism the abandonment of cultural goals but a rigid adherence to the legitimate means of attaining those goals. It is the opposite of innovation (as a response to structural strain); the game is played according to the rules despite defeat.

Rituals rules that govern how people must behave when in the presence of the sacred. These rules may take the form of instructions detailing the appropriate context, the roles of various participants, acceptable attire, and the precise wording of chants, songs, and prayers.

Role the behavior expected of a status in relation to another status.

Role conflict a predicament in which the expectations associated with two or more roles in a role set are contradictory.

Role set the array of roles associated with every status.

Role strain a predicament in which contradictory or conflicting expectations are associated with the role that a person is occupying.

Role-taking stepping outside the self and viewing its appearance and behavior imaginatively from an outsider's perspective.

Routinized charisma a situation in which the community must establish procedures, rules, and traditions to regulate the members' conduct, to recruit new members, and to ensure the orderly transfer of power.

Sacramental religion religions in which the sacred is sought in places, objects, and actions believed to house a god or a spirit.

Sacred all phenomena that are regarded as extraordinary and that inspire in believers deep and absorbing sentiments of awe, respect, mystery, and reverence.

Sample a portion of cases from a particular population.

Sampling frame a complete list of every case in the population.

Sanctions reactions of approval and disapproval to behavior and appearances. Sanctions can be positive or negative, formal or informal.

Scapegoat a person or a group that is assigned blame for conditions that cannot be controlled, that threaten a community's sense of well-being, or that shake the foundations of a trusted institution.

Schooling a program of formal and systematic instruction that takes place primarily in a classroom but also includes extracurricular activities and out-of-classroom assignments.

Scientific method an approach to data collection guided by the assumptions that knowledge about the world is acquired through the senses (is observable) and that the truth of the knowledge is confirmed by other persons making the same observations.

Scientific revolution a condition that occurs when enough people in the community break with an old paradigm and change the nature of their research in favor of the incompatible new paradigm.

Search engine tool that allows internet users to submit a keyword(s) to identify topics on which they want to find information; the search engine identifies Web sites and corresponding URLs.

Secondary sex characteristics physical traits not essential to reproduction (breast development, voice quality, distribution of facial and body hair, and skeletal form) that result from the action of so-called male hormones (androgen) and female hormones (estrogen). Although testes produce androgen and ovaries produce estrogen, the adrenal cortex produces androgen and estrogen in both sexes.

Secondary sources data that has been collected by other researchers for some other purpose.

Secret deviants people who have broken the rules, but whose violation goes unnoticed, or, if it is noticed, goes unpunished.

Sect a small community of believers led by a lay ministry, with no formal hierarchy or official governing body to oversee the various religious gatherings and activities.

Secularization a process by which religious influences on thought and behavior are reduced.

Selective perception the process whereby prejudiced persons notice only those behaviors or events that support their stereotypes about an outgroup.

Self-administered questionnaire a set of questions given or mailed to respondents, who read the instructions and fill in the answers themselves.

Self-fulfilling prophecy a concept that begins with a false definition of a situation. The false definition is assumed to be accurate, and people behave as if the definition were true. In the end the misguided behavior produces responses that confirm the false definition.

Sellers persons who conduct one or more of the extensive transactions (producing, advertising, shipping, storing, selling) needed to sell a commodity or service.

Semi-peripheral economies the moderately diversified economies of moderately wealthy countries.

Sexist ideologies the ideologies that justify one sex's social, economic, and political dominance over the other.

Sexual property "the relatively permanent claim to exclusive sexual rights over a particular person" (Collins 1971, p. 7).

Sick role a term coined by sociologist Talcott Parsons to represent the rights and obligations accorded people when they are sick.

Significant others people or characters who are important in a person's life—important in the sense that they have considerable influence on self-evaluation and encourage a person to behave in a certain manner.

Significant symbols words, gestures, and other learned signs that are used to convey a meaning that is the same for both the communicator and the recipient. Particularly important significant symbols are language and symbolic gestures.

Simultaneous-independent inventions situations in which the same invention is created by two or more persons working independently of one another at about the same time (sometimes within a few days or months).

Situational factors forces outside an individual's control, such as environmental conditions or bad luck.

Social action behavior or actions that people take in response to others.

Social change any alteration, modification, or transformation of social phenomena over a specified period.

Social emotions internal bodily sensations that people experience in relationships with others.

Social identity the category to which a person belongs and the qualities that others believe, rightly or wrongly, to be "ordinary and natural" (Goffman 1963, p. 2) for a member of that category.

Social interactions events involving at least two people who communicate through language and symbolic gestures to affect one another's behavior and thinking.

Socialization a complex, lifelong process of learning about the social world. Socialization begins immediately after birth and continues throughout life. It is a process through which newcomers develop their human capacities, acquire a unique personality and identity, and internalize the norms, values, beliefs, and language needed to participate in the larger society.

Social movements organized, deliberate efforts by believers to transform, reform, or replace some element of culture, and in the process to convert nonbelievers to their position.

Social promotion passing students from one grade to another on the basis of age rather than academic competency.

Social relativity the view that ideas, beliefs, and behavior are to a large extent a product of time and place.

Social status a position in a system of social relationships or a social structure.

Social stratification a systematic, nonrandom process by which people in a society are ranked according to a scale of social worth and are awarded unequal amounts of income, wealth, prestige, and power on the basis of that ranking.

Social structure two or more people interacting and interrelating in expected ways, regardless of the unique personalities involved.

Society a group of people living in a given territory who share a culture and who interact with people of that territory more than with people of another territory.

Sociological imagination the ability to connect seemingly impersonal and remote historical forces to the most basic incidents of an individual's life. The sociological imagination enables people to distinguish between personal troubles and public issues.

Sociological theory a set of principles and definitions that tell how societies operate and how people relate to one another.

Sociology the scientific study of the causes and consequences of human interaction.

Solidarity the ties that bind people to one another.

Spurious correlation a correlation that is coincidental or accidental.

State (1) a political entity recognized by foreign governments, with a civilian and military bureaucracy to carry out its policies, to enforce its rules, and to regulate other activities within its borders; (2) a governing body organized to manage and control specified activities of people living in a given territory.

Stationary pyramids population pyramids that are characteristic of most developed nations and that are similar to constrictive pyramids except that all of the age cohorts in the population are roughly the same size and fertility is at replacement level. (See also **population pyramid**.)

Status group a plurality of persons held together by virtue of a common lifestyle, formal education, family background, or occupation and "by the level of social esteem and honor accorded them by others" (Coser 1977, p. 229).

Status system classification of achievements resulting in popularity, respect, and acceptance into the crowd, as opposed to disdain, discouragement, and disrespect.

Status value a situation in which persons who possess one category of a characteristic (white skin versus brown skin, blond hair versus dark hair) are believed to be and are treated as more valuable or worthy than persons who possess other categories (Ridgeway 1991).

Stereotypes exaggerated and inaccurate generalizations about people who are members of an outgroup.

Stigmas statuses that are deeply discrediting in the sense that they overshadow all other statuses that a person occupies.

Streaming the arranging of middle school and high school students into instructional groups according to similarities in past academic performance and/or on standardized test scores.

Structural strain a condition that occurs when the valued goals have no clear boundaries, when it is not clear whether the legitimate means that society provides will lead to the goals, and when the legitimate opportunities for meeting the goals are closed to a significant portion of the population; anomie.

Structured interview an interview in which the wording and the sequence of questions are predetermined and cannot be altered during the course of the interview.

Subcultures groups that share in some parts of the dominant culture but that have their own distinctive values, norms, language, or material culture.

Subjective secularization a decrease in the number of people who view the world and their place in it from a religious perspective.

Symbol any kind of physical phenomenon—a word, an object, a taste—to which people assign a meaning or value.

Symbolic gestures extraverbal cues that include tone of voice, inflection, facial expression, posture, and other body movements or positions that convey meaning from one person to another.

Technological determinist someone who believes that human beings have no free will and are controlled entirely by their material innovations.

Theory a framework that can be used to comprehend and explain events.

Theory of the demographic transition a model that outlines historical changes in birthrates and death rates among the mechanized rich countries and the factors underlying those changes. Some demographers have theorized that a country's birthrates and death rates are linked to its level of industrial or economic development.

This-worldly asceticism a belief that people are instruments of divine will and that their activities are determined and directed by God.

Total fertility the average number of children women bear over their lifetime.

Total institutions settings in which people surrender control of their lives, voluntarily or involuntarily, to an administrative staff and in which they (as inmates) carry out daily activities in the presence of other inmates.

Tracking the arranging of middle school and high school students into instructional groups according to similarities in past academic performance and/or on standardized test scores.

Trained incapacity the inability to respond to new and unusual circumstances or to recognize when official rules and procedures are outmoded or no longer applicable.

Transformative powers of history the dramatic consequences of important historical events on people's thinking and behavior.

Troubles private matters that can be explained in terms of personal characteristics (motivation level, mood, personality, or ability) or immediate relationships with others (family members, friends, acquaintances, or coworkers). The resolution of a trouble, if it can be resolved, lies in changing an individual's character or immediate relationships.

Uniform Resource Locator (URL) an internet document's address.

Unprejudiced discriminators fair-weather liberals; persons who believe in equal opportunity but engage in discriminatory behaviors because it is to their advantage to do so or because they fail to consider the discriminatory consequences of some of their actions.

Unstructured interview an interview that is flexible and open-ended; the question–answer sequence is spontaneous and resembles a conversation.

Upward mobility a change in social class that corresponds to a gain in rank or prestige.

Urbanization an increase in the number of cities and in the proportion of the population living in cities.

Urban underclass the ghetto poor, or a "heterogeneous grouping of families and individuals in the inner city that are outside the mainstream of the American occupational system and that consequently represent the very bottom of the economic hierarchy" (Wilson 1983, p. 80).

Validity the degree to which an operational definition measures what it claims to measure.

Values general conceptions about what is good, right, appropriate, worthwhile, and important with regard to modes of conduct and states of existence.

Variable any trait or characteristic that can vary or have more than one category. (For example, gender is a variable with two categories: male and female.)

Vertical mobility a change in class status that corresponds to a gain or loss in rank or prestige.

Voluntary minorities racial and ethnic groups that come to a country expecting to improve their way of life.

White-collar crime "crime committed by persons of respectability and high social status in the course of their occupations" (Sutherland and Cressey 1978, p. 44).

Witch-hunt a campaign or purge launched to reach subversive elements on the pretext of investigating and correcting activities that undermine a group or a country.

World-economy an economy in which economic transactions transcend national boundaries.

World Wide Web one of several internet-based services that facilitate the exchange of text-, video-, and audio-based information.

References

Preface

Katz, Yvonne, and Gay Chedester. 1992. "Redefining Success: Public Education in the 21st Century." *Community Services CATALYST* 22 (3). gopher://borg.lib.vf.edu//00/catalyst/v22n3/katz.v22n3.

Chapter 1

Klapp, Orrin E. 1986. *Overload and Boredom: Essays on the Quality of Life in the Information Society.* New York: Greenwood Press.

Lewis, Peter H. 1996. "When On-Line Service Cannot Be Counted On." *The New York Times* (June 24):C1.

Steinberg, Steve. 1996. "Seek and Ye Shall Find (Maybe)." *Wired* (May):108–14+.

Chapter 2

Abercrombie, Nicholas, Stephen Hill, and Bryan S. Turner. 1988. *The Penguin Dictionary of Sociology.* New York: Penguin.

Bardis, Panos D. 1980. "Sociology as a Science." *Social Science Journals* 55(3):141–80.

Berger, Peter. 1963. *Invitation to Sociology: A Humanistic Perspective.* New York: Anchor.

Boden, Deirdre, Anthony Giddens, and Harvey L. Molotch. 1990. "Sociology's Role in Addressing Society's Problems Is Undervalued and Misunderstood in Academe." *The Chronicle of Higher Education* (February 21):B1+.

Charyn, Jerome. 1978. "Black Diamond." *The New York Review of Books* (August 17):41.

Coser, Lewis A. 1977. *Masters of Sociological Thought*, ed. R. K. Merton. New York: Harcourt Brace Jovanovich.

Freund, Julien. 1968. *The Sociology of Max Weber.* New York: Random House.

Gordon, John Steele. 1989. "When Our Ancestors Became Us." *American Heritage* (December): 106–21.

Gould, Stephen Jay. 1981. *The Mismeasure of Man.* New York: Norton.

Henslin, James M., ed. 1993. *Down to Earth Sociology: Introductory Readings.* New York: Free Press.

Lengermann, Patricia M. 1974. *Definitions of Sociology: A Historical Approach.* Columbus, OH: Merrill.

Martineau, Harriet. [1837] 1968. *Society in America*, ed. and abrg. S. M. Lipset. Gloucester, MA: Peter Smith.

Miller, S. M. 1963. *Max Weber: Selections from His Work.* New York: Crowell.

Mills, C. Wright. 1959. *The Sociological Imagination.* New York: Oxford University Press.

The New Columbia Encyclopedia. 1975. "Steamship." New York: Columbia University Press.

Ornstein, Robert, and Paul Ehrlich. 1989. *New World New Mind.* New York: Touchstone Books.

Zuboff, Shoshana. 1988. *In the Age of the Smart Machine.* New York: Basic Books.

Chapter 3

Bearden, Tom. 1993. "Focus: Help Wanted." Interview with anonymous Denver woman on her use of undocumented worker for childcare. "MacNeil/Lehrer Newshour" (transcript no. 4548). New York: WNET.

Blumer, Herbert. 1962. "Society as Symbolic Interaction." In *Human Behavior and Social Processes*, ed. A. Rose. Boston: Houghton Mifflin.

Carver, Terrell. 1987. *A Marx Dictionary.* Totowa, NJ: Barnes & Noble.

CommerceNet/Nielsen. 1995. "The CommerceNet/Nielsen Demographics Survey." http://www.commerce.net:8000/information/surveys/execsum/exec_s.

Crawford, Jack. 1995. *Renaissance Two: Second Coming of the Printing Press.* http://lincoln.ac.nz/reg/ferenaiss2.htm.

Erlanger, Steven. 1996. "Americans in Desert Wonder How Long They'll Stay." *The New York Times* (June 29):Y5.

Gans, Herbert. 1972. "The Positive Functions of Poverty." *American Journal of Sociology* 78: 275–89.

Halberstam, David. 1986. *The Reckoning.* New York: Morrow.

Internet Society. 1996. "Growth in New IP Networks 2Q93-3Q93." http://info.isoc.org/images/tokyo8.gif.

Kilborn, Peter T. 1990. "When Plant Shuts Down, Retraining Laid-Off Workers Is Toughest Job Yet." *The New York Times* (April 23):A12.

———. 1992. "Tide of Migrant Labor Tells of a Law's Failure." *The New York Times* (November 4):A9.

Koenenn, Connie. 1992. "The Power of Pulling Purse Strings." *Los Angeles Times* (December 1):E1.

Hall, Edward T. 1992. *An Anthropology of Everyday Life.* New York: Doubleday.

Hamilton, Virginia. 1988. *In the Beginning: Creation Stories from Around the World.* New York: Harcourt Brace Jovanovich.

Marx, Karl. [1888] 1961. "The Class Struggle." Pp. 529–35 in *Theories of Society*, ed. by T. Parsons, E. Shils, K. D. Naegele, and J. R. Pitts. New York: Free Press.

———. 1848. *Manifesto of the Communist Party.* http://csf.Colorado.Edu/psn/marx/Archive/1848-cm/cm.html.

Mead, George H. 1934. *Mind, Self and Society.* Chicago: University of Chicago Press.

Merton, Robert K. 1967. "Manifest and Latent Functions." Pp. 73–137 in *On Theoretical Sociology: Five Essays, Old and New.* New York: Free Press.

Rodale, Jerome Irving. 1986. *The Synonym Finder*, rev. L. Urdang and N. LaRoche. New York: Warner.

Rose, Kenneth J. 1988. *The Body in Time.* New York: Wiley.

Tuleja, Tad. 1987. *Curious Customs: The Stories Behind 296 Popular American Rituals.* New York: Harmony Books.

Tumin, Melvin. 1964. "The Functionalist Approach to Social Problems." *Social Problems* 12:379–88.

Chapter 4

Cameron, William B. 1963. *Informal Sociology.* New York: Random House.

Gregg, Alan. 1989. Quoted on pp. 48–55 in *Science and the Human Spirit*, ed. R. D. White. Belmont, CA: Wadsworth.

Joseph, Michael. 1982. *The Timetable of Technology.* London: Marshal Editions.

Katzer, Jeffrey, Kenneth H. Cook, and Wayne W. Crouch. 1991. *Evaluating Information: A Guide for Users of Social Science Research*, 3rd ed. New York: McGraw-Hill.

Klapp, Orrin E. 1986. *Overload and Boredom: Essays on the Quality of Life in the Information Society.* New York: Greenwood Press.

Lucky, Robert W. 1985. "Message by Light Wave." *Science* (November):112–13.

Marsa, Linda. 1992. "Scientific Fraud." *Omni* (June):39+.

Michael, Donald. 1984. "Too Much of a Good Thing?: Dilemmas of an Information Society." *Technological Forecasting and Social Change* 25(4):347–54.

Schonberg, Harold C. 1981. "Sumo—Embodies Ancient Rituals." *The New York Times*:B9.

Singleton, Royce A., Jr., Bruce C. Straits, and Margaret Miller Straits. 1993. *Approaches to Social Research*, 2nd ed. New York: Oxford University Press.

Smith, Joel. 1991. "A Methodology for the Twenty-First Century Sociology." *Social Forces* 70(1):1–17.

Thayer, John E., III. 1983. "Sumo." Pp. 270–74 in *Kodansha Encyclopedia of Japan*, Vol. 7. Tokyo: Kodansha.

Thomas, Bill. 1992. "King Stacks." *Los Angeles Times Magazine* (November 15):31+.

Chapter 5

Agar, Michael. 1994. *Language Shock/Understanding the Culture of Conversation*. New York: Morrow.

Anderson, Barbara Gallatin. 1971. "Adaptive Aspects of Culture Shock." *American Anthropologist* 73: 1121–25.

Barnlund, Dean. 1994. "Communication in a Global Village." Pp. 26–36 in *Intercultural Communication: A Reader*, ed. L. A. Samovar and R. E. Porter. Belmont, CA: Wadsworth.

Bateson, Mary Catherine. 1968. "Insight in a Bicultural Context." *Philippine Studies* 16: 605–21.

Behrangi, Samad. 1994. Quoted in "International Rural Education Teacher and Literary Critic: Samad Behrangi's Life and Thoughts." *Journal of Global Awareness* 21(1):27–35.

Berkhofer, Robert F., Jr. 1978. *The White Man's Indian: Images of the American Indian from Columbus to the Present*. New York: Knopf.

Berreby, David. 1995. "Unabsolute Truths: Clifford Geertz." *The New York Times Magazine* (April 9):44–47.

Berry, Michael. 1995. "Curse-Cultural Communication." *Word* 16(1):8.

Benedict, Ruth. 1976. Quoted on p. 14 in *The Person: His and Her Development Throughout the Life Cycle*, by Theodor Lidz. New York: Basic Books.

Breton, Raymond. 1967. "Institutional Completeness of Ethnic Communities and the Personal Relations of Immigrants." *American Journal of Sociology* 70:193–205.

Brown, Rita Mae. 1988. *Rubyfruit Jungle*. New York: Bantam Books.

Burke, James. 1978. *Connections*. Boston: Little, Brown.

Cambridge International Dictionary of English. 1995. New York: Cambridge University Press.

Clifton, James A. 1989. *Being and Becoming Indian: Biographical Studies of North American Frontiers*. Chicago: Dorsey.

Dunn, Ashley. 1995. "Skilled Asians Leaving U.S. for High-Tech Jobs at Home." *The New York Times* (February 21):A11.

Frank, Lawrence. 1948. "World Order and Cultural Diversity." Pp. 389–95 in *Society as the Patient*. New Brunswick, NJ: Rutgers University Press.

Geertz, Clifford. 1995a. *After the Fact: Two Countries, Four Decades, One Anthropologist*. Cambridge, MA: Harvard University Press.

———. 1995b. Quoted in "Unabsolute Truths: Clifford Geertz." *The New York Times Magazine* (April 9):44–47.

Gordon, Emily Fox. 1995. "Faculty Brat: A Memoir." *Boulevard* 10(1-2):1–17.

Gordon, Steven L. 1981. "The Sociology of Sentiments and Emotion." Pp. 562–92 in *Social Psychology Sociological Perspectives*, ed. M. Rosenberg and R. H. Turner. New York: Basic Books.

Gullick, C.J.M.R. 1990. "Expatriate British Executives and Culture Shock." *Studies in Third World Societies*, 42 (Cross-Cultural Management and Organizational Culture).

Hannerz, Ulf. 1986. "Theory in Anthropology: Small Is Beautiful? The Problem of Complex Cultures." *Comparative Studies in Society and History* 28:362–67.

———. 1990. "Cosmopolitans and Locals in World Culture." *Theory, Culture & Society* 7: 237–51.

———. 1992. *Cultural Complexity: Studies in the Social Organization of Meaning*. New York: Columbia University Press.

———. 1993. "The Withering Away of the Nation? An Afterword." *Ethnos* 3(3-4):377–91.

Herskovits, Melville J. 1948. *Man and His Works: The Science of Cultural Anthropology*. New York: Knopf.

Hochschild, Arlie R. 1976. "The Sociology of Feeling and Emotion: Selected Possibilities." Pp. 280–307 in *Another Voice*, ed. M. Millman and R. Kanter. New York: Octagon.

———. 1979. "Emotion Work, Feeling Rules, and Social Structure." *American Journal of Sociology* 85:551–75.

Hughes, Everett C. 1984. *The Sociological Eye: Selected Papers.* New Brunswick, NJ: Transaction Books.

Hunter, Victor. 1986. "Closure and Commencement: The Stress of Finding Home." Pp. 179–80 in *The Cultural Transition: Human Experience and Social Transformation in the Third World and Japan,* ed. M. I. White and S. Pollak. Boston: Routledge & Kegan.

Ingram, Erik. 1992. "Water Use Continues to Decline: Bay Area Districts Report Record Savings." *San Francisco Chronicle* (July):A15.

Kempley, Rita. 1994. "Jodie Foster: In Control." *The Washington Post* (December 25):G6.

Kim, Bo-Kyung, and Kevin Kirby. 1996. Personal correspondence (April 25).

Kluckhohn, Clyde. 1949. *Mirror for Man: Anthropology and Modern Life.* New York: McGraw-Hill.

Kracke, Waud. 1987. "Encounter with Other Cultures: Psychological and Epistemological Aspects." *Ethos* 15(1):58–81.

Kristof, Nicholas D. 1995. "When Doctor Won't Tell Cancer Patients the Truth." *The New York Times* (February 25):Y6.

———. 1996. "Rappers' Credo: No Sex Please! We're Japanese." *The New York Times* (January 29):A4.

Kroeber, A. L., and Clyde Kluckhohn. 1952. *Culture: A Critical Review of Concepts and Definitions.* New York: Vintage.

Lamb, David. 1987. *The Arabs: Journeys Beyond the Mirage.* New York: Random House.

Levinson, David. 1991. "Preface." Pp. xvii–xxi in *Encyclopedia of World Cultures, Volume 1: North America,* ed. T. J. O'Leary and D. Levinson. Boston: G. K. Hall.

Lidz, Theodore. 1976. *The Person: His and Her Development Throughout the Life Cycle.* New York: Basic Books.

Linton, Ralph. 1936. *The Study of Man: An Introduction.* New York: Appleton-Century-Crofts.

Liu, Hsein-Tung. 1994. "Intercultural Relations in an Emerging World Civilization." *Journal of Global Awareness* 2(1):48–53.

Mahmood, Cynthia K., and Sharon Armstrong. 1992. "Do Ethnic Groups Exist? A Cognitive Perspective on the Concept of Cultures." *Ethnology* 31(1):1–14.

Marshall, Tyler. 1995. "7 Nations in Europe Union Open Borders." *The New York Times* (March 27):A1.

Mendelsohn, Harold. 1964. "Listening to the Radio." Pp. 239–49 in *People, Society, and Mass Communications,* ed. L. A. Dexter and D. M. White. London: Collier-Macmillan.

Moran, Robert T. 1987. "Cross-Cultural Contact: What's Funny to You May Not Be Funny to Other Cultures." *International Management* 42 (July/August):74.

Murphy, Dean. 1994. "New East Europe Retailers Told to Put on a Happy Face." *Los Angeles Times* (November 26):A1.

Nasaw, David. 1991. "The Stuff That Made Us What We Are." *The New York Times Book Review* (December 15):10.

National Public Radio. "Morning Edition." 1995. "How Americans Perceive the Japanese and Vice Versa." (February 10).

———. 1994a. "High-Tech TV Cameras Akin to Fountain of Youth." (December 5).

———. 1994b. "France Gives Birth to Multi-Platinum Rapper." (December 5).

Peterson, Mark. 1977. "Some Korean Attitudes Toward Adoption." *Korea Journal* 17(12):28–31.

Reader, John. 1988. *Man on Earth.* Austin: University of Texas Press.

Redfield, Robert. 1962a. "The Universally Human and the Culturally Variable." Pp. 439–53 in *Human Nature and the Study of Society: The Papers of Robert Redfield,* Vol. 1, ed. M. P. Redfield. Chicago: University of Chicago.

———. 1962b. "Anthropological Understanding of Man." Pp. 453–69 in *Human Nature and the Study of Society: The Papers of Robert Redfield,* Vol. 1, ed. M. P. Redfield. Chicago: University of Chicago.

Rohner, Ronald P. 1984. "Toward a Conception of Culture for Cross-Cultural Psychology." *Journal of Cross-Cultural Psychology* 15(2):111–38.

Rohter, Larry. 1994. "Battle over Patriotism Curriculum." *The New York Times* (May 15):Y12.

Rokeach, Milton. 1973. *The Nature of Human Values.* New York: Free Press.

Rosenfeld, Jeffrey P. 1987. "Barking Up the Right Tree." *American Demographics* (May):40–43.

Sacks, Oliver. 1989. *Seeing Voices: A Journey in the World of the Deaf.* Los Angeles: University of California Press.

Schoenberger, Karl. 1992. "Moving Between 2 Worlds." *Los Angeles Times* (July 12):A1+.

Smalley, William A. 1963. "Culture Shock, Language Shock, and the Shock of Self-Discovery." *Practical Anthropology* (March/April):49–56.

Sobie, Jane Hipkins. 1986. "The Cultural Shock of Coming Home Again." Pp. 95–102 in *The Cultural Transition: Human Experience and Social Transformation in the Third World and Japan*, ed. M. I. White and S. Pollak. Boston: Routledge & Kegan.

Sumner, William Graham. 1907. *Folkways*. Boston: Ginn.

Taniuchi, Lois. 1986. "Cultural Continuity in an Educational Institution: A Case Study of the Suzuki Method of Music Instruction." Pp. 113–40 in *The Cultural Transition: Human Experience and Social Transformation in the Third World and Japan*, ed. M. I. White and S. Pollak. Boston: Routledge & Kegan.

U.S. Bureau of the Census. 1994. "Nonimmigrants Admitted, by Class of Admission: 1984–1991." *Statistical Abstract of the United States, 1993–94*. Washington, DC: U.S. Government Printing Office.

Visser, Margaret. 1989. "A Meditation on the Microwave." *Psychology Today* (December):38–42.

Wallace, Charles P. 1994. "Singapore Affirms Flogging of American." *Los Angeles Times* (April 1):A5.

Weiss, Lowell. 1995. "Speaking in Tongues." *The Atlantic Monthly* (June):36–42.

Werkman, Sidney L. 1986. "Coming Home: Adjustment of Americans to the United States After Living Abroad." Pp. 5–18 in *Cross-Cultural Reentry: A Book of Readings*, ed. C. N. Austin. Abilene, TX: Abilene Christian University.

White, Leslie A. 1949. *The Science of Culture*. New York: Farrar, Straus & Cudahy.

Yeh, May. 1991. "A Letter." *Amerasia* 17(2):1–7.

Chapter 6

Bunuel, Luis. 1985. Quoted on p. 22 in *The Man Who Mistook His Wife for a Hat and Other Clinical Tales*, by Oliver Sacks. New York: Summit Books.

Cooley, Charles Horton. 1909. *Social Organization*. New York: Scribner.

———. 1961. "The Social Self." Pp. 822–28 in *Theories of Society: Foundations of Modern Sociological Theory*, ed. T. Parsons, E. Shils, K. D. Naegele, and J. R. Pitts. New York: Free Press.

Corsaro, William A. 1985. *Friendship and Peer Culture in the Early Years*. Norwood, NJ. Ablex.

Coser, Lewis A. 1992. "The Revival of the Sociology of Culture: The Case of Collective Memory." *Sociological Forum* 7(2):365–73.

Davis, Kingsley. 1940. "Extreme Isolation of a Child." *American Journal of Sociology* 45:554–65.

———. 1947. "Final Note on a Case of Extreme Isolation." *American Journal of Sociology* 3(5):432–37.

Delgado, Jose M. R. 1970. Quoted on p. 170 in "Brain Researcher Jose Delgado Asks—'What Kind of Humans Would We Like to Construct?'" *The New York Times Magazine* (November 15):46+.

Dyer, Gwynne. 1985. *War*. New York: Crown.

Faris, Ellsworth. 1964. "The Primary Group: Essence and Accident." Pp. 314–19 in *Sociological Theory: A Book of Readings*, ed. L. A. Coser and B. Rosenberg. New York: Macmillan.

Figler, Stephen K., and Gail Whitaker. 1991. *Sport and Play in American Life*. Dubuque, IA: Brown.

Goffman, Erving. 1961. *Asylums: Essays on the Social Situation of Mental Patients and Other Inmates*. New York: Anchor.

Halbwachs, Maurice. 1980. *The Collective Memory*, trans. F. J. Ditter, Jr., and V. Y. Ditter. New York: Harper & Row.

Hellerstein, David. 1988. "Plotting a Theory of the Brain." *The New York Times Magazine* (May 22):17+.

Kagan, Jerome. 1988a. Interview on "The Mind." PBS.

———. 1988b. Quoted on pp. 21–22 in *The Mind*, by R. M. Restak. New York: Bantam Books.

———. 1989. *Unstable Ideas: Temperament, Cognition, and Self*. Cambridge, MA: Harvard University Press.

Mannheim, Karl. 1952. "The Problem of Generations." Pp. 276–322 in *Essays on the Sociology of Knowledge*, ed. P. Kecskemeti. New York: Oxford University Press.

Mead, George Herbert. 1934. *Mind, Self and Society*. Chicago: University of Chicago Press.

Merton, Robert K. 1976. *Sociological Ambivalence and Other Essays*. New York: Free Press.

Montgomery, Geoffrey. 1989. "Molecules of Memory." *Discover* (December):46–55.

"Nova." 1986. "Life's First Feelings." (February 11). Boston: WGBH.

Ornstein, Robert, and Richard F. Thompson. 1984. *The Amazing Brain.* Boston: Houghton Mifflin.

Penfield, Wilder, and P. Perot. 1963. "The Brain's Record of Auditory and Visual Experience: A Final Summary and Discussion." *Brain* 86. 595–696.

Piaget, Jean. 1923. *The Language and Thought of the Child,* trans. M. Worden. New York: Harcourt, Brace & World.

———. 1929. *The Child's Conception of the World,* trans. J. Tomlinson and A. Tomlinson. Savage, MD: Rowan & Littlefield.

———. 1932. *The Moral Judgement of the Child,* trans. M. Worden. New York: Harcourt, Brace & World.

———. 1946. *The Child's Conception of Time,* trans. A. J. Pomerans. London: Routledge & Kegan Paul.

———. 1967. *On the Development of Memory and Identity.* Worcester, MA: Clark University Press.

Restak, Richard M. 1988. *The Mind.* New York: Bantam Books.

Rose, Peter I., Myron Glazer, and Penina M. Glazer. 1979. "In Controlled Environments: Four Cases of Intensive Resocialization." Pp. 320–38 in *Socialization and the Life Cycle,* ed. P. I. Rose. New York: St. Martin's Press.

Rosenthal, Elisabeth. 1989. "Mystery on Arrival." *Discover* (December):78–82.

Rubinstein, Danny. 1988. "The Uprising: Reporter's Notebook." *Present Tense* 15:22–25.

———. 1991. *The People of Nowhere: The Palestinian Vision of Home,* trans. T. Friedman. New York: Random House.

Sacks, Oliver. 1989. *Seeing Voices: A Journey into the World of the Deaf.* Los Angeles: University of California Press.

Satterly, D. J. 1987. "Jean Piaget (1896–1980)." Pp. 621–22 in *The Oxford Companion to the Mind,* ed. R. I. Gregory. Oxford: Oxford University Press.

Spitz, Rene A. 1951. "The Psychogenic Diseases in Infancy: An Attempt at Their Etiological Classification." Pp. 255–78 in *The Psychoanalytic Study of the Child,* Vol. 27, ed. R. S. Eissler and A. Freud. New York: Quadrangle.

Theodorson, George A., and Achilles G. Theodorson. 1979. *A Modern Dictionary of Sociology.* New York: Barnes & Noble.

Townsend, Peter. 1962. Quoted on pp. 146–47 in *The Last Frontier: The Social Meaning of Growing Old,* by Andrea Fontana. Beverly Hills, CA: Sage.

Chapter 7

Barr, David. 1990. "What Is AIDS? Think Again." *The New York Times* (December 1):Y15.

Bloor, Michael, David Goldberg, and John Emslie. 1991. "Research Note: Ethnostatics [sic] and the AIDS Epidemic." *The British Journal of Sociology* 42(1):131–38.

Brooke, James. 1987. "In Cradle of AIDS Theory, a Defensive Africa Sees a Disguise for Racism." *The New York Times* (November 19):B13.

———. 1988a. "In Africa, Tribal Hatreds Defy the Borders of State." *The New York Times* (August 28):E1.

———. 1988b. "Mobutu's Village Basks in His Glory." *The New York Times* (September 9):Y4.

Clark, Matt, with Stryker McGuire. 1980. "Blood Across the Border." *Newsweek* (December 29):61.

Clarke, Thurston. 1988. *Equator: A Journey.* New York: Morrow.

Colby, Ron. 1986. Quoted in "Did Media Sensationalize Student AIDS Case?" by John McGauley. *Editor and Publisher* 119:19.

Conrad, Joseph. 1971. *Heart of Darkness,* rev. and ed. R. Kimbrough. New York: Norton.

De Cock, Kevin M., and Joseph B. McCormick. 1988. "Correspondence: Reply to HIV Infection in Zaire." *New England Journal of Medicine* 319(5):309.

Doyal, Lesley, with Imogen Pennell. 1981. *The Political Economy of Health.* Boston: South End Press.

Durkheim, Emile. [1933] 1964. *The Division of Labor in Society,* trans. G. Simpson. New York: Free Press.

The Economist. 1981. "America the Blood Bank." (October 17):87.

———. 1983. "Vein Hopes, Mainline Profits." (January 22):63–64.

Fox, Renée. 1988. *Essays in Medical Sociology: Journeys into the Field.* New Brunswick, NJ: Transaction Books.

"Frontline." 1993. "AIDS, Blood, and Politics." Boston: WGBH Educational Foundation and Health Quarterly.

Giese, Jo. 1987. "Sexual Landscape: On the Difficulty of Asking a Man to Wear a Condom." *Vogue* 177 (June):227+.

Goffman, Erving. 1959. *The Presentation of Self in Everyday Life.* New York: Anchor.

———. 1963. *Stigma: Notes on the Management of Spoiled Identity.* Englewood Cliffs, NJ: Prentice-Hall.

Grmek, Mirkod. 1990. *History of AIDS: Emergence and Origin of a Modern Pandemic,* trans. R. C. Maulitz and J. Duffin. Princeton, NJ: Princeton University Press.

Grover, Jan Zita. 1987. "AIDS: Keywords." *October* 43:17–30.

Halberstam, David. 1986. *The Reckoning.* New York: Morrow.

Hiatt, Fred. 1988. "Tainted U.S. Blood Blamed for AIDS' Spread in Japan." *The Washington Post* (June 23):A29.

Hilts, Philip J. 1988. "Dispelling Myths About AIDS in Africa." *Africa Report* 33:27–31.

Hunt, Charles W. 1989. "Migrant Labor and Sexually Transmitted Diseases: AIDS in Africa." *Journal of Health and Social Behavior* 30: 353–73.

Hurley, Peter, and Glenn Pinder. 1992. "Ethics, Social Forces, and Politics in AIDS-Related Research: Experience in Planning and Implementing a Household HIV Seroprevalence Survey." *The Milbank Quarterly* 70(4):605–28.

Irwin, Kathleen. 1991. "Knowledge, Attitudes and Beliefs About HIV Infection and AIDS Among Healthy Factory Workers and Their Wives, Kinshasa, Zaire." *Social Science and Medicine* 32(8):917–30.

Johnson, Diane, and John F. Murray, M.D. 1988. "AIDS Without End." *The New York Review of Books* (August 18):57–63.

Kaptchuk, Ted, and Michael Croucher, with the BBC. 1986. *The Healing Arts: Exploring the Medical Ways of the World.* New York: Summit.

Kerr, Dianne L. 1990. "AIDS Update: Ryan White's Death." *Journal of School Health* 60(5):237–38.

Kolata, Gina. 1989. "AIDS Test May Fail to Detect Virus for Years, Study Finds." *The New York Times* (June 1):Y1.

Kornfield, Ruth. 1986. "Dr., Teacher, or Comforter?: Medical Consultation in a Zairian Pediatrics Clinic." *Culture, Medicine and Psychiatry* 10:367–87.

Kramer, Reed. 1993. "Ties That Bind: Pressure Points Considered to Back Change in Zaire." *Africa News* (March 8–21):2.

Kramer, Staci D. 1988. "The Media and AIDS." *Editor and Publisher* 121:10–11, 43.

Krause, Richard. 1993. Quoted on p. xii in *A Dancing Matrix: Voyage Along the Viral Frontier,* by Robin Marantz Henig. New York: Knopf.

Lasker, Judith N. 1977. "The Role of Health Services in Colonial Rule: The Case of the Ivory Coast." *Culture, Medicine and Psychiatry* 1: 277–97.

Lippmann, Walter. 1976. "The World Outside and the Pictures in Our Heads." Pp. 174–81 in *Drama in Life: The Uses of Communication in Society,* ed. J. E. Combs and M. W. Mansfield. New York: Hastings House.

Liversidge, Anthony. 1993. "Heresy!: 3 Modern Galileos." *Omni* (June):43–51.

Mahler, Halfdan. 1989. Quoted on p. 91 in *AIDS and Its Metaphors,* by Susan Sontag. New York: Farrar, Straus & Giroux.

Mannheim, Karl. 1952. "The Problem of Generations." Pp. 276–322 in *Essays on the Sociology of Knowledge,* ed. P. Kecskemeti. New York: Oxford University Press.

McNeill, William H. 1976. *Plagues and People.* New York: Anchor.

Meltzer, Milton. 1960. *Mark Twain: A Pictorial Biography.* New York: Bonanza.

Merton, Robert K. 1957. *Social Theory and Social Structure.* Glencoe, IL: Free Press.

Noble, Kenneth B. 1989. "More Zaire AIDS Cases Show Less Underreporting." *The New York Times* (December 26):J4.

The Panos Institute. 1989. *AIDS and the Third World.* Philadelphia: New Society.

Parsons, Talcott. 1975. "The Sick Role and the Role of the Physician Reconsidered." *Milbank Memorial Fund Quarterly: Health and Society* 53(1):257–78.

Peretz, S. Michael. 1984. "Providing Drugs to the Third World: An Industry View." *Multinational Business* 84 (Spring):20–30.

Postman, Neil. 1985. *Amusing Ourselves to Death.* New York: Penguin.

Shilts, Randy. 1987. *And the Band Played On: Politics, People, and the AIDS Epidemic.* New York: St. Martin's Press.

Sontag, Susan. 1989. *AIDS and Its Metaphors.* New York: Farrar, Straus & Giroux.

Stolberg, Sheryl. 1992. "New AIDS Definition to Increase Tally." *Los Angeles Times* (December 31):A1+.

Swenson, Robert M. 1988. "Plagues, History, and AIDS." *The American Scholar* 57:183–200.

Thomas, William I., and Dorothy Swain Thomas. [1928] 1970. *The Child in America*. New York: Johnson.

Tuchman, Barbara W. 1981. *Practicing History: Selected Essays*. New York: Ballantine Books.

Turnbull, Colin M. 1961. *The Forest People*. New York: Simon & Schuster.

———. 1962. *The Lonely African*. New York: Simon & Schuster.

———. 1965. *Wayward Servants*. New York: Doubleday.

———. 1983. *The Human Cycle*. New York: Simon & Schuster.

U.S. Bureau for Refugee Programs. 1988. *World Refugee Report*. Washington, DC: U.S. Government Printing Office.

U.S. Bureau of the Census. 1992a. *U.S. Exports and General Imports by Hormonized Commodity by Country* (Report no. FT947/91-A). Washington, DC: U.S. Government Printing Office.

———. 1992b. *Statistical Abstract of the United States*, 112th ed. Washington, DC: U.S. Government Printing Office.

———. 1992. "AIDS Knowledge and the Attitudes for January–March 1991: Provisional Data from the National Health Interview Survey." *Advance Data* No. 216 (August 21). Washington, DC: U.S. Government Printing Office.

U.S. General Accounting Office. 1987. *AIDS: Information of Global Dimensions and Possible Impacts*. Washington, DC: U.S. Government Printing Office.

Watson, William. 1970. "Migrant Labor and Detribalization." Pp. 38–48 in *Black Africa: Its Peoples and Their Cultures Today*, ed. J. Middleton. London: Collier-Macmillan.

Whitaker, Jennifer Seymour. 1988. *How Can Africa Survive?* New York: Harper & Row.

Chapter 8

Abercrombie, Nicholas, Stephen Hill, and Bryan S. Turner. 1988. *The Penguin Dictionary of Sociology*. New York: Penguin.

Aldrich, Howard E., and Peter V. Marsden. 1988. "Environments and Organizations." Pp. 361–92 in *Handbook of Sociology*, ed. N. J. Smelser. Newbury Park, CA: Sage.

Barnet, Richard J., and Ronald E. Müller. 1974. *Global Reach: The Power of the Multinational Corporations*. New York: Simon & Schuster.

Blau, Peter M. 1974. *On the Nature of Organizations*. New York: Wiley.

Blau, Peter M., and Richard A. Schoenherr. 1973. *The Structure of Organizations*. White Plains, NY: Longman.

Castleman, Barry. 1986. Quoted in "The Dilemmas of Advanced Technology for the Third World," by Rashid A. Shaikh. *Technology Review* 89 (April):62.

Chenevière, Alain. 1987. *Vanishing Tribes*. Garden City, NY: Doubleday.

Clark, Andrew. 1993. "Learning the Rules of Global Citizenship: Transnationals Need to Meet Their Challenges, or Be Overwhelmed by Them." *The World Paper* (February):5.

Coser, Lewis A. 1977. *Masters of Sociological Thought*. New York: Harcourt Brace.

Crossette, Barbara. 1989. "New Delhi Prepares Attempt to Control Pervasive Pollution." *The New York Times* (July 4):Y24.

Engler, Robert. 1985. "Many Bhopals: Technology Out of Control." *The Nation* 240 (April): 488–500.

Freund, Julien. 1968. *The Sociology of Max Weber*. New York: Random House.

Keller, George M. 1986. "International Business and the National Interest." *Vital Speeches of the Day* (December 1):124–28.

Kennedy, Paul. 1993. *Preparing for the Twenty-First Century*. New York: Random House.

Khan, Rahat Nabi. 1986. "Multinational Companies and the World Economy: Economic and Technological Impact." *Impact of Science on Society* 36(141):15–25.

Lengermann, Patricia M. 1974. *Definitions of Sociology: A Historical Approach*. Columbus, OH: Merrill.

Lepkowski, Wil. 1985. "Chemical Safety in Developing Countries: The Lessons of Bhopal." *Chemical and Engineering News* 63:9–14.

McIntosh-Fletcher, W. Thomas. 1990. "When Two Cultures Meet." *Twin Plant News: The Magazine of the Maquiladora Industry* 5 (April): 32–33.

Michels, Robert. 1962. *Political Parties*, trans. E. Paul and C. Paul. New York: Dover.

Moskowitz, Milton. 1987. *The Global Marketplace.* New York: Macmillan.

National Public Radio. "Morning Edition." 1990. "Global Corporations" (tape/cassette no. 900820). (August 20).

The New York Times. 1989. "Rain Forest Worth More If Uncut, Study Says" (July 4):Y24.

Reich, Robert B. 1988. "Corporation and Nation." *The Atlantic Monthly* (May):76–81.

———. 1990. Quoted on p. 84 in "Calhoun County Goes Global," by Jan Bowermaster. *The New York Times Magazine* (December 2):58+.

Sekulic, Dusko. 1978. "Approaches to the Study of Informal Organization." *Sociologija* 20(1):27–43.

Shabecoff, Philip. 1985. "Tangled Rules on Toxic Hazards Hamper Efforts to Protect Public." *The New York Times* (November 27):A1+.

Snow, Charles P. 1961. *Science and Government.* Cambridge, MA: Harvard University Press.

Standke, Klaus-Heinrich. 1986. "Technology Assessment: An Essentially Political Process." *Impact of Science on Society* 36(141):65–76.

U.S. General Accounting Office. 1978. *U.S. Foreign Relations and Multinational Corporations: What's the Connection?* Washington, DC: U.S. Government Printing Office.

Veblen, Thorstein. 1933. *The Engineers and the Price System.* New York: Viking Press.

Wald, Matthew L. 1990. "Where All That Gas Goes: Drivers' Thirst for Power." *The New York Times* (November 21):A1, C17.

Weber, Max. 1947. *The Theory of Social and Economic Organization,* ed. and trans. A. M. Henderson and T. Parsons. New York: Macmillan.

Wexler, Mark N. 1989. "Learning from Bhopal." *The Midwest Quarterly* 31(1):106–29.

Young, T. R. 1975. "Karl Marx and Alienation: The Contributions of Karl Marx to Social Psychology." *Humboldt Journal of Social Relations* 2(2):26–33.

Zuboff, Shoshana. 1988. *In the Age of the Smart Machine: The Future of Work and Power.* New York: Basic Books.

Chapter 9

Becker, Howard S. 1963. *Outsiders: Studies in the Sociology of Deviance.* New York: Free Press.

———. 1973. "Labelling Theory Reconsidered." Pp. 177–212 in *Outsiders: Studies in the Sociology of Deviance.* New York: Free Press.

Belkin, Lisa. 1990. "Airport Anti-Drug Nets Snare Many People Fitting 'Profiles.'" *The New York Times* (March 20):A1+.

Best, Joel. 1989. *Images of Issues: Typifying Contemporary Social Problems.* New York: Aldine de Gruyter.

Chambliss, William. 1974. "The State, the Law, and the Definition of Behavior as Criminal or Delinquent." Pp. 7–24 in *Handbook of Criminology,* ed. D. Glaser. Indianapolis, IN: Bobbs-Merrill.

Collins, Randall. 1982. *Sociological Insight: An Introduction to Nonobvious Sociology.* New York: Oxford University Press.

Cowell, Alan. 1992. "Strike Hits Tobacco Industry, and 13 Million Italians Suffer." *The New York Times* (November 19):A5.

Durkheim, Emile. [1901] 1982. *The Rules of Sociological Method and Selected Texts on Sociology and Its Method,* ed. S. Lukes, trans. W. D. Halls. New York: Free Press.

Erikson, Kai T. 1966. *Wayward Puritans.* New York: Wiley.

Fox, James Alan, and Jack Levin. 1990. "Inside the Mind of Charles Stuart." *Boston Magazine* (April):66–70.

Gould, Stephen Jay. 1990. "Taxonomy as Politics: The Harm of False Classification." *Dissent* (Winter):73–78.

Henriques, Diana B. 1993. "Great Men and Tiny Bubbles: For God, Country and Coca-Cola." *The New York Times Book Review* (May 23):13.

Kitsuse, John I. 1962. "Societal Reaction to Deviant Behavior: Problems of Theory and Method." *Social Problems* 9 (Winter):247–56.

Kometani, Foumiko. 1987. "Pictures from Their Nightmare." *The New York Times Book Review* (July 19):9–10.

Lemert, Edwin M. 1951. *Social Pathology.* New York: McGraw-Hill.

Merton, Robert K. 1957. *Social Theory and Social Structure.* Glencoe, IL: Free Press.

Monkerud, Don. 1990. "Blurring the Lines: Androgyny on Trial." *Omni* (October):81–86+.

National Council for Crime Prevention in Sweden. 1985. *Crime and Criminal Policy in Sweden* (Report no. 19). Stockholm: Liber Distribution.

Ramos, Francisco Martins. 1993. "My American Glasses." Pp. 1–10 in *Distant Mirrors: America as a Foreign Culture,* by Philip R. DeVita and

James D. Armstrong. Belmont, CA: Wadsworth.

Reinarman, Craig, and Harry G. Levine. 1989. "The Crack Attack: Politics and Media in America's Latest Drug Scare." Pp. 115–38 in *Images of Issues: Typifying Contemporary Social Problems*, ed. J. Best. New York: Aldine de Gruyter.

Simmons, J. L., with Hazel Chambers. 1965. "Public Stereotypes of Deviants." *Social Problems* 3(2):223–32.

Spector, Malcolm, and J. I. Kitsuse. 1977. *Constructing Social Problems*. Menlo Park, CA: Cummings.

Sutherland, Edwin H., and Donald R. Cressey. 1978. *Principles of Criminology*, 10th ed. Philadelphia: Lippincott.

Tannenbaum, Frank. 1938. *Crime and the Community*. New York: Ginn.

Chapter 10

Angelou, Maya. 1987. "Intra-Racism." Interview on the "Oprah Winfrey Show" (Journal Graphics transcript no. W172):2.

Berreman, Gerald D. 1972. "Race, Caste, and Other Invidious Distinctions in Social Stratification." *Race* 13(4):385–414.

Boudon, Raymond, and François Bourricaud. 1989. *A Critical Dictionary of Sociology*, sel. and trans. P. Hamilton. Chicago: University of Chicago Press.

Chass, Murray. 1992. "A Zillionaire at the Bat." *International Herald Tribune* (February 28):16.

———. 1993. "25 Men on a Team and 7 Figures Per Man." *The New York Times* (April 11):S4.

Coser, Lewis A. 1977. *Masters of Sociological Thought*, 2nd ed., ed. R. K. Merton. New York: Harcourt Brace Jovanovich.

Davis, Kingsley, and Wilbert E. Moore. 1945. "Some Principles of Stratification." Pp. 413–25 in *Sociological Theory: A Book of Readings*, ed. L. A. Coser and B. Rosenberg. New York: Macmillan.

Eiseley, Loren. 1990. "Man: Prejudice and Personal Choice." Pp. 640–943 in *The Random House Encyclopedia*, 3rd ed. New York: Random House.

Finnegan, William. 1986. *Crossing the Line: A Year in the Land of Apartheid*. New York: Harper & Row.

Franklin, John Hope. 1990. Quoted in "That's History, Not Black History," by Mark Mcgurl. *The New York Times Book Review* (June 3):13.

"The Freedom Charter." 1990. From *One Nation, One Country*, by Nelson Mandela. *The Phelps-Stokes Fund* 4 (May):47–51.

"Frontline." 1985. "A Class Divided" (transcript no. 309). Boston: WGBH Educational Foundation.

Gerth, Hans, and C. Wright Mills. 1954. *Character and Social Structure: The Psychology of Social Institutions*. London: Routledge & Kegan Paul.

Jencks, Christopher. 1990. Quoted in "The Rise of the 'Hyperpoor,'" by David Whitman. *U.S. News & World Report* (October 15):40–42.

Lee, Spike. 1989. Quoted in "He's Got to Have It His Way," by Jeanne McDowell. *Time* (July 17): 92–93.

Lock, Margaret. 1993. "The Concept of Race: An Ideological Construct." *Transcultural Psychiatric Research Review* 30:203–27.

Lopez, Ian F. Haney. 1994. "Some Observations on Illusion, Fabrication, and Choice." *Harvard Civil Rights–Civil Liberties Law Review* 29: 1–61.

Loy, John W., and Joseph F. Elvogue. 1971. "Racial Segregation in American Sport." *International Review of Sport Sociology* 5:5–24.

Marx, Karl. 1909. *Capital: A Critique of Political Economy*, Vol. 3, ed. F. Engels, trans. E. Untermann. Chicago: Kerr.

———. [1895] 1976. *The Class Struggles in France 1848–1850*. New York: International.

Medoff, Marshall H. 1977. "Positional Segregation and Professional Baseball." *International Review of Sport Sociology* 12:49–56.

Merton, Robert K. 1958. *Social Theory and Social Structure*. New York: Free Press.

Mukherjee, Bharati. 1990. Quoted on pp. 3–10 in *Bill Moyers, A World of Ideas II: Public Opinion from Private Citizens*, ed. A. Tucher. New York: Doubleday.

O'Hare, William P., and Brenda Curry-White. 1992. "Demographer's Page: Is There a Rural Underclass?" *Population Today* 20(3):6–8.

Ridgeway, Cecilia. 1991. "The Social Construction of Status Value: Gender and Other Nominal Characteristics." *Social Forces* 70(2): 367–86.

Ross, Edward Alsworth. [1908] 1929. *Social Psychology: An Outline and Source Book*. New York: Macmillan.

Thomas, Isiah. 1987. Quoted in "The Coloring of Bird," by Ira Berkow. *The New York Times* (June 2):D27.

Tumin, Melvin M. 1953. "Some Principles of Stratification: A Critical Analysis." *American Sociological Review* 18:387–94.

Wacquant, Loic J. D. 1989. "The Ghetto, the State, and the New Capitalist Economy." *Dissent* (Fall):508–20.

Weber, Max. [1947] 1982. "Status Groups and Classes." Pp. 69–73 in *Classes, Power, and Conflict: Classical and Contemporary Debates*, ed. A. Giddens and D. Held. Los Angeles: University of California Press.

———. [1947] 1985. "Social Stratification and Class Structure." Pp. 573–76 in *Theories of Society: Foundations of Modern Sociological Theory*, ed. T. Parsons, E. Shils, K. D. Naegele, and J. R. Pitts. New York: Free Press.

Wilson, Francis, and Mamphela Ramphele. 1989. *Uprooting Poverty: The South African Challenge*. New York: Norton.

Wilson, William Julius. 1983. "The Urban Underclass: Inner-City Dislocations." *Society* 21: 80–86.

———. 1987. *The Truly Disadvantaged: The Inner City, the Underclass, and Public Policy*. Chicago: University of Chicago Press.

Wirth, Louis. [1945] 1985. "The Problem of Minority Groups." Pp. 309–15 in *Theories of Society: Foundations of Modern Sociological Theory*, ed. T. Parsons, E. Shils, K. D. Naegele, and J. R. Pitts. New York: Free Press.

Yeutter, Clayton. 1992. "When 'Fairness' Isn't Fair." *The New York Times* (March 24):A13.

Chapter 11

Alba, Richard D. 1992. "Ethnicity." Pp. 575–84 in *Encyclopedia of Sociology*, Vol. 2, ed. E. F. Borgatta and M. L. Borgatta. New York: Macmillan.

Anson, Robert Sam. 1987. *Best Intentions: The Education and Killing of Edmund Perry*. New York: Random House.

Atkins, Elizabeth. 1991. "For Many Mixed-Race Americans, Life Isn't Simply Black or White." *The New York Times* (June 5):B8.

Burns, Jim. 1993. "Fusion Cooking from the Pacific Rim." *American Visions* 8(5):36–38.

Bustamante, Jorge A. 1993. "Mexico-Bashing: A Case Where Words Could Hurt." *Los Angeles Times* (August 13):B7.

Carver, Terrell. 1987. *A Marx Dictionary*. Totowa, NJ: Barnes & Noble.

Castles, Steven, and Godula Kosack. 1985. *Immigrant Workers and Class Structure in Western Europe*, 2nd ed. New York: Oxford University Press.

Cornell, Stephen. 1990. "Land, Labour and Group Formation: Blacks and Indians in the United States." *Ethnic and Racial Studies* 13(3): 368–88.

Crapanzano, Vincent. 1985. *Waiting: The Whites of South Africa*. New York: Random House.

Davis, F. James. 1978. *Minority–Dominant Relations: A Sociological Analysis*. Arlington Heights, IL: AHM.

Emde, Helga. 1992. "An 'Occupation Baby' in Postwar Germany." Pp. 101–11 in *Showing Our Colors: Afro-German Women Speak Out*, ed. M. Opitz, K. Oguntoye, and D. Schultz. Amherst: University of Massachusetts Press.

Goffman, Erving. 1963. *Stigma: Notes on the Management of Spoiled Identity*. Englewood Cliffs, NJ: Prentice-Hall.

Gordon, Milton M. 1978. *Human Nature, Class, and Ethnicity*. New York: Oxford University Press.

Gould, Stephen Jay. 1981a. "The Politics of Census." *Natural History* 90(1):20–24.

———. 1981b. *The Mismeasure of Man*. New York: Norton.

Hacker, Andrew. 1992. *Two Nations: Black and White, Separate, Hostile, Unequal*. New York: Scribner.

Houston, Velin Hasu. 1991. "The Past Meets the Future: A Cultural Essay." *Amerasia Journal* 17(1):53–56.

Levin, Jack, and Jack McDevitt. 1993. *Hate Crimes: The Rising Tide of Bigotry and Bloodshed*. New York: Plenum.

Lieberman, Leonard. 1968. "The Debate over Race: A Study in the Sociology of Knowledge." *Phylon* 39 (Summer):127–41.

"MacNeil/Lehrer Newshour." 1991. New York: WNET.

Merton, Robert K. 1957. *Social Theory and Social Structure*. New York: Free Press.

———. 1976. "Discrimination and the American Creed." Pp. 189–216 in *Sociological Ambivalence and Other Essays*. New York: Free Press.

National Public Radio. "All Things Considered." 1990. "Prejudice Puzzle" (September 13).

Light, Ivan. 1990. Quoted in Daniel Goleman, "As Bias Crime Seems to Rise, Scientists Study Roots of Racism." *The New York Times* (May 29):B5+.

Ogbu, John U. 1990. "Minority Status and Literacy in Comparative Perspective." *Daedalus* 119(2):141–68.

Opitz, May. 1992a. "In Search of My Father (from a Conversation with Ellen Wiedenroth)." Pp. 172–77 in *Showing Our Colors: Afro-German Women Speak Out*, ed. M. Opitz, K. Oguntoye, and D. Schultz. Amherst: University of Massachusetts Press.

———. 1992b. "Recapitulation and Outlook." Pp. 228–33 in *Showing Our Colors: Afro-German Women Speak Out*, ed. M. Opitz, K. Oguntoye, and D. Schultz. Amherst: University of Massachusetts Press.

Rawley, James A. 1981. *The Transatlantic Slave Trade: A History*. New York: Norton.

Reynolds, Larry T. 1992. "A Retrospective on 'Race': The Career of a Concept." *Sociological Focus* 25(1):1–14.

Smokes, Saundra. 1992. "A Lifetime of Racial Rage Control Snaps with a Telephone Call." *The Cincinnati Post* (May 13):A14.

Stark, Evan. 1990. "The Myth of Black Violence." *The New York Times* (July 18):A21.

Steele, Shelby. 1990. "A Negative Vote on Affirmative Action." *The New York Times Magazine* (May 13):46–49+.

Walton, Anthony. 1989. "Willie Horton and Me." *The New York Times Magazine* (August 20):52+.

Webster, Yehundi O. 1992. *The Racialization of America*. New York: St. Martin's Press.

Wiedenroth, Ellen. 1992. "What Makes Me So Different in the Eyes of Others?" Pp. 165–77 in *Showing Our Colors: Afro-German Women Speak Out*, ed. M. Opitz, K. Oguntoye, and D. Schultz. Amherst: University of Massachusetts Press.

Wirth, Louis. 1945. "The Problem of Minority Groups." Pp. 347–72 in *The Science of Man*, ed. R. Linton. New York: Columbia University Press.

Chapter 12

Alderman, Craig, Jr. 1990. "10 February 1989 Memo for Mr. Peter Nelson." P. 108 in *Gays in Uniform: The Pentagon's Secret Reports*, ed. K. Dyer. Boston: Alyson.

Almquist, Elizabeth M. 1992. Review of "Gender, Family, and Economy: The Triple Overlap." *Contemporary Sociology* 21(3):331–32.

Anspach, Renee R. 1987. "Prognostic Conflict in Life-and-Death Decisions: The Organization as an Ecology of Knowledge." *Journal of Health and Social Behavior* 28(3):215–31.

Anthias, Floya, and Nira Yuval-Davis. 1989. "Introduction." Pp. 1–15 in *Woman-Nation-State*, ed. N. Yuval-Davis and F. Anthias. New York: St. Martin's Press.

Baumgartner-Papageorgiou, Alice. 1992. *My Daddy Might Have Loved Me: Student Perceptions of Differences Between Being Male and Being Female*. Denver: Institute for Equality in Education.

Bem, Sandra Lipsitz. 1993. *The Lenses of Gender: Transforming the Debate on Sexual Inequality*. Binghamton, NY: Vail-Ballou.

Boroughs, Don L. 1990. "Valley of the Doll?" *U.S. News & World Report* (December 3):56–59.

Burns, John F. 1992. "Canada Moves to Strengthen Sexual Assault Law." *The New York Times* (February 21):B9.

Collins, Randall. 1971. "A Conflict Theory of Sexual Stratification." *Social Problems* 19(1):3–21.

Cordes, Helen. 1992. "What a Doll! Barbie: Materialistic Bimbo or Feminist Trailblazer." *Utne Reader* (March/April):46, 50.

Davis, F. James. 1979. *Understanding Minority-Dominant Relations: Sociological Contributions*. Arlington Heights, IL: AHM.

Dewhurst, Christopher J., and Ronald R. Gordon. 1993. Quoted in "How Many Sexes Are There?" *The New York Times* (March 12):A15.

Doherty, Jake. 1993. "Conference to Focus on Plight of Wartime 'Comfort Women.'" *Los Angeles Times* (February 20):B3.

Fagot, Beverly, Richard Hagan, Mary Driver Leinbach, and Sandra Kronsberg. 1985. "Differential Reactions to Assertive and Communicative Acts of Toddler Boys and Girls." *Child Development* 56(6):1499–505.

Fairstein, Linda A. 1993. *Sexual Violence: Our War Against Rape*. New York: Morrow.

Fausto-Sterling, Anne. 1993. "How Many Sexes Are There?" *The New York Times* (March 12):A15.

Ferrante, Joan. 1988. "Biomedical Versus Cultural Constructions of Abnormality: The Case of Idiopathic Hirsutism in the United States." *Culture, Medicine, and Psychiatry* 12:219–38.

Garb, Frances. 1991. "Secondary Sex Characteristics." Pp. 326–27 in *Women's Studies Encyclopedia, Vol. 1: Views from the Sciences*, ed. H. Tierney. New York: Bedrick.

Geschwender, James A. 1992. "Ethgender, Women's Waged Labor, and Economic Mobility." *Social Problems* 39(1):1–16.

Grady, Denise. 1992. "Sex Test of Champions." *Discover* (June):78–82.

Hall, Edward T. 1959. *The Silent Language*. New York: Doubleday.

Henry, Jules. 1963. *Culture Against Man*. New York: Random House.

Hoon, Shim Jae. 1992. "Haunted by the Past." *Far Eastern Economic Review* (February 6):20.

Johnson, G. David, Gloria J. Palileo, and Norma B. Gray. 1992. "'Date Rape' on a Southern Campus: Reports from 1991." *SSR* 76(2):37–44.

Jones, Ann. 1994. "Change from Within." *The Women's Review of Books* 11(4):14.

Kifner, John. 1994. "Bosnian Serbs Order General Mobilization for 'Conclusion of War.'" *The New York Times* (February 1):A4.

Kolata, Gina. 1992. "Track Federation Urges End to Gene Test for Femaleness." *The New York Times* (February 12):A1, B1.

Komarovsky, Mirna. 1991. "Some Reflections on the Feminist Scholarship in Sociology." Pp. 1–25 in *Annual Review of Sociology*, Vol. 17, ed. W. R. Scott and J. Blake. Palo Alto, CA: Annual Reviews.

Koss, Mary P., Christine A. Gidycz, and Nadine Wisniewski. 1987. "The Scope of Rape: Incidence and Prevalence of Sexual Aggression and Victimization in a National Sample of Higher Education Students." *Journal of Consulting and Clinical Psychology* 55(2):162–70.

Lemonick, Michael D. 1992. "Genetic Tests Under Fire." *Time* (February 24):65.

Lewin, Tamar. 1993. "At Bases, Debate Rages over Impact of New Gay Policy." *The New York Times* (December 24):A1+.

Mills, Janet Lee. 1985. "Body Language Speaks Louder Than Words." *Horizons* (February): 8–12.

Morawski, Jill G. 1991. "Femininity." Pp. 136–39 in *Women's Studies Encyclopedia, Volume 1: Views from the Sciences*, ed. H. Tierney. New York: Bedrick.

Morgenson, Gretchen. 1991. "Barbie Does Budapest." *Forbes* (January 7):66–69.

Pion, Allison. 1993. "Accessorizing Ken." *Origins* (November):8.

Rank, Mark R. 1989. "Fertility Among Women on Welfare: Incidence and Determinants." *American Sociological Review* 54(4):296–304.

Roiphe, Katie. 1993. *The Morning After: Sex, Fear, and Feminism on Campus*. New York: Little, Brown.

Schaller, Jane Green, and Elena O. Nightingale. 1992. "Children and Childhoods: Hidden Casualties of War and Civil Unrest." *Journal of the American Medical Association* 268(5):642–44.

Segal, Lynne. 1990. *Slow Motion: Changing Masculinities, Changing Men*. London: Virago.

Shweder, Richard A. 1994. "What Do Men Want? A Reading List for the Male Identity Crisis." *The New York Times Book Review* (January 9):3+.

Solinger, Rickie. 1992. *Wake Up Little Susie: Single Pregnancy and Race Before Roe v. Wade*. New York: Routledge.

Sturdevant, Saundra Pollock, and Brenda Stoltzfus. 1992. *Let the Good Times Roll: Prostitution and the U.S. Military in Asia*. New York: New Press.

Tattersall, Ian. 1993. "Focus—All in the Family. (Homosapien Exhibit at New York's American Museum of Natural History)."

Tierney, Helen. 1991. "Gender/Sex." P. 153 in *Women's Studies Encyclopedia, Volume 1: Views from the Sciences*, ed. H. Tierney. New York: Bedrick.

U.S. Department of Commerce. 1989. *Child Support and Alimony, 1989*. Current Population Reports, Series P-60. Washington, DC: Government Printing Office.

U.S. Department of Defense. 1990. "DOD Directive 1332.14." P. 19 in *Gays in Uniform: The Pentagon's Secret Reports*, ed. K. Dyer. Boston: Alyson.

Williams, Lena. 1993. "Pregnant Teen-agers Are Outcasts No Longer." *The New York Times* (December 2):B1+.

Chapter 13

Adams, Bert. 1968. *Kinship in an Urban Setting*. Chicago: Markham.

Allan, Graham. 1977. "Sibling Solidarity." *Journal of Marriage and the Family* 9(1):177–83.

Behnam, Djamshid. 1990. "An International Inquiry into the Future of the Family: A UNESCO Project." *International Social Science Journal* 42:547–52.

Bell, Daniel. 1989. "The Third Technological Revolution." *Dissent* (Spring):164–76.

Berelson, Bernard. 1978. "Prospects and Programs for Fertility Reduction: What? Where?" *Population and Development Review* 4:579–616.

Bernardo, Felix M. 1967. "Kinship Interaction and Communications Among Space-Age Migrants." *Journal of Marriage and the Family* 29(3):541–54.

Boccaccio, Giovanni. [1353] 1984. "The Black Death." Pp. 728–40 in *The Norton Reader: An Anthology of Expository Prose*, 6th ed., ed. A. M. Eastman. New York: Norton.

Brown, Lester R. 1987. "Analyzing the Demographic Trap." *State of the World 1987: A Worldwatch Institute Report on Progress Toward a Sustainable Society*. New York: Norton.

Burke, B. Meredith. 1989. "Ceausescu's Main Victims: Women and Children." *The New York Times* (January 16):Y15.

Davis, Kingsley. 1984. "Wives and Work: The Sex Role Revolution and Its Consequences." *Population and Development Review* 10(3): 397–417.

Durning, Alan B. 1990. "Ending Poverty." Pp. 135–53 in *State of the World 1990: A Worldwatch Institute Report on Progress Toward a Sustainable Society*, ed. L. Starke. New York: Norton.

Dychtwald, Ken, and Joe Flower. 1989. *Age Wave: The Challenges and Opportunities of an Aging America*. Los Angeles: Tarcher.

Eckholm, Erik. 1990. "An Aging Nation Grapples with Caring for the Frail." *The New York Times* (March 27):A1+.

Glascock, Anthony P. 1982. "Decrepitude and Death Hastening: The Nature of Old Age in Third World Societies (Part I)." *Studies in Third World Societies* 22:43–66.

Goldenberg, Sheldon. 1987. *Thinking Sociologically*. Belmont, CA: Wadsworth.

Gutis, Philip S. 1989. "Family Redefines Itself, and Now the Law Follows." *The New York Times* (May 28):B1.

Johansson, S. Ryan. 1987. "Status Anxiety and Demographic Contraction of Privileged Populations." *Population and Development Review* 13(3):439–70.

Leigh, Geoffrey. 1982. "Kinship Interaction over the Family Life Span." *Journal of Marriage and the Family* 41(1):197–208.

Lengermann, Patricia M. 1974. *Definitions of Sociology: A Historical Approach*. Columbus, OH: Merrill.

Lewin, Tamar. 1990. "Strategies to Let Elderly Keep Some Control." *The New York Times* (March 28):A1+.

Light, Ivan. 1983. *Cities in World Perspective*. New York: Macmillan.

Litwak, Eugene. 1960. "Geographic Mobility and Extended Family Cohesion." *American Sociological Review* 25:385–94.

Malthus, Thomas R. [1798] 1965. *First Essay on Population*. New York: Augustus Kelley.

Olshansky, S. Jay, and A. Brian Ault. 1986. "The Fourth Stage of the Epidemiologic Transition: The Age of Delayed Degenerative Diseases." *The Milbank Quarterly* 64(3):355–91.

Omran, Abdel R. 1971. "The Epidemiologic Transition: A Theory of the Epidemiology of Population Change." *The Milbank Quarterly* 49(4): 509–38.

Parsons, Talcott. 1966. "The Kinship System of the Contemporary United States." Pp. 177–96 in *Essays in Sociological Theory*, rev. ed. New York: Free Press.

Pullman, Thomas W. 1992. "Population." Pp. 1499–1507 in *Encyclopedia of Sociology*, Vol. 3, ed. E. F. Borgatta and M. L. Borgatta. New York: Macmillan.

Riesman, David, with Nathan Glazer and Reuel Denney. 1977. *The Lonely Crowd: A Study of the Changing American Character*, abrg. ed. New Haven, CT: Yale University Press.

Rock, Andrea. 1990. "Can You Afford Your Kids?" *Money* (July):88–99.

Rusinow, Dennison. 1986. "Mega-Cities Today and Tomorrow: Is the Cup Half Full or Half Empty?" *UFSI Reports* 12.

Sagan, Carl. 1978. *Murmurs of Earth: The Voyager Interstellar Record*. New York: Random House.

Skidmore, Thomas E. 1993. "Bi-racial U.S.A. vs. Multi-racial Brazil: Is the Contrast Still Valid?" *Journal of Latin American Studies* 25:373–86.

Soldo, Beth J., and Emily M. Agree. 1988. "America's Elderly." *Population Bulletin* 43(3):5+.

Sorel, Nancy Caldwell. 1984. *Ever Since Eve: Personal Reflections on Childbirth*. New York: Oxford University Press.

Stockwell, Edward G., and H. Theodore Groat. 1984. *World Population: An Introduction to Demography*. New York: Franklin Watts.

Stone, Robyn, Gail Lee Cafferata, and Judith Sangl. 1987. "Caregivers of the Frail Elderly: A National Profile." *The Gerontologist* 27(5): 616–26.

Stub, Holger R. 1982. *The Social Consequences of Long Life*. Springfield, IL: Thomas.

Targ, Dena B. 1989. "Feminist Family Sociology: Some Reflections." *Sociological Focus* 22(3): 151–60.

United Nations. 1983. *World Population Trends and Policies: 1983 Monitoring Report,* Vol. 1. New York: United Nations.

U.S. Bureau of the Census. 1947. *Statistical Abstract of the United States, 1947.* Washington, DC: U.S. Government Printing Office.

———. 1989. *World Population Profile: 1989.* Washington, DC: U.S. Government Printing Office.

———. 1991. *World Population Profile: 1991.* Washington, DC: U.S. Government Printing Office.

———. 1993. *Statistical Abstract of the United States, 1993.* Washington, DC: U.S. Government Printing Office.

van de Kaa, Dirk J. 1987. "Europe's Second Demographic Transition." *Population Bulletin* 42(1): 1–59.

Watkins, Susan C., and Jane Menken. 1985. "Famines in Historical Perspective." *Population and Development Review* 11(4):647–75.

Zelditch, Morris. 1964. "Family, Marriage, and Kinship." Pp. 680–733 in *Handbook of Modern Sociology,* ed. R. E. L. Faris. Chicago: Rand McNally.

Chapter 14

Alpert, Bracha. 1991. "Students' Resistance in the Classroom." *Anthropology and Education Quarterly* 22(4):350–66.

Barrett, Michael J. 1990. "The Case for More School Days." *The Atlantic Monthly* (November):78–106.

Bettelheim, Bruno, and Karen Zeland. 1981. *On Learning to Read: The Child's Fascination with Meaning.* New York: Knopf.

Bloom, Benjamin S. 1981. *All Our Children Learning: A Primer for Parents, Teachers and Other Educators.* New York: McGraw-Hill.

The Book of the States, 1992–93 Edition. 1992. Lexington, KY: Council of State Governments.

Botstein, Leon. 1990. "Damaged Literacy: Illiteracies and American Democracy." *Daedalus* 119(2):55–84.

Boyer, Ernest. 1986. "Forum: How Not to Fix the Schools." *Harper's* (February):39–51.

Celis, William III. 1992. "A Texas-Size Battle to Teach Rich and Poor Alike." *The New York Times* (February 12):B6.

———. 1993a. "International Report Card Shows U.S. Schools Work." *The New York Times* (December 9):A1+.

———. 1993b. "Study Finds Rising Concentration of Black and Hispanic Students." *The New York Times* (December 14):A1+.

Cetron, Marvin. 1988. "Forum: Teach Our Children Well." *Omni* 10(11):14.

Clements, Marcelle. 1992. "Fear of Reading." *The New York Times* (May 18):A11.

Cohen, David K., and Barbara Neufeld. 1981. "The Failure of High Schools and the Progress of Education." *Daedalus* (Summer):69–89.

Coleman, James S. 1960. "The Adolescent Subculture and Academic Achievement." *American Journal of Sociology* 65:337–47.

———. 1966. *Equality of Educational Opportunity.* Washington, DC: U.S. Government Printing Office.

———. 1977. "Choice in American Education." Pp. 1–12 in *Parents, Teachers, and Children: Prospects for Choice in American Education.* San Francisco: Institute for Contemporary Studies.

Coleman, James S., John W. C. Johnstone, and Kurt Jonassohn. 1961. *The Adolescent Society.* New York: Free Press.

Csikszentmihalyi, Mihaly. 1990. "Literacy and Intrinsic Motivation." *Daedalus* 119(2):115–40.

Danner, Mark D. 1986. "Forum: How Not to Fix the Schools." *Harper's* (February):39–51.

Dorris, Michael. 1989. *The Broken Cord.* New York: Harper & Row.

Durkheim, Emile. 1961. "On the Learning of Discipline." Pp. 860–65 in *Theories of Society: Foundations of Modern Sociological Theory,* Vol. 2, ed. T. Parsons, E. Shils, K. D. Naegele, and J. R. Pitts. New York: Free Press.

———. 1968. *Education and Sociology,* trans. S. D. Fox. New York: Free Press.

Early, Margaret. 1987. Pp. 785–98 in *Streamers Workbook.* Chicago: Harcourt Brace Jovanovich.

Erickson, Frederick. 1984. "School Literacy, Reasoning, and Civility: An Anthropologist's Perspective." *Review of Educational Research* 54(4):525–46.

Foster, Jack D. 1991. "The Role of Accountability in Kentucky's Education Reform Act of 1990." *Education Leadership* (February):34–36.

Gardner, John W. 1984. *Excellence: Can We Be Equal and Excellent Too?* New York: Norton.

Gisi, Lynn Grover. 1985. "How States Are Reforming Public Education." *USA Today* 113(2478): 76–78.

Guzzardi, Walter, Jr. 1976. "The Uncertain Passage from College to Job." *Fortune* (January): 126–27, 168–72.

Hakim, Joy. 1993. Interview on National Public Radio, "Morning Edition" (June 2):4–7.

Hallinan, M. T. 1988. "Equality of Educational Opportunity." Pp. 249–68 in *Annual Review of Sociology*, Vol. 14, ed. W. R. Scott and J. Blake. Palo Alto, CA: Annual Reviews.

Henry, Jules. 1965. *Culture Against Man.* New York: Random House.

Hirsch, E. D., Jr. 1989. "The Primal Scene of Education." *The New York Review of Books* (March 2):29–35.

Horn, Miriam. 1987. "The Burgeoning Educational Underclass." *U.S. News & World Report* (May 18):66–67.

Lapointe, Archie E., Nancy A. Mead, and Gary W. Phillips. 1989. *A World of Differences: An International Assessment of Mathematics and Science.* Princeton, NJ: Educational Testing Service.

Lightfoot, Sara Lawrence. 1988. "Bill Moyers' World of Ideas" (transcript no. 123). New York: Public Affairs Television.

Los Angeles Times. 1993. "Gifted Students Found Unchallenged." (November 5):A39.

Luria, A. R. 1979. *The Making of Mind: A Personal Account of Soviet Psychology*, ed. M. Cole and S. Cole. Cambridge, MA: Harvard University Press.

Lynn, Richard. 1988. *Educational Achievement in Japan: Lessons for the West.* London: Macmillan.

Merton, Robert K. 1957. *Social Theory and Social Structure.* Glencoe, IL: Free Press.

The National Commission on Excellence in Education. 1983. *A Nation at Risk: The Imperative for Educational Reform.* Washington, DC: U.S. Government Printing Office.

National Endowment for the Humanities. 1991. *National Tests: What Other Countries Expect Their Students to Know.* Washington, DC: U.S. Government Printing Office.

National Science Board. 1991. *Science and Engineering Indicators 1991.* Washington, DC: U.S. Government Printing Office.

Oakes, Jeannie. 1985. *Keeping Track: How Schools Structure Inequality.* Binghamton, NY: Vail-Ballou.

———. 1986a. "Keeping Track, Part 1: The Policy and Practice of Curriculum Inequality." *Phi Delta Kappan* 67 (September):12–17.

———. 1986b. "Keeping Track, Part 2: Curriculum Inequality and School Reform." *Phi Delta Kappan* 67 (October):148–54.

O'Connor, John J. 1990. "Critic's Notebook: How TV Sends Mixed Messages About Education." *The New York Times* (September 13):B1.

Ouane, Adama. 1990. "National Languages and Mother Tongues." *UNESCO Courier* (July): 27–29.

Potter, J. Hasloch, and A. E. W. Sheard. 1918. *Catechizings for the Church and Sunday Schools*, 2nd series. London: Skeffington.

Ramirez, Francisco, and John W. Meyer. 1980. "Comparative Education: The Social Construction of the Modern World System." Pp. 369–99 in *Annual Review of Sociology*, Vol. 6, ed. A. Inkeles, N. J. Smelser, and R. H. Turner. Palo Alto, CA: Annual Reviews.

Reich, Robert. 1993. Quoted in "Skipping School," by Bettijane Levine. *Los Angeles Times* (July 3):E1+.

Remlinger, Connie, and Debra Vance. 1989. "Business Enlisting in War on Illiteracy." *The Kentucky Post.* Special Supplement on Literacy. (October 19):2.

Resnick, Daniel P. 1990. "Historical Perspectives on Literacy and Schooling." *Daedalus* 119(2): 15–32.

Resnick, Daniel P., and Lauren B. Resnick. 1989. "Varieties of Literacy." Pp. 171–206 in *Social History and Issues in Human Consciousness*, ed. A. E. Barnes and P. N. Stearns. New York: New York University Press.

Resnick, Lauren B. 1990. "Literacy in School and Out." *Daedalus* 119(2):169–86.

Richardson, Lynda. 1994. "More Schools Are Trying to Write Textbooks Out of the Curriculum." *The New York Times* (January 31):A1+.

Rohlen, Thomas P. 1986. "Japanese Education: If They Can Do It, Should We?" *The American Scholar* 55:29–43.

Rush, Benjamin. 1966. Quoted in "Forming the National Character," by David Tyack. *Harvard Educational Review* 36:29–41.

Sanchez, Claudio. 1993. Interview on National Public Radio, "Morning Edition" (December 8): 11–13.

Schlack, Lawrence B. 1992. "Letter: School Days." *World Monitor* (November):5.

Shelley, Kristina J. 1992. "The Future of Jobs for College Graduates." *Monthly Labor Review* (July):13–21.

Sowell, Thomas. 1981. *Ethnic America: A History.* New York: Basic Books.

Stevenson, Harold W., Shin-ying Lee, and James W. Stigler. 1986. "Mathematics Achievement of Chinese, Japanese, and American Children." *Science* 231.693–99.

Thomas, William I., and Dorothy Swain Thomas. [1928] 1970. *The Child in America.* New York: Johnson.

Thomson, Scott D. 1989. "Report Card USA: How Much Do Americans Value Schooling?" *NASSP Bulletin* 73(519):51–67.

Tyack, David, and Elisabeth Hansot. 1981. "Conflict and Consensus in American Public Education." *Daedalus* (Summer):1–43.

Tyler, Ralph. 1974. Quoted in "Divergent Views on the Schools: Some Optimism Justified," by Alan C. Purves. *The New York Times* (January 16):C74.

U.S. Bureau of the Census. 1982. *Illiteracy: 1982.* Washington, DC: U.S. Government Printing Office.

———. 1993. *Statistical Abstract of the United States*, 113th ed. Washington, DC: U.S. Government Printing Office.

U.S. Department of Education. 1992a. *International Mathematics and Science Assessments: What Have We Learned?* Washington, DC: U.S. Government Printing Office.

———. 1992b. *International Education Comparisons.* Washington, DC: U.S. Government Printing Office.

———. 1993a. *Occupational and Educational Outcomes of Recent College Graduates 1 Year After Graduation: 1991.* Washington, DC: U.S. Government Printing Office.

———. 1993b. *Adult Literacy in America: A First Look at the Results of the National Literacy Survey.* Washington, DC: U.S. Government Printing Office.

Webster, Noah. 1966. Quoted in "Forming the National Character," by David Tyack. *Harvard Educational Review* 36:29–41.

White, Robert. 1989. "Building a Better Classroom." *The Cincinnati Post* (September 18):A1+.

Chapter 15

Abercrombie, Nicholas, and Bryan S. Turner. 1978. "The Dominant Ideology Thesis." *British Journal of Sociology* 29(2):149–70.

Alston, William P. 1972. "Religion." Pp. 140–45 in *The Encyclopedia of Philosophy*, Vol. 7, ed. P. Edwards. New York: Macmillan.

Aron, R. 1969. Quoted on p. 204 in *The Sociology of Max Weber*, by Julien Freund. New York: Random House.

Berger, Peter L. 1967. *The Sacred Canopy: Elements of a Sociological Theory of Religion.* New York: Doubleday.

Bush, George H. 1991. State of the Union Address by the President of the United States (January 29).

Caplan, Lionel. 1987. "Introduction: Popular Conceptions of Fundamentalism." Pp. 1–24 in *Studies in Religious Fundamentalism*, ed. L. Caplan. Albany: State University of New York Press.

Christianity Today. 1986. "Letters" (October 17):6.

Coles, Robert. 1990. *The Spiritual Life of Children.* Boston: Houghton Mifflin.

Durkheim, Emile. [1915] 1964. *The Elementary Forms of the Religious Life*, 5th ed., trans. J. W. Swain. New York: Macmillan.

———. 1951. *Suicide: A Study in Sociology*, trans. J. A. Spaulding and G. Simpson. New York: Free Press.

———. 1984. Quoted in *Durkheim's Sociology of Religion*, by W. S. F. Pickering. London: Routledge & Kegan Paul.

Ebersole, Luke. 1967. "Sacred." P. 613 in *A Dictionary of the Social Sciences*, ed. J. Gould and W. L. Kolb. New York: UNESCO.

Esposito, John L. 1986. "Islam in the Politics of the Middle East." *Current History* (February):53–57, 81.

———. 1992. *The Islamic Threat: Myth or Reality?* New York: Oxford University Press.

Gallup, George, Jr., and Jim Castelli. 1989. *The People's Religion: American Faith in the 90s.* New York: Macmillan.

Haddad, Yvonne. 1991. Interview with Bill Moyers on "Images of God in the Arab World." PBS.

Hammond, Phillip E. 1976. "The Sociology of American Civil Religion: A Bibliographic Essay." *Sociological Analysis* 37(2):169–82.

Hourani, Albert. 1991. *A History of the Arab Peoples*. Cambridge, MA: Belknap Press.

Lechner, Frank J. 1989. "Fundamentalism Revisited." *Society* (January/February):51–59.

Lifton, Robert Jay. 1969. *Boundaries: Psychological Man in Revolution*. New York: Touchstone Books.

Lifton, Robert Jay, and Nicholas Humphrey. 1984. *In a Dark Time: Images of Survival*. Cambridge, MA: Harvard University Press.

Mead, George Herbert. 1940. *Mind, Self and Society*, 3rd ed. Chicago: University of Chicago Press.

Merton, Robert K. 1957. *Social Theory and Social Structure*. Glencoe, IL: Free Press.

National Public Radio. 1984. "Black Islam." *The World of Islam*. Washington, DC: NPR.

The New York Times. 1991. "Iraqi Message: 'Duty' Fulfilled." (February 26):Y1.

Nottingham, Elizabeth K. 1971. *Religion: A Sociological View*. New York: Random House.

Pickering, W. S. F. 1984. *Durkheim's Sociology of Religion*. London: Routledge & Kegan Paul.

Robertson, Roland. 1987. "Economics and Religion." Pp. 1–11 in *The Encyclopedia of Religion*, ed. M. Eliade. New York: Macmillan.

Smart, Ninian. 1976. *The Religious Experience of Mankind*. New York: Scribner.

Stark, Rodney, and William S. Bainbridge. 1985. *The Future of Religion: Secularization, Revival and Cult Formation*. Berkeley, CA: University of California Press.

Stavenhagen, Rodolfo. 1991. "Ethnic Conflicts and Their Impact on International Society." *International Social Science Journal* (February):117–32.

Turner, Bryan S. 1974. *Weber and Islam: A Critical Study*. Boston: Routledge & Kegan Paul.

Van Doren, Charles L. 1991. *A History of Knowledge: Past, Present, and Future*. New York: Carol.

Watchtower Bible and Tract Society. 1987. "Life in a Peaceful New World." Brooklyn: Watchtower.

Weber, Max. 1922. *The Sociology of Religion*, trans. E. Fischoff. Boston: Beacon Press.

———. 1958. *The Protestant Ethic and the Spirit of Capitalism*, 5th ed., trans. T. Parsons. New York: Scribner.

The World Almanac and Book of Facts 1991. 1990. New York: Pharos Books.

The World Treasury of Modern Religious Thought. 1990. Ed. Jaroslav Pelikan and Clifton Fadiman. Boston: Little, Brown.

Yinger, J. Milton. 1971. *The Scientific Study of Religion*. New York: Macmillan.

Zangwill, O. L. 1987. "Isolation Experiments." Pp. 393–94 in *The Oxford Companion to the Mind*, ed. R. L. Gregory. New York: Oxford University Press.

Chapter 16

Ash, Timothy Garton. 1989. *The Uses of Adversity: Essays on the Fate of Central Europe*. New York: Random House.

Barnet, Richard J. 1990. "Reflections: Defining the Moment." *The New Yorker* (July):45–60.

Boudon, Raymond, and François Bourricaud. 1989. *A Critical Dictionary of Sociology*, sel. and trans. P. Hamilton. Chicago: University of Chicago Press.

Broad, William J. 1991. "Nuclear Designers from East and West Plan Bomb Disposal." *The New York Times* (December 17):B5+.

Calonius, Erik. 1991. "Smart Moves by Quality Champs." *Fortune* (Spring/Summer):24–28.

Colihan, Jane, and Robert J. T. Joy. 1984. "Military Medicine." *American Heritage* (October/November):65.

Coser, Lewis A. 1973. "Social Conflict and the Theory of Social Change." Pp. 114–22 in *Social Change: Sources, Patterns, and Consequences*, ed. E. Etzioni-Halevy and A. Etzioni. New York: Basic Books.

Currie, Elliott, and Jerome H. Skolnick. 1988. *America's Problems: Social Issues and Public Policy*, 2nd ed. Boston: Little, Brown.

Dahrendorf, Ralf. 1973. "Toward a Theory of Social Conflict." Pp. 100–13 in *Social Change: Sources, Patterns, and Consequences*, 2nd ed., ed. E. Etzioni-Halevy and A. Etzioni. New York: Basic Books.

Erikson, Kai T. 1971. "Sociology and the Historical Perspective." Pp. 61–77 in *The Sociology of the Future*, ed. W. Bell and J. A. Mau. New York: Russell Sage Foundation.

Galbraith, John K. 1958. *The Affluent Society*. Boston: Houghton Mifflin.

Gould, Stephen Jay. 1990. Interview on "MacNeil/Lehrer Newshour" (January 1). New York: WNET.

Guttman, Robert J. 1990. "Interview with John Georgas: The Coca-Cola Company." *Europe* (July/August):16.

Hacker, Andrew. 1971. "Power to Do What?" Pp. 134–46 in *The New Sociology: Essays in Social Science and Social Theory in Honor of C. Wright Mills*, ed. I. L. Horowitz. New York: Oxford University Press.

Halberstam, David. 1986. *The Reckoning.* New York: Morrow.

Holusha, John. 1989. "Eastern Europe: Its Lure and Hurdles." *The New York Times* (December 18): Y25+.

Kuhn, Thomas S. 1975. *The Structure of Scientific Revolutions.* Chicago: University of Chicago Press.

Life. 1945. "Advertisement: Millions of Military Telephones." (August 10):3.

Mandelbaum, Maurice. 1977. *The Anatomy of Historical Knowledge.* Baltimore: Johns Hopkins University Press.

Martel, Leon. 1986. *Mastering Change: The Key to Business Success.* New York: Simon & Schuster.

Marx, Karl. [1881] 1965. "The Class Struggle." Pp. 529–35 in *Theories of Society*, ed. T. Parsons, E. Shils, K. D. Naegele, and J. R. Pitts. New York: Free Press.

Michnik, Adam. 1990. "The Moral and Spiritual Origins of Solidarity." Pp. 239–50 in *Without Force or Lies: Voices from the Revolution of Central Europe in 1989–90*, ed. W. M. Brinton and A. Rinzler. San Francisco: Mercury House.

Miller, Judith. 1991. "Displaced in the Gulf War: 5 Million Refugees." *The New York Times* (June 16):E3.

Mills, C. Wright. 1959. *The Sociological Imagination.* New York: Oxford University Press.

———. 1963. "The Structure of Power in American Society." Pp. 23–38 in *Power, Politics and People: The Collected Essays of C. Wright Mills*, ed. I. L. Horowitz. New York: Oxford University Press.

———. 1973. "The Sources of Societal Power." Pp. 123–30 in *Social Change: Sources, Patterns, and Consequences*, 2nd ed., ed. E. Etzioni-Halevy and A. Etzioni. New York: Basic Books.

Norman, Donald A. 1988. Quoted in "Management's High-Tech Challenge." *Editorial Research Report* (September 30):482–91.

Ogburn, William F. 1968. "Cultural Lag as Theory." Pp. 86–95 in *William F. Ogburn on Culture and Social Change*, 2nd ed., ed. O. D. Duncan. Chicago: University of Chicago Press.

Oppenheimer, Robert. 1986. Quoted on p. 11 of *The Making of the Bomb*, by Richard Rhodes. New York: Touchstone.

O'Sullivan, Anthony. 1990. "Eastern Europe." *Europe* (September):21–22.

Rawe, Dick. 1991. "P&G Expands Entry in Eastern Europe." *The Cincinnati Post* (June 20):B10.

Reich, Jens. 1989. Quoted on p. 20 of "People of the Year." *Newsweek* (December 25):18–25.

Schneider, Keith. 1990. "Uranium Miners Inherit Dispute's Sad Legacy." *The New York Times* (January 9):Y1+.

Stevenson, Richard W. 1991. "Northrop Settles Workers' Suit on False Missile Tests for $8 Million." *The New York Times* (June 25):A7.

Van Evera, Stephen. 1990. "The Case Against Intervention." *The Atlantic Monthly* (July): 72–80.

Wald, Matthew L. 1989. "Finding a Burial Place for Nuclear Wastes Grows More Difficult." *The New York Times* (December 5):Y19+.

Wallerstein, Immanuel. 1984. *The Politics of the World-Economy: The States, the Movements and the Civilizations.* New York: Cambridge University Press.

Weber, Max. 1947. *The Theory of Social and Economic Organization*, trans. A. M. Henderson and T. Parsons. Glencoe, IL: Free Press.

White, Leslie A. 1949. *The Science of Culture: A Study of Man and Civilization.* New York: Grove Press.

Key Concepts

Bem, Sandra Lipsitz. 1993. *The Lenses of Gender: Transforming the Debate on Sexual Inequality.* Binghamton, NY: Vail-Ballou.

Collins, Randall. 1971. "A Conflict Theory of Sexual Stratification." *Social Problems* 19(1): 3–21.

Coser, Lewis A. 1977. *Masters of Sociological Thought*, 2nd ed., ed. R. K. Merton. New York: Harcourt Brace Jovanovich.

Gerth, Hans, and C. Wright Mills. 1954. *Character and Social Structure: The Psychology of Social Institutions.* London: Routledge & Kegan Paul.

Goffman, Erving. 1963. *Stigma: Notes on the Management of Spoiled Identity.* Englewood Cliffs, NJ: Prentice-Hall.

Hammond, Phillip E. 1976. "The Sociology of American Civil Religion: A Bibliographic Essay." *Sociological Analysis* 37(2):169–82.

Ridgeway, Cecilia. 1991. "The Social Construction of Status Value: Gender and Other Nominal Characteristics." *Social Forces* 70(2): 367–86.

Robertson, Roland. 1987. "Economics and Religion." Pp. 1–11 in *The Encyclopedia of Religion*, ed. M. Eliade. New York: Macmillan.

Sutherland, Edwin H., and Donald R. Cressey. 1978. *Principles of Criminology*, 10th ed. Philadelphia: Lippincott.

Wilson, William Julius. 1983. "The Urban Underclass: Inner-City Dislocations." *Society* 21: 80–86.

CREDITS

The numbers in parentheses after each entry are the page numbers on which the material appears.

From "Faculty Brat: A Memoir," by Emily Fox Gordon, *Boulevard* 10 (1–2):1–17. Copyright © 1994 by Opojaz, Inc. Reprinted by permission. (70–71)

From *Culture Against Man* by Jules Henry. Copyright © 1963 Random House, Inc. Reprinted by permission of the publisher. (306–307)

From *The Making of Mind: A Personal Account of Soviet Psychology* by A. R. Luria, Cambridge, MA: Harvard University Press. Copyright © 1979 by the President and Fellows of Harvard College. Reprinted by permission of the publisher. (314–315)

From *And the Band Played On: Politics, People and the AIDS Epidemic* by Randy Shilts. Copyright © 1987 by Randy Shilts. Reprinted by permission of St. Martin's Press. (114–115)

Excerpts from *In the Age of the Smart Machine: The Future of Work and Power* by Shoshana Zuboff. Copyright © 1988 by Basic Books, Inc. Reprinted by permission of Basic Books, a division of HarperCollins Publishers, Inc. (14, 15, 155–157)

URL Index

Abortion
  gopher://gopher.undp.org/00/ungophers/
  popin/wdtrends/charts.asc
  Document: World Abortion Policies

African National Congress
  gopher://gopher.anc.org.za/00/anc/history/
  75years.87
  Document: Advance to Power—75 Years of
  Struggle

AIDS
  http://gopher.hivnet.org:70/0/hivtext1/
  aids101
  Document: An Introduction to AIDS

  http://www.nectec.or.th/users/craig/
  hiv-aids.htm
  Document: The Relationship Between HIV
  and AIDS

  CDC definition of
  http://www.safersex.org/hiv/
  howisaidstransmitted.html
  Document: How is AIDS Transmitted?

  prevention
  http://www.urban.org/periodcl/
  prr25_2c.htm
  Document: Why Teenagers Do Not Use
  Condoms

  social-human issues related to
  gopher://gopher.niaid.nih.gov/00/aids/nca/
  Working%20Group%20on%
  20Social-Human%20Issues
  Document: Working Group on Social-
  Human Issues

  theories of
  ftp://ftp.cs.berkeley.edu/ucb/sprite/www/
  theories/africa1
  Document: Myths of AIDS and Sex

  ftp://ftp.cs.berkeley.edu/ucb/sprite/www/
  theories/africa2
  Document: Myths of AIDS and Sex

  http://www.visions.net/zaire.html
  Document: Medissage In Zaire: The
  Biodiversity Emergency Team

  treatment of
  http://gpawww.who.ch/whademo/
  approach.htm

Document: New Approach to Fighting AIDS
http://www.talamasca.org/avatar/alt-healing.html
Document: Alternative Healing

underdiagnosing
http://www.indiana.edu/~aids/news/news1.html#hiv
Document: HIV Infection and AIDS in Rural America
http://www.aoa.dhhs.gov/aoa/pages/agepages/aids.html
Document: National Institute on Aging Age Page: HIV, AIDS, and Older Adults

Amnesty International
http://www.io.org/amnesty/
Document: Amnesty International

Apartheid
http://curry.edschool.virginia.edu/go/multicultural/edusa/Socioecon.html
Document: Socio-Economic Conditions of South African Children

Assimilation/absorption
http://etext.lib.virginia.edu/etcbin/browse-mixed-new?id=DawHave&tag=public&images/modeng&data=/texts/englishmodeng/parsed
Document: Have We Failed with the Indian?

Attribution theory
ftp://ftp.cs.berkeley.edu/ucb/sprite/www/theories/pages.html
Path: Follow any paths.
Document: AIDS Theories

Backstage
http://www.nmia.com/~mdibble/Japan2.html
Document: Japanese Hemophiliacs Suffering from HIV Infection

Bhopal
http://www.essential.org/monitor/hyper/mm1294.html
Document: Remembering Bhopal

Birthrates
gopher://gopher.undp.org/00/ungophers/popin/wdtrends/fertilit
Document: New Fertility Declines in Sub-Saharan Africa and South-Central Asia

Blood
blood products and
http://www.access.digex.net/%7Enpc/
Document: Directory of News Sources
http://www.aabb.org/docs/receive.html
Document: Receiving a Blood Transfusion: What Every Patient Should Know
http://www.nmia.com/~mdibble/japan.html
Document: HIV Transmission Through Blood Products
http://www.webcom.com/~lef/texts/fda-lem-nov95.html
Document: Offshore Drug Update
http://www.web-depot.com/hemophilia/archives/iom_summary.html
Document: Institute of Medicine Report: HIV and the Blood Supply: An Analysis of Crisis Decision Making

indirect interaction between donors and receivers
http://www.crossnet.org/biomed/bio-fact.html
Document: Biomedical Services 1995–96

Body image
gopher://gopher.cc.columbia.edu:71/00/publications/women/wh27
Document: Body Image and "Eating Disorders"

Capitalism
http://www.ocf.berkeley.edu/~shadab/capit-2.html
Document: Capitalism: Frequently Asked Questions (Theory)
http://www.ocf.berkeley.edu/~shadab/capit-3.html
Document: Capitalism: Frequently Asked Questions (Practice)

critique of
http://members.gnn.com/dgude/capital.htm
Document: The Truth About American Capitalism
http://members.gnn.com/dgude/capital.htm#Caste
Document: The American Caste System

Characteristics, ascribed
http://www.cdinet.com/Rockefeller/Briefs/brief8.html
Document: Racial Discrimination in Hiring

Charisma, routinized
http://www.igc.apc.org/elaw/asia/india/icela.html
Document: Goldman Prize Winner: M. C. Mehta

Charismatic authority
http://www.scetv.org/scetv/over2.html
Document: South Carolina Voices: Lessons from the Holocaust

Child care, cost of
http://www.census.gov/population/socdemochild/file1.dat
Document: Child Care Expenditures

http://www.census.gov/cgi-bin/print_hit_bold.p1/pub/Press-Release/cb95-182.txt
Document: Child Care Costs Greater Burden for the Poor

Child support
http://info-sys.home.vix.com/free/index.html
Document: The Fathers' Rights and Equality Exchange

Claims making
on death row
http://www.gbiz.com/odell/
Document: Innocent on Death Row

on illegal substances
http://www.acsp.uic.edu/lib/ussc/chapter8.htm
Document: Findings, Discussions, and Recommendations

http://tt.dx.com/tobacco/Misc/ash.kids.clinton.contest.html
Document: Hey, Kids, Help Make History and Maybe Even Win $1000

Cognitive development, Piaget's theory of
http://www.fishnet.net/~pparents/johnson1.html
Document: What Is Normal Development?

Collective memory
http://www.saed.kent.edu/Architronic/v2n2/v2n2.05.html
Document: Ritual and Monument

College, access to
ftp://stats.bls.gov/pub/news.release/hsgec.txt
Document: College Enrollment and Work Activity of 1995 High School Graduates

gopher://INET.ed.gov:12002
Path: ERIC.scr; enter ED355251
Document: National Assessments in Europe and Japan

Communist Manifesto
gopher://wiretap.spies.com/00/Library/Classic/manifesto.txt
Document: Manifesto of the Communist Party

Computers, access to
http://www.cs.cmu.edu/afs/cs.cmmu.edu/user/bam/www/number.html#ComputerInHomeWork
Document: Computer Use at Home and at Work

Conflict
as agent of change
http://www.commerce.net/work/pilot/nielsen_96/exec.html
Document: The CommerceNet/Nielsen Internet Demographics Survey

http://scuba.uwsuper.edu/~rwhiffen/web-intro/history/welcome.html
Document: Ten Minutes of Internet History

http://www.arpa.mil/news.html
Document: DARPA News

http://www.socool.com/socool/news/protest.html
Document: Internet Day of Protest

structural origins of
http://www.enn.com/feature/fe042296/feature1.htm
Document: Chernobyl 10 Years After: The Nightmare Accelerates

Conservation, water
http://www.uswaternews.com/archive/96/conserv/consort.html
Consortium Promotes Washing Machines That Use Less Water, Energy

http://www.innovativ.com/waterwise.html
Document: Be Water-Wise: 7 Ways in 7 Days

http://ianrwww.unl.edu/ianr/waterctr/wctriv.html
Document: Water Trivia

http://www.mbnet.mb.ca/wpgwater/welcome.html
Document: Welcome to the Waterfront—Winnipeg's Water Conservation Information Source

Correlation, causation, prediction, and
http://www2.ncsu.edu/ncsu/pams/stat/info/jse/v2n2/datasets.rossman.html
Document: Televisions, Physicians, and Life Expectancy

Crime and punishment, international comparisons of
http://lcweb2.loc.gov/frd/country.html
Document: Army Area Handbook Access

Crimestoppers
http://www.nj.com/crimestoppers/morriscounty/index.html
Document: New Jersey Online: Morris County CrimeStoppers

http://www.C-S-I.org/
Document: Crimestoppers International

http://www.matsu-crimestoppers.org/
Document: Matsu Crime Stoppers Inc.

http://www.fbi.gov/
Document: Federal Bureau of Investigation Home Page

Cultural flow

diffusion and

http://www.teleport.com/~napoleon/louiscreole/intro.html
Document: Overview of Louisiana Creole

http://www.pangea.ca/~trade
Document: The Global Connections Project

http://www.carleton.ca/npsia/cfpj/john.html
Document: Culture and Foreign Policy

http://www.perseus.tufts.edu/GreekScience/Students/Ellen/EarlyGkAstronomy.html
Document: Early Greek Astronomy

state control over

http://www.usia.gov/abtusia/factsht.htm
Document: About the U.S. Information Agency and the U.S. Information Service

Cultural relativism
http://www.law.indiana.edu/glsj/vol1/green.html
Document: Cultural Identities and Global Political Economy from an Anthropological Vantage Point

as applied to poopets

http://www.poopets.com/cgi-bin/dbml.exe?Template=/mall/poopets/abouttewks.dbm&sid=1&cid=-1406890265
Document: About Poopets

Culture
http://www.loc.gov/folklife/cwc.html
Document: American Folklife a Commonwealth of Cultures

http://www.wsu.edu:8000/vcwsu/commons/topics/culture/culture-definitions/whose-text.html
Document: The Culture Debate in the U.S.: Whose Culture Is This Anyway?

adaptive

http://www.magpage.com/~tdoherty/futures.html
Document: Futures Research: New Technologies for the 21st Century

historical and geographical forces in shaping

http://www.uswaternews.com/archive/96/conserv/swdrou.html
Document: Emergency Water Conservation Measures Implemented in Drought-Stricken Southwest

http://www.uswaternews.com/archive/96/conserv/albuq.html
Document: Albuquerque Saves a Billion Gallons in '95 Usage

issues in

http://www2.uchicago.edu/jnl-pub-cult
Document: Public Culture Home Page

Data
http://www.census.gov.
Document: U.S. Census Bureau

http://lcweb2.loc.gov/wpaintro/wpahome.html
Document: Life History Manuscripts from the Folklore Project

Data sets, analysis of
http://www2.ncsu.edu/ncsu/pams/stat/info/jse
Document: Journal of Statistical Education

Death penalty, as negative sanction
http://www.best.com/~mlacabe/amnesty/info/eng/dpfacts.html
Document: Facts and Figures on the Death Penalty

Demography
ftp://coombs.anu.edu.au/coombspapers/coombsarchives/demography/what-is-demography.txt
Document: What Is Demography?

Demographic characteristics, children and teenagers
> gopher://ra.reeusda.gov/11/CYFER-net/statistics/Kids_Count/kidscnt94
> Document: Kids Count Data Book, 1994

Deviance

relative nature of
> http://www.lib.virginia.edu/journals/EH/EH37/Murphy.html
> Document: The Advertising of Installment Plans

secret
> http://www.ojp.usdoj.gov/bjs/
> Document: Bureau of Justice Statistics Crime and Victim Statistics

selective character of
> http://www.ca-probate.com/molest.htm
> Document: Child Molester Hotline Number

> http://www.acsp.uic.edu/lib/ussc/chapter1.htm
> Document: United States Sentencing Commission—Cocaine Federal Sentencing Policy

Discrimination
> http://www.baclaw.com/#Jury
> Document: Jury Award Statistics

> http://www.getnet.com/silent/hatecrimes.html
> Document: Phoenix Police Department: Hate

individual
> http://wiretap.spies.com/ftp.items/Library/Humor/Jokes/offensiv.jok
> Document: Are You Sure You Want to Read This?

institutional
> http://www.bmj.com
> Path: /BMJ/Medicine and Global Survival/Medicine and Global Survival: Current Issue
> Document: Whither Nuremberg?: Medicine's Continuing Nazi Heritage

> http://www.law.cornell.edu/uscode/42/2000e-2.html
> Document: Unlawful Employment Practices

> http://hcs.harvard.edu/~perspy/may95/187cons.html
> Document: The Aftermath of Prop 187: Licensing Human Rights Abuses Against Racial Minorities

> gopher://wiretap.spies.com/00/Gov/US-Docs/civil91.act
> Document: The Civil Rights Act of 1991

> http://wiretap.spies.com/ftp.items/Gov/Treaties/Treaties/racial.un
> Document: International Convention on the Elimination of All Forms of Racial Discrimination

unemployment and
> http://www.cdinet.com/Rockefeller/Briefs/brief9.html
> Document: Employment Discrimination Against Hispanics

> http://www.cdinet.com/Rockefeller/Briefs/brief8.html
> Document: Racial Discrimination in Hiring

> http://epn.org/sage/rstill.html
> Document: "Soft" Skills and Race: An Investigation of Black Men's Employment

Divorce
> http://www.divorce-online.com/ther_art/tract1.html
> Document: Parental Rights vs. What Is Right

Domestic violence
> gopher://www.ojp.usdoj.gov/00/bjs/press/spousfac.pr
> Document: Husbands Convicted More Often Than Wives for Spouse Murder

> gopher://www.ojp.usdoj.gov/00/bjs/press/femvied.pr
> Document: Women Usually Victimized by Offenders They Know

Downsizing, effects of income on
> http://www.census.gov/ftp/pub/hhes/laborfor/dewb9092/jobturntab.html
> Document: Distribution of Average Weekly Earnings Before and After Job Turnover

Dropout rate
> gopher://INET.ed.gov:12002
> Path: ERIC.src; enter ED386515
> Document: School Dropouts: New Information About an Old Problem

Dubois, W. E. B.
> http://www.gms.ocps.k12.fl.us/biopage/dubois.html
> Document: W. E. B. Dubois

> http://www.swarthmore.edu/SocSeilHistory/Dubois.html
> Document: The Souls of Black Folk

> http://ezinfo.ucs.indiana.edu/~jgreen/home16.html
> Document: W. E. B. Dubois

http://www.calpoly.edu/~clor/bois.html
Document: W. E. B. Dubois

http://www.cc.columbia.edu/acis/bartleby/dubois/0.html
Document: The Forethought

http://www.cc.columbia.edu/acis/bartleby/dubois/15.html
Document: The After-thought

Durkheim, Emile

http://www.lang.uiuc.edu/RelSt/Durkheim/DurkheimHome.html
Document: Durkheim Home Page

Ecclesiae, Saudi Arabia as example of

gopher://dosfan.lib.uic.edu/OF-1%3a8210%3aSaudi%20Arabia%20Country
Document: Saudi Arabia Country Commercial Guide

Economy, information-based

http://www.educom.edu/educom.review/review.96/jan.feb/varion.html
Document: The Information Economy: How Much Will Two Bits Be Worth in the Future?

Education,
assessment of

gopher://gopher.ed.gov:10000/00/tab/assess/naep/math/readme
Document: National Assessment of Educational Progress 1992 Mathematics Assessment

http://www.ed.gov/bulletin/summer1994/writnaep.html
Document: How Well Do Students Write? Can They Persuade?

enrollment levels and

http://www.census.gov/cgi-bin/print_hit_bold.pl/pub/Press-Release/cb94-177.txt
Document: Changes in School Enrollment Levels of Nation's Race and Hispanic Origin Groups

functions of

gopher://INET.ed.gov:12002
Path: ERIC.src; enter ED325659
Document: Employability—The Fifth Basic Skill

funding for

gopher://INET.ed.gov:12002
Path: ERIC.src; enter ED350717 (if number doesn't work, plug in financial equity)
Document: Financial Equity in the Schools

hidden curriculum in

http://www.ed.gov/bulletin/fall1993/fallread.html
Document: The Word on Reading

income and

http://www.census.gov/cgi-bin/print_hit_bold.pl/pub/Press-Release/cb94-113.txt
Document: College Degree Can Earn Millions over a Lifetime

ftp://stat.bls.gov/pub/news.release/hsgec.txt
Document: 1995 College Enrollment and Work Activity of High School Graduates

international rankings in

gopher://INET.ed.gov:12002
Path: ERIC.src; enter ED328604
Document: The International Association for the Evaluation of Educational Achievement

news about

http://www.utopia.com/mailings/reportcard/
Document: Daily Report Card—Index

reform in

gopher://gopher.ed.gov:10001/00/initiatives/goals/legislation/g2k-1
Document: National Education Goals

http://www.oecdwash.org
Document: Washington Center Home Page

statistics on

http://www.census.gov/cgi-bin/print_hit_bold.pl/pub/Press-Release/cb94-34.txt
Document: Nation Sets Record for Educational Attainment, Farm Population, Other Areas

http://chronicle.merit.edu/.almanac/.almdem2.html
Document: Demographics

http://chronicle.merit.edu/.almanac/.almmon6.html
Document: Money

successes in

gopher://INET.ed.gov:12002
Path: ERIC.src; enter ED378665
Document: What's Right with Schools?

Educational attainment, among immigrants

ftp://ftp.census.gov/pub/press-release/cb94-01.txt
Document: New Racial and Ethnic Information May Debunk Stereotypes

Emoticons
> http://www.organic.com/1800collect/
> Emoticons/index.html
> Document: Emoticons

Environment, youths' reaction to
> http://www.unicef.org/voy/
> Document: Voices of Youth Home Page

Environmental crime
> gopher://justice2.usdoj.gov/00/fbi/May95/
> may3.txt
> Document: Traditional Policing and Environmental

Ethnic categories
> http://www.census.gov/td/stf3/
> append_b.html#hispanic
> Document: Summary Table 3 Technical Documents

Ethnicity
> http://www.census.gov/td/stf3/
> append_b.html
> Document: Definition of Subject Characteristics

Ethnocentrism
> http://www.lawrence.edu/dept/
> anthropology/classics_april96.html
> Document: East Africa: Colonized Minds

> http://www.univie.ac.at/voelkerkunde/
> theoretical-anthropology/godina.html
> Document: What Is Wrong with the Concept of Human (Social) Evolution?

Euthanasia
> http://www.biol.tsukuba.ac.jp/~macer/
> EJ63/EJ63D.html
> Document: The First Euthanasia Court Case in China—Cong Yali, M.D.

Externality costs
> gopher://justice2.usdoj.gov/ORO-4947-/
> press_releases/previous/May96/233.enr
> Document: Builder of Vast Northeastern Gas Pipeline Pleads Guilty

Family
> *coalition structures within*
>
> http://www.uiowa.edu/~grpproc/crisp/
> crisp.1.3.html
> Document: Latest Issue/The Coalition Structure of the Four-Person Family
>
> *nontraditional*
>
> http://sunsite.unc.edu/gaylaw/files/
> coleman.pag
> Document: The Registration of Names of Family Associations
>
> http://www.census.gov/population/
> socdemo/hh-fam/his9.prn
> Document: Average Number of Own Children Under 18 per Family, by Type of Family: 1955 to Present
>
> http://www.census.gov/cgi-bin/
> print_hit_bold.pl/pub/Press-Release/
> cb94-121.txt
> Document: Half of Nation's Children Live in Non-Traditional Families
>
> *student school performance and*
>
> http://www.rand.org/publications/RB/
> RB8009
> Document: Student Performance and the Changing American Family

Femininity
> gopher://gopher.cc.columbia.edu:71/00/
> publications/women/wh27
> Document: Body Image and "Eating Disorders"

Feminism
> http://www.cis.ohio-state.edu/hypertext/
> faq/usenet/feminism/info/faq.html
> Document: The alt.feminism FAQ (monthly posting)
>
> http://www.rochester.edu/SBA/declare.html
> Document: The Declaration of Sentiments

Fertility
> gopher://gopher.undp.org/00/ungophers/
> popin/wdtrends/agespec.tab
> Document: Age-Specific Fertility

Fetal alcohol syndrome
> http://www.niaaa.nih.gov/publications/
> aa13.htm
> Document: Fetal Alcohol Syndrome

Foreign-born individuals
> http://www.census.gov/ftp/pub/population/
> socdemo/foreign/foreign_rpt.html
> Document: Foreign-Born Report Highlights

Fundamentalism, religious
> http://www.ais.org/bsb/Herald/Previous/
> 495/fundamentalism.html
> Document: What Does Fundamentalism Really Mean?

Gangs, mothers against
> http://www.winternet.com/~jannmart/
> nkcmag.html
> Document: Mothers Against Gangs

Gender, fluidity of
  http://ezinfo.ucs.indiana.edu/~mberz/ttt/articles/rights
  Document: The International Bill of Gender Rights

  http://garnet.berkeley.edu:4248/Fantasies_of.html
  Document: Fantasies of Straight Men About Gays in the Military

Gender inequality, in wages and income
  http://www.dol.gov/dol/wb/public/wb_pubs/wagegap2.htm
  Document: Earnings Differences Between Women and Men

  http://www.dol.gov/dol/wb/public/wb_pubs/wwmf1.htm
  Document: Women Who Maintain Families

  http://www.cdinet.com/Rockefeller/FemMasc/wpb6.html
  Document: Comparable Worth: Theories and Evidence

Gender issues, youths' perspective on
  http://ux641a12.unicef.org/voy/past/voyI/tindex.html
  Document: Messages from Youth Indexed by Topic

Gender nonconformists
  http://ezinfo.ucs.indiana.edu/~mberz/ttt/articles/crossing-line.html
  Document: Crossing the Line

Gender-schematic decisions, in education
  http://www.ed.gov/pubs/Cond0fEd_95/ovw3.html
  Document: Progress in the Achievement and Attainment of Women

Genocide, cultural
  http://daedalus.ee.ic.ac.uk/cyprus/Cyprus_Problem/destruction2.html
  Document: Destruction of the Cultural Identity of the Occupied Area

Glass ceiling
  gopher://gopher.etext.org/11/Politics/Womens.Studies/GenderIssues/GlassCeiling
  Document: Various Documents

Goals 2000
  http://www.utopia.com/mailings/reportcard/DAILY.REPORT.CARD113.html#Index
  Document: Goals 2000: A Politically Charged Battleground

Groups, primary
  http://www.psych.med.umich.edu/web/aacap/factsFam/fmlymove.htm
  Document: Children and Family Moves

Health care, role of faith in
  http://www.interaccess.com/ihpnet/alpha
  Document: Expanding the Public Health Envelope Through Faith Community

Health issues
  http://www.cdc.gov/nchswww/nchshome.htm
  Document: Centers for Disease Control

High school graduates, college enrollment among
  ftp://stats.bls.gov/pub/news.release/hsgec.txt
  Document: College Enrollment and Work Activity of 1995 High School Graduates

Housing, quality of
  http://www.census.gov/ftp/pub/hhes/housing/ahs/tab2-11.html
  Document: Why Move?

  http://www.census.gov/ftp/pub/hhes/housing/ahs/tab2-3.html
  Document: Big Homes?

  http://www.census.gov/ftp/pub/hhes/housing/ahs/tab2-2.html
  Document: Good Buildings?

Human rights, youths' perspective on
  http://www.unicef.org/voy/past/voyI/tget.cgi?hum
  Document: Messages About Human Rights

Hypertext
  http://www.teleport.com/~cdeemer/essay.html
  Document: What Is Hypertext?

  http://www.rtvf.nwu.edu/telecine/Tele-vol3.html#feature
  Document: Turning the Page of Journalism

Immigrants, educational needs of
  http://www.rand.org/publications/MR/MR103/MR103.html
  Document: Newcomers in American Schools

Immigration, undocumented
  http://www.law.indiana.edu/glsj/vol2/calavita.html
  Document: U.S. Immigration Policy: Contradictions and Projections for the Future

Income
  distribution by gender
  http://www.census.gov/ftp/pub/hhes/laborfor/dewb9193/tableb.html
  Document: Dynamics of Economic Well-Being: Labor Force, 1991 to 1993

  operational definitions of
  http://www.census.gov/cdrom/lookup
  Document: 1990 Census Lookup

  http://www.census.gov/td/stf5/append_b.html#Income
  Document: Income in 1989

Industrialization
  effect on chocolate easter eggs of
  http://www.cadbury.co.uk/facts.htm
  Document: The Effects of Industrialization: The Case of Chocolate Easter Eggs

  effect on paper making of
  http://www.ipst.edu/amp/machine.html
  Document: The Advent of the Paper Machine

Inequality, gender
  http://gopher.census.gov:70/ls/Bureau/Pr/Subject/Income/cb95-129.txt
  Document: Female-Householders Most Likely to Stay Poor

Infant footprints, as measure of identity
  gopher://justice2.usdoj.gov
  Path: /Federal Bureau of Investigations/November94
  Document: Best Foot Forward: Infant Footprints for Personal Identification

Information, dirty
  http://sun.soci.niu.edu/sssi/papers/dirty.data
  Document: Dirty Information and Clean Conscience: Communication Problems in Studying "Bad Guys"

  http://www.educom.edu/educom.review/review.96/mar.apr/peters.html
  Document: Raison d'Net: Are Your Ready for That Thing Called "Change"?

Information infrastructure, global
  gopher://198.80.36.82/00s/usa/media/global/global.txt
  Document: Toward a Global Information Infrastructure

Information technology
  http://www.beacham.com/essay_contents_848.html
  Document: Information Technology: Essays and Technology Criticism

Innovations
  basic
  http://www.newciv.org/GIB/BOV/BV-2.HTML
  Document: The History of Social Inventions

  improving
  http://ourworld.compuserve.com/homepages/invent/
  Document: The Inventor's Home Page

Institution, total
  gopher://server.gdn.org/00/Human_Rights_Reports/Human_Rights_Watch/Child_Soldiers_in_Sudan
  Document: Human Rights Watch/Africa

  gopher://server.gdn.org/00/Human_Rights_Reports/Human_Rights_Watch/Child_Soldiers_in_Liberia
  Document: Human Rights Watch

Institutionalization, long-term effects of
  http://www.cyfc.umn.edu/Adoptinfo/institutionalization.html
  Document: The Long-Term Effects of Institutionalization on the Behavior of Children from Eastern Europe and the Former Soviet Union

Internet
  http://info.isoc.org/speeches/interop-tokyo.html
  Document: The Present and the Future of the Internet: Five Faces

  http://info.isoc.org/papers/truth.html
  Document: Truth and the Internet

  http://www.sjcoe.k12.ca.us/employees.html
  Document: Employee Internet Agreement

  http://heg-school.aw.com/bc/is/bclink/bclink1/Rules.html
  Document: On-line Survival Tips

  as "deity"
  http://www.cemetery.org/about.html
  Document: World Wide Cemetery

  functions and dysfunctions of
  http://www.columbia.edu/~rh120/
  Document: Netizens: An Anthology: Preface

gopher://borg.lib.vt.edu/00/catalyst/v22n3/katz.v22n3
Document: Redefining Success: Public Education in the 21st Century

http://www.islandnet.com/~rcarr/oddy.html#mentornumbers
Document: Oddysey, Volume 3,2, May 1995

as second coming of printing press

http://www.lincoln.ac.nz/reg/futures/renaiss2.htm
Document: Renaissance Two: Second Coming of the Printing Press

users of

http://future.sri.com/vals/vals-survey.results.html
Document: Exploring the World Wide Web Population's Other Half

Internment, in WWII

http://www.freedom-server.co.uk/Japanese-Internment.html
Document: The Freedom Server-Japanese Internment

http://www.netzone.com/~adjacobs/
Document: European American Internment

Intersexed

http://www.isna.org/FAQ.html
Document: Frequently Asked Questions

http://www.isna.org/pamphlets/recommendations.html
Document: Recommendations for Treatment of Intersexed Infants and Children

Interviews

http://www.myna.com/~davidck/gotlib.htm
Document: David Gotlib on the Banks of the Charles

Labeling theory

http://www.educom.edu/educom.review/review.96/jan.feb/sliwa.html
Document: Cyber-Cops: Angels on the Net

Labor, exploitation of

http://www.idbsu.edu/surveyrc/Staff/jaynes/marxism/fairwage.htm
Document: A Fair Day's Wage for a Fair Day's Work

Language

http://www.teleport.com/~napoleon/louisianafrenchcreole/introtolouisianafrenchcreo.html
Document: Louisiana Creole

http://www.travlang.com/languages/indextext.html
Document: Foreign Languages for Travelers

http://mendel.mbb.sfu.ca/berg/breden.lab/dotw.html
Document: Days of the Week

cultural context of

http://www.comenius.com/idiom/index.html
Document: The Weekly Idiom

Liberation theology

gopher://server.gdn.org/00/Miscellaneous_Items/World_HR
Document: The World Human Rights Movement: Reflections and Proposed Next Steps

Literacy

gopher://INET.ed.gov:12002
Path: ERIC.src; enter ED372664
Document: Estimating Literacy in the Multilingual U.S.

new forms of

http://www.educom.edu/educom.review/review.96/mar.apr/shapiro.html
Document: Information Literacy as a Liberal Art

Longevity, consequences of

http://www.census.gov/cgi-bin/print_hit_bold.pl/Press-Release/cb95-90.txt
Document: The Aging of the U.S. Population

Looking-glass self

http://www.dnai.com/children/media/content_study.html
Document: The Reflection on the Screen: Television's Image of Children

Malthus, Thomas

http://socserv2.socsci.mcmaster.ca/~econ/ugcm/3ll3/malthus/population
Document: An Essay on the Principles of Population

Martineau, Harriet

http://miso.wwa.com/~jej/martinh.html
Document: Harriet Martineau

Marx, Karl

http://csf.Colorado.EDU/psn/marx/Archive/Interviews/1879int1.htm
Document: Interview with Karl Marx

capitalism and
http://csf.colorado.edu/psn/marx/Archive/1849-WLC/wlc9.txt
Document: Karl Marx: Wage-Labor and Capital

critique of
http://home.sol.no/hansom/papers/960506.htm
Document: A Review of Reviews: Making Sense of Marx

Friedrich Engels and
http://csf.Colorado.EDU/psn/marx/index1.htm (for nongraphic browsers)
Document: The Marx-Engels Internet Archives

http://csf.colorado.edu/psn/marx (for graphic browsers)
Document: The Marx-Engels Internet Archives

influence of
http://www.idbsu.edu/surveyrc/Staff/jaynes/marxism/websites.htm
Document: Other Marxist-Related Sites on the Web

Material culture, nonmaterial culture and
http://www.biol.tsukuba.ac.jp/~macer/EJ54E.html
Document: Bioethics and Japanese Culture: Brain Death, Patients' Rights

Mechanization

paper-making and
http://www.ipst.edu/amp/inventn.html
Document: The Invention of Paper

http://www.ipst.edu/amp/forerun.html
Document: Forerunners of Paper

steam locomotive and
http://www.arc.umn.edu/wes/misc/other.html
Document: Other Web Sites Providing Steam Locomotive Information

http://www-cse.ucsd.edu/users/bowdidge/railroad/rail-groups.html
Document: Museums and Historical Societies

Media, censored stories in
http://censored.sonoma.edu/ProjectCensored/Stories1994.html
Document: Top Censored News Stories of 1994

http://censored.sonoma.edu/ProjectCensored/Stories1995.html
Document: Top Censored News Stories of 1995

Migration
http://www.census.gov/population/socdemo/migration/tab-a-3.txt
Document: Annual Inmigration, Outmigration, and Net Migration for Metropolitan Areas: 1985–1994

http://www.census.gov/ftp/pub/population/socdemo/migration/net-mig.txt
Document: Immigrants, Outmigrants, and Net Migration for States

international
gopher://gopher.undp.org/00/ungophers/popin/wdtrends/inttab
Document: South-to-North Migration Flows

Mingling, everyday
http://www.intrnet.net/~jbailey/suzuki/whois-ss.html
Document: Who Is Shinichi Suzuki?

Misinformation, role of television in
http://www.law.indiana.edu/fclj/v47/no1/jrhodes.html
Document: Even My Own Mother Couldn't Recognize Me

Mores, euthanasia and
http://www.biol.tsukuba.ac.jp/~macer/EJ63/EJ63D.html
Document: The First Euthanasia Court Case in China

Mortality
http://www.census.gov/ftp/pub/statab/ranks/pg05.txt
Document: Birth and Infant Mortality Rates in the United States by State and Area in 1992

child
gopher://gopher.undp.org/00/ungophers/popin/wdtrends/child
Document: Child Mortality Estimates

plagues and
http://jefferson.village.virginia.edu/osheim/intro.html
Document: Plague and Public Health in Renaissance Europe

Multinational corporations
 http://pathfinder.com/
 @@oz00hwuapotknqeg/fortune/magazine/
 1995/950807/global500/revenue.html
 Document: Fortune Global 500

 http://www.essential.org/monitor
 Document: Multinational Monitor

 effect of on community
 http://www.igc.apc.org/elaw/
 update_summer_95.html#hyundai
 Document: Hyundai Plans Chip Factory for Eugene

 most admired
 http://pathfinder.com/
 @@oz00hwuapotknqeg/fortune/magazine/
 specials/most admired/comebacks.html
 Document: Fortune Magazine's Most Admired Corporations

 worst
 http://www.essential.org/monitor/hyper/
 mm1294.html#topten
 Document: The Ten Worst Corporations of 1994

Nature, as factor in alcoholism
 http://www.niaaa.nih.gov/publications/
 aa18.htm
 Document: The Genetics of Alcoholism

Netizens
 http://www.columbia.edu/~hauben/
 netbook/
 Document: Neitzens: On the History and Impact of the Net

 rights of
 http://www.columbia.edu/~rh120/
 netizen-rights.txt
 Document: Proposed Declaration of the Rights of Netizens

Norms
 gopher:// liberty.uc.wlu.edu/00/library/
 human/eashum/nameconv
 Document: Non-Western Naming Conventions

Nurture, as factor in alcoholism
 http://home.navisoft.com/aapa/aca.htm
 Document: Adult Children of Alcoholics

Occupation, gender segregation by
 gopher://una.hh.lib.umich.edu/00/census/
 summaries/eeous
 Document: Detailed Occupation by Sex

Occupational structure, projected changes in
 ftp://stats.bls.gov/pub/news.release/
 ecopro.txt
 Document: BLS Press Release: New 1994–2005 Employment Projections

Operational definitions
 http://www2.ncsu.edu/ncsu/pams/stat/info
 /jse/v3n3/datasets.dawson.html
 Document: The "Unusual Episode" Data Revisited

 examples of
 http://stats.bls.gov/flsfaqs.htm
 Document: FLS Frequently Asked Questions

 http://stats.bls.gov/csxgloss.htm
 Document: Consumer Expenditure Surveys Glossary

Organizations
 dark side of
 http://epawww.ciesin.org/national/
 epacoop.html
 Document: Integration and Use of Information Regarding the Human Dimensions of Environmental Change

 http://www.esential.org/monitor/hyper/
 list.html
 Document: List of Back Issues of Multinational Monitor

 informal dimension of
 http://iww.org/labor/sabotage/S2.html
 Document: General Forms of Sabotage

 http://www.oshadata.com/fsoihu.htm
 Document: It's Confirmed—OSHA Inspectors Are Human

 http://www.oshadata.com/fssy.htm
 Document: Seek and Ye Shall Not Find

Outgroup, elderly as
 gopher://justice2.usdoj.gov/00/fbi/
 January95/jan6.txt
 Document: Focus on Training: Teaching Officers to Serve Seniors

 gopher://justice2.usdoj.gov/00/fbi/
 January95/jan2.txt
 Document: Police Practice: Horseplay Brings Officers Closer to Community

Paradigm shift, in education
 gopher://borg.lib.vt.edu/00/catalyst/v22n3/
 katz.v22n3
 Document: Redefining Success: Public Education in the 21st Century

Perception, selective
   gopher://gopher.etext.org/00/Politics/Fourth.World/Americas/manifest.txt
   Document: Reflections on Race and Manifest Destiny

Plagues
   http://jefferson.village.virginia.edu/osheim/intro.html
   Document: Plagues and Public Health in Renaissance Europe
   http://www.ento.vt.edu/IHS/plagueHistory.html
   Document: Plagues after the Roman Empire

Population
   by country
   gopher://gopher.undp.org/00/ungophers/popin/wdtrends/pop1994
   Document: Population Figures for Countries of the World

   world milestones in
   gopher://gopher.undp.org/00/ungophers/popin/wdtrends/mileston
   Document: World Population Milestones

Population growth
   http://www.census.gov/population/estimates/metro-city/metal95.txt
   Document: National Population Projections

   in Asian and Eastern European successor states
   gopher://gopher.undp.org/00/ungophers/popin/wdtrends/transit
   Document: The Demography of Countries with Economies in Transition

Poverty
   federal guidelines on
   http://www.os.dhhs.gov/poverty/poverty.htm
   Document: HHS Poverty Guidelines

   gender and
   http://www.yahoo.com/society_and_culture/GenderIssues
   Path: /Gender&Society/Feminism/Gender Bias Causes Poverty
   Document: Gender Bias Causes Poverty

   public definitions of
   http://www.cdinet.com/Rockefeller/Briefs/brief23.html
   Document: Where the American Public Would Set the Poverty Line

   white
   http://www.cdinet.com/Rockefeller/Briefs/brief28.html
   Document: White Poverty in America

Preschoolers, low-income
   http://www.cdinet.com/Rockefeller/Briefs/brief35.html
   Document: Changing the Odds for Low-Income Preschoolers

Race
   federal definition of
   http://www.census.gov/td/stf3/append_b.html#RACE
   Document: Definitions of Subject Characteristics: Race
   http://ftp.fedworld.gov/pub/omb/re.fr2
   Document: Standards for the Classification of Federal Data on Race and Ethnicity
   http://www.vrx.net/aar/educate5.html
   Document: When Racial Categories Make No Sense

   history of concept of
   http://www.grapevine-sys.com/~newworld/racehist.html
   Document: The Islamic Solution to the Issue of "Race"

Race and ethnicity
   book reviews on
   gopher://rs6000.cmp.ilstu.edu/h0/depts/polisci/COURSES/POS302/AAwelcome
   Document: Race and Ethnicity Seminar Book Reviews

   definitions and concepts of
   http://www.inform.umd.edu:8080/EdRes/Topic/Diversity/Reference/Diversity_Dictionary/
   Document: Diversity Dictionary

Racial categories
   personal experiences with
   http://www.webcom.com/intvoice/letters1.html
   Document: Letters/Voices to the Editor
   http://www.webcom.com/intvoice/letters3.html
   Document: Letters/Voices to the Editor

Racially mixed people
   http://www.webcom/intvoice/rights.html
   Document: Bill of Rights for Racially Mixed People

Racist ideology, examples of
http://grid.let.rug.nl/~welling/usa/documents/jefslav_note.html
Document: Thomas Jefferson on Slavery

gopher://gopher.etext.org/00/Politics/Fourth.World/Americas/manifest.txt
Document: Reflections on Race and Manifest Destiny

Rape
gopher://justice2.usdoj.gov/00/fbi/January95/jan3.txt
Document: The Art of Interrogating Rapists

Religion
descriptions of
http://www.kosone.com/people/ocrt/var_rel.htm
Document: Descriptions of 49 Religions, Faith Groups, and Ethical Systems

general resources in
http://www.pitts.emory.edu/ptl_rel-std.html
Document: Pitts Theology Library: Religious Studies Resources on the Internet

http://www.pitts.emory.edu/boblist.html
Document: Internet Lists Related to Topics in Religion

media coverage of
http://www.missouri.edu/~c676747/religion/religion.html
Document: Religion, Journalism, and the Internet

mystical
http://www.realtime.net/~rlp/dwp/mystic/index.html
Document: Mysticism in World Religions

sacramental
http://www.kosone.com/people/ocrt/nataspir.htm
Document: Native American Spirituality

science vs.
http://www.myna.com/~davidck/hawking.htm
Document: David Cherniack Films: Transcripts—Stephen Hawking

Religious Freedom Restoration Act (RFRA)
http://northshore.shore.net/rf/theact.html
Document: Full Text of the Religious Freedom Restoration Act

http://northshore.shore.net/rf/nowl.html
Document: NOWL Panel Discussion on RFRA

Religious terms, glossary of
http://www.kosone.com/people/ocrt/glossary.htm
Document: Glossary of Confusing Religious Terms

Religious tolerance
http://www.kosone.com/people/ocrt
Document: Ontario Centre for Religious Tolerance

Research, guide to
http://www.dnai.com/~children/report_guide.html
Document: The Report Card Guide

Resilience model
http://www.ncrel.org/ncrel/sdrs/cityschl/city1_1b.htm
Document: Resilience Research: How Can It Help City Schools?

http://www.ncrel.org/ncrel/sdrs/cityschl/city1_1c.htm
Document: Funds of Knowledge: A Look at Luis Moll's Research into Hidden Family Resources

Resocialization, examples of
http://www.shore.net/~tcfraser/blurblst.htm
Document: Blurbs from the Big Book: Chapter 5

http://www.ilinkgn.net/commercl/author/news.htm
Document: News from the Front Lines of Hospice Care

Rule enforcement, selective characteristics of
http://www.acsp.uic.edu/lib/ussc/chapter1.htm
Document: Background and Methodology

Rule makers and enforcers
http://www.ojp.usdoj.gov/bjs/
Document: Bureau of Justice Statistics

http://gopher.usdoj.gov/bop/facts.html
Document: U.S. Bureau of Prisons Fact Sheet

http://www.law.cornell.edu/supct/supct.table.html
Document: U.S. Supreme Court Decisions

Sacred phenomena
  http://www.kosone.com/people/ocrt/holy_day.htm
  Document: Holy Days, Seasonal Days of Religious Celebration, Etc.

Sampling frame, generating
  http://sun.soci.niu.edu/~sssi/papers/ralphw.txt
  Document: Strategies for Identifying and Interviewing "Deviant" Informants

Sanctions, formal
  gopher://wiretap.spies.com/00/Gov/US-State/compcrime.tx
  Path: Government Docs/various US State Laws/Texas Computer Crime Statutes
  Document: Texas Computer Crime Statutes

Schools, rural
  gopher://INET.ed.gov:12002
  Path: ERIC.src; enter ED317332
  Document: Small Schools: An International Overview

Scientific method, political elements in
  http://szocio.tgi.bme.hu/replika/hozzaszol2.html
  Document: Acquired Immune Deficiency Syndrome in Social Science in Eastern Europe: The Colonization of Eastern Europe Social Science

Secularization
  objective
  http://www.kosone.com/people/ocrt/prayer.htm
  Document: Prayer in Public School

  gopher://justice2.usdoj.gov/00/fbi/June95/7june.txt
  Document: Freedom of Religion and Law Enforcement Employment

  subjective
  http://www.Trinity.Edu/~mkearl/never.html
  Document: You Never Have to Die!

  http://www.interaccess.com/ihpnet/andrews
  Document: The Forgotten Player in Health Care Reform: Organized Religion

Segregation, school
  gopher://INET.ed.gov:12002
  Path: ERIC.src; enter ED316616
  Document: Hispanic Education in America: Separate and Unequal

Self-fulfilling prophecy, deficit model as
  http://www.ncrel.org/ncrel/sdrs/cityschl/city1_1a.htm
  Document: Who Are Today's City Kids? Beyond the "Deficit Model"

Sex ratio
  http://www.census.gov/population/projection-extract/nation/npas9600.asc
  Document: Resident Population of the United States

Sex segregation
  http://www.ul.cs.cmu.edu/books/sex_segregation/sex.htm
  Document: Segregation in the Workplace

Shilts, Randy
  http://qrd.rdrop.com/qrd/aids/obits/shilts.obituary-SFChron
  Document: Randy Shilts Obituary

Skills for Youth
  gopher://gopher.undp.org/00/ungophers/popin/unfpa/speeches/1995/youth95.gen
  Document: Geneva Youth Forum

Social action, as applied to Taj Mahal
  http://www.lonely planet.com/dest/ind/nor.htm#taj
  Document: Taj Mahal

Social change
  http://www.hotwired.com/special/unabom/principles.html
  Document: Some Principals of History

  http://www.samizdat.com/global.html
  Document: Global Competition and the Long Road to General Prosperity

Social class, as measured by sources of income
  http://www.propertyguide.com/steps.html
  Path: Step Three: Home Buying Power

Social emotions
  laughter as
  http://rampages.onramp.net/~ejunkins/
  Document: Laughter Therapy Homepage

  toward animals
  http://www.biol.tsukuba.ac.jp/~macer/EJ51A.html
  Document: Editorial—Why a New Journal?

  http://www.primenet.com/meggie/bridge.htm
  Document: Informational Dog-Related Web Sites

Socialization

in daycare

gopher://gopher-cyfernet.mes.umn.edu:4242/00/ChildCare/ChildDevel/cynet15
Document: Preschooler Development

gopher://gopher-cyfernet.mes.umn.edu:4242/00/ChildCare/ChildDevel/cynet17
Document: Toddler Development

gopher://gopher-cyfernet.mes.umn.edu:4242/00/ChildCare/Curriculum/cynet21
Document: Play Activities

http://people.delphi.com/punkyhaake/preschool.htm
Document: Preschool for Day Care

gender

http://www.toysrus.com/bestsell.html
Document: Toys at the Top

lifelong

gopher://borg.lib.vt.edu/00/catalyst/v21n4/cross.v21n4
Document: The Renaissance in Adult Learning

Social mobility

http://www.census.gov/ftp/pub/hhes/housing/ahs/tab2-11.html
Document: Why Move?

Social movements

http://www.gospelcom.net/ibs/who.html#HISTORIC
Document: A Historical Perspective

Social structure

http://www.lm.com/~rs7717/wilmerdi.html
Document: An Historical Small Town Treasure: A View from Our Front Porch

Society, as object of worship

http://shamash.nysernet.org/pirchei/death.html#III
Document: Part III: Proper Attitude at a Jewish Funeral

Statistical records of performance, problems with

http://www.oshadata.com/fsdr.htm
Document: The Defeat of Repeat or "What's in a Name?"

Status groups, defined by neighborhood characteristics

http://propertyguide.com/steps.html#Step2
Document: Step 2: Selecting a New Neighborhood

Stigma, associated with sex research

http://math.ucsd.edu/~weinrich/theScientist2.html#part 2
Document: SEX: Still a Bad Word for Some People

Stigmatization

http://www.pathcom.com/%7Efreedom/hf/upfront19.html
Document: Up Front, Issue 19

gopher://gopher.umsl.edu/00/library/govdocs/armyahbs/aahb4/aahb0269
Document: Employment and Labor Relations

gopher://una.hh.lib.umich.edu/00/census/summaries/eeous
Document: Equal Employment Opportunity File

Structural changes, downward mobility attributed to

ftp://stats.bls.gov/pub/news.release/demdat.txt
Document: Bureau of Labor Statistics Releases 1992 Demographic Data Book for State and Large Metropolitan Areas

Structural strain, groups vulnerable to

http://www.acsp.uic.edu/lib/ussc/chapter4.htm#sectg
Document: Role of Youth and Women in Crack Cocaine Distribution

Subcultures

Korean

http://www.koma.org/ktown_15.html
Document: All About Koreatown

http://www.koma.org/koreatown.html
Document: Welcome to Koreatown

punk rock

gopher://alishaw.ucsb.edu/00/.thresholds/.tvc8/.histpunk/.steigner.txt
Document: Do-It-Yourself Punk Rock and Hardcore

Symbolic gestures

http://www.law.pitt.edu/hibbitts/re_mem.htm
Document: Re-membering Law: Legal Gesture in the Past, Present, and Future

Symbolic interaction

http://sun.soci.niu.edu/~sssi/papers/papers.html
Document: Society for the Study of Symbolic Interaction

Teachers, problems facing
    http://www.utopia.com/mailings/
    reportcard/
    DAILY.REPORT.CARD170.html#Index4
    Document: Something Old, Something New:
    American Teacher Survey

Technological determinism
    http://www.aber.ac.uk/~dgc/tdet01.html
    Document: Technological or Media
    Determinism: Introduction

    http://www.aber.ac.uk/~dgc/tdet02.html
    Document: Technological or Media
    Determinism: Technology-Led Theories

Television, violence on
    http://www.mediascope.org/mediascope
    Document: National Television Violence
    Study

Tests, misuses of
    gopher://INET.ed.gov:12002
    Path: ERIC.src; enter ED315429
    Document: Five Common Misuses of Tests

Transformative powers of history
    http://www.digitalcentury.com/encyclo/
    update/print.html
    Document: Printing: History and
    Development

Travel, disease and
    http://www.cdc.gov/ncidod/EID/vol1no2/
    wilson.htm
    Document: Travel and the Emergence of
    Infectious Diseases

    http://www.cdc.gov/ncidod/EID/vol2no1/
    fritz.htm
    Document: Surveillance for Pneumonic
    Plague in the United States During an
    International Emergency: A Model for
    Control of Imported Emerging Diseases

Urbanization, world
    gopher://gopher.undp.org/00/ungophers/
    popin/wdtrends/urban
    Document: World Urbanization Prospects

    http://www.census.gov/population/
    socdemo/migration/net-mig.txt
    Document: Inmigration, Outmigration, and
    Net Migration Between 1985 and 1990

Values, uniqueness of Asian
    http://ifrm.glocom.ac.jp/DOC/k02.001.html#5
    Document: Can "Asian Values" Be Unique?

Victimization, power, ideology and
    http://english-www.hss.cmu.edu/bs/23/
    newitz.html
    Document: Myth of the Million Man March

Violence, against women
    http://edie.cprost.sfu.ca/gcnet/
    ISS4-05d.html
    Document: India Bans the Use of Sex
    Screening Tests

    http://cdie.cprost.sfu.ca/gcnet/
    ISS4-05e.html
    Document: Egypt Against Female
    Circumcision

    gopher://csf.Colorado.Edu/
    OR149488-153662-/feminist/Femisa/
    95/dec95
    Document: Rural Indian Women Protest
    Rape Verdict

Wages, part-time and temporary
    ftp://stats.bls.gov/pub/news.release/
    occomp.txt
    Document: New Survey Reports on Wages
    and Benefits for Temporary Help Services
    Workers

Witch-hunts
    http://liquid2-sun.mit.edu/FAQ.html
    Document: Witch-Hunt Frequently Asked
    Questions

Women
    U.N. efforts to empower
    gopher://gopher.undp.org/00/ungophers/
    popin/unfpa/speeches/1995/summngo.asc
    Document: 180 Days–180 Ways Women's
    Action Campaign

    work, quality of life, and
    http://www.hec.ohio-state.edu/famlife/
    bulletin/volume.1/bull13a.htm
    Document: Multiple Roles and Women's
    Mental Health

Women's suffrage
    http://lcweb2.loc.gov/ammem/
    rbnawsahtml/nawstime.html
    Document: One Hundred Years Toward
    Suffrage: An Overview

Women workers, characteristics of
    gopher://english.hss.cmu.edu/
    00ftp%3aEnglish.Server%3aGender%
    3aFacts%20on%20Working%20Women
    Document: 20 Facts on Women Workers

Worker displacement
: http://stats.bls.gov/pub/news.release/disp.txt
Document: Worker Displacement During Mid-1990s

http://stats.bls.gov/ceschick.htm
Document: Why Did Employment Expand in Poultry Processing Plants?

Workplace, fatalities/illnesses/injuries in
: ftp://stats.bls.gov/pub/news.release/osh2.txt
Document: Characteristics of Injuries and Illnesses Resulting in Absences from Work, 1994

ftp://stats.bls.gov/pub/news.release/osh.txt
Document: Workplace Injuries and Illnesses in 1994

ftp://stats.bls.gov/pub/news.release/cfoi.txt
Document: National Census of Fatal Occupational Injuries

Work-related training, rewards of
: http://www1.ifs.org.uk/research/education/workrelatedtraining.htm
Document: Work-Related Training Provides Big Rewards for Those Who Get It

World conflicts
: http://www.emory.edu/CARTER_CENTER/demo.htm#conres
Document: Conflict Resolution Update

World economy, strategies to realize
: gopher://gopher.undp.org/00/ungophers/unctad/efficiency/columbus
Document: Columbus Ministerial Declaration on Trade Efficiency

World Fact Book
: http://www.odci.gov/cia/publications/95fact/index.html
Document: 1995 World Factbook

World system theory
: http://cil.andrew.cmu.edu/projects/World_History/Wall.html
Document: The Development of a World Economic System

http://www.worldbank.org/html/extpb/wdr95/WDRENG.html
Document: Workers in an Integrating World

Yanomami, effects of degradation and population pressures on
: gopher://gopher.etext.org/00/Politics/Fourth.World/Americas/yanomami.txt
Document: Yanomami in Peril

Zaire
: http://www.odci.gov/cia/publications/95fact/cg.html
Document: Zaire

Index

Abbreviations for International Organizations and Groups, 396
Ability grouping, 310
Abortion, government policies regarding, 277
Absorption assimilation, 216–218
Academe This Week, 370
Academic achievement
  adolescent subculture and, 320–321
  family background and, 317–319
Acculturation, 216, 217
Achieved characteristics, 188
Achieved status, 126
Acquired immunodeficiency virus (AIDS). See HIV/AIDS
Active adaptation, 108
Adaptive culture, 352
Administration for Children and Families Press Releases, 383
Adolescent subcultures, 320–321
Adoption, 68–69
Advanced market economies, 242
Advanced Research Project Agency (ARPA), 359
Affectional action, 19
Affirmative action, 225, 226
African Americans. See also Racism; Racist ideologies
  college recruitment and, 226
  as crime victims, 221
  school integration and, 318
Agar, Michael, 80
Alcoholism, 95–96
Alderman, Craig, Jr., 251
Aleuts, 91
Alienation, 163–164
American Association for the Advancement of Science, 46
American Sociological Association, 382
Amish, 91
Amnesty International Home Page, 391
Amnesty International UK Press Releases, 384
Anderson, Barbara Gallatin, 84, 85
Anderson, Elijah, 21
Anderson, Warren, 162
Androgen, 236, 237
Annual per capita consumption of energy, 264
Anomaly, 353
Anomie, 182
Anspach, Renee R., 248, 249
Anthias, Floya, 252, 253, 255
Apartheid, 195
Armstrong, Carla, 100
Armstrong, Sharon, 61
ARPANET (Advanced Research Projects Agency), 30

ARTFL Project: Roget's Thesaurus Search Form, 392–393
Asceticism, this-worldly, 340
Ascribed characteristics
  experiment involving, 188–189
  explanation of, 187–188
Ascribed status, 126
Assimilation
  absorption, 216–218
  discrimination and, 222–226
  explanation of, 216
  melting pot, 218–219
  prejudice and stereotyping and, 221–222
  racial ideologies and, 219–220
  stratification theory and barriers to, 219–227. See also Stratification theory
Asylums: Essays on the Social Situation of Mental Patients and Other Inmates (Goffman), 110
Attitude receptional assimilation, 216
Attribution theory
  assumptions of, 134–135
  dispositional factors and, 134–136
  HIV/AIDS and, 135–138
Authority
  charismatic, 354
  explanation of, 354
  legal-rational, 356
Automate, 156
Automobiles, 28–29

Background Notes on the Countries of the World, 373
Back stage, 132–133
Barnard/Columbia Women's Handbook, 237
Barrett, Michael, 295
Bartlett's Familiar Quotations, 393
Basic innovations, 350. See also innovations
Bateson, Mary Catherine, 86–87
Baumgartner-Papageorgiou, Alice, 238
Baumrucker, Steven, 111
Bearden, Tom, 34
Becker, Howard, 173, 175, 177
Becquerel, Henri, 350
Behavior receptional assimilation, 216
Behrangi, Samad, 88
Belgium, imperialism in Africa, 121
Beliefs, 63
Bem, Sandra Lipsitz, 171, 238
Ben-Yehuda, Eliezer, 71
Berelson, Bernard, 270
Berger, Peter, 23
The Berkeley Sociology Center, 383

Berlin West Africa Conference, 121
Berners-Lee, Tim, 2
Best, Joel, 179
The Best of the Best of the Web, 401
Bettelheim, Bruno, 308, 309
Bibliographic Formats for Citing Electronic Information, 395
Birthrates
  age-sex composition and, 272
  demographic transition theory and, 265–269
  government policies and, 277
  in labor-intensive poor countries, 270–271
  patterns in, 265
Black Plague, 267
Blau, Peter, 158–161
Blood banks, 132–134
Bloor, Michael, 137
Bonacich, Philip, 48–49
Bonda, 149
Boudon, Raymond, 203
Boundaries, of culture, 62
Bourgeoisie, 18, 32
Bourricaud, François, 203
Boyer, Ernest, 304
Brain, human, 96
Breadwinner system, 282–284
Brewer, Wayne, 178
A Brief Guide to State Facts, 376
Briefings, 383–386
Brown, Rita Mae, 73
Brown v. Board of Education, 317
Browsers, 3
Bruce, Terry, 180
Buddhism, 328
Bureaucracy, 152–153
Bureau of Indian Affairs Press Releases, 384
Bureau of Labor Statistics, 245
Burke, James, 80
Bush, George, 289, 332–333, 338
Buyers, 79

Calculators On-Line, 396
Calendar Generator, 396
Calhoun, John C., 220
California Molester Identification Line, 174
Calvinism, capitalism and, 340
Capital: A Critique of Political Economy (Marx), 203, 204
Capitalism
  core, peripheral, and semi-peripheral economy and, 365–366
  development of modern, 339
  explanation of, 349, 362
  role in global economy of, 363–364
  world system theory and, 362–363, 366

453

Caplan, Lionel, 344
Career Connections, 399
Career guides, 399–400
Career Magazine, 399
CareerPath, 399
Career Shop's Resume, Job and Employment Site, 399
The Carter Center, 392
Caste system
  apartheid and, 193
  class vs., 193–194
  explanation of, 192–193
  interrelations between class and, 197–200
Castles, Stephen, 221–222
Catechisms, 298
Ceausescu, Nicolai, 277
Census Bureau
  ethnic categories used by, 213–214
  racial categories of, 211
Census Bureau Press Releases, 384
Centers for Disease Control, 132, 138, 390
Chambliss, William J., 21
Charismatic leaders, 349, 354–355
Cherniack, David, 52
Child rearing, sexual stratification and, 243
Children, status of, 280–281
Child support, 250
Christianity, 327
Chromosomes, sex, 256
Church, 329–330
City Destinations by Text Express, 377
City-level information, 377
City Net, 377
CITYSCHOOLS, 379
City Schools (North Central Regional Educational Laboratory), 313
Civil assimilation, 216
Civil religion, 332–333
Civil Rights Act of 1964, 224, 316–317, 343
Civil Rights Act of 1991, 226
Claims makers
  examples of, 179–180
  explanation of, 178
Classless society, 337
Class system. See also Social class
  explanation of, 192, 193–194
  interrelations between caste and, 197–200
  in United States, 195–197
The Class Struggles in France 1848–1850 (Marx), 203–205
Clinton, Bill, 65, 289
CNN Interactive, 388
Coca-Cola Company, 364
Cocaine, legalization of, 181
Cochran, Floyd, 224
Cognitive development stages, 108–109
Cohort, 271
Cold War, 30
Cole, Robert, 325, 326
Coleman, James S., 316–321

Collective memory, 100
College
  availability of, 300–301
  coping with, 398–399
Collins, Randall, 171, 241, 242, 244
The Communist Manifesto (Marx & Engels), 17, 203–204
Computers
  culture flow and, 80–81
  technological advances in, 48
Concepts, 48–49
Concrete operational stage, 108–109
"Condition of Education–1995," 239
Condom use, 130–131, 136
Conflict
  consequences of, 358–359
  explanation of, 349
  role, 127
  structural origins of, 359–361
Conflict perspective
  critique of, 35
  on Internet, 35–36
  overview of, 32–35
  religion and, 337–339
  symbolic interactionist perspective vs., 38
Conformists, 174
Conformity, 168, 185
Confucianism, 327
Conger, Stuart, 350
Connotation, 68
Constitution, U.S., First Amendment, 343
Constrictive pyramids, 272
Constructionist approach
  claims makers and, 179–180
  dangers of legal and illegal substances and, 180–181
  explanation of, 179
Consumer Information Center, 390
Content, 115, 125–129
Context, 115, 140–142
Control variables, 58
Cooley, Charles Horton, 103, 107, 112
Cool Site Winners, 400
Core economies, 364–365
Corporate crime, 178
Correlation, 57, 58
Correlation coefficient, 57
Coser, Lewis, 358
Costa Rica, 270
Countercultures, 90
Country Destinations by Text Express, 373
Country Health Profiles for the Americas Only, 373
Country-level information, 373–375
Country List, 373
Country Reports on Economic Policy and Trade, 374
Country Studies, 374
County Business Patterns, 377
County-level information, 377–378
County Population Profiles, 378

Cover Stories from Previous Congressional Quarterly Weekly Report Stories, 379
Cox, Melissa, 2, 7
Crawford, Jack, 31
Cressey, Donald R., 183, 184
Crime
  corporate, 178
  hate, 223–224
  meaning of, 171
  social class and, 177–178
  white-collar, 178
Csikszentmihalyi, Mihaly, 322
Cuba, 135
Cults, 332
Cultural diffusion
  explanation of, 76–78
  in marketplace, 79
  in social movements, 79–81
  through mingling, 78
  through state officials and masses, 78–79
Cultural Immersion, 397
Cultural lag, 352–353
Cultural relativism
  example of, 86–88
  explanation of, 86–87
  function of, 88–89
Culture
  adaptive, 352
  explanation of, 60–62
  geographical and historical forces in, 66–67
  home, 81–85
  material, 62–63
  nonmaterial, 63–65
  relationship between material and nonmaterial, 75–76
  sub-, 89–90
  as tool for problems of living, 71–75
  transmission of, 67–71
  women as transmitters of, 253
Culture shock
  explanation of, 83
  responses to, 84–85
  in reverse, 85–86
Currency Converter, 396
Curriculum
  formal and hidden, 306–307
  for reading, 307–310
  variations in, 301–302
Curry-White, Brenda, 208
CUSI, 4
Cyber Angels, 173

Dahrendorf, Ralf, 359
Daily news, 388–389
Darwin, Charles, 378
Das Kapital (Marx & Engels), 17
Data. See also Scientific method
  analysis of, 57–58
  explanation of, 42
  honesty in research, 46
  organization and distribution of, 43–44
  poor-quality, 44–45
Data collection, methods of, 49, 51–54

Data on Poverty, 371
Data on the Elderly, 371
Data on the Net, 383
Date rape, 255–257
Davis, Kingsley, 96, 97, 99, 200, 201, 281–285
Dawes, H. L., 217
Dawson, Robert J. McG., 54, 56
Dearth of feedback, 44–45
Death rates
 age-sex composition and, 272, 273
 demographic transition and, 265–269
 in labor-intensive poor countries, 270
 patterns in, 265
DeBerry, Jennifer, 2, 5
Deemer, Charles, 43
Demographic gap, 268
Demographic transition
 explanation of, 265–266
 in labor-intensive poor countries, 269–276
 stages in, 266–269
Demographic trap, 270
Demography, 265
Denomination, religious, 330–331
Denotation, 68
Department of Defense, 390
Department of Education, 390
Department of Housing and Urban Development, 390
Department of Justice, 390
Department of Labor, 390
Department of Veteran Affairs Press Releases, 384
Dependent variables, 54
Descourt, Jacques, 236
Deviance
 explanation of, 168
 functions of, 171–172
 labeling theory and, 173–177
 pure, 174
 secret, 175
 variations in, 169
 as violation of norms, 169–171
Dictionaries, 393
Differential association theory, 184
Diffusion, 76. *See also* Cultural diffusion
The Disability Rag, 381–382
Discoveries, 350
Discrimination
 explanation of, 222
 institutionalized vs. individual, 224–226
 unprejudiced vs. prejudiced, 223
Disenchantment of world, 150
Dispositional factors
 attributing cause to, 135–138
 explanation of, 134
Division of labor
 explanation of, 117
 industrialization and, 281–282, 284
Divorce
 effects between former spouse and children of, 285

length of marriage and, 279
women in labor force and, 284
Documents in the News, 386
Doherty, Thomas, 352
Dolphin Project (Earth Island Institute), 35
Domain name system, 3–4
Dominant group, 215, 226
Dorris, Michael, 304
Doubling time, 264
Downward mobility, 194
Dramaturgical model
 blood banks and, 132–134
 explanation of, 130
 impression management and, 130–131
 staging behavior and, 132
Drugs, legal and illegal, 180–181
Dubois, W. E. B., 22
Durkheim, Emile
 deviance and, 171–173
 education and, 291
 influence of, 20
 profile of, 18
 religion and, 325–329, 332–336
 solidarity and, 117, 119
Dysfunctions
 explanation of, 29
 latent, 29, 30
 manifest, 29, 30

Earth Island Institute, Dolphin Project, 35
Easton's Bible Dictionary, 393
Ecclesiae, 330
Ecola's Newsstand, 388
Economics
 interplay between religion and, 339–341
 sexual stratification and, 241–243
Economic Statistics Briefing Room, 387–388
Economy
 core, 364–365
 peripheral, 365
 semi-peripheral, 366
 world, 363
Education
 availability of college, 300–301
 classroom environment and, 305–316
 crisis in, 290
 curriculum and, 301–302, 306–307
 formal, 291
 funding for, 302–303
 historical background of, 297–299
 illiteracy and, 292–297
 informal, 290–291
 meaning of, 290–291
 problems faced by teachers and, 316
 purpose and value of, 304–305
 reforms in, 288–289
 social context of, 316–322
 social functions of, 291–292
 to solve social problems, 303–304
 tests and, 314–315
 tracking and, 301, 302, 310–313

Education Attainment–Historical Tables, 371
Education Policy Analysis Archives, 379
E-law, 165
Elderly individuals
 caregivers for, 280
 increase in, 279
 minimal interaction with, 99
Electronic books, 278–279
Electronic journals and newsletters, 379–382
Elements of Style, 395
Elliot, Jane, 188–189
Emigration, 273
Emotions
 cultural formulas for, 72–75
 social, 240
Encyclopedia of World Cultures (Levinson), 61–62
Encyclopedias, 393–394
Encyclopedia Smithsonian, 393
Energy conservation, 66
Engels, Friedrich, 17, 33, 203, 378
English as a Second Language Home Page, 395
English language instruction, 299
Engrams, 99–100
Environmental Defense Fund, 35
Environmental Protection Agency (EPA), 164, 390
Environmental Protection Agency (EPA) Journal, 379
Equality of Educational Opportunity (Coleman Report), 316
ERIC Digests, 398–400
Erickson, Frederick, 314
Erikson, Kai, 173, 176
Esposito, John L., 345, 346
Essential Information, Inc., 165
Established sects, 331–332
Estrogen, 236, 237
Ethgender, 252
Ethnic blending, 214
Ethnicity
 Census Bureau data on, 213–214
 determinants of, 215
Ethnocentrism
 effects of, 81–82
 explanation of, 81
 reverse, 82–83
Everyday mingling, 78–79
Excite, 4
Expansive pyramids, 272
Expert knowledge, decision making and, 160–161
External costs, 151

Facade of legitimacy
 conflict theory and, 34
 example of capitalist use of, 34–35
 explanation of, 33
Fagot, Beverly, 246
Fairstein, Linda A., 256–257
Falsely accusation
 examples of, 176–177
 explanation of, 175
 situations for likely, 176

Families
  academic achievement and, 317–319
  consequences of life span on, 278–280
  definitions of, 260–261
  effect of geographic mobility on, 278
  effect of industrialization on, 276–278
  effects of urbanization on, 281
  effects of women in labor force on, 281–285
  Industrial Revolution and, 262–263
  key episodes in, 261
  overview of, 259–260
  status of children in, 280–281
Famine, 267
Farad, W. D., 339
Farrell, Thomas, 333
Faye, Michael, 65
Federal Bureau of Investigation (FBI) Law Enforcement Bulletin, 379–380
Federal Government Agencies, 390
Federal Writers' Project (Works Project Administration), 54
Feedback, dearth of, 44–45
Feeling rules, 73, 240
Femininity, 237
Feminism, in United States, 234–235
Fertility rate
  decline in, 268–269
  thresholds associated with decline in, 270–271
  welfare recipients and, 254
  women in labor force and, 283
Fetal alcohol syndrome, 304
Fiber-optic cables, 43
Finding an E-mail Address, 396
Finding Data on the Internet: Links to Potential Story Data, 394
First Amendment to Constitution, 343
Folkways, 64, 170
Ford, Henry, 204, 350
Foreign newspapers, 389
Formal curriculum, 306
Formal dimensions, of organizations, 155
Formal education, 291. *See also* Education
Formal operational stage, 109
Formal sanctions, 171
Fortified households, 242
Foster, Jodie, 65
Franklin, John Hope, 198
F.R.E.E. (Fathers' Rights and Equality Exchange), 250
Free Internet Encyclopedia, 394
Freeman, Jayne, 296
Frisians, 61–62
Functional illiteracy, 292. *See also* Illiteracy
Functionalist perspective, 334–336
  critique of, 28–29, 201–203
  on internet, 30–32

Merton and, 29–30
  overview of, 26–28
  stratification from, 200–201
  symbolic interactionist perspective vs., 38–39
Functions
  explanation of, 26
  manifest and latent, 29, 30
Fundamentalism
  complexity of, 344–345
  explanation of, 344
  Islamic, 345–346
Funding College, 398

Gaines, Patricia, 2, 5
Games, 106
Gans, Herbert, 27
Geertz, Clifford, 88–89
Gender
  date rape and, 255–257
  explanation of, 233
  organization of social life around, 238–241
  as social construct, 237–238
Gender expectations
  overview of, 245–246
  sexist ideologies and, 250–252
  situational constraints and, 247–250
  socialization and, 246–247
Gender nonconformists, 251
Gender polarization, 238–241
Gender-schematic decisions, 239
General Electric, 34, 35
Generalizability, 57
Generalized others, 106
General references, 393–394
Geographic mobility, family life and, 278
Gibson, Josh, 23
Gidycz, Christine A., 255
Global Child Health News and Review Online Newspaper, 380
Global organizations, 391–392
Global Stewardship Network, 380
Global village, 117
Goals 2000 Act, 289
Goddard, Joel, 120
Goffman, Erving, 110, 112, 126, 130, 131, 133, 227, 228
Gordon, Emily Fox, 70
Gordon, Milton M., 216, 217
Gotlib, David, 49
Gould, Stephen Jay, 23, 180, 181, 220
Government and the Information Superhighway, 386
Government Resources Via the Web—National, State, and Local, 390–391
Grammar Handbook, 395
Gray, Norma B., 255
Grief
  explanation of, 72
  in Philippines, 87, 88
Groups
  explanation of, 101

  ingroups and outgroups, 102–103
  primary, 101–102
Grusky, Oscar, 48–49
Guardian Angels, 173
A Guide for Writing Research Papers Based on Modern Language Association Documentation, 395
Guinet, Paul, 206
Gutenberg printing press, 14–15

Hacker, Andrew, 228
Hafford, Mary, 61
Hair, cultural standards for body, 237
Haiti, 135
Hakin, Joy, 315
Halberstam, David, 120
Hall, Edward T., 38
Hannerz, Ulf, 78
Hate crime, 223, 224. *See also* Crime
Hauben, Michael, 32
Hauben, Rhonda, 32
Havel, Vaclav, 360
Hawthorne effect, 53
Heads of State and Heads of Government, 374
Health-care industry, division of labor in, 248–249
Henry, Jules, 257, 306, 307, 309
Henslin, James M., 20–21
Hermaphrodites, 235, 236
Hetrick, Ron, 208
Hewlett-Packard, 146
Hibbitts, Bernard, 104
Hidden curriculum, 306, 307
Hidden rape, 255. *See also* Date rape; Rape
Hinduism, as mystical religion, 328
Hirsch, E. D., Jr., 302
Hirschman, Charles, 214
Hispanic Online, 380
Historical documents, 391
Historical Letters, Documents, Essays, and Speeches, 391
Historical Newspapers, 380
HIV/AIDS
  blood banks and, 132–134
  claims makers and, 179–180
  classification issues related to, 137–138
  determining who is infected with, 138–139
  individual knowledge about, 140–141
  origin and transmission of, 116, 123–125, 135–137
  reduction techniques for, 130–131
  social interaction during illness from, 114–115
  staging behavior and, 132
Hixson, Lindsay, 2, 6–7
Holley, Jeannette, 214
Homeless Fact Sheets, 371
Homosexuality, in military, 251

Households
  fortified, 242
  sexual stratification and responsibilities in, 243
HTML (hypertext markup language), 3
Huber, Ryan, 2, 6
Hughes, Everett, 82
Human immunodeficiency virus (HIV). See HIV/AIDS
Human Rights, 374
Human Rights Watch Children's Rights Project, 111
Hunger, cultural formulas for, 72
Hussein, Saddham, 338
Hypertext, 43, 44
Hypothesis
  explanation of, 54
  testing, 55

Identification assimilation, 216
Ideologies
  conflict theory and, 33
  explanation of, 329
  racist, 219–220
  sexist, 250–252
Idiom, 68
Illiteracy
  foreign education systems and, 294–296
  functional, 292
  schools and, 294
  in United States, 292–293
Immigration, 273
Immigration to the United States in 1994, 371
Impression management, 130–131
Improving innovations, 350, 351. See also innovations
Income, inequality in, 195–197
Independent variables, 54
Indiana Journal of Global Legal Studies, 380
Individual discrimination, 224
Industrialization
  decline of breadwinner system and, 282, 283
  demographic transition theory and, 265, 268–270
  family life and, 262–263
  Industrial Revolution and, 15, 16
  in mechanized rich vs. labor-intensive poor countries, 263–264
  sociological perspective on, 16
Industrial Revolution
  explanation of, 13, 14
  families and, 262–263, 282
Infant mortality, 264
Infants, Children, and Teenagers, 371–372
Infants, evaluation of, 248–249
Informal dimensions, of organizations, 153–154
Informal education, 290–291. See also Education
Informal sanctions, 171
Informate, 156
Information, 42, 44

Information explosion
  effects of, 44–46
  explanation of, 42
  overview of, 43–44
Ingroups, 102–103
In-migration, 275
Innovations
  explanation of, 349–351
  rate of change and, 351–352
  as response to structural strain, 185
Institutionalized discrimination
  explanation of, 224, 225
  forms of, 225–226
  gender and, 249–250
Institutionally complete, subcultures as, 91
Institutions, total, 110
Instrumental action, 19–20
The Interface Database, 272
Intergenerational mobility, 194
Internalization, 95
Internal migration, 275–276
International Agencies and Information, 370
International Demographic Data, 374
International Labor Organization, 391
International Labor Organization Press Releases, 384
International migration, 273–274
International Monetary Fund (IMF) Press Releases, 384
International Red Cross/Red Crescent, 392
International Sociological Association, 382
The International Monetary Fund, 392
Internet
  access to, 5
  accuracy and value of documents on, 45
  conflict perspective on, 35–36
  explanation of, 2
  functionalist perspective on, 30–32
  history of, 30
  perspectives on, 25
  student suggestions for use of, 5–7
  student views of, 1
  symbolic interactionist perspective on, 39–40
Internet home library
  career guides, 399–400
  city-level information, 377
  coping with college, 398–399
  country-level information, 373–375
  county-level information, 377–378
  daily news, 388–389
  dictionaries, thesauruses, and quotations, 392–393
  electronic books, 278–279
  electronic journals and newsletters, 379–382

  encyclopedias and general references, 393–394
  foreign newspapers, 389
  global organizations, 391–392
  handy reference guides, 396–397
  historical documents, 391
  population profile of United States, 370
  press releases and briefings, 383–386
  resources of interest to sociologists, 382–383
  social issues, 386–387
  specific population groups, 370–373
  state-level information, 376
  statistical sources, 387–388
  study abroad, 397–398
  U.S. government agencies, 390–391
  world information, 369–370
  World Wide Web, 400–401
  writing resources, 395
Internet Resources for Sociology, 383
Interracial Voice, 212, 380
Intersexed individuals
  explanation of, 235–236
  as gender nonconformists, 251
Intersex Society of North America, 235, 236
Interviews
  explanation of, 51
  types of, 51–52
Intragenerational mobility, 194
Inventions, simultaneous-independent, 351–352
Involuntary minorities, 218
Irwin, Kathleen, 131
Islam
  fundamentalism and, 345–346
  as official religion, 330
  as sacramental religion, 327
  sects of, 331
Isolation, 97–98
Issues, 9
The Issues Page, 386

Jacobson, Lenore, 312, 313
Jarvis, Edward, 220
Jealousy, 72
Jefferson, Thomas, 219, 298
Jobs site, 400
JobTrack, 400
Johansson, S. Ryan, 280–281
Johnson, Diane Clark, 109
Johnson, G. David, 255, 256
Jones, Robert, 35
Journal of Statistics Education Information Service, 380–381
Judaism, 327
Judith Bower's Law Lists, 394

Kagwahiv culture, 84
Keller, George, 146–147
Kentucky Education and Reform Act (KERA), 303
Kilborn, Peter, 34, 35
Kim, Bong Hwan, 83

Kitsuse, John, 173
Klapp, Orin, 1, 44, 45
Kondo, Makoto, 78
Koop, C. Everett, 180
Kornfield, Ruth, 129
Kosack, Godula, 221–222
Koss, Mary P., 255, 256
Kracke, Waud, 84
Kuhn, Thomas, 353

Labeling theory
  conformists and deviants and, 174–175
  explanation of, 173–174
  falsely accused individuals and, 175–177
Labor-intensive poor countries
  demographic transition in, 269–276
  explanation of, 264
  peripheral economies in, 365
Language, culture and, 68–70
Language (reference guide), 396
Latent functions, 29, 30
Laughter, 74
Law enforcement, 177–178
Lawrence, Cecile Ann, 213
Leaders, 354–355
LeBon, Gustave, 378
Lee, Shin-ying, 295
Lee, Spike, 198
Legislation. *See also* specific laws
  institutional discrimination due to, 225
  role of religion addressed in, 343
Lemert, Edwin, 173
Lengermann, Patricia M., 20
Levin, Jack, 224
Levinson, David, 61
Liberation theology, 338
Library of Congress Foreign Newspapers, 389
Library of Congress Home Page, 394
Life changes
  effect of sex on, 241
  explanation of, 187
  importance of ascribed characteristics to, 188
Life expectancy
  family life and, 278–280
  male vs. female, 240
  women in labor force and, 283
Lifton, Robert Jay, 333–334
Light, Ivan, 224
Links, 3
List of American Universities Home Pages, 398
Literacy, 293. *See also* Illiteracy
Local internet access providers, 5
Looking-glass self, 107
Los Angeles Times, 389
Love, 72
Low-technology tribal societies, 241–242
Luria, A. R., 314
Lycos, 4

Mahmood, Cynthia K., 61
Main On-Line Books Page, 378
Malan, D. F., 193
Malcolm X, 339
Malthus, Thomas, 267, 378
Manifest Destiny, 219–220
Manifest functions, 29, 30
Marital assimilation, 216, 217
Market, 79
Market economies, 242
Marketplace, 79
Marquart, James, 53
Marriage length, 278–279
Martineau, Harriet, 21–22
Marx, Karl
  alienation and, 162, 163
  capitalism and, 362
  conflict theory and, 32, 33
  electronic books by, 379
  influence of, 20
  profile of, 17–18
  religion and, 337, 338
  social class and, 203–205
Masculinity, 237
Material culture
  explanation of, 62–63
  relationship between nonmaterial and, 75–76
Mathematical Notation, Weights, and Measures, 397
Mbuti, 118–119
McDevitt, Jack, 224
Mead, George Herbert, 36, 103, 105, 107, 112, 113
Meaning
  denotative and connotative, 68
  symbolic interactionist perspective and, 36–38
Mechanical solidarity
  explanation of, 117–118
  in Zaire, 118–119
Mechanization
  conflict theory and, 32–33
  explanation of, 14
Mechanized rich countries, 364–365
Media Watchdog, 386
Mehta, M. C., 355
Melting pot assimilation, 218
Memory, 99–100
Mennheim, Karl, 100
Merrill, William F., 257
Merton, Robert K., 29–30, 181–183, 222–224, 336
Metzroth, Jane, 263
Michael, Donald, 42–43
Michels, Robert, 161
Michnik, Adam, 360
Microwave ovens, 75–76
Mid South Sociological Association, 382
Migration
  elements involved in, 277
  explanation of, 273
  internal, 275–276
  international, 273–274
  net, 271

Military
  role of women in, 254
  sexist ideologies in, 251–252
Military units, 101–102
Mills, C. Wright, 8–11, 356, 357
Mills, Janet Lee, 246
Minority groups
  characteristics of, 215–216
  voluntary vs. involuntary, 210
Mixed contacts, 227–229
Mobility. *See* Geographic mobility
Mobutu, Sese, 122
Modern capitalism, 339
Mokhiber, Russell, 144, 145
The Monster Magazine List, 382
Moore, Wilbert, 200, 201
Mores, 64–65, 170
Morioka, Masahiro, 75
Morrow, Jennifer, 244
Mortality crises
  examples of, 267, 268
  explanation of, 266
Mukherjee, Bharati, 194
Multinational corporations. *See also* Organizations
  criteria for evaluating, 165–166
  explanation of, 145–146
  nature of, 146–147
  responsibilities of, 151–152
Multinational Monitor, 144
Mystical religions, 327–328

Nader, Ralph, 144
National Aeronautics and Space Administration, 390
National Commission on Education, 288
National Crime Victimization Survey, 221
National Institute on Alcohol Abuse and Alcoholism, 304
National Library of Australia Internet Site, 374
National Literacy Act of 1991, 293
National PopClock from the U.S. Bureau of the Census, 370
National Television Violence Study, 184
A Nation at Risk (National Commission on Education), 288–289
Nation of Islam, 339
Native American spirituality, as sacramental religion, 327
NATO Press Releases, 384–385
Nature, 95
Navajo, 38
Negative sanctions, 170
Nepal, 270
New Jour, 381
News from Reuters Online, 389
New Sites on the Web, 400
Newspaper Listing—Worldwide, 389
Newspapers, foreign, 389
New York Times, 389
1995 CIA World Factbook, 375
Nonconformists, gender, 251
Nondiscriminators, 222, 223

Nonhouseholder class, 242
Nonmaterial culture
　explanation of, 63–65
　relationship between material and, 75–76
Nonparticipant observation, 52
Nonprejudiced nondiscriminators, 223
Normals
　contact between stigmatized and, 227–229
　explanation of, 227
Norms
　explanation of, 64–65
　origin of, 67
　types of, 169–170
Norris, Laureen, 2, 6
North Central Regional Educational Laboratory, 313
North Central Sociological Association, 382
Nuclear explosions, 333–334
Nurture, 95

Oakes, Jeannie, 310, 311
Objective secularization, 343–344
Objectivity, in research, 46
Obligations, associated with roles, 127
Observation, 52–53
Occupational Safety and Health Administration (OSHA), 158
Occupational structure, poverty and, 207–209
Occupations
　functional importance of, 200–201
　inequality in, 196–197
Ogburn, William F., 352
O'Hare, William P., 208
Old World Amish, 91
Oligarchy, in organizations, 161–162
OneWorld Online Home Page, 375
Online Dictionaries, 393
On-Line Resources for Writers, 395
Ontario Centre for Religious Tolerance, 324
Operational definitions, 55–56
Oppression, 337
Organic solidarity, 119–120
Organization for Economic Cooperation and Development homepage, 289, 290
Organization for Economic Cooperation and Development Press Releases, 385
Organizations. *See also* Multinational corporations
　alienation within, 162–166
　bureaucracy and, 152–153
　decision-making obstacles within, 159–162
　explanation of, 145
　formal vs. informal dimensions of, 153–155
　means-to-end thought and action and, 149–150

　performance statistics of, 158–159
　political parties as, 206
　rationalization and, 150–152
　religious, 329–330
　social action and, 148–149
　trained incapacity in, 155–157
　two faces of, 147
Outgroups, 102–103
Out-migration, 275

Pacific Sociological Association, 382
Palileo, Gloria J., 255
Paradigms, 353
Parent Network for Post-Institutionalized Children, 98
Parks, Rosa, 230
Parsons, Talcott, 127
Participant observation, 52–53
Patient-physician interactions, cultural variations in, 128–129
Patino, Maria José Martinez, 236
Per capita income, 264
Perelman, Eliezer, 71
Performance, statistical records of, 158–159
Peripheral economies, 365
Persons of Indian Ancestry, 272
Piaget, Jean, 107, 108
Pickering, W. S. F., 335
Play, 105
Political parties, 206
Population
　explanation of, 49
　profile of United States, 370
　of specific groups, 370–373
　study of, 261
Population 1900–1990, 378
Population for the Countries of the World, 375
Population growth, calculation of, 271–272
Population Profile of the U.S.: 1995, 370
Population pyramids, 271, 272
Population Reference Bureau Releases, 385
Populi, 381
Positive checks, 267
Positive sanctions, 170
Postman, Neil, 140
Postmodern Culture, 381
Poverty
　functions of, 27–28
　occupational structure and, 207–209
Poverty Clock, 375
Powell, Colin, 213
Power elite, 356–358
Predestination, 340
Pregnancy, 253
Prejudice, 221–222
Prejudiced discriminators, 223
Prejudiced nondiscriminators, 223
Preoperational stage, 108
Preparing Your Student for College, 398

Press releases, 383–386
Press Releases by Country, 385
Primary groups, 101–102
Primary sex characteristics, 235, 236
Prison Legal News, 381
Prison Populations, 272
Private households, 242
Probabilistic models, 58
Production, means of, 17
Profane
　explanation of, 328–329
　fundamentalism and, 345
Professionalization, 160
Proletariat
　conflict theory and, 32–33
　explanation of, 18
Proposition 187 (California), 225
Protestantism, 345
Pseudo hermaphrodites, 235
Pull factors, 273
Pulp mills, 155–157
Pure deviants, 174
Push factors, 273

Questionnaires, self-administered, 51
Quotations, 393
Quotations Home Page, 393

Race
　complexity of defining, 213
　evaluation of accomplishments and effects of, 198
Race and Ethnicity Standards for the Classification of Federal Data on Race and Ethnicity, 272
Racial classification
　and idea of race, 213–215
　problems with, 190–191, 211–212
Racism, 219–220
Racist ideologies, 219, 220
Rack, Julie, 2, 6
Ramos, Francisco Martins, 170
Random samples
　to determine number of AIDS-infected people, 139
　explanation of, 49
　procedure for, 50
Rank, Mark R., 254
Rape
　date, 255–257
　treatment of women following, 244
　wartime, 245
Rap music, 77–78
Rask, Margrethe, 114–115, 122–124, 126, 129, 141
Raspberry, William, 226
Rationalization
　explanation of, 148–150
　means-to-end logic and, 149–150
　multinationals and, 151–152
　negative side of, 150
Reading instruction, 307–310
Reagan administration, 288

Reality construction. *See also* Social interaction
  explanation of, 139–140
  television and, 140–142
Rebellion, 185
Reentry shock, 85
Reference Center of the Internet Public Library, 394
Reference guides, 393–394, 396–397
Reflective thinking, 104
Register of Leading Social Sciences Electronic Journals, 381
Reich, Robert, 165, 166, 296
Religion
  civil, 332
  communities of worshipers and, 329–332
  conflict perspective on, 337–339
  Durkheim on, 325–329, 332–336
  functionalist perspective on, 334–336
  fundamentalism and, 344–346
  interplay between economics and, 339–341
  meaning of, 325–326
  mystical, 327–328
  prophetic, 327, 328
  rituals attached to, 329
  sacramental, 327, 328
  sacred and profane and, 326–329
  secularization and, 341–344
  sociological perspective on, 324–325
Religious Freedom Restoration Act of 1993, 343
Representative samples, 49
Research, 42, 46
Research design, 49
Research methods, 42, 46. *See also* Scientific method
Resnick, Daniel, 299
Resnick, Lauren B., 294
Resocialization, 109–111
Retreatism, 185
Reverse ethnocentrism, 82–83
Revolutionary ideas, 353
Reynolds, Larry T., 220
Rhodes, Jane, 141
Richardson, Keith, 151
Rights, associated with roles, 127
Ritualism, as response to structural strain, 185
Rituals, religious, 329
Rohter, Larry, 81
Roiphe, Katie, 256
Rokeach, Milton, 63–64
Role conflict, 127
Role expectations, 127, 128
Roles
  cultural variations in, 128–129
  explanation of, 126
  individual behavior and, 127–128
  sick, 127
Role set, 127
Role strain, 127
Role-taking, 105–107
Rosenthal, Elisabeth, 111
Rosenthal, Robert, 312, 313

Ross, E. A., 199
Rossman, Allan J., 58
Routinized charisma, 355
Rubenstein, Danny, 103
Rubyfruit Jungle (Brown), 73–74
Rule enforcers, 177–178
Rule makers, 177–178
Rural-to-rural migration, 276
Rural-to-urban migration, 275–276
Rush, Benjamin, 298, 299

Sabotage, 154
Sacramental religions, 327
Sacred
  explanation of, 326, 328
  fundamentalism and, 345
  ideas about what is, 327
  rituals to preserve what is, 329
Said, Abdul Aziz, 338
Samples, 49–50
Sampling frame, 50
Sanctions, 170–171
Saskatchewan NewStart, 350
Schengen Agreement, 79
Schindler's List, 89
Schoefield, L., 343–344
Schoenherr, Richard, 159–161
Scholarships and Fellowships, 398
School Enrollment—Historical Tables, 272
Schooling, 291
Schroeder, Patricia, 251
Scientific method
  analyzing data and drawing conclusions in, 57–59
  choosing design and collecting data in, 49–57
  defining topic for investigation in, 47–48
  explanation of, 46–47
  identifying core concepts in, 48–49
  reviewing literature in, 48
Scientific revolution, 353
Search engines, 4–5
Search for an Area Code, 397
Search the White House Virtual Library, 391
Secondary sex characteristics
  explanation of, 236–237
  gender nonconformity and, 251
Secondary sources, 53–54
Secret deviants, 175
Sects, 331–332
Secularization
  explanation of, 341
  objective, 343–344
  subjective, 341–342
Selective perception, 221
Self-administered questionnaires, 51
Self-awareness, 104–105
Self-fulfilling prophecy, 312–313
Sellers, 79
Semester at Sea's Home Page on CampusNET, 397
Semi-peripheral economies, 366
Sensorimotor stage, 108
Sex
  as biological concept, 235–237

  gender vs., 237
  life changes and effect of, 241
Sex characteristics
  primary, 235, 236
  secondary, 236–237
Sexist ideologies
  characteristics of, 250–251
  effects of, 251–252
  explanation of, 250
  in military, 251
Sexual property, 241, 242
Sexual relationships
  ethnicity and, 253
  gender polarization and, 240
Sexual stratification
  access to agents of violence control and, 244–245
  economic arrangements and, 241–243
  explanation of, 241
  household and child-rearing responsibilities and, 243
  workplace hazards and, 245
Shilts, Randy, 136
Shumaker, Richard, 8
Sick roles, 127
Significant others, 105–106
Significant symbols, 104
Simmons, J. L., 168, 169
Simpson, Carole, 198
Simpson, Richard L., 201–203
Simultaneous-independent inventions, 351–352
Situational factors, 134, 247
Slavery, 219, 220
Smoking, 180–181
Snow, Charles, 161
Social action, 19, 148
Social change
  Capitalism and, 362–366
  causes and consequences of, 349–350
  charismatic leaders as, 354–355
  conflict and, 358–361
  explanation of, 348
  innovations and, 350–354
  legal-rational authority and, 356–358
Social class. *See also* Class system
  conflict theory and, 32–33
  Weber and, 205–207
Social contact
  effects of lack in, 97–98
  importance of, 97, 99
Social control, mechanisms of, 168, 170–171
Social emotions
  cultural formulas for, 72–75
  explanation of, 240
Social identity, 227
Social interaction
  attribution theory and, 134–138
  content of, 125–129
  context of, 117–120
  dramaturgical model of, 130–134
  explanation of, 115
Social issues, 386–387

Socialization
  cognitive development and, 107–109
  explanation of, 94–95, 246
  gender expectations and, 246–247
  looking-glass self and, 107
  memory and, 99–100
  nature and nurture and, 95–96
  resocialization and, 109–111
  role of groups in, 101–103
  role of social contact in, 96–99
  role-taking and, 105–107
  self-awareness and, 104–105
  unpredictable elements of, 112–113
Social mobility, 194
Social movements, 79–81
Social relativity, 11–12
Social Statistics Briefing Room, 388
Social status, 125–126
Social stratification. *See* Stratification
Social structure, 125, 248
Society for Applied Sociology, 382
Society for the Study of Symbolic Interaction (SSSI): Papers of Interest, 381
Sociological imagination
  explanation of, 9
  nature of interaction and, 15–16
  nature of work and, 14–15
  transformative powers of history and, 12–13, 20
  troubles, issues, and, 9–11
Sociological Research Online, 381
Sociological theory, 25–26
Sociologists, resources for, 382–383
Sociology
  evolution of discipline of, 20–23
  explanation of, 17
  focus of, 17–20
Sociology Departments, 383
Sociology.net:Sociology on the Internet, 2
Solaar, M. C., 77
Solidarity
  explanation of, 18
  mechanical, 117–119
  organic, 119–120
  within societies, 117
Solinger, Rickie, 253
Sources, secondary, 53–54
South Africa, 193
Southern Sociological Association, 382
Spencer, Herbert, 333
Spitz, Rene, 96, 98, 99
Sports
  adolescent interest in, 321
  caste and class systems of stratification in, 198–199
  recruitment for college, 226
  sex tests and, 256
SRI International, 40
Staging behavior, 132
Standard of living, 243
Standke, Klaus-Heinrich, 150
Stanton, Elizabeth Cady, 234
Starr, Paul D., 215
Starting Point, 4

State
  cultural diffusion through, 78–79
  explanation of, 252
State and Local Governments, 376
State-level information, 376
State of the World Indicators, 369
State Profiles 1993—by the Small Business Association, 376
State Rankings, 376
The State of the World's Children, 387
Stationary pyramids, 272
Statistical Abstract Frequently Requested Population Tables, 370
Statistical Abstract Frequently Requested Tables, 388
Statistical Briefs, 272
Statistical Resources on the Web, 388
Statistical sources, 387–388
Status, 125, 126
Status group, 206
Status system, 320
Status value, 188
Steele, Shelby, 225
Stereotypes, 221–222
Stevenson, Harold W., 295, 316
Stewart, Jacob, 2, 6
Stigler, James W., 295
Stigmas, 126
Stigmatized individuals
  contact between normals and, 227–229
  explanation of, 227
  responses by, 229–230
Stoltzfus, Brenda, 254–255
Stratification
  achieved and ascribed characteristics and, 187–190
  assimilation and, 219–227
  explanation of, 187, 241
  functionalist view of, 200–203
  mixed systems of, 197–200
  occupational structure and, 207–209
  open and closed systems of, 194–197
  racial categories and, 190–192
  sexual, 241–245
Stratification theory
  assimilation and, 219
  discrimination and, 222–224
  institutionalized vs. individual discrimination and, 224–226
  prejudice and stereotyping and, 221–222
  racist ideologies and, 219–220
Structural assimilation
  achievement of, 217–218
  connection between marital and, 217
  explanation of, 216
Structural strain, 182, 185
Structured interviews, 51
Stuart, Charles, 175–176
Stub, Holger R., 268, 278, 279
Studds, Gerry, 251
Study abroad, 397–398

Study Abroad Home Page, 397
Subcultures
  adolescent, 320–322
  examples of, 91
  explanation of, 89–90
Subjective secularization, 341–342
Summer Institute of Linguistics, 80
Suntans, 37–38
Surdevant, Saundra Pollock, 254–255
Sutherland, Edwin H., 183, 184
Suzuki, Shinichi, 77, 78
Symbolic gestures, 104
Symbolic interactionist perspective
  critique of, 39
  on internet, 39–40
  overview of, 36–39
  self-development and, 104–107
Symbols
  explanation of, 36–37, 65
  ingroup or outgroup clash over, 103
  shared, 38
  significant, 104

Tan, Peter K. W., 69
Tannenbaum, Frank, 173
Tattersall, Ian, 240
Technological determinism, 352–353
Technology, 150–151
Teenage pregnancy, 253
Telecommunications, 43
Television
  reality construction and, 140–142
  as source of misinformation, 141–142
Tests, 314–315
Textbooks, 298–299
Theory, 25–26
Thesauruses, 392–393
This Week's Magazine, 389
This-worldly asceticism, 340
Thomas, Dorothy, 134
Thomas, Isiah, 198
Thomas, Jim, 53
THOMAS: Legislative Information on the Internet, 387
Thomas, William, 134
Thomson, Scott, 295–296
Thorne, Barrie, 21
Thorton, Alice Wandesworth, 263
TimeDaily, 389
Tocqueville, Alexis de, 379
Today in History, 389
Today's Press Releases from the White House, 385
Top 25 American Indian Tribes for the United States, 272
Top City Rankings, 377
Top Web Sites, 401
Total institutions, 110
Townsend, Peter, 96, 99
Toys, gender and, 246–247
Tracking
  effects of, 310–312
  explanation of, 310
  teacher expectations and, 312–313

Tracking *(continued)*
  use of, 301, 302
Traditional action, 19
Trained incapacity, 155–157
Transformative powers of history, 12–13, 20
Transportation modes, 15–16
Travel Guides, 377
Tribal societies, 241–242
Troubles, 8–9
Tumin, Melvin M., 201–203
Turnbull, Colin, 118, 119, 134–135
Tyson, Brady, 338

Unemployment, labor migrants and, 222
Uniform Resource Locators (URLs), 3–4
United Nations, 263–264, 277
United Nations High Commissioner for Refugees, 387
United Nations Home Page, 392
United States
  class system in, 195–197
  death rates in, 273
  education in, 297–305. *See also* Education
  feminism in, 234–235
  occupational structure and poverty in, 207–209
  physical contact between same-sexed persons in, 240–241
  rape laws in, 244
United States Census, 273
United States Information Agency (USIA), 79
United States Postal Service Zip Code Lookup and Address Information, 397
Unprejudiced discriminators, 223
Unstructured interviews, 51–52
Up Front (Heritage Front), 229
Upward mobility, 194
Urbanization, 268, 281
Urban underclass, 208
U.S. Bureau of the Census, 390
U.S. Department of Education Publications, 382
U.S. government agencies, 390–391
U.S. Historical Documents, 391
U.S. International Trade Statistics Current and Past Press Release File, 385
U.S. National Debt Clock, 388
U.S. State Fact Sheets, 376
USA Statistics in Brief: Part 1, 388
USA Today, 389
Usenet FAQs, 394

Validity, 56
VALS segmentation system, 40
Value-rational action, 19, 20
Values
  explanation of, 63–64
  language and, 69
  origin of, 67
  women as transmitters of, 253
Variables
  control, 58
  correlated, 57
  explanation of, 54
  identification of, 54–55
Vaughn, Angela, 2, 6
Veblen, Thorstein, 155, 379
Vertical mobility, 194
Veterans (Male and Female), 273
Violence control, 244–245
Voluntary minorities, 218
Voluntary Service Overseas, 397–398

Wacquant, Loic J. D., 208
Wallerstein, Immanuel, 362–363
War, 254
Washington Post, 389
Webcrawler, 4, 45
Weber, Max
  organizations and, 147, 148
  profile of, 19–20
  religion and, 325, 339–340
  social class and, 205–207
Webster, Cynthia, 48–49
Webster, Noah, 298, 299
Webster's English Dictionary, 393
"The Weekly Idiom" (Comenius), 68
Weisheit, Ralph, 50
Westinghouse Air Brake Company, 8, 9
White, Leslie, 351, 352
White, Ryan, 140–141
White-collar crime, 178
Wilmerding, Pennsylvania, 8–10
Wilson, Mary E., 123
Wilson, William Julius, 208
Wirth, Louis, 215, 216
Wisniewski, Nadine, 255
Witch-hunts, 176
Women. *See also* Gender; Gender expectations
  access to agents of violence control for, 244
  as caregivers for elderly, 280
  effect of multiple roles on, 284
  employment opportunities for, 284
  in labor force, 281–284
  as participants in struggles, 254–255
  as reproducers, 252–253
  as sexual property, 241
  as signifiers of ethnic and racial differences, 254
  standard of living for divorced, 243
  as transmitters of social values and culture, 253
  in workplace, 247–248
Work, 14–15
Workplace
  alienation in, 163–164
  hazards in, 245
  women in, 247–248
Works Project Administration (WPA), 53–54
World Bank, 263–264, 392
World Bank Press Releases, 385
World Constitutions, 375
World Demographics, 369
World economy, 363
World Factbook, 369
World Health Organization, 385–386
World Health Organization WWW Home Page, 392
World information, 369–370
World List Servers, 375
World PopClock from the U.S. Bureau of the Census, 375
World Resources Institute, 387
World Resources Institute News Release, 386
Worldspeaker, 387
World system theory, 362–363, 366
The World Council of Churches, 392
World Trade Organization (WTO) Press Releases, 386
World Watch Institute, 248
World Wide Web
  access to, 5
  explanation of, 2–3
  sites on, 400–401
WPA Federal Writers' Project, 54
Writing resources, 395
WWW Virtual Library, 383
Wycliff Bible Translators, 80

Yahoo, 4, 5
Yali, Cong, 63
Yanomami, 276
Yasuda, Yukuo, 133
Yinger, J. Milton, 339
Yuval-Davis, Nira, 252, 253, 255

Zaire
  Belgian imperialism in, 121–122
  blood banks and, 133
  health care in, 128–129
  HIV/AIDS and, 114, 116, 123–125, 135, 136, 138
  HIV risk reduction techniques in, 130–131
  independence of, 122–123
  mechanical solidarity in, 118–119
  raw materials in, 117, 122
Zeland, Karen, 308, 309
Zimbabwean Male Motivation Project, 271
Zuboff, Shoshana, 155–157